Lecture Notes in Computer Scie

T0237964

Commenced Publication in 1973
Founding and Former Series Editors:
Gerhard Goos, Juris Hartmanis, and Jan van Leeuwen

Ioannis G. Tollis   Maurizio Patrignani (Eds.)

Graph Drawing

16th International Symposium, GD 2008
Heraklion, Crete, Greece, September 21-24, 2008
Revised Papers

 Springer

Volume Editors

Ioannis G. Tollis
University of Crete
Department of Computer Science
71409 Heraklion, Crete, Greece
and
Institute of Computer Science
Foundation for Research and Technology-Hellas (FORTH)
Science and Technology Park of Crete, 71110 Heraklion, Crete, Greece
E-mail: tollis@ics.forth.gr

Maurizio Patrignani
Università Roma Tre
Dip. Informatica e Automazione
Via della Vasca Navale, 79, 00146 Rome, Italy
E-mail: patrigna@dia.uniroma3.it

Library of Congress Control Number: Applied for

CR Subject Classification (1998): G.2, F.2, I.3, E.1

LNCS Sublibrary: SL 1 – Theoretical Computer Science and General Issues

ISSN   0302-9743

ISBN   978-3-642-00218-2 Springer Berlin Heidelberg New York

springer.com

© Springer-Verlag Berlin Heidelberg 2009

Typesetting: Camera-ready by author, data conversion by Scientific Publishing Services, Chennai, India
Printed on acid-free paper   SPIN: 12609796   06/3180   5 4 3 2 1 0

Preface

The 16th International Symposium on Graph Drawing (GD 2008) was held in Hersonissos, near Heraklion, Crete, Greece, September 21-24, 2008, and was attended by 91 participants from 19 countries.

In response to the call for papers the Program Committee received 83 submissions, each describing original research and/or a system demonstration. Each submission was reviewed by at least three Program Committee members and the reviewer's comments were returned to the authors. Following extensive discussions, the committee accepted 31 long papers and 8 short papers. In addition, 10 posters were accepted and displayed at the conference site. Each poster was granted a two-page description in the conference proceedings.

Two invited speakers, Jesper Tegnér from Karolinska Institute (Monday) and Roberto Tamassia from Brown University (Tuesday), gave fascinating talks during the conference. Professor Tegnér focused on the challenges and opportunities posed by the discovery, analysis, and interpretation of biological networks to information visualization, while Prof. Tamassia showed how graph drawing techniques can be used as an effective tool in computer security and pointed to future research directions in this area.

Following what is now a tradition, the 15th Annual Graph Drawing Contest was held during the conference, also including a Graph Drawing Challenge to the conference attendees. A report is included in the conference proceedings.

Many people contributed to the success of GD 2008. First of all, special thanks to the authors of submitted papers, demos, and posters. Many thanks to the members of the Program Committee and the external referees who worked diligently to select only the best of the submitted papers. The Organizing Committee worked tirelessly in the months leading to the crucial final four days: Emilio Di Giacomo was a great Publicity Chair; Theano Apostolidi, Kiriaki Kaiserli, Maria Prevelianaki, and Vassilis Tsiaras carried a large part of the work regarding local organization and management of the conference. Also, many thanks to the student volunteers who helped in many ways during the conference.

The conference was organized and supported by the Institute of Computer Science (ICS)-FORTH and the Computer Science Department of the University of Crete. GD 2008 also received generous support from our sponsors: Tom Sawyer Software (Gold Sponsor), and OTE, ILOG, and Virtual Trip (Silver Sponsors).

The 17th International Symposium on Graph Drawing (GD 2009) will be held September 23-25, 2009 in Chicago, USA, co-chaired by David Eppstein and Emden R. Gansner.

November 2008

Ioannis G. Tollis
Maurizio Patrignani

Organization

Steering Committee

Franz Josef Brandenburg	University of Passau
Ulrik Brandes	University of Konstanz
Giuseppe Di Battista	Università Roma Tre
Peter Eades	NICTA and University of Sydney
David Eppstein	University of California, Irvine
Hubert de Fraysseix	CAMS-CNRS
Seokhee Hong	NICTA and University of Sydney
Giuseppe Liotta	Università degli Studi di Perugia
Takao Nishizeki	Tohoku University
Pierre Rosenstiehl	CAMS-CNRS
Roberto Tamassia	Brown University
Ioannis G. Tollis	FORTH-ICS and University of Crete
Emden R. Gansner	AT&T Labs

Program Committee

Ulrik Brandes	University of Konstanz
Walter Didimo	Università degli Studi di Perugia
Peter Eades	NICTA and University of Sydney
David Eppstein	University of California, Irvine
Robert Gentleman	Fred Hutchinson Cancer Research Center
Seok-Hee Hong	NICTA and University of Sydney
Michael Kaufmann	Universität Tübingen
Stephen Kobourov	University of Arizona
Yehuda Koren	AT&T Labs
Jan Kratochvíl	Charles University
Kwan-Liu Ma	University of California, Davis
Henk Meijer	Roosevelt Academy
Kazuyuki Miura	Fukushima University
Tamara Munzner	University of British Columbia
János Pach	City College and Courant Institute
Maurizio Patrignani	Università Roma Tre
Natasa Przulj	University of California, Irvine
Antonios Symvonis	National Technical University of Athens
Ioannis G. Tollis	FORTH-ICS and University of Crete (Chair)
Stephen K. Wismath	University of Lethbridge

Organizing Committee

Theano Apostolidi   FORTH-ICS
Emilio Di Giacomo   Università degli Studi di Perugia
                                 (Publicity Chair)
Kiriaki Kaiserli   FORTH-ICS
Maurizio Patrignani   Università Roma Tre (Co-chair)
Maria Prevelianaki   FORTH-ICS
Ioannis G. Tollis   FORTH-ICS and University of Crete (Co-chair)
Vassilis Tsiaras   FORTH-ICS and University of Crete

Contest Committee

Ugur Dogrusoz   Bilkent University and Tom Sawyer Software
Christian A. Duncan   Lousiana Tech Univesity
Carsten Gutwenger   Dortmund University of Technology
Georg Sander   ILOG Deutschland GmbH (Chair)

External Referees

Patrizio Angelini   Francesco Giordano   Barbara Pampel
Yasuhito Asano   Luca Grilli   Charis Papadopoulos
Melanie Badent   Kyle Hambrook   Christian Pich
Michael A. Bekos   Yifan Hu   Maurizio Pizzonia
Carla Binucci   Weidong Huang   Katerina Potika
Krists Boitmanis   Konstantinos Kakoulis   Md. Saidur Rahman
Pier Francesco Cortese   Sven Kosub   Zeqian Shen
Giuseppe Di Battista   Irini Koutaki   Martin Siebenhaller
Emilio Di Giacomo   Martin Mader   Janet M. Six
A. Estrella-Balderrama   Jiří Matoušek   Csaba D. Tóth
Daniel Fleischer   Tamara Mchedlidze   Francesco Trotta
J. Joseph Fowler   Christopher Muelder   Vassilis Tsiaras
Fabrizio Frati   Uwe Nagel   Xiao Zhou
Emden R. Gansner   Shin-Ichi Nakano   Katharina A. Zweig
Markus Geyer   Pietro Palladino

Sponsoring Institutions

Organized and Supported by

Gold Sponsor

Silver Sponsors

Table of Contents

Posters

Graph Drawing Contest

Networks in Biology – From Identification, Analysis to Interpretation

Jesper Tegnér

Institutionen för Medicin
Karolinska Universitetssjukhuset
Solna, Stockholm
jesper.tegner@ki.se

Abstract. Over the last decade networks has become a unifying language in biology. Yet we are only in the beginning of understanding their significance for biology and their medical applications. I will talk about the diversity of biological networks composed either of genes, proteins, metabolites, or cells and the associated methods for finding these graphs in the data. Next I will provide an overview of different methods of analysis and what kind of insights that have been obtained. During the talk I will highlight current challenging problems requiring computational skills with respect to identification, analysis, algorithms, visualization and software.

I.G. Tollis and M. Patrignani (Eds.): GD 2008, LNCS 5417, p. 1, 2009.
© Springer-Verlag Berlin Heidelberg 2009

Graph Drawing for Security Visualization*

Roberto Tamassia[1], Bernardo Palazzi[1,2,3], and Charalampos Papamanthou[1]

[1] Brown University, Department of Computer Science, Providence, RI, USA
{rt,bernardo,cpap}@cs.brown.edu
[2] Roma TRE University, Rome, Italy
palazzi@dia.uniroma3.it
[3] ISCOM Italian Ministry of Economic Development-Communications, Rome, Italy

Abstract. With the number of devices connected to the internet growing rapidly and software systems being increasingly deployed on the web, security and privacy have become crucial properties for networks and applications. Due the complexity and subtlety of cryptographic methods and protocols, software architects and developers often fail to incorporate security principles in their designs and implementations. Also, most users have minimal understanding of security threats. While several tools for developers, system administrators and security analysts are available, these tools typically provide information in the form of textual logs or tables, which are cumbersome to analyze. Thus, in recent years, the field of security visualization has emerged to provide novel ways to display security-related information so that it is easier to understand. In this work, we give a preliminary survey of approaches to the visualization of computer security concepts that use graph drawing techniques.

1   Introduction

As an increasing number of software applications are web-based or web-connected, security and privacy have become critical issues for everyday computing. Computer systems are constantly being threatened by attackers who want to compromise the privacy of transactions (e.g., steal credit card numbers) and the integrity of data (e.g., return a corrupted file to a client). Therefore, computer security experts are continuously developing methods and associated protocols to defend against a growing number and variety of attacks. The development of security tools is an ongoing process that keeps on reacting to newly discovered vulnerabilities of existing software and newly deployed technologies.

Both the discovery of vulnerabilities and the development of security protocols can be greatly aided by visualization. For example, a graphical representation of a complex multi-party security protocol can give experts better intuition of its execution and security properties. In current practice, however, computer security analysts read through

---

* This work has been presented at the 2008 Symposium on Graph Drawing in a invited talk dedicated to the memory of Paris C. Kanellakis, a prominent computer scientist and Brown faculty member who died with his family in an airplane crash in December 1995. His unbounded energy and outstanding scholarship greatly inspired all those who interacted with him.

I.G. Tollis and M. Patrignani (Eds.): GD 2008, LNCS 5417, pp. 2–13, 2009.
© Springer-Verlag Berlin Heidelberg 2009

large logs produced by applications, operating systems, and network devices. The visual inspection of such logs is quite cumbersome and often unwieldy, even for experts. Motivated by the growing need for automated visualization methods and tools for computer security, the field of *security visualization* has recently emerged as an interdisciplinary community of researchers with its own annual meeting (VizSec).

In this paper, we give a preliminary survey of security visualization systems that use graph drawing methods. Thanks to their versatility, graph drawing techniques are one of the main approaches employed in security visualization. Indeed, not only computer networks are naturally modeled as graphs, but also data organization (e.g., file systems) and vulnerability models (e.g., attack trees) can be effectively represented by graphs. In the rest of this paper, we specifically overview graph drawing approaches for the visualization of the following selected computer security concepts:

1. *Network Monitoring.* Monitoring network activity and identifying anomalous behavior, such as unusually high traffic to/from certain hosts, helps identifying several types of attacks, such as intrusion attempts, scans, worm outbreaks, and denial of service.

2. *Border Gateway Protocol (BGP).* BGP manages reachability between hosts in different autonomous systems, i.e., networks under the administrative control of different Internet Service Providers. Understanding the evolution of BGP routing patterns over time is very important to detect and correct disruptions in Internet traffic caused by router configuration errors or malicious attacks.

3. *Access Control.* Access to resources on a computer system or network is regulated by policies and enforced through authentication and authorization mechanisms. It is critical to protect systems not only from unauthorized access by outside attackers but also from accidental disclosure of private information to legitimate users. Access control systems and their associated protocols can be very complex to manage and understand. Thus, it is important to have tools for analyzing and specifying policies, identifying the possibility of unauthorized access, and updating permissions according to desired goals.

4. *Trust Negotiation.* Using a web service requires an initial setup phase where the client and server enter into a negotiation to determine the service parameters and cost by exchanging credentials and policies. Trust negotiation is a protocol that protects the privacy of the client and server by enabling the incremental disclosure of credentials and policies. Planning and executing an effective trust negotiation strategy can be greatly aided by tools that explore alternative scenarios and show the consequences of possible moves.

5. *Attack Graphs.* A typical strategy employed by an attacker to compromise a system is to follow a path in a directed graph that models vulnerabilities and their dependencies. After an initial successful attack to a part of a system, an attacker can exploit one vulnerability after the other and reach the desired goal. Tools for building and analyzing attack graphs help computer security analysts identify and fix vulnerabilities.

In Table 1, we show the graph drawing methods used by the systems surveyed in this paper.

Table 1. Graph drawing methods used in the security visualization systems surveyed in this paper

	Force-Directed	Layered	Bipartite	Circular	Treemap	3D
Network Monitoring	[9, 12, 14, 21]		[1, 4, 24]	[20]		
BGP	[19]			[19]		[18]
Access Control		[13]			[10]	
Trust Negotiation		[23]				
Attack Graphs		[16, 17]				

2   Network Monitoring

Supporting Intrusion Detection by Graph Clustering and Graph Drawing [21]. In this paper, the authors use a combination of force-directed drawing, graph clustering, and regression-based learning in a system for intrusion detection (see Fig. 1(a)). The system consists of modules for the following functions: packet collection, graph construction and clustering, graph layout, regression-based learning, and event generation.

The authors model the computer network with a graph where the nodes are computers and the edges are communication links with weight proportional to the network traffic on that link. The clustering of the graph is performed with a simple iterative method. Initially, every node forms its own cluster. Next, nodes join clusters that already have most of their neighbors. A force-directed approach is used to place clusters and nodes within the clusters. Since forces are proportional to the weights of the edges, if there is a lot of communication between two hosts, their nodes are placed close to each other. Also, in the graph of clusters, there is an edge between clusters A and B if there is at least one edge between some node of cluster A and some node of cluster B. The layout of the graph of clusters and of each cluster are computed using the classic force-directed spring embedder method [6].

Various features of the clustered graph (including statistics on the node degrees, number of clusters, and internal/external connectivity of clusters) are used to describe the current state of network traffic and are summarized by a feature vector. Using test traffic samples and a regression-based strategy, the system learns how to map feature vectors to intrusion detection events. The security analyst is helped by the visualization of the clustered graph in assessing the severity of the intrusion detection events generated by the system.

Graph-Based Monitoring of Host Behavior for Network Security [12]. In this paper, the authors show how to visualize the evolution over time of the volume and type of network traffic using force-directed graph drawing techniques (see Fig. 1(b)). Since there are different types of traffic protocols (HTTP, FTP, SMTP, SSH, etc.) and multiple time periods, this multi-dimensional data set is modeled by a graph with two types of nodes: *dimension nodes* represent traffic protocols and *observation nodes* represent the state of a certain host in a given time interval. Edges are also of two types: *trace edges* link observation nodes of consecutive time intervals and *attraction edges* link observation nodes with dimension nodes and have weight proportional to the traffic of that type.

The layout of the above graph is computed starting with a fixed placement of the dimension nodes and then executing a modified version of the Fruchterman-Reingold

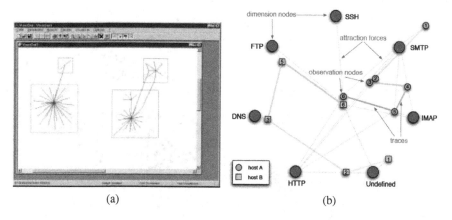

Fig. 1. (a) Force-directed clustered drawing for intrusion detection (thumbnail of image from [21]). (b) Evolution of network traffic over time (thumbnail of image from [12]): dimension nodes represent types of traffic and observation nodes represent the state of a host at a given time.

force-directed algorithm [8] that aims at achieving uniform edge lengths. The authors show how intrusion detection alerts can be associated with visual patterns in the layout of the graph.

A Visual Approach for Monitoring Logs [9]. This paper (see Fig. 2(a)) presents a technique to visualize log entries obtained by monitoring network traffic. The log entries are basically vectors whose elements correspond to features of the network traffic, including origin IP, destination IP, and traffic volume. The authors build a weighted similarity graph for the log entries using a simple distance metric for two entries given by the sum of the differences of the respective elements. The force-directed drawing algorithm of [3] is used to compute a drawing of the similarity graph of the entries.

A Visualization Methodology for Characterization of Network Scans [14]. This work considers network scans, often used as the preliminary phase of an attack. The authors develop a visualization system that shows the relationships between different network scans (see Fig. 2(b)). The authors set up a graph where each node represents a scan and the connection between them is weighted according to some metric (similarity measure) that is defined for the two scans. Features taken into consideration for the definition of the similarity measure include the origin IP, the destination IP and the time of the connection. To avoid displaying a complete graph, the authors define a minimum weight threshold, below which edges are removed. The LinLog force directed layout method [15] is used for the visualization of this graph. In the drawing produced, sets of similar scans are grouped together, thus facilitating the visual identification of malicious scans.

VisFlowConnect: NetFlow Visualizations of Link Relationships for Security Situational Awareness [24]. In this work, the authors apply a simple bipartite drawing technique to provide a visualization solution for network monitoring and intrusion detection (see Fig. 3(a)). The nodes, representing internal hosts and external domains, are placed on three vertical lines. The external domains that send traffic to some internal host are

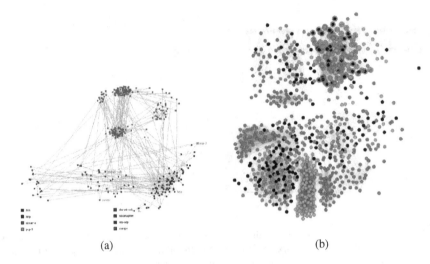

         (a)   (b)

Fig. 2. (a) Similarity graph of log entries (thumbnail of image from [9]). (b) Similarity graph of network scans (thumbnail of image from [14]).

placed on the left line. The domains of the internal hosts are placed on the middle line. The external domains that receive traffic from some internal host are placed on the right line. Each edge represents a network flow, which is a sequence of related packets transmitted from one host to another host (e.g., a TCP packet stream). Basically, the layout represents a tripartite graph. The vertical ordering of the domains along each line is computed by the drawing algorithm with the goal of minimizing crossings.

The tool uses a slider to display network flows at various time intervals and provides three views. In the global view, the entire tripartite graph is displayed to show all the communication between internal and external hosts. In the internal view and domain view, the tool isolates certain parts of the network, such as internal senders and internal receivers, and correspondingly displays a bipartite graph. The domain view and internal view are easier to analyze and provide more details on the network activity being visualized but on the other hand, the global view produces a high-level overview of the network flows. The authors apply the tool in various security-related scenarios, such as virus outbreaks and denial-of-service attacks.

Home-Centric Visualization of Network Traffic for Security Administration [1]. In this paper the authors use a matrix display combined with a simple graph drawing method in order to visualize the traffic between domains in network and external domains (see Fig. 3(b)). To visualize the internal network, the authors use a square matrix: each entry of the matrix corresponds to a host of the internal network. External hosts are represented by squares placed outside the matrix, with size proportional to the traffic sent or received. Straight-line edges represent traffic between internal and external hosts and can be colored to denote the predominant direction of the traffic (outgoing, incoming, or bidirectional). The placement of the squares arranges hosts of the same class A, B

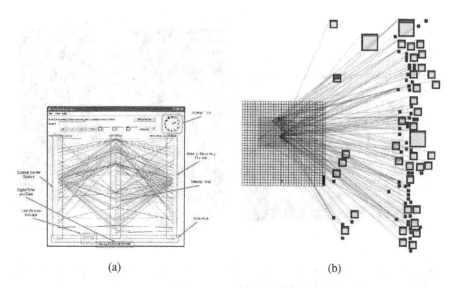

(a)   (b)

Fig. 3. (a) Global view of network flows using a tripartite graph layout: nodes represent external domains (on the left and right) and internal domains (in the middle) and edges represent network flows (packet streams) between domains (thumbnail of image from [24]). **(b)** Visualization of internal vs. external hosts using a matrix combined with a straight-line drawing. Internal hosts correspond to entries of the matrix while external hosts are drawn as squares placed around the matrix. The size of the square for an external host is proportional to the amount of traffic from/to that host (thumbnail of image from [1]).

or C network along the same vertical line and attempts to reduce the number of edge crossings. Further details on the type of traffic can be also displayed in this tool. For example, vertical lines inside each square indicate ports with active traffic. This system can be used to visually identify traffic patterns associated with common attacks, such as virus outbreaks and network scans.

EtherApe: A Live Graphical Network Monitor Tool [20]. This tool shows traffic captured on the network interface (in a dynamic fashion) or optionally reads log files like PCAP (Fig. 4(a)). A simple circular layout places the hosts around a circle and represents network traffic between hosts by straight-line edges between them. Each protocol is distinguished by a different color and the width of an edge shows the amount of traffic. This tool allows to quickly understand the role of a host in the network and the changes in traffic patterns over time. Beyond the graphical representation, it is also possible to display detailed traffic statistics of active ports.

RUMINT [4]. This system (named after RUMor INTelligence) is a free tool for network and security visualization (Fig. 4(b)). It takes captured traffic as input and visualizes it in various unconventional ways. The most interesting visualization related to graph drawing is a parallel plot that allows one to see at a glance how multiple packet fields are related. An animation feature allows to analyze various trends over time.

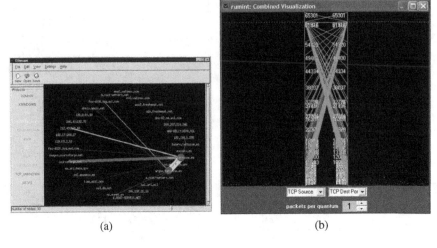

(a)   (b)

Fig. 4. (a) Traffic monitoring with Etherape (thumbnail of image from [20]). **(b)** Visualization of an NMAP scan with RUMINT (thumbnail of image from [4]).

3   Border Gateway Protocol

BGP Eye: A New Visualization Tool for Real-Time Detection and Analysis of BGP Anomalies [19]. In this paper, the authors present a visualization tool, called *BGP Eye*, that provides a real-time status of BGP activity with easy-to-read layouts (Fig. 5). BGP Eye is a tool for root-cause analysis of BGP anomalies. Its main objective is to track the healthiness of BGP activity, raise an alert when an anomaly is detected, and indicate its most probable cause. BGP Eye allows two different types of BGP dynamics visualization: *internet-centric view* and *home-centric view*. The internet-centric view studies the activity among ASes (autonomous systems) in terms of BGP events exchanged. The home-centric view has been designed to understand the BGP behavior from the perspective of a specific AS. The inner ring contains the routers of the customer AS and the outer ring contains their peer routers, belonging to other ASes. In the outer layer, the layout method groups together routers belonging to the same AS and uses a placement algorithm for the nodes to reduce the distance between connected nodes.

VAST: Visualizing Autonomous System Topology [18]. This tool (Fig. 6(a)) uses 3D straight-line drawings to display the BGP interconnection topology of ASes with the goal of allowing security researchers to extract quickly relevant information from raw routing datasets. VAST employs a quad-tree to show per-AS information and an octo-tree to represent relationships between multiple ASes. Routing anomalies and sensitive points can be quickly detected, including route leakage events, critical Internet infrastructure and space hijacking incidents. The authors have also developed another tool, called *Flamingo*, that uses the same graphical engine as VAST but is used for real-time visualization of network traffic.

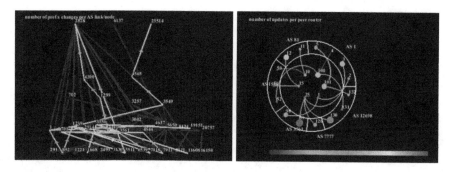

Fig. 5. Internet-centric view and home-centric view in BGP Eye (thumbnails of images from [19])

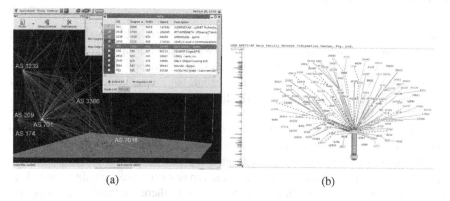

(a)   (b)

Fig. 6. (a) Some large autonomous systems in the internet visualized with VAST (thumbnail of image from [18]). **(b)** In BGPlay, nodes represent autonomous systems and paths are sequences of autonomous systems to be traversed to reach the destination (thumbnail of image from [5]).

BGPlay: A system for visualizing the interdomain routing evolution [5]. BGPlay and *iBGPlay* (Fig. 6(b)) provide animated graphs of the BGP routing announcements for a certain IP prefix within a specified time interval. Both visualization tools are targeted to Internet service providers. Each nodes represents an AS and paths are used to indicate the sequence of ASes needed to be traversed to reach a given destination. BGPlay shows paths traversed by IP packets from several probes spread over the Internet to the chosen destination (prefix). iBGPlay shows data privately collected by one ISP. The ISP can obtain from iBGPlay visualizations of outgoing paths from itself to any destination. The drawing algorithm is a modification of the force-directed approach that aims at optimizing the layout of the paths.

4   Access Control

Information Visualization for Rule-based Resource Access Control [13]. In this paper, the authors provide a visualization solution for managing and querying rule-based

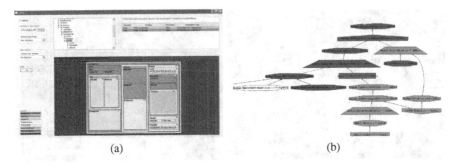

(a)   (b)

Fig. 7. (**a**) Visualization of permissions in the NTFS file system with TrACE (thumbnail of image from [10]). (**b**) Drawing of the trust-target graph generated by a trust negotiation session (thumbnail of image from [23]).

access control systems. They develop a tool, called RubaViz, which makes it easy to answer questions like "What group has access to which files during what time duration?". RubaViz constructs a graphs whose nodes are subjects (people or processes), groups, resources, and rules. Directed edges go from subjects/groups to rules and from rules to resources to display allowed accesses. The layout is straight-line and upward.

Effective Visualization of File System Access-Control [10]. This paper presents a tool, called TrACE, for visualizing file permissions in the NTFS file system (Fig. 7(a)). TrACE allows a user or administrator to gain a global view of the permissions in a file system, thus simplifying the detection and repair of incorrect configurations leading to unauthorized accesses. In the NTFS file system there are three types of permissions: (a) *explicit* permissions are set by the owner of each group/user; (b) *inherited* permissions are dynamically inherited from the explicit permissions of the ancestor folders; and (c) *effective* permissions are obtained by combining the explicit and inherited permissions. The tool uses a treemap layout [11] to draw the file system tree and colors the tiles with a palette denoting various access levels. The size of a tile indicates how much the permissions of a folder/file differ from those of its parent and children. Advanced properties, such as a break of inheritance at some folder, are also graphically displayed. The tool makes is easy to figure out which explicit and inherited permissions of which nodes affect the effective permissions of a given node in the file system tree.

5   Trust Negotiation

Visualization of Automated Trust Negotiation [23]. In this paper, the authors use a layered upward drawing to visualize automated trust negotiation (ATN) (Fig. 7(b)). In a typical ATN session, the client and server engage in a protocol that results in the collaborative and incremental construction of a directed acyclic graph, called trust-target graph, that represents credentials (e.g., a proof that a party has a certain role in an organization) and policies indicating that the disclosure of a credential by one party is subject to the prior disclosure of a set of credentials by the other party [22]. A tool based on the Grappa system [2], a Java port of Graphviz [7], is used to generate successive drawings of the trust-target graph being constructed in an ATN session.

6 Attack Graphs

Multiple Coordinated Views for Network Attack Graphs [16] This paper describes a tool for visualizing attack graphs (Fig. 8). Given a network and a database of known vulnerabilities that apply to certain machines of the network, one can construct a directed graph where each node is a machine (or group of machines) and an edge denotes how a successful attack on the source machine allows to exploit a vulnerability on the destination machine. Since attack graphs can be rather large and complex, it is essential to use automated tools to analyze them. The tool presented in this paper clusters machines in order to reduce the complexity of the attack graph (e.g., machines that belong to the same subnet may be susceptible to the same attack). The Graphviz tool [7] is used to produce a layered drawing of the clustered attack graph. Similar layered drawings for attack graphs are proposed in [17].

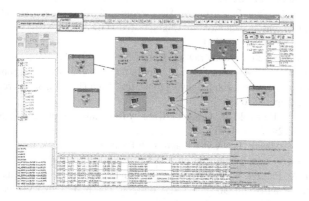

Fig. 8. Visualization of an attack graph (thumbnail of image from [16])

7 Conclusions

In this paper, we have presented a preliminary survey of security visualization methods that use graph drawing techniques. The growing field of security and privacy offers many opportunities to graph drawing researchers to develop new drawing methods and tools. In computer and network security applications, the input to the visualization system is often a large multidimensional and temporal data set. Moreover, the layout needs to support color, labels, variable node shape/size and edge thickness. In most of the security visualization papers we have reviewed, either simple layout algorithms have been implemented (e.g., spring embedders) or open-source software has been used (e.g., Graphviz). In order to make a larger collection of sophisticated graph drawing techniques available to computer security researchers, it is important for the graph drawing community to develop and distribute reliable software implementations.

Acknowledgments

This work was supported in part by the U.S. National Science Foundation under grants IIS–0713403 and CCF–0830149, by the Kanellakis Fellowship at Brown University, and by the Italian Ministry of Research under grant RBIP06BZW8.

References

[1] Ball, R., Fink, G.A., North, C.: Home-centric visualization of network traffic for security administration. In: Proc. Workshop on Visualization and Data Mining for Computer Security (VIZSEC/DMSEC), pp. 55–64 (2004)

[2] Barghouti, N.S., Mocenigo, J., Lee, W.: Grappa: A GRAPh PAckage in Java. In: DiBattista, G. (ed.) GD 1997. LNCS, vol. 1353, pp. 336–343. Springer, Heidelberg (1997)

[3] Chalmers, M.: A linear iteration time layout algorithm for visualising high-dimensional data. In: Proc. Conference on Visualization (VIS), pp. 127–132 (1996)

[4] Conti, G.: Security Data Visualization. No Starch Press, San Francisco (2007), http://www.rumint.org

[5] Di Battista, G., Mariani, F., Patrignani, M., Pizzonia, M.: Bgplay: A system for visualizing the interdomain routing evolution. In: Liotta, G. (ed.) GD 2003. LNCS, vol. 1353, pp. 295–306. Springer, Heidelberg (2003)

[6] Eades, P.: A heuristic for graph drawing. Congr. Numer. 42, 149–160 (1984)

[7] Ellson, J., Gansner, E.R., Koutsofios, L., North, S.C., Woodhull, G.: Graphviz and dynagraph - static and dynamic graph drawing tools. In: Graph Drawing Software, pp. 127–148. Springer, Heidelberg (2003)

[8] Fruchterman, T., Reingold, E.: Graph drawing by force-directed placement. Softw. – Pract. Exp. 21(11), 1129–1164 (1991)

[9] Girardin, L., Brodbeck, D.: A visual approach for monitoring logs. In: Proc. of USENIX Conference on System Administration (LISA), pp. 299–308 (1998)

[10] Heitzmann, A., Palazzi, B., Papamanthou, C., Tamassia, R.: Effective visualization of file system access-control. In: Goodall, J.R., Conti, G., Ma, K.-L. (eds.) VizSec 2008. LNCS, vol. 5210, pp. 18–25. Springer, Heidelberg (2008)

[11] Johnson, B., Shneiderman, B.: Tree maps: A space-filling approach to the visualization of hierarchical information structures. In: Proc. Conference on Visualization (VIS), pp. 284–291 (1991)

[12] Mansmann, F., Meier, L., Keim, D.: Graph-based monitoring of host behavior for network security. In: Proc. Visualization for Cyper Security (VIZSEC), pp. 187–202 (2007)

[13] Montemayor, J., Freeman, A., Gersh, J., Llanso, T., Patrone, D.: Information visualization for rule-based resource access control. In: Proc. of Int. Symposium on Usable Privacy and Security (SOUPS) (2006)

[14] Muelder, C., Ma, K.L., Bartoletti, T.: A visualization methodology for characterization of network scans. In: Proc. Visualization for Cyber Security (VIZSEC) (2005)

[15] Noack, A.: An energy model for visual graph clustering. In: Liotta, G. (ed.) GD 2003. LNCS, vol. 1353, pp. 425–436. Springer, Heidelberg (2003)

[16] Noel, S., Jacobs, M., Kalapa, P., Jajodia, S.: Multiple coordinated views for network attack graphs. In: Proc.Visualization for Cyber Security (VIZSEC), pp. 99–106 (2005)

[17] Noel, S., Jajodia, S.: Managing attack graph complexity through visual hierarchical aggregation. In: Proc. Workshop on Visualization and Data Mining for Computer Security (VIZSEC/DMSEC), pp. 109–118 (2004)

[18] Oberheide, J., Karir, M., Blazakis, D.: VAST: Visualizing autonomous system topology. In: Proc. Visualization for Cyber Security (VIZSEC), pp. 71–80 (2006)

[19] Teoh, S.T., Ranjan, S., Nucci, A., Chuah, C.N.: BGP Eye: a new visualization tool for real-time detection and analysis of BGP anomalies. In: Proc. Visualization for Cyber Security (VIZSEC), pp. 81–90 (2006)

[20] Toledo, J.: Etherape: a live graphical network monitor tool, `http://etherape.sourceforge.net`

[21] Tölle, J., Niggermann, O.: Supporting intrusion detection by graph clustering and graph drawing. In: Debar, H., Mé, L., Wu, S.F. (eds.) RAID 2000. LNCS, vol. 1907. Springer, Heidelberg (2000)

[22] Winsborough, W.H., Li, N.: Towards practical automated trust negotiation. In: Proc. Workshop on Policies for Distributed Systems and Networks (POLICY), pp. 92–103 (2002)

[23] Yao, D., Shin, M., Tamassia, R., Winsborough, W.H.: Visualization of automated trust negotiation. In: Proc. Visualization for Cyber Security (VIZSEC), pp. 65–74 (2005)

[24] Yin, X., Yurcik, W., Treaster, M., Li, Y., Lakkaraju, K.: VisFlowConnect: Netflow visualizations of link relationships for security situational awareness. In: Proc. Workshop on Visualization and Data Mining for Computer Security (VizSEC/DMSEC), pp. 26–34 (2004)

Succinct Greedy Graph Drawing
in the Hyperbolic Plane

David Eppstein[1] and Michael T. Goodrich[2]

Computer Science Department, University of California, Irvine, USA
[1] http://www.ics.uci.edu/~eppstein/
[2] http://www.ics.uci.edu/~goodrich/

Abstract. We describe a method for producing a greedy embedding of any n-vertex simple graph G in the hyperbolic plane, so that a message M between any pair of vertices may be routed by having each vertex that receives M pass it to a neighbor that is closer to M's destination. Our algorithm produces *succinct* drawings, where vertex positions are represented using $O(\log n)$ bits and distance comparisons may be performed efficiently using these representations.

1   Introduction

Viewing network routing as an algorithmic problem, we are given an n-vertex graph G representing a communication network, where each vertex in G is a computational agent, and the edges in G represent communication channels. The routing problem is to set up an efficient means to support message passing between the vertices in G.

There is a recent non-traditional approach to solving the routing problem, which can be viewed as new and exciting application of graph drawing. In this new approach, called *geometric routing* [2,7,10,11,12] or *geographic routing* [8], the graph G is drawn in a geometric metric space $\mathcal{S}$ in the standard way, so that vertices are drawn as points in $\mathcal{S}$ and each edge is drawn as the loci of points along the shortest path between its two endpoints. For example, if $\mathcal{S}$ is the Euclidean plane, $\mathbf{R}^2$, then edges would be drawn as straight line segments in this approach. Routing is then performed by having any vertex v holding a message destined for a node w use a simple policy involving only the coordinates of v and w and the coordinates and topology of v's neighbors to determine the neighbor of v to which v should forward the message. It is important to note that even in applications where the vertices of G come with pre-defined geometric coordinates (e.g., GPS coordinates of smart sensors), the drawing of G need not take these coordinates into consideration.

Perhaps the simplest routing policy imaginable is the *greedy* one:

- If a vertex v receives a message M with destination w, v should forward M to any neighbor of v in G that is closer than v to w.

We are interested in this paper in *greedy* drawings of arbitrary graphs, that is, drawings for which greedy routing is always successful. Unfortunately, greedy routing doesn't always work. For example, it is not uncommon for geometric graph embeddings

I.G. Tollis and M. Patrignani (Eds.): GD 2008, LNCS 5417, pp. 14–25, 2009.

to have "lakes" and "voids" that make greedy routing impossible in some cases [17]. Indeed, in any fixed-dimensional Euclidean space, a star with sufficiently many leaves cannot be embedded so that all paths are greedy. Thus, in order to find greedy drawing schemes for arbitrary connected graphs, we must consider drawings in non-Euclidean spaces.

Following Papadimitriou and Ratajczak [17], we say that a *distance decreasing path* from v to w in a geometric embedding of G is a path $(v_1, v_2, \ldots, v_k)$ such that $v = v_1$, $w = v_k$, and $d(v_i, w) > d(v_{i+1}, w)$, for $i = 1, 2, \ldots, k - 1$. A *greedy embedding* of a graph G in a geometric metric space $\mathcal{S}$ is a drawing of G in $\mathcal{S}$ such that a distance decreasing path exists between every pair of vertices in G.

Prior Related Work. Early papers on geometric routing include work by Bose *et al.* [2], who extract a planar subgraph of G, embed it, and then route a message from v to w by marching around the faces intersected by the line segment vw using a subdivision traversal algorithm of Kranakis *et al.* [9]. Karp and Kung [7] introduce a hybrid scheme, which combines a greedy routing strategy with face routing. Similar hybrid schemes were subsequently studied by several other researchers [10,11,12].

Rao *et al.* [18] introduce the idea of drawing a graph using virtual coordinates and doing a pure greedy routing strategy with that drawing, although they make no theoretical guarantees. Papadimitriou and Ratajczak [17] continue this line of work on greedy drawings, studying greedy schemes that are guaranteed to work, and they conjecture that Euclidean greedy drawings exist for any graph containing a 3-connected planar spanning subgraph. They present a greedy drawing algorithm for embedding 3-connected planar graphs in $\mathbf{R}^3$ based on a specialization of Steinitz's Theorem for circle packings, albeit with a non-standard metric. Dhandapani [4] provides an existence proof that two-dimensional Euclidean greedy drawings of triangulations are always possible, but he does not provide a polynomial-time algorithm to find them. Chen *et al.* [3] study methods for producing two-dimensional Euclidean greedy drawings for graphs containing power diagrams, and Lillis and Pemmaraju [14] provide similar methods for graphs containing Delaunay triangulations. It is not clear whether either of these greedy drawings in Euclidean spaces run in polynomial time, however. Nevertheless, Leighton and Moitra [13] have recently given a polynomial-time algorithm for producing two-dimensional Euclidean greedy drawings of 3-connected planar graphs. The corresponding two-dimensional problem for greedy drawings of arbitrary graphs in non-Euclidean geometries also has a solution, in that Kleinberg [8] provides a polynomial-time algorithm for embedding any graph in the hyperbolic plane so as to allow for greedy routing using the standard metric for hyperbolic space.

The Importance of Succinctness. Unfortunately, all of the algorithms mentioned above for producing greedy embeddings, including the hyperbolic-space solution of Kleinberg [8] and the Euclidean-space solution of Leighton and Moitra [13], contain a hidden drawback that makes them ill-suited for the motivating application of geometric routing. Namely, each of the greedy embeddings mentioned above use vertex coordinates with representations requiring $\Omega(n \log n)$ bits in the worst case. Thus, these greedy approaches to geometric routing have the same space usage as traditional routing table approaches. Since the *raison d'être* for greedy embeddings is to improve and simplify traditional routing schemes, if embeddings are to be useful for geometric routing purposes, they should be *succinct*, that is, they should use vertices with representations having a number of bits that is polylogarithmic in n and the should allow for efficient distance comparisons using these representations.

We are, in fact, not the first to make this observation. Muhammad [16] specifically addresses succinctness, observing that a method based on extracting a planar subgraph of the routing network G and performing a hybrid greedy/face-routing algorithm in this embedding can be implemented using only $O(\log n)$ bits for each vertex coordinate, since planar graphs can be drawn in $O(n) \times O(n)$ grids [5,20]. For non-Euclidean spaces, Maymounkov [15] provides a greedy drawing method for three-dimensional hyperbolic space using vertices that can be represented with $O(\log^2 n)$ bits. His work leaves open the existence of succinct greedy embeddings for two-dimensional non-Euclidean spaces, however, as well as whether there are succinct non-Euclidean greedy embeddings that use only $O(\log n)$ bits per vertex.

Our Results. In this paper, we settle both questions of whether there are succinct greedy embeddings in two-dimensional non-Euclidean spaces and whether the vertices in such embeddings can be represented using an asymptotically optimal number of bits. In particular, we show that any n-vertex connected graph can be drawn in the hyperbolic plane with coordinates that can be represented using $O(\log n)$ bits so as to support greedy geometric routing between any pair of vertices, using a standard distance metric for hyperbolic space. Our scheme is constructive, runs in polynomial time, and allows the distance between any two vertices to be calculated efficiently from our representation of their coordinates. In addition, our greedy drawing scheme is based on the combination of a number of graph drawing and data structuring techniques.

2   Autocratic Weight-Balanced Trees

One of the new data structuring techniques we use in our greedy drawing scheme is a data structure that we call *autocratic weight-balanced binary trees*. These are first and foremost weight-balanced binary trees, which store weighted items at their leaves so that the depth of each item of weight w_i is $O(\log W/w_i)$, where W is the sum of all weights. Just as important, however, is that they are autocratic, by which we mean that the distance from any leaf v to any other leaf w is strictly greater than the distance from the root to w, where tree distance is measured by simple path length. Of course, this autocratic property implies that such binary trees are not proper, in that we allow for some internal nodes in such trees to have only one child. The challenge, of course, is to have a structure that is both autocratic and weight-balanced.

It turns out that there is a fairly simple method for turning any weight-balanced binary tree into an autocratic weight-balanced tree. So suppose we are given an ordered collection of k items with weights $\{w_1, w_2, \ldots, w_k\}$, such that each $w_i \geq 1$. If we store these items at the leaves of a binary tree T, we say that T is *weight-balanced* if the depth of each item i is $O(\log W/w_i)$, where $W = \sum_i w_i$. There are several existing schemes for producing a weight-balanced binary tree so that an inorder listing of the items stored at its leaves preserves the given order (e.g., see [6]).

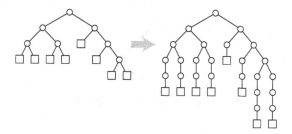

Fig. 1. Converting a weight-balanced binary tree into an autocratic weight-balanced binary tree

Suppose, then, that T is such an ordered weight-balanced tree, and let r denote the root of T. To convert T into an autocratic weight-balanced tree, T', we replace the edge connecting each leaf v to its parent with a path of length $1 + d_T(r, \text{parent}(v))$, where $d_T(v, w)$ denotes the length of the path from v to w in the tree T. That is, we insert a number of "dummy" nodes between each leaf and its parent that is equal to the depth of its parent. (See Fig. 1.)

This transformation increases the depth of each leaf in T by less than a factor of two and it keeps the depth of all other nodes in T unchanged. Thus, if the depth of a leaf storing item i in T was previously at most $c \log W/w_i$, for some constant c, then the depth of the corresponding leaf in T' is less than $2c \log W/w_i$, which is still $O(\log W/w_i)$. Given that T was weight-balanced, this implies that T' is a weight-balanced tree. More importantly, we have the following lemma.

Lemma 1. *The above transformation of a weight-balanced tree T produces an autocratic weight-balanced tree T'.*

Proof. We have already observed that the tree T' is weight-balanced. So we have yet to show that T' is autocratic. First, observe that, by a simple induction argument, if u is an ancestor in T of a leaf v, then in T' we have the following:

$$d_{T'}(u, v) = d_T(r, v) + d_T(u, v) - 1.$$

In particular, $d_{T'}(r, v) = 2d_T(r, v) - 1$. Let v and w be two leaves in T'. Furthermore, let u be the least common ancestor of v and w in T'. Then

$$
\begin{aligned}
d_{T'}(v, w) &= d_{T'}(u, v) + d_{T'}(u, w) \\
&= d_T(r, v) + d_T(u, v) - 1 + d_T(r, w) + d_T(u, w) - 1 \\
&= 2d_T(r, w) + 2d_T(u, v) - 2 \\
&\geq 2d_T(r, w) > d_{T'}(r, w).
\end{aligned}
$$

Thus, T' is an autocratic weight-balanced tree.   □

Therefore, we have a way of constructing for any ordered set of weighted items an autocratic weight-balanced tree for that set. We will use such data structures as auxiliary components in the structures we discuss next.

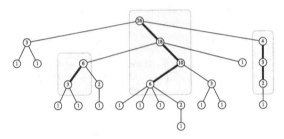

Fig. 2. The heavy path decomposition of a tree. Three heavy paths are shown; the remaining 17 nodes form degenerate length-0 heavy paths. Nodes are labelled with their subtree sizes.

3   Heavy Path Decompositions

Let T be a rooted ordered tree of arbitrary degree and depth having n nodes. Sleator and Tarjan [21] describe a scheme, which we call the *heavy path decomposition*, for decomposing T into a hierarchical collection of paths (see also [19] for an alternative path decomposition scheme with similar properties). Their scheme works as follows. For each node v in T, let $n(v)$ denote the number of descendents in the subtree rooted at v, including v itself. For each child-to-parent edge, $e = (v, w)$ in T, label e as a *heavy* edge if $n(v) > n(w)/2$. Otherwise, label e as a *light* edge. Connected components of heavy edges form paths, called *heavy paths*, which may in turn have many incident light edges. As a degenerate case, we also consider the zero-length path consisting of a single node in T incident only to light edges as a heavy path.

Note that the size of a subtree at least doubles every time we traverse a light edge from a child to a parent. (See Fig. 2.) Thus, if we compress every heavy path in T to a single "super" node, preserving the relative order of the nodes, then we define a tree, Z, of depth $O(\log n)$. Of course, the nodes in Z can have arbitrary degree. Nevertheless, for data structuring purposes, following Alstrup et al. [1], we may replace each vertex v in Z having d children $v_1, v_2, \ldots, v_d$ with a weight-balanced binary tree that uses the $n(v_i)$ values as weights. The useful property of this substitution is that any leaf-to-root path P in the resulting binary tree, Z'', will have length $O(\log n)$, since the lengths of the subpaths of P in the weight-balanced binary trees traversed in P form a telescoping sum that adds up to $O(\log n)$.

In our case, we use autocratic weight-balanced binary trees for the substitutions of high-degree super nodes in Z, so as to define a binary tree of depth $O(\log n)$. This construction will prove essential for our greedy embedding scheme. Before we present this geometric embedding, however, we first present a combinatorial greedy embedding in a completely contrived metric space, which we will subsequently show how to turn into a greedy embedding in the hyperbolic plane using the standard hyperbolic metric.

4   Greedy Embeddings in the Dyadic Tree Metric Space

Let G be a graph with n vertices and m edges for which we wish to construct a succinct greedy embedding. We show in this section how to produce a combinatorial greedy embedding in a contrived space we call the *dyadic tree metric space*.

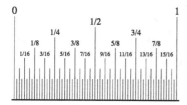
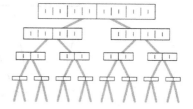

Fig. 3. The dyadic rational numbers (left) and a schematic of the dyadic tree metric space (right)

We may consider the infinite binary tree, $\mathcal{B}$, to be an abstract metric space, in which the distance between any two tree nodes is just the number of edges on the shortest path between them. But there is another natural metric that can be formed on the same tree by embedding it into the *dyadic rational numbers* (Fig. 3, left), that is, rational numbers with denominators that are powers of two. Let f be the map from $\mathcal{B}$ to the open interval $(0, 1)$ that maps the root of the tree to $1/2$, and that maps the children of a node x at level i of the tree to $f(x) \pm 2^{-i-2}$; thus, the children of the root map to the dyadic rational numbers $1/4$ and $3/4$, the grandchildren of the root map to $1/8, 3/8, 5/8, 7/8$, and so on. We define the *dyadic metric* on $\mathcal{B}$ as the metric in which the distance between two tree nodes x and y is $|f(x) - f(y)|$.

We will show that any graph may be greedily embedded into an ad-hoc metric space that combines these two tree metrics; we call it the *dyadic tree metric space*. A point in this space is represented by a pair (x, y), where x and y are nodes in the infinite binary tree, $\mathcal{B}$, and where x must be an ancestor of y (possibly equal to y itself). We define the distance between two points (x, y) and (x', y') in the dyadic tree metric space to be the sum of the tree distance between x and x' and of the dyadic distance $|f(y) - f(y')|$. The dyadic tree metric space can be represented as an infinite binary tree representing the x coordinates of each of its points, in which each tree node contains an interval of dyadic rational numbers; this interval of numbers is split into two halves at the two children of each node. This representation is depicted in Fig. 3, right.

Our embedding begins with us finding a spanning tree T of G, choosing a root arbitrarily, and producing a heavy path decomposition of T. For technical reasons we require that each node in a nontrivial heavy path of the decomposition have at least one child that is not in the path; we add dummy nodes to T if necessary, after forming the path decomposition, to ensure that this is true.

We orient the light edges for each heavy path P so that they are all on the same side of P and we orient the light edges incident upon the same vertex. We then compress each heavy path into a super node, using the orientation of edges around the vertices of each heavy path to determine the ordering of children for each node in the resulting tree, Z. If a super node in Z is the right child of its parent, we make the left-to-right ordering of children be the same as the ordering from parent to child in the heavy path; if, on the other hand, it is the left child of its parent, we make the left-to-right ordering of children be the same as the ordering from child to parent in the heavy path.

Next, we form groups of the nodes in Z that have the same parent in T. We form a weight-balanced binary tree for each these groups. Furthermore, within each group, we form a weight-balanced binary tree of the nodes in the group. Concatenating these

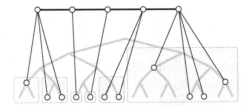

Fig. 4. Our two-level weight-balanced strategy for placing the children of the nodes on a heavy path. The groups of children for each heavy path node are assigned to subtrees in a weight-balanced way (gray shaded areas), and then within each subtree the individual children are placed using a second level of weight balancing. The third step of child placement, in which we make the subtree between the root (representing the heavy path) and its children autocratic, is not shown.

two levels of weight-balanced trees forms a single weight-balanced tree connecting the node in Z to each of its children; we apply the transformation described earlier to make this tree autocratic. The first three steps, in which we form a weight-balanced tree of the groups and a weight-balanced tree within each group, and then concatenate these two levels of weight-balanced trees to form a single binary tree for all children of the node in Z, are depicted in Fig. 4.

This construction of an autocratic weight-balanced tree for each node in Z can be used to embed Z into the infinite binary tree, $\mathcal{B}$. The root of Z may be placed at the root of $\mathcal{B}$, and the children of each node v in Z are placed under that node in the positions of $\mathcal{B}$ corresponding to their positions in the autocratic weight-balanced tree constructed for v. We observe that, in this way, all nodes of Z are placed at most $O(\log n)$ levels deep; for, due to the weight balancing, the distance in $\mathcal{B}$ between any node w and its parent v is proportional to the difference in the logarithms of the weights of the subtrees rooted at v and w, and along any path of Z these differences add in a telescoping series to $O(\log n)$.

We have embedded Z into the infinite binary tree, $\mathcal{B}$; hence, we are now ready to embed T itself into the dyadic tree metric. To do so, we must determine a pair (x, y) of coordinates for any node v of T; both x and y must be nodes of $\mathcal{B}$, and x must be an ancestor of y. The x coordinate of v is simply the node of $\mathcal{B}$ at which the heavy path of v is placed. The y coordinate of v is the least common ancestor in $\mathcal{B}$ of the placements of all the children of v. This calculation is the reason we required v to have at least one child; for leaf nodes of T, we instead set $y = x$. Due to our two-level weight balancing strategy, two nodes of T that belong to the same heavy path (and that therefore share the same x coordinate) will have different y coordinates, for their children will be placed within disjoint subtrees of the infinite binary tree, $\mathcal{B}$.

Lemma 2. *The above embedding of T into the dyadic tree metric space is greedy.*

Proof. Any directed path in T consists of edges that, when translated into the dyadic tree metric space, have three types: edges from a node to the parent heavy path in Z, edges within a heavy path, and edges from a node to a child heavy path in Z. We must show that edges of each type lead to a node that is closer to the terminus of the path.

For the edges that go from a node to the parent heavy path or to a child heavy path, this is straightforward: the contribution of the x-coordinates to the distance to the terminus decreases by one at each step, due to the autocratic property of our weight-balanced trees, more than offsetting any possible increase in the contribution of the y-coordinates.

For the edges that remain within a heavy path, the x coordinates remain unchanged and do not lead to any increase or decrease of the distance to the terminus. The y coordinates are linearly ordered by the map f from infinite binary tree nodes to dyadic rationals, and our weight-balanced trees were chosen to be consistent with this linear ordering; therefore, any step along the heavy path, either towards a node of the path that is the ancestor of the terminus or towards the topmost node of the path and the edge leading to the parent node in Z, decreases the distance to the terminus.   □

5   Succinct Greedy Embedding in the Hyperbolic Plane

We have shown that any tree T (and any graph G by choosing a spanning tree of G) may be greedily and succinctly embedded into a dyadic tree metric space. To complete our greedy embedding, it remains to show that this space may be embedded, independently of our original graph (but depending on a parameter D determined by the number of vertices of the graph), into the hyperbolic plane in such a way that the greedy property of the embedding of T is preserved. That is, although the distances themselves in the hyperbolic plane may differ from those in the dyadic tree metric space, composing our embedding of T into the dyadic tree metric space with our embedding of the dyadic tree metric space should yield a greedy embedding of T into the hyperbolic plane.

Due to the existence of this embedding, we may reinterpret the succinct coordinates computed for the embedding of a graph into the dyadic tree metric space as also being coordinates for a subset of points in the hyperbolic plane.

Our overall strategy will be to embed the infinite binary tree, $\mathcal{B}$, into the hyperbolic plane in such a way that any edge has length $D + O(1)$ and crosses a *buffer zone* of width D, bounded by two hyperbolic lines (Fig. 5). The buffer zones for different edges will be disjoint from each other. Thus, any two nodes of the tree that have tree distance k units apart will have hyperbolic distance at least Dk (because any path between the two nodes must cross k buffer zones) and at most $(D+O(1))k$ (there exists a path following tree edges with that length). In our application, all tree paths will have $O(\log n)$ edges; thus, by choosing $D = \Omega(\log n)$ we may guarantee that the order relation between any two distinct tree distances remains unchanged by this hyperbolic embedding. Any point (x, y) of the dyadic tree metric will be placed near the embedding of tree node x,

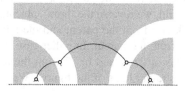

Fig. 5. Disjoint buffer zones of width D are crossed by each edge of an embedding of $\mathcal{B}$ into the hyperbolic plane, so that tree distance and hyperbolic distance closely approximate each other

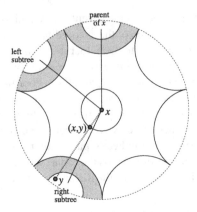

Fig. 6. Top-down placement of node x of $\mathcal{B}$ and point (x, y) of the dyadic tree metric space into the hyperbolic plane, shown in a Poincaré disk model centered at x

and this placement will ensure the greediness of any edge whose endpoints belong to different paths of our heavy path decomposition.

Next, we place nodes of the infinite binary tree, $\mathcal{B}$, into the hyperbolic plane, with the buffer zones described above. Although this placement is conceptual rather than algorithmic, we may view it as being performed in a top down traversal of the tree, so that when node x is placed we will already know the location of its parent, the buffer zone separating x from its parent, and a line connecting it to its parent and on which it must be placed. We place x itself on this line in such a way that the boundary of the parental buffer zone forms one of the seven sides of an ideal regular heptagon—a figure in the hyperbolic plane formed by seven lines that are asymptotic to each other but never intersect, such that the angle subtended by each line as viewed from x is equal. Figure 6 shows this placement, in a Poincaré disk model of the hyperbolic plane centered at x; the parental buffer zone is the topmost shaded region in the figure and the vertical line through x is the one connecting it to its parent node. The large arcs depict hyperbolic lines forming the heptagon described above.

In the case where x is the right child of its parent, so that the upper nodes of the heavy path represented by x have children in its left subtree and the lower nodes of the heavy path have children in the right subtree, shown in the figure, we place the left subtree within the halfplane bounded by the heptagon side one step counterclockwise from the parent, and the right subtree within the halfplane bounded by the heptagon side three steps counterclockwise from the parent, as shown in the figure. In the case where x is its parent's left child, we reverse the figure, placing the right subtree within the halfplane one step clockwise from the parent and the left subtree within the halfplane three steps clockwise from the parent. In either case, we draw lines connecting x to its child nodes, at angles of $2\pi/7$ and $6\pi/7$ from the angle of the line connecting x to its parent (the solid straight lines of the figure). We use the heptagon edges as the outer boundaries of buffer zones between x and its children, and we set the inner boundaries of the buffer zones to be hyperbolic lines perpendicular to the lines connecting x to its children, at distance D from the outer boundaries of the buffer zones.

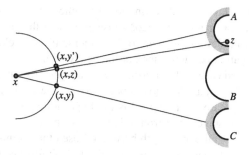

Fig. 7. Illustration for proof of greediness of our embedding (not to scale)

With this information determined, we may continue to place the children of x in the same way.

We are finally ready to describe the mapping of the dyadic tree metric space into the hyperbolic plane. Recall that each point of the dyadic tree metric space consists of a pair (x, y) where x and y are nodes of $\mathcal{B}$, x a parent of y. We draw small circles of equal radius centered at each point where we have placed a node of $\mathcal{B}$—the precise radius is unimportant as long as it is small enough that the circles are disjoint from the buffer zones. Then, given a point (x, y) of the dyadic tree metric space, we draw a hyperbolic line segment from x to y (the dotted straight line in the figure), and place (x, y) at the point where this line segment intersects the circle centered at x. In the case $x = y$, which happens in our construction only for leaves, we instead place (x, x) at the point where the line segment from x to its parent intersects the circle centered at x.

Theorem 1. *For sufficiently large values of D, the embedding of G formed by composing the embedding from G into the dyadic tree metric space and the embedding of the dyadic tree metric space into the hyperbolic plane is greedy.*

Proof. We show that, for every edge e of the chosen spanning tree, and every possible terminus v of a path using e, that traveling along e reduces the distance to the terminus. We assume that the starting endpoint of e is placed at point (x, y) of the dyadic tree metric, the ending endpoint is placed at point (x', y'), and that these points are mapped as described above to the hyperbolic plane. We distinguish several cases.

First, if $x \neq x'$, let $k = O(\log n)$ be the tree distance from x' to the destination. Then, due to the autocratic property of our weight-balanced placement of heavy paths into the dyadic tree metric, x is at tree distance at least $k + 1$ from the destination. As discussed above, due to the buffer zones of our construction, (x, y) is at hyperbolic distance at least $(k + 1)D$ from the destination, while (x', y') is at hyperbolic distance at most $k(D + O(1))$. By choosing D sufficiently large (a constant times $\log n$), we can guarantee that the former distance is larger than the latter and that this step is greedy.

Second, if $x = x'$ and the eventual destination also has the same value of x, the result follows from the fact that our embedding places the nodes of any heavy path consecutively over an arc of less than half of a circle. Such an embedding is greedy for any path, no matter how the nodes are distributed within the arc.

Third, if $x = x'$ and the eventual destination is reached via the parent of x, the step is greedy for the same reason as in the second case: the nodes that are mapped to x form a heavy path placed in order along an arc of less than half the circle, with the node of the arc closest to the parent being the apex of the heavy path.

The most complicated case is the fourth: $x = x'$ and the eventual destination z has x'' as a proper descendant of x. The closest point to z on the circle surrounding x onto which (x, y) and (x', y') are both mapped is the hyperbolic point represented by the coordinates (x, z); the distance to z from other points on the circle can be calculated as a monotonic function of the arc length between those other points and (x, z). Thus, moving around the circle towards (x, z) is a greedy step. Unfortunately, the point (x, z) may not be a node of the heavy path; rather, the node of the heavy path from which z descends may be some other nearby point (x, y''). We must show that any step along the heavy path towards this point is greedy.

In most cases, it is straightforward to show that this step is greedy: a step around the circle towards (x, z) is also a step towards (x, y''), which as we have argued immediately above is greedy. The only possible exception occurs when $y' = y''$ and when the true closest point on the circle to z, that is, (x, z), lies on the arc of the circle between y and y'. In this case we must show that (x, y') and (x, z) are closer in arc length than (x, y) and (x, z), for then the greediness of the step will follow from the monotonicity of the distance to z as a function of arc length.

Let $\hat{y}$ be the least common ancestor in the binary tree of the two disjoint subtrees containing y and y'. Let A be the inner boundary of the buffer zone adjacent to $\hat{y}$ that contains y', let C be the inner boundary of the buffer zone adjacent to $\hat{y}$ that contains y, and let B be the edge of the regular ideal heptagon adjacent to $\hat{y}$ that separates A from C. Figure 7 illustrates this notation. These three hyperbolic lines may not be symmetrically placed relative to x, due to the asymmetry of the placement of the two subtrees relative to the parent at each node x. However, the distances from x to A and to C are within $O(1)$ of each other, and B is closer to x by a distance of $D - O(1)$. It is a basic property of hyperbolic geometry that the angle that an object subtends, as viewed from a fixed point of view x, is inversely proportional to an exponential function of the distance of the object from x. Thus, B will subtend an angle, as viewed from x, that is larger than the angles subtended by A and C by a factor exponential in $D - O(1)$. In particular, for sufficiently large D (larger than some fixed constant, a weaker requirement than the one above that $D = \Omega(\log n)$), both A and C will subtend smaller angles than the angle subtended by B. Then, any point behind line A, and in particular the point z, will form an arc from (x, y') to (x, z) that is shorter than the arc from (x, y) to (x, z). The greediness of the step from (x, y) to (x, y') follows from the monotonicity of the distance to z as a function of arc length. $\square$

References

1. Alstrup, S., Lauridsen, P.W., Sommerlund, P., Thorup, M.: Finding cores of limited length. In: Rau-Chaplin, A., Dehne, F., Sack, J.-R., Tamassia, R. (eds.) WADS 1997. LNCS, vol. 1272, pp. 45–54. Springer, Heidelberg (1997)
2. Bose, P., Morin, P., Stojmenović, I., Urrutia, J.: Routing with Guaranteed Delivery in Ad Hoc Wireless Networks. Wireless Networks 6(7), 609–616 (2001)

3. Chen, M.B., Gotsman, C., Wormser, C.: Distributed computation of virtual coordinates. In: Proc. 23rd Symp. Computational Geometry (SoCG 1997), pp. 210–219 (1997)

4. Dhandapani, R.: Greedy drawings of triangulations. In: Proc. 19th ACM-SIAM Symp. Discrete Algorithms (SODA 2008), pp. 102–111 (2008)

5. de Fraysseix, H., Pach, J., Pollack, R.: How to draw a planar graph on a grid. Combinatorica 10(1), 41–51 (1990)

6. Gilbert, E.N., Moore, E.F.: Variable-Length binary encodings. Bell System Tech. J. 38, 933–968 (1959)

7. Karp, B., Kung, H.T.: GPSR: greedy perimeter stateless routing for wireless networks. In: Proc. 6th ACM Mobile Computing and Networking (MobiCom), pp. 243–254 (2000)

8. Kleinberg, R.: Geographic Routing Using Hyperbolic Space. In: Proc. 26th IEEE Int. Conf. Computer Communications (INFOCOM 2007), pp. 1902–1909. IEEE Press, Los Alamitos (2007)

9. Kranakis, E., Singh, H., Urrutia, J.: Compass routing on geometric networks. In: Proc. 11th Canad. Conf. Computational Geometry (CCCG), pp. 51–54 (1999)

10. Kuhn, F., Wattenhofer, R., Zhang, Y., Zollinger, A.: Geometric ad-hoc routing: of theory and practice. In: Proc. 22nd ACM Symp. Principles of Distributed Computing (PODC), pp. 63–72 (2003)

11. Kuhn, F., Wattenhofer, R., Zollinger, A.: Asymptotically optimal geometric mobile ad-hoc routing. In: Proc. 6th ACM Discrete Algorithms and Methods for Mobile Computing and Communications (DIALM), pp. 24–33 (2002)

12. Kuhn, F., Wattenhofer, R., Zollinger, A.: Worst-Case optimal and average-case efficient geometric ad-hoc routing. In: Proc. 4th ACM Symp. Mobile Ad Hoc Networking & Computing (MobiHoc), pp. 267–278 (2003)

13. Leighton, T., Moitra, A.: Some results on greedy embeddings in metric spaces. In: Proc. 49th IEEE Symp. Foundations of Computer Science (FOCS) (2008)

14. Lillis, K.M., Pemmaraju, S.V.: On the Efficiency of a Local Iterative Algorithm to Compute Delaunay Realizations. In: McGeoch, C.C. (ed.) WEA 2008. LNCS, vol. 5038, pp. 69–86. Springer, Heidelberg (2008)

15. Maymounkov, P.: Greedy Embeddings, Trees, and Euclidean vs. Lobachevsky Geometry. M.I.T (manuscript, 2006),
http://pdos.csail.mit.edu/~petar/papers/
maymounkov-greedy-prelim.pdf

16. Muhammad, R.B.: A distributed geometric routing algorithm for ad hoc wireless networks. In: Proc. IEEE Conf. Information Technology (ITNG), pp. 961–963 (2007)

17. Papadimitriou, C.H., Ratajczak, D.: On a conjecture related to geometric routing. Theor. Comput. Sci. 344(1), 3–14 (2005)

18. Rao, A., Ratnasamy, S., Papadimitriou, C.H., Shenker, S., Stoica, I.: Geographic routing without location information. In: Proc. 9th Int. Conf. Mobile Computing and Networking (MobiCom 2003), pp. 96–108. ACM, New York (2003)

19. Schieber, B., Vishkin, U.: On finding lowest common ancestors: simplification and parallelization. SIAM J. Comput. 17(6), 1253–1262 (1988)

20. Schnyder, W.: Embedding planar graphs on the grid. In: Proc. 1st ACM-SIAM Sympos. Discrete Algorithms, pp. 138–148 (1990)

21. Sleator, D.D., Tarjan, R.E.: A data structure for dynamic trees. J. Comp. and Sys. Sci. 26(3), 362–391 (1983)

An Algorithm to Construct
Greedy Drawings of Triangulations*

Patrizio Angelini[1], Fabrizio Frati[1], and Luca Grilli[2]

[1] Dipartimento di Informatica e Automazione - Roma Tre University
{angelini,frati}@dia.uniroma3.it
[2] Dipartimento di Ingegneria Elettronica e dell'Informazione - Perugia University
luca.grilli@diei.unipg.it

Abstract. We show an algorithm to construct greedy drawings of every given triangulation.

1 Introduction

In a *greedy routing* setting, a node forwards packets to a neighbor that is *closer* to the destination's geographic location. Different distance metrics define different meanings for the word "closer", and consequently define different routing algorithms for the packet delivery. The most used and studied metric is of course the *Euclidean distance*.

The efficiency of the greedy routing algorithms strongly relies on the geographic coordinates of the nodes. This is a drawback of such algorithms, for the following reasons: (i) Nodes of the network have to know their locations, hence they have to be equipped with GPS devices, which are expensive and increase the energy consumption of the nodes; (ii) geographic coordinates are independent of the network obstructions, i.e. obstacles making the communication between two close nodes impossible, and, more in general, they are independent of the network topology; this could lead to situations in which the communication fails because a *void* has been reached, i.e., the packet has reached a node whose neighbors are all farther from the destination than the node itself.

A brilliant solution to such weaknesses has been proposed by Rao *et al.* who in [9] proposed a scheme in which nodes decide *virtual coordinates* and then apply the greedy routing algorithm relying on such coordinates rather than on the real geographic ones. Since virtual coordinates do not need to reflect the nodes actual positions, they can be suitably chosen to guarantee that the greedy routing algorithm delivers packets with high probability. Experiments have shown that such an approach strongly improves the reliability of greedy routing [9,8]. Further, it has been proved that virtual coordinates guarantee greedy routing to work for every connected topology when they can be chosen in the hyperbolic plane [5], and that some modifications of the routing algorithm guarantee that Euclidean virtual coordinates can be chosen so that the packet delivery always succeeds [1], even if the coordinates need to be locally computed [2].

* Work partially supported by MUR under Project "MAINSTREAM: Algoritmi per strutture informative di grandi dimensioni e data streams" and by the Italian Ministry of Research, Grant number RBIP06BZW8, project FIRB "Advanced tracking system in intermodal freight transportation".

I.G. Tollis and M. Patrignani (Eds.): GD 2008, LNCS 5417, pp. 26–37, 2009.

Subsequent to the Rao *et al.* paper [9], an intense research effort has been devoted to determine on which network topologies the Euclidean greedy routing with virtual coordinates is guaranteed to work. From a graph-theoretic point of view, the problem is as follows: Which are the graphs that admit a *greedy embedding*, i.e., a straight-line drawing Γ such that, for every pair of nodes u and v, there exists a *distance-decreasing path* in Γ? A path $(v_0, v_1, \ldots, v_m)$ is distance-decreasing if $d(v_i, v_m) < d(v_{i-1}, v_m)$, for $i = 1, \ldots, m$. In [8] Papadimitriou and Ratajczak conjectured the following:

Conjecture 1. (Papadimitriou and Ratajczak [8]) Every triconnected planar graph admits a greedy embedding.

Papadimitriou and Ratajczak showed that $K_{k,5k+1}$ has no greedy embedding, for $k \geq 1$. As a consequence, both the triconnectivity and the planarity are necessary, because there exist planar non-triconnected graphs, such as $K_{2,11}$, and non-planar triconnected graphs, such as $K_{3,16}$, that do not admit any greedy embedding. Further, they observed that, if a graph G has a greedy embedding, then any graph containing G as a spanning subgraph has a greedy embedding. It follows that Conjecture 1 extends to all graphs which are spanned by a triconnected planar graph. Related to such an observation, they proved that every triconnected graph not containing a $K_{3,3}$-minor has a triconnected planar spanning subgraph.

For a few classes of triconnected planar graphs the conjecture is easily shown to be true, for example graphs with a *Hamiltonian path* and *Delaunay Triangulations*. At SODA'08 [3], Dhandapani proved the conjecture for the first non-trivial class of triconnected planar graphs, namely he showed that every *triangulation* admits a greedy embedding. The proof of Dhandapani is probabilistic, namely the author proves that among all the *Schnyder drawings* of a triangulation [10], there exists a drawing which is greedy. Although such a proof is elegant, relying at the same time on an old Combinatorial Geometry theorem, known as the *Knaster-Kuratowski-Mazurkievicz Theorem* [6], and on standard Graph Drawing techniques, as the *Schnyder realizers* [10] and the *canonical orderings* of a triangulation [4], it does not lead to an embedding algorithm.

In this paper we show an algorithm for constructing greedy drawings of triangulations. The algorithm relies on a different and maybe more intuitive approach with respect to the one used in [3]. We define a simple class of graphs, called *binary cactuses*, and we provide an algorithm to construct a greedy drawing of any binary cactus. Finally, we show how to find, for every triangulation, a binary cactus spanning it. It is clear that the previous statements imply an algorithm for constructing greedy drawings of triangulations. Namely, consider any triangulation G, apply the algorithm to find a binary cactus S spanning G, and then apply the algorithm to construct a greedy drawing of S. As already observed, adding edges to a greedy drawing leaves the drawing greedy, hence S can be augmented to G, obtaining the desired greedy drawing of G.

Theorem 1. *Given a triangulation G, there exists an algorithm to compute a greedy drawing of G.*

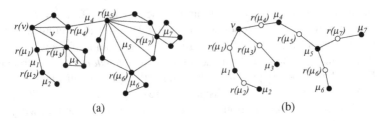

Fig. 1. (a) A binary cactus S. (b) The block-cutvertex tree of S. White (resp. black) circles represent C-nodes (resp. B-nodes).

2   Preliminaries

A graph G is *connected* if every pair of vertices of G is connected by a path. A *cutvertex* is a vertex whose removal increases the number of connected components of G. A connected graph is *biconnected* if it has no cutvertices. The maximal biconnected subgraphs of a graph are its *blocks*. Each edge of G falls into a single block of G, while cutvertices are shared by different blocks. The *block-cutvertex tree*, or BC-tree, of a connected graph G is a tree with a B-node for each block of G and a C-node for each cutvertex of G. Edges in the BC-tree connect each B-node μ to the C-nodes associated with the cutvertices in the block of μ.

The BC-tree of G may be thought as rooted at a specific block ν. When the BC-tree $\mathcal{T}$ of a graph G is rooted at a certain block ν, we denote by $G(\mu)$ the subgraph of G induced by all vertices in the blocks contained in the subtree of $\mathcal{T}$ rooted at μ. In a rooted BC-tree $\mathcal{T}$ of a graph G, for each B-node μ we denote by $r(\mu)$ the cutvertex of G parent of μ in $\mathcal{T}$. If μ is the root of $\mathcal{T}$, i.e., $\mu = \nu$, then we let $r(\mu)$ denote any non-cutvertex node of the block associated with μ. In the following, unless otherwise specified, each considered BC-tree is meant to be rooted at a certain B-node ν such that the block associated with ν has at least one vertex $r(\nu)$ which is not a cutvertex. It is not difficult to see that such a block exists in every planar graph.

A *rooted triangulated binary cactus S*, in the following simply called *binary cactus*, is a connected graph such that (see Fig 1): (i) the block associated with each B-node of $\mathcal{T}$ is either an edge or a *triangulated cycle*, i.e., a cycle $(r(\mu), u_1, u_2, \ldots, u_h)$ triangulated by the edges from $r(\mu)$ to each of $u_1, u_2, \ldots, u_h$; (ii) every cutvertex is shared by exactly two blocks of S.

A *planar drawing* of a graph is a mapping of each vertex to a distinct point of the plane and of each edge to a Jordan curve between its endpoints such that no two edges intersect except, possibly, at common endpoints. A planar drawing of a graph determines a circular ordering of the edges incident to each vertex. Two drawings of the same graph are *equivalent* if they determine the same circular ordering around each vertex. A *planar embedding* is an equivalence class of planar drawings. A planar drawing partitions the plane into topologically connected regions, called *faces*. The unbounded face is the *outer face*. The outer face of a graph G is denoted by $f(G)$. A *chord* of a graph G is an edge connecting two non-adjacent vertices of $f(G)$. A graph together with a planar embedding and a choice for its outer face is called *plane graph*. A plane graph is a *triangulation* when all its faces are triangles. A plane graph is *internally-triangulated*

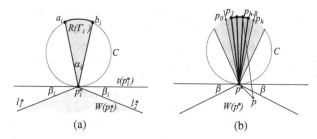

Fig. 2. (a) Illustration for Properties 1–3 of Γ. (b) Base case of the algorithm. The light and dark shaded region represents $R(\Gamma)$ (the angle of $R(\Gamma)$ at p^* is α). The dark shaded region represents the intersection of $W(p^*, \alpha/2)$ with the circle delimited by C.

when all its internal faces are triangles. An *outerplane* graph is a plane graph such that all its vertices are incident to the outer face. A *Hamiltonian cycle* of a graph G is a simple cycle passing through all vertices of G. Notice that a biconnected outerplane graph has only one Hamiltonian cycle, the one delimiting its outer face.

3 Greedy Drawing of a Binary Cactus

In this section, we give an algorithm to compute a greedy drawing of a binary cactus S. Such a drawing is constructed by a bottom-up traversal of the BC-tree $\mathcal{T}$ of S.

Consider the root μ of a subtree of $\mathcal{T}$ corresponding to a block of S, consider the k children of μ, which correspond to cutvertices of S, and consider the children of such cutvertices, say $\mu_1, \mu_2, \ldots, \mu_k$. Notice that each C-node child of μ is parent of exactly one B-node μ_i of $\mathcal{T}$, by definition of binary cactus. For each $i = 1, \ldots, k$, inductively assume to have a drawing Γ_i of $S(\mu_i)$ satisfying the following properties. Let α_i and β_i be any two angles less than $\pi/4$ such that $\beta_i \geq \alpha_i$. Refer to Fig. 2.a.

- *Property 1.* Γ_i is a greedy drawing.
- *Property 2.* Γ_i is entirely contained inside a region $R(\Gamma_i)$ delimited by an arc (a_i, b_i) of a circumference C and by two segments (p_i^*, a_i) and (p_i^*, b_i), such that p_i^* is a point of C and the diameter through p_i^* cuts (a_i, b_i) in two arcs of the same length. The angle $\widehat{a_i p_i^* b_i}$ is α_i.
- *Property 3.* Consider the tangent $t(p_i^*)$ to C in p_i^*. Consider two half-lines l_1^* and l_2^* incident to p_i^*, lying on the opposite part of C with respect to $t(p_i^*)$, and forming angles equal to β_i with $t(p_i^*)$. Denote by $W(p_i^*)$ the wedge centered at p_i^*, delimited by l_1^* and l_2^*, and not containing C. Then, for every vertex v in $S(\mu_i)$ and for every point p internal to $W(p_i^*)$, a distance-decreasing path $(v = v_0, v_1, \ldots, v_l = r(\mu_i))$ from v to $r(\mu_i)$ exists in Γ_i such that $d(v_j, p) < d(v_{j-1}, p)$ for $j = 1, \ldots, l$.

In the base case, block μ has no child. Denote by $(r(\mu) = u_0, u_1, \ldots, u_{h-1})$ the block of S corresponding to μ. If $h = 2$, i.e., μ corresponds to an edge, draw such an edge as a vertical segment, with u_1 above u_0. A region $R(\Gamma_i)$ can be easily constructed, for every angles α and β, with $\beta \geq \alpha$, satisfying the above properties. If $h > 2$, i.e., μ corresponds to a triangulated cycle of S, place $r(\mu)$ at any point p^* and consider a wedge $W(p^*, \alpha/2)$ that has an angle equal to $\alpha/2$, that is incident to $r(\mu)$, and that is bisected by the vertical half-line incident to $r(\mu)$ and directed upward (see Fig. 2.b). Denote by p'_a and p'_b the intersection points of the half-lines delimiting $W(p^*, \alpha/2)$ with a circumference C through $r(\mu)$, properly intersecting the border of $W(p^*, \alpha/2)$ twice. Denote by A the arc of C between p'_a and p'_b not containing p^*. Consider points $p'_a = p_0, p_1, \ldots, p_h = p'_b$ on A such that the distance between any two consecutive points p_i and p_{i+1} is the same. Place vertex u_i at point p_i, for $i = 1, 2, \ldots, h-1$.

We show that the constructed drawing Γ satisfies Property 1. Consider any two vertices u_i and u_j, with $i < j$. If $i = 0$, then u_0 and u_j are joined by an edge, which provides a distance-decreasing path among them. Otherwise, we claim that path $(u_i, u_{i+1}, \ldots, u_j)$ is distance-decreasing. In fact, for each $l = i, i+1, \ldots, j-2$, angle $\widehat{u_l u_{l+1} u_j}$ is greater than $\pi/2$, because triangle (u_l, u_{l+1}, u_j) is inscribed in less than half a circumference with u_{l+1} as middle point. Hence, (u_l, u_j) is the longest side of triangle (u_l, u_{l+1}, u_j) and $d(u_{l+1}, u_j) < d(u_l, u_j)$ follows. Drawing Γ satisfies Property 2 by construction. In order to prove that Γ satisfies Property 3, we have to show that, for every vertex u_i, with $i \geq 1$, and for every point p in $W(p^*)$, $d(u_0, p) < d(u_i, p)$. However, angle $\widehat{pp^*p_i}$ is at least $\beta + (\frac{\pi}{2} - \frac{\alpha}{4})$, which is more than $\pi/2$. It follows that segment $\overline{pp_i}$ is the longest side of triangle (p, p^*, p_i), thus proving that $d(u_0, p) < d(u_i, p)$.

Now suppose μ is a node of $\mathcal{T}$ having k children. We show how to construct a drawing Γ of $S(\mu)$ satisfying Properties 1–3 with parameters α and β. Denote by $(r(\mu) = u_0, u_1, \ldots, u_{h-1})$ the block of S corresponding to μ. Consider any circumference C with center c. Let p^* be the point of C with smallest y-coordinate. Consider wedges $W(p^*, \alpha)$ and $W(p^*, \alpha/2)$ with angles α and $\alpha/2$, respectively, incident to p^* and such that the diameter of C through p^* is their bisector. Region $R(\Gamma)$ is the intersection region of $W(p^*, \alpha)$ with the closed circle delimited by C.

Consider a circumference C' with center c intersecting the two lines delimiting $W(p^*, \alpha/2)$ in two points p'_a and p'_b such that angle $\widehat{p'_a c p'_b} = 3\alpha/2$. Denote by p' the intersection point between C' and (c, p^*). Observe that angle $\widehat{p'_a p' p'_b} = 3\alpha/4$. Denote by A the arc of C' delimited by p'_a and p'_b not containing p'. Consider points $p'_a = p_0, p_1, \ldots, p_h = p'_b$ on A such that the distance between any two consecutive points p_i and p_{i+1} is the same. Observe that, for each $i = 0, 1, \ldots, h-1$, angle $\widehat{p_i c p_{i+1}} = \frac{3\alpha}{2h}$.

First, we draw the block of S corresponding to μ. As in the base case, place vertex $u_0 = r(\mu)$ at p^* and, for $i = 1, 2, \ldots, h-1$, place u_i at point p_i. Recursively construct a drawing Γ_i of $S(\mu_i)$ satisfying Properties 1–3 with $\alpha_i = \frac{3\alpha}{16h}$ and $\beta_i = \frac{3\alpha}{8h}$.

We are going to place each drawing Γ_i of $S(\mu_i)$ together with the drawing of the block of S corresponding to μ, thus obtaining a drawing Γ of $S(\mu)$. Not all h nodes u_i are cutvertices of S. However, with a slight abuse of notation, we suppose that block $S(\mu_i)$ has to be placed at node u_i. Refer to Fig 3. Consider point p_i and its "neighbors" p_{i-1} and p_{i+1}. Consider lines $t(p_{i-1})$ and $t(p_{i+1})$ tangent to C' through p_{i-1} and p_{i+1}, respectively. Further, consider circumferences C_{i-1} and C_{i+1} centered at p_{i-1} and p_{i+1}, respectively, and passing through p_i. Moreover, consider lines h_{i-1} and h_{i+1} through p_i and tangent to C_{i-1} and C_{i+1}, respectively. For each point p_i, consider two half-lines t_1^i and t_2^i incident to p_i, cutting C' twice, and forming angles $\beta_i = \frac{3\alpha}{8h}$ with $t(p_i)$. Denote by $W(p_i)$ the wedge delimited by t_1^i and t_2^i and containing c.

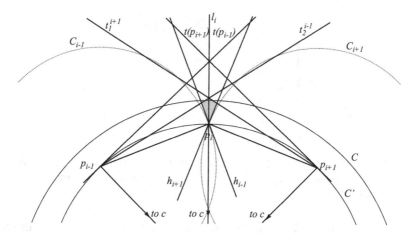

Fig. 3. Lines and circumferences in the construction of Γ. The shaded region is R_i.

We place Γ_i inside the bounded region R_i intersection of the half-plane H^{i-1} delimited by h_{i-1} and not containing C_{i-1}, of the half-plane H^{i+1} delimited by h_{i+1} and not containing C_{i+1}, of $W(p_{i-1})$, of $W(p_{i+1})$, and of the circle delimited by C.

First, we show that R_i is "large enough" to contain Γ_i, namely we claim that there exists an isosceles triangle T that has an angle larger than $\alpha_i = \frac{3\alpha}{16h}$ incident to p_i and that is completely contained in R_i. Such a triangle will have the further feature that the angle incident to p_i is bisected by the half-line l_i incident to c and passing through p_i.

Lines h_{i-1} and h_{i+1} are both passing through p_i; we prove that they have different slopes and we compute the angles they form at p_i. Line h_{i-1} forms an angle of $\pi/2$ with segment $\overline{p_{i-1}p_i}$; angle $\widehat{cp_ip_{i-1}}$ is equal to $\frac{\pi}{2} - \frac{3\alpha}{4h}$, since $\widehat{p_icp_{i-1}} = \frac{3\alpha}{2h}$ and since triangle (p_{i-1}, c, p_i) is isosceles. Hence, the angle delimited by h_{i-1} and l_i is $\pi - \pi/2 - (\frac{\pi}{2} - \frac{3\alpha}{4h}) = \frac{3\alpha}{4h}$. Analogously, the angle between l_i and h_{i+1} is $\frac{3\alpha}{4h}$. Hence, the intersection of H^{i-1} and H^{i+1} is a wedge $W(p_i, h_{i-1}, h_{i+1})$ centered at p_i, with an angle of $\frac{3\alpha}{2h}$, and bisected by l_i. We claim that each of t_2^{i-1} and t_1^{i+1} cuts the border of $W(p_i, h_{i-1}, h_{i+1})$ twice. The angle between $t(p_{i-1})$ and $\overline{p_{i-1}p_i}$ is $\frac{3\alpha}{4h}$, namely the angle between $t(p_{i-1})$ and $\overline{cp_{i-1}}$ is $\pi/2$, and angle $\widehat{cp_{i-1}p_i}$ is $\frac{\pi}{2} - \frac{3\alpha}{4h}$. The angle between $t(p_{i-1})$ and t_2^{i-1} is $\beta_i = \frac{3\alpha}{8h}$ by construction. Hence, the angle between t_2^{i-1} and $\overline{p_{i-1}p_i}$ is $\frac{3\alpha}{4h} - \frac{3\alpha}{8h} = \frac{3\alpha}{8h}$. Since the slope of both h_{i-1} and h_{i+1} with respect to $\overline{p_{i-1}p_i}$ is greater than $\frac{3\alpha}{8h}$ and less than $\pi - \frac{3\alpha}{8h}$, namely the slope of h_{i-1} and h_{i+1} with respect to $\overline{p_{i-1}p_i}$ is $\frac{\pi}{2}$ and $\frac{\pi}{2} + \frac{3\alpha}{2h}$, respectively (notice that $\alpha \le \pi/4$ and $h \ge 2$), then t_2^{i-1} intersects both h_{i-1} and h_{i+1}. It can be analogously proved that t_1^{i+1} intersects h_{i-1} and h_{i+1}. It follows that the intersection of H^{i-1}, H^{i+1}, $W(p_{i-1})$, and $W(p_{i+1})$ contains a triangle T as required by the claim (the angle of T incident to p_i is $\frac{3\alpha}{2h}$). Considering circumference C does not invalidate the existence of T, since C is concentric with C' and has a bigger radius, hence T can be chosen sufficiently small so that it completely lies inside C.

Now Γ_i can be placed inside T, by scaling Γ_i down till it fits inside T (see Fig. 4.a). The scaling always allows Γ_i to be placed inside T, since the angle of $R(\Gamma_i)$ incident to p is $\alpha_i = \frac{3\alpha}{16h}$, that is smaller than the angle of T incident to p_i, which is $\frac{3\alpha}{2h}$. In

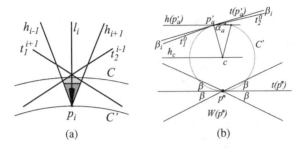

Fig. 4. (a) Placement of Γ inside R_i. Region $R(\Gamma)$ is the darkest, triangle T is composed of $R(\Gamma)$ and of the second darkest region, R_i is composed of T and of the light shaded region. (b) Illustration for the proof of Lemma 1.

particular, we choose to place Γ_i inside T so that l_i bisects the angle of $R(\Gamma_i)$ incident to p_i. This concludes the construction of Γ. We have the following lemmata.

Lemma 1. *The closed wedge $W(p^*)$ is completely contained inside the open wedge $W(p_i)$, for each $i = 0, 1, \ldots, h$.*

Proof: Consider any point p_i. Observe that p_i is contained inside the wedge $\overline{W}(p^*)$ obtained by reflecting $W(p^*)$ with respect to $t(p^*)$. Namely, p_i is contained inside $W(p^*, \alpha/2)$, which is in turn contained inside $\overline{W}(p^*)$, since $\alpha/2 < \pi - 2\beta$, as a consequence of the fact that $\pi/4 > \beta \geq \alpha$. Hence, in order to prove the lemma, it suffices to show that the absolute value of the slope of each of t_1^i and t_2^i is less than the absolute value of the slope of the half-lines delimiting $W(p^*)$. Such latter half-lines form angles of β, by construction, with the x-axis.

The slope of t_1^i can be computed by summing up the slope of t_1^i with respect to $t(p_i)$ with the slope of $t(p_i)$. The former slope is equal to $\beta_i = \frac{3\alpha}{8h}$, by construction. Recalling that $t(p_i)$ is the tangent to A in p_i, the slope of $t(p_i)$ is bounded by the maximum among the slopes of the tangents to points of A. Such a maximum is clearly achieved at p'_a and p'_b and is equal to $3\alpha/4$. Namely, refer to Fig. 4.b and consider the horizontal lines $h(c)$ and $h(p'_a)$ through c and p'_a, respectively, that are traversed by radius (c, p'_a). Such a radius forms angles of $\pi/2$ with $t(p'_a)$; hence, the slope of $t(p'_a)$, that is equal to the angle between $t(p'_a)$ and $h(p'_a)$, is $\pi/2$ minus the angle α_a between $h(p'_a)$ and (c, p'_a). Angle α_a is the alternate interior of the angle between $h(c)$ and (c, p'_a), which is complementary to the half of angle $\widehat{p'_a c p'_b}$, which is equal to $3\alpha/2$, by construction. It follows that α_a is equal to $\frac{\pi}{2} - \frac{3\alpha}{4}$ and the slope of $t(p'_a)$ is $\frac{3\alpha}{4}$.

Hence, the slope of t_1^i is at most $\frac{3\alpha}{4} + \frac{3\alpha}{8h}$, which is less than α, since $h \geq 2$, and hence less than β. Analogously, the slope of t_2^i is less than β, and the lemma follows. $\square$

Corollary 1. *Point p^* is inside the open wedge $W(p_i)$, for each $i = 1, 2, \ldots, h$.*

Lemma 2. *For every pair of indices i and j such that $1 \leq i < j \leq k$, the drawing of $S(\mu_j)$ is contained inside $W(p_i)$ and the drawing of $S(\mu_i)$ is contained inside $W(p_j)$.*

Proof: If $S(\mu_i)$ and $S(\mu_j)$ are consecutive, i.e., the cutvertices parents of $S(\mu_i)$ and $S(\mu_j)$ are u_i and u_j and $j = i+1$, then the statement is true by construction. Suppose

$S(\mu_i)$ and $S(\mu_j)$ are not consecutive. Consider the triangle T_i delimited by (p^*, p_i), by t_2^i, and by the line through p^* and p_b'. T_i contains the triangle delimited by (p^*, p_{i+1}), by t_2^{i+1}, and by the line through p^* and p_b', which in turn contains the triangle delimited by (p^*, p_{i+2}), by t_2^{i+2}, and by the line through p^* and p_b'. Repeating such an argument shows that T_i contains the triangle T_{j-1} delimited by (p^*, p_{j-1}), by t_2^{j-1}, and by the line through p^* and p_b'. By construction, Γ_j lies inside T_{j-1}, and the lemma follows. $\square$

We prove that the constructed drawing Γ satisfies Properties 1–3.

Property 1. We show that, for every pair of vertices w_1 and w_2, there exists a distance-decreasing path between them in Γ. If both w_1 and w_2 are internal to the same graph $S(\mu_i)$, the property follows by induction. If one of w_1 and w_2, say w_1, is $r(\mu)$ and the other one, say w_2, is a node in $S(\mu_i)$ then, by Property 3, there exists a distance-decreasing path $(w_2 = v_0, v_1, \ldots, v_l = r(\mu_i))$ from w_2 to $r(\mu_i)$ such that, for every point p in $W(p_i)$, $d(v_j, p) < d(v_{j-1}, p)$, for $j = 1, 2, \ldots, l$. By Corollary 1, p^* is contained inside $W(p_i)$. Hence path $(w_2 = v_0, v_1, \ldots, v_l = r(\mu_i), w_1 = r(\mu))$ is a distance-decreasing path between w_1 and w_2. If w_1 belongs to $S(\mu_i)$ (possibly $w_1 = u_i$) and w_2 belongs to $S(\mu_j)$ (possibly $w_2 = u_j$) then suppose, w.l.o.g., that $j > i$. We show the existence of a distance-decreasing path $\mathcal{P}$ in Γ, composed of three subpaths $\mathcal{P}_1, \mathcal{P}_2$, and $\mathcal{P}_3$. By Property 3, Γ_j is such that there exists a distance-decreasing path $\mathcal{P}_1 = (w_1 = v_0, v_1, \ldots, v_l = r(\mu_i))$ from w_1 to $r(\mu_i)$ such that, for every point p in $W(p_i)$, $d(v_j, p) < d(v_{j-1}, p)$, for $j = 1, 2, \ldots, l$. By Lemma 2, drawing Γ_j, and hence vertex w_2, is contained inside $W(p_i)$, hence path $\mathcal{P}_1$ decreases the distance from w_2 at every vertex. Path $\mathcal{P}_2 = (u_i = r(\mu_i), u_{i+1}, \ldots, u_j = r(\mu_j))$ is easily shown to decrease the distance from w_2 at every vertex. In fact, for each $l = i, i+1, \ldots, j-2$, angle $\widehat{u_l u_{l+1} u_j}$ is greater than $\pi/2$, because triangle (u_l, u_{l+1}, u_j) is inscribed in less than half a circumference with u_{l+1} as middle point. Angle $\widehat{u_l u_{l+1} w_2}$ is strictly greater than $\widehat{u_l u_{l+1} u_j}$, hence it is the biggest angle in triangle (u_l, u_{l+1}, w_2) and $d(u_{l+1}, w_2) < d(u_l, w_2)$ follows. By induction, there exists a distance-decreasing path $\mathcal{P}_3$ from $r(\mu_j)$ to w_2, thus obtaining a distance-decreasing path $\mathcal{P}$ from w_1 to w_2.

Property 2. Such a property holds for Γ by construction.

Property 3. Consider any node v in $S(\mu_i)$ and any point p internal to $W(p^*)$. By Lemma 1, p is internal to $W(p_i)$. By induction, there exists a distance-decreasing path $(v = v_0, v_1, \ldots, v_l = r(\mu_i))$ such that $d(v_j, p) < d(v_{j-1}, p)$, for $j = 1, 2, \ldots, l$. Hence, path $(v = v_0, v_1, \ldots, v_l = r(\mu_i), v_{l+1} = r(\mu))$ is a distance-decreasing path such that $d(v_j, p) < d(v_{j-1}, p)$, for $j = 1, 2, \ldots, l+1$, if and only if $d(r(\mu), p) < d(r(\mu_i), p)$. However, angle $\widehat{pr(\mu)r(\mu_i)}$ is at least $\beta + (\frac{p_i}{2} - \frac{3\alpha}{8})$, which is more than $\pi/2$. Hence, $(p, r(\mu_i))$ is the longest side of triangle $(p, r(\mu), r(\mu_i))$, thus proving that $d(r(\mu), p) < d(r(\mu_i), p)$, and Property 3 holds for Γ.

When the induction on $\mathcal{T}$ is performed with $\mu = \nu$, we obtain a greedy drawing of S, thus proving the following:

Theorem 2. *There exists an algorithm that constructs a greedy drawing of any binary cactus.*

4   Spanning a Triangulation with a Binary Cactus

In this section we prove the following theorem:

Theorem 3. *Given a triangulation G, there exists a spanning subgraph S of G such that S is a binary cactus.*

Consider any triangulation G. We are going to construct a binary cactus S spanning G. First, we outline the algorithm to construct S. Such an algorithm has several steps. At the first step, we choose a vertex u incident to $f(G)$ and we construct a triangulated cycle C_T composed of u and all its neighbors. We remove u and its incident edges from G, obtaining a biconnected internally-triangulated plane graph G^*. At the beginning of each step after the first one, we suppose to have already constructed a binary cactus S whose vertices are a subset of the vertices of G (at the beginning of the second step, S coincides with C_T), and to have a set $\mathcal{G}$ of subgraphs of G (at the beginning of the second step, G^* is the only graph in $\mathcal{G}$). Each of such subgraphs is biconnected, internally-triangulated, has an outer face whose vertices already belong to S, and has internal vertices. All such internal vertices do not belong to S and each vertex of G not belonging to S is internal to a graph in $\mathcal{G}$. Only one of the graphs in $\mathcal{G}$ may have chords (at the beginning of the second step, G^* is such a graph). During each step, we perform the following two actions: (1) We partition the only graph G_C of $\mathcal{G}$ with chords, if any, into several biconnected internally-triangulated chordless plane graphs; we remove G_C from $\mathcal{G}$ and we add to $\mathcal{G}$ all graphs with internal vertices into which G_C has been partitioned; (2) we choose a graph G_i from $\mathcal{G}$, we choose a vertex u incident to the outer face of G_i and already belonging to exactly one block of S, and we add to S a block composed of u and of all its neighbors internal to G_i. We remove u and its incident edges from G_i, obtaining a biconnected internally-triangulated plane graph G_i^*. We remove G_i from $\mathcal{G}$ and we add G_i^* to $\mathcal{G}$. The algorithm stops when $\mathcal{G}$ is empty.

Now we give the details of the above outlined algorithm. At the first step of the algorithm, choose any vertex u incident to $f(G)$. Consider the neighbors $(u_1, u_2, \ldots, u_l)$ of u in clockwise order around it. Since G is a triangulation, $C = (u, u_1, u_2, \ldots, u_l)$ is a cycle. Let C_T be the triangulated cycle obtained by adding to C the edges connecting u to its neighbors. Let $S = C_T$. Remove vertex u and its incident edges from G, obtaining a biconnected internally-triangulated graph G^*. If G^* has no internal vertex, then all the vertices of G belong to S and we have a binary cactus spanning G. Otherwise, let $\mathcal{G} = \{G^*\}$. For each graph $G_i \in \mathcal{G}$, consider the vertices incident to $f(G_i)$. Each of such vertices can be either *forbidden for G_i* or *assigned to G_i*. A vertex w is forbidden for G_i if the choice of not introducing in S any new block incident to w and spanning a subgraph of G_i has been done. Conversely, a vertex w is assigned to G_i if a new block incident to w and spanning a subgraph of G_i could be introduced in S. For example, w is forbidden for G_i if there exist two blocks of S sharing w as a cutvertex. At the end of the first step of the algorithm, choose any two vertices incident to $f(G^*)$ as the only forbidden vertices for G^*. All other vertices incident to $f(G^*)$ are assigned to G^*. At the beginning of the i-th step, with $i \geq 2$, we assume that each of the following holds:

- *Invariant A:* Graph S is a binary cactus spanning all and only the vertices that are not internal to any graph in $\mathcal{G}$.

- *Invariant B:* Each graph in $\mathcal{G}$ is biconnected, internally-triangulated, and has internal vertices.
- *Invariant C:* Only one of the graphs in $\mathcal{G}$ may have chords.
- *Invariant D:* No internal vertex of a graph $G_i \in \mathcal{G}$ belongs to a graph $G_j \in \mathcal{G}$.
- *Invariant E:* For each graph $G_i \in \mathcal{G}$, all vertices incident to $f(G_i)$ are assigned to G_i, except for two vertices, which are forbidden.
- *Invariant F:* Each vertex v incident to the outer face of a graph in $\mathcal{G}$ is assigned to at most one graph $G_v \in \mathcal{G}$. The same vertex is forbidden for all graphs $\overline{G}_v \in \mathcal{G}$ such that v is incident to $f(\overline{G}_v)$ and $\overline{G}_v \neq G_v$.
- *Invariant G:* Each vertex assigned to a graph in $\mathcal{G}$ belongs to exactly one block of S.

Such invariants clearly hold after the first step of the algorithm.

Action 1: If all graphs in $\mathcal{G}$ are chordless, go to Action 2. Otherwise, by Invariant C, only one of the graphs in $\mathcal{G}$, say G_C, may have chords. We use such chords to partition G_C into k biconnected, internally-triangulated, chordless graphs G_C^j, with $j = 1, 2, \ldots, k$. Consider the biconnected outerplane subgraph O_C of G_C induced by the vertices incident to $f(G_C)$. To each internal face f of O_C delimited by a cycle c, a graph G_C^j is associated such that G_C^j is the subgraph of G_C induced by the vertices of c or inside c. Before replacing G_C with graphs G_C^j in $\mathcal{G}$, we show how to decide which vertices incident to the outer face of a graph G_C^j are assigned to G_C^j and which vertices are forbidden for G_C^j. Since each graph G_C^j is univocally associated with a face of O_C (namely the face of O_C delimited by the cycle that delimits $f(G_C^j)$), in the following we assign vertices to the faces of O_C and we forbid vertices for the faces of O_C, meaning that if a vertex is assigned to (forbidden for) a face of O_C delimited by a cycle c then it is assigned to (resp. forbidden for) graph G_C^j whose outer face is delimited by c.

We want to assign the vertices incident to $f(O_C)$ to faces of O_C so that the following properties are satisfied. *Property 1:* No forbidden vertex is assigned to any face of O_C. *Property 2:* No vertex is assigned to more than one face of O_C; *Property 3:* Each face of O_C has exactly two incident vertices which are forbidden for it; all other vertices of the face are assigned to it.

By Invariant E, G_C has two forbidden vertices. We construct an assignment of vertices to faces of O_C in some steps. Let p be the number of chords of O_C. Consider the Hamiltonian cycle O_C^0 of O_C, and assign all vertices of O_C^0, but for the two forbidden vertices, to the only internal face of O_C^0. At the i-th step, $1 \leq i \leq p$, we insert into O_C^{i-1} a chord of O_C, obtaining a graph O_C^i. This is done so that Properties 1–3 are satisfied by O_C^i (with O_C^i instead of O_C). After all p chords of O_C have been inserted, $O_C^p = O_C$, and we have an assignment of vertices to faces of O_C satisfying Properties 1–3. Properties 1–3 are clearly satisfied by the assignment of vertices to faces of O_C^0. Inductively assume Properties 1–3 are satisfied by the assignment of vertices to faces of O_C^{i-1}. Let (u_a, u_b) be the chord that is inserted at the i-th step. Chord (u_a, u_b) partitions a face f of O_C^{i-1} into two faces f_1 and f_2. By Property 3, two vertices u_1^* and u_2^* incident to f are forbidden for it and all other vertices incident to f are assigned to it. For each face of O_C^i different from f_1 and f_2, assign and forbid vertices as in the same face in O_C^{i-1}. Assign and forbid vertices for f_1 and f_2 as follows.

- If vertices u_a and u_b are the same vertices of u_1^* and u_2^*, assign to each of f_1 and f_2 all vertices incident to it, except for u_a and u_b. No forbidden vertex has been assigned to any face of O_C^i (Property 1). Vertices u_a and u_b have not been assigned to any face. All vertices assigned to f belong to exactly one of f_1 and f_2 and so they have been assigned to exactly one face (Property 2). The only vertices of f_1 (resp. of f_2) not assigned to it are u_a and u_b, while all other vertices are assigned to such a face (Property 3).

- If vertices u_a and u_b are both distinct from u_1^* and u_2^* and both u_1^* and u_2^* are in the same of f_1 and f_2, say in f_1, assign to f_1 all vertices incident to it, except for u_1^* and u_2^*, and assign to f_2 all vertices incident to it, except for u_a and u_b. No forbidden vertex has been assigned to any face of O_C^i (Property 1). Vertices u_a and u_b have been assigned to exactly one face. All other vertices assigned to f belong to exactly one of f_1 and f_2 and so they have been assigned to exactly one face (Property 2). The only vertices of f_1 (resp. of f_2) not assigned to it are u_1^* and u_2^* (resp. u_a and u_b), while all other vertices are assigned to such a face (Property 3).

- If vertices u_a and u_b are both distinct from u_1^* and u_2^* and one of u_1^* and u_2^*, say u_1^*, is in f_1 while the other one, say u_2^*, is in f_2, assign to f_1 all vertices incident to it, except for u_1^* and u_a, and assign to f_2 all vertices incident to it, except for u_2^* and u_b. No forbidden vertex has been assigned to any face of O_C^i (Property 1). Vertices u_a and u_b have been assigned to exactly one face. All other vertices assigned to f belong to exactly one of f_1 and f_2 and so they have been assigned to exactly one face (Property 2). The only vertices of f_1 (resp. of f_2) not assigned to it are u_1^* and u_a (resp. u_2^* and u_b), while all other vertices are assigned to such a face (Property 3).

- If one of vertices u_1^* and u_2^* coincides with one of u_a and u_b, say u_1^* coincides with u_a, and the other one, say u_2^*, is in one of f_1 and f_2, say in f_1, assign to f_1 all vertices incident to it, except for u_2^* and u_a, and assign to f_2 all vertices incident to it, except for u_a and u_b. No forbidden vertex has been assigned to any face of O_C^i (Property 1). Vertex u_a has not been assigned to any face and vertex u_b has been assigned to exactly one face. All other vertices assigned to f belong to exactly one of f_1 and f_2 and so they have been assigned to exactly one face (Property 2). The only vertices of f_1 (resp. of f_2) not assigned to it are u_2^* and u_a (resp. u_a and u_b), while all other vertices are assigned to such a face (Property 3).

Graph G_C is removed from $\mathcal{G}$. All graphs G_C^j having internal vertices are added to $\mathcal{G}$. It is easy to see that Invariants A–G are satisfied after Action 1.

Action 2: After Action 1 all graphs in $\mathcal{G}$ are chordless. There is at least one graph G_i in $\mathcal{G}$, otherwise the algorithm would have stopped before Action 1. By Invariant B, G_i has internal vertices. Choose any vertex u incident to $f(G_i)$ and assigned to G_i. Since G_i is biconnected and has internal vertices, $f(G_i)$ has at least three vertices. Since each graph in $\mathcal{G}$ has at most two forbidden vertices (by Invariant E), a vertex u assigned to G_i exists. Consider all the neighbors $(u_1, u_2, \ldots, u_l)$ of u internal to G_i, in clockwise order around u. Since G is biconnected, chordless, internally triangulated, and has internal vertices, then $l \geq 1$. If $l = 1$ then let C_T be edge (u, u_1). Otherwise, let C_T be the triangulated cycle obtained by adding to cycle $(u, u_1, u_2, \ldots, u_l)$ the edges connecting u to its neighbors. Add C_T to S. Remove u and its incident edges from G_i, obtaining a graph G_i^*. Assign to G_i^* all vertices incident to $f(G_i^*)$, except for the two vertices forbidden for G_i. Remove G_i from $\mathcal{G}$ and insert G_i^*, if it has internal vertices, in $\mathcal{G}$. It is easy to see that Invariants A–G are satisfied after Action 2.

When the algorithm stops, i.e., when there is no graph in $\mathcal{G}$, by Invariant A graph S is a binary cactus spanning all vertices of G, hence proving Theorem 3.

5 Conclusions

In this paper we have shown an algorithm for constructing greedy drawings of triangulations. The algorithm relies on two main results. The first one states that every binary cactus admits a greedy drawing. The second result, that may be of its own interest, is that, for every triangulation G, there exists a binary cactus S spanning G.

After this paper was submitted, the authors realized that a slight modification of the two main arguments, presented in Sect. 3 and 4, proves Conjecture 1. Namely, it can be shown that every triconnected planar graph can be spanned by a rooted *non-triangulated* binary cactus, i.e. a connected graph such that the block associated with each B-node of $\mathcal{T}$ is either an edge or a cycle and every cutvertex is shared by exactly two blocks. A greedy drawing of such a graph can be constructed by the drawing algorithm presented for rooted *triangulated* binary cactuses (the proof that the drawings constructed by the algorithm are greedy is slightly more involved due to the absence of edges $(r(\mu), u_i)$, for $i = 2, 3, \cdots, h - 2$). However, two reviewers of our paper made us aware that the conjecture has been positively settled by Leighton and Moitra in a paper to appear at FOCS'08 [7]. The approach used by Leighton and Moitra is surprisingly similar to ours.

References

1. Ben-Chen, M., Gotsman, C., Gortler, S.J.: Routing with guaranteed delivery on virtual coordinates. In: CCCG 2006 (2006)
2. Ben-Chen, M., Gotsman, C., Wormser, C.: Distributed computation of virtual coordinates. In: Erickson, J. (ed.) SoCG 2007, pp. 210–219. ACM Press, New York (2007)
3. Dhandapani, R.: Greedy drawings of triangulations. In: Huang, S.-T. (ed.) SODA 2008, pp. 102–111. SIAM, Philadelphia (2008)
4. de Fraysseix, H., Pach, J., Pollack, R.: How to draw a planar graph on a grid. Combinatorica 10(1), 41–51 (1990)
5. Kleinberg, R.: Geographic routing using hyperbolic space. In: INFOCOM 2007, pp. 1902–1909. IEEE, Los Alamitos (2007)
6. Knaster, B., Kuratowski, C., Mazurkiewicz, C.: Ein beweis des fixpunktsatzes fur n dimensionale simplexe. Fundamenta Mathematicae 14, 132–137 (1929)
7. Leighton, T., Moitra, A.: Some results on greedy embeddings in metric spaces. In: FOCS 2008 (2008)
8. Papadimitriou, C.H., Ratajczak, D.: On a conjecture related to geometric routing. Theor. Comput. Sci. 344(1), 3–14 (2005)
9. Rao, A., Papadimitriou, C.H., Shenker, S., Stoica, I.: Geographic routing without location information. In: Johnson, D.B., Joseph, A.D., Vaidya, N.H. (eds.) MOBICOM 2003, pp. 96–108. ACM Press, New York (2003)
10. Schnyder, W.: Embedding planar graphs on the grid. In: SODA 1990, pp. 138–148. SIAM, Philadelphia (1990)

Crossing and Weighted Crossing Number of Near-Planar Graphs

Sergio Cabello[1,*] and Bojan Mohar[2,**,***]

[1] Department of Mathematics, FMF, University of Ljubljana
sergio.cabello@fmf.uni-lj.si
[2] Department of Mathematics, Simon Fraser University, Burnaby, B.C. V5A 1S6
mohar@sfu.ca

Abstract. A nonplanar graph G is near-planar if it contains an edge e such that $G - e$ is planar. The problem of determining the crossing number of a near-planar graph is exhibited from different combinatorial viewpoints. On the one hand, we develop min-max formulas involving efficiently computable lower and upper bounds. These min-max results are the first of their kind in the study of crossing numbers and improve the approximation factor for the approximation algorithm given by Hliněný and Salazar (Graph Drawing GD 2006). On the other hand, we show that it is NP-hard to compute a weighted version of the crossing number for near-planar graphs.

1 Introduction

Crossing number minimization is one of the fundamental optimization problems in the sense that it is related to various other widely used notions. Besides its mathematical interest, there are numerous applications, most notably those in VLSI design [1,8,9], in combinatorial geometry and even in number theory, see, e.g, [16]. We refer to [10,15] and to [18] for more details about diverse applications of this important notion.

A nonplanar graph G is *near-planar* if it contains an edge e such that $G - e$ is planar. Such an edge e is called a *planarizing edge*. It is easy to see that near-planar graphs can have arbitrarily large crossing number. However, it seems that computing the crossing number of near-planar graphs should be much easier than in unrestricted cases. This is supported by a less known, but particularly interesting result of Riskin [14], who proved that the crossing number of a 3-connected cubic near-planar graph G can be computed easily as the length of a shortest path in the geometric dual graph of the planar subgraph $G - x - y$, where

---

* Supported in part by the Slovenian Research Agency, project J1-7218 and program P1-0297.

** Supported in part by the ARRS, Research Program P1-0297, by an NSERC Discovery Grant, and by the Canada Research Chair Program.

*** On leave from IMFM & FMF, Department of Mathematics, University of Ljubljana, 1000 Ljubljana, Slovenia.

I.G. Tollis and M. Patrignani (Eds.): GD 2008, LNCS 5417, pp. 38–49, 2009.

$xy \in E(G)$ is the edge whose removal yields a planar graph. Riskin asked if a similar correspondence holds in more general situations, but this was disproved by Mohar [13] (see also [5]). Another relevant paper about crossing numbers of near-planar graphs was published by Hliněný and Salazar [6].

In this paper we show that several generalizations of Riskin's result are indeed possible. We provide efficiently computable upper and lower bounds on the crossing number of near-planar graphs in a form of min-max relations. These relations can be extended to the non-3-connected case and even to the case of weighted edges. As far as we know, these results are the first of their kind in the study of crossing numbers. It is shown that they generalize and improve some known results and we foresee that generalizations and further applications are possible.

On the other hand, we show that computing the crossing number of weighted near-planar graphs is NP-hard. This discovery is a surprise and brings more questions than answers.

Drawings and crossings. A *drawing* of a graph G is a representation of G in the Euclidean plane $\mathbb{R}^2$ where vertices are represented as distinct points and edges by simple polygonal arcs joining points that correspond to their endvertices. A drawing is *clean* if the interior of every arc representing an edge contains no points representing the vertices of G. If interiors of two arcs intersect or if an arc contains a vertex of G in its interior we speak about crossings of the drawing. More precisely, a *crossing* of a drawing $\mathcal{D}$ is a pair $(\{e, f\}, p)$, where e and f are distinct edges and $p \in \mathbb{R}^2$ is a point that belongs to interiors of both arcs representing e and f in $\mathcal{D}$. If the drawing is not clean, then the arc of an edge e may contain in its interior a point $p \in \mathbb{R}^2$ that represents a vertex v of G. In such a case, the pair $(\{v, e\}, p)$ is also referred to as a *crossing* of $\mathcal{D}$.

The number of crossings of $\mathcal{D}$ is denoted by $\mathrm{cr}(\mathcal{D})$ and is called the crossing number of the drawing $\mathcal{D}$. The *crossing number* $\mathrm{cr}(G)$ of a graph G is the minimum $\mathrm{cr}(\mathcal{D})$ taken over all clean drawings $\mathcal{D}$ of G. When each edge e of G has a weight $w_e \in \mathbb{N}$, the weighted crossing number $\mathrm{wcr}(\mathcal{D})$ of a clean drawing $\mathcal{D}$ is the sum $\sum w_e \cdot w_f$ over all crossings $(\{e, f\}, p)$ in $\mathcal{D}$. The *weighted crossing number* $\mathrm{wcr}(G)$ of G is the minimum $\mathrm{wcr}(\mathcal{D})$ taken over all clean drawings $\mathcal{D}$ of G. Of course, if all edge-weights are equal to 1, then $\mathrm{wcr}(G) = \mathrm{cr}(G)$.

We shall discuss both, the weighted and unweighted crossing number. Most of the results are treated for the general weighted case. However, some results hold only in the unweighted case or are too technical to state for the weighted case. For a graph we shall assume that it is unweighted (i.e., all edge-weights are equal to 1) unless stated explicitly or when it is clear from the context that it is weighted.

A clean drawing $\mathcal{D}$ with $\mathrm{cr}(\mathcal{D}) = 0$ is also called an *embedding* of G. By a *plane graph* we refer to a planar graph together with a fixed embedding in the Euclidean plane. We shall identify a plane graph with its image in the plane.

Dual and facial distances. Let G_0 be a plane graph and let x, y be two of its vertices. A simple (polygonal) arc $\gamma : [0, 1] \to \mathbb{R}^2$ is an (x, y)-*arc* if $\gamma(0) = x$ and

$\gamma(1) = y$. If $\gamma(t)$ is not a vertex of G_0 for every t, $0 < t < 1$, then we say that γ is *clean*. For an (x, y)-arc γ we define the crossing number of γ with G_0 as

$$\mathtt{cr}(\gamma, G_0) = |\{t \mid \gamma(t) \in G_0 \text{ and } 0 < t < 1\}|. \tag{1}$$

This definition extends to the weighted case as follows. If the graph G_0 is weighted and the edge xy realized by an (x, y)-arc γ also has weight w_{xy}, then each crossing of γ with an edge e contributes $w_{xy} \cdot w_e$ towards the value $\mathtt{cr}(\gamma, G_0)$, and each crossing $(\{v, xy\}, p)$ of xy with a vertex of G_0 contributes 1 (independently of the edge-weights).

Using this notation, we define the *dual distance*

$$d^*(x, y) = \min\{\mathtt{cr}(\gamma, G_0) \mid \gamma \text{ is a clean } (x, y)\text{-arc}\}.$$

We also introduce a similar quantity, the *facial distance* between x and y:

$$d'(x, y) = \min\{\mathtt{cr}(\gamma, G_0) \mid \gamma \text{ is an } (x, y)\text{-arc}\}.$$

It should be observed at this point that the value $d'(x, y)$ is independent of the weights – since all weights are integers, we can replace each crossing of an edge with a crossing through an incident vertex and henceforth replace weight contributions simply by counting the number of crossings.

Let $G_{x,y}^*$ be the geometric dual graph of $G_0 - x - y$. Then $d^*(x, y)$ is equal to the distance in $G_{x,y}^*$ between the two vertices corresponding to the faces of $G_0 - x - y$ containing x and y. Of course, in the weighted case the distances are determined by the weights of their dual edges. This shows that $d^*(x, y)$ can be computed in linear time by using known shortest path algorithms for planar graphs. Similarly, one can compute $d'(x, y)$ in linear time by using the vertex-face incidence graph (see [12]).

Clearly, $d'(x, y) \leq d^*(x, y)$. Note that d^* and d' depend on the embedding of G_0 in the plane. However, if G_0 is (a subdivision of) a 3-connected graph, then this dependency disappears since G_0 has essentially a unique embedding. To compensate for this dependence, we define $d_0^*(x, y)$ (and $d_0'(x, y)$) as the minimum of $d^*(x, y)$ (resp. $d'(x, y)$) taken over all embeddings of G_0 in the plane.

Overview of results. The following proposition is clear from the definition of d^*:

Proposition 1. *If G_0 is a weighted planar graph and $x, y \in V(G_0)$, then* $\mathtt{cr}(G_0 + xy) \leq d_0^*(x, y)$.

This result shows that the value $d_0^*(x, y)$ is of interest. Gutwenger, Mutzel, and Weiskircher [5] provided a linear-time algorithm to compute $d_0^*(x, y)$. In Sect. 2 we study $d_0^*(x, y)$ from a combinatorial point of view and obtain a min-max characterization that results very useful.

Riskin [14] proved the following strengthening of Proposition 1 in a special case when G_0 is 3-connected and cubic:

Theorem 1 ([14]). *If G_0 is a 3-connected cubic planar graph, then*

$$\mathrm{cr}(G_0 + xy) = d_0^*(x, y).$$

Riskin asked in [14] if Theorem 1 extends to arbitrary 3-connected planar graphs. One of the authors [13] has shown that this is not the case: for every integer k, there exists a 5-connected planar graph G_0 and two vertices $x, y \in V(G_0)$ such that $\mathrm{cr}(G_0 + xy) \leq 11$ and $d_0^*(x, y) \geq k$. See also Gutwenger, Mutzel, and Weiskircher [5] for an alternative construction.

However, several extensions of Theorem 1 are possible, and some of them are presented throughout this paper. In particular, we show how to deal with graphs that are not 3-connected, and what happens when we allow vertices of arbitrary degrees.

Theorem 2. *If G_0 is a weighted planar graph and $x, y \in V(G_0)$, then*

$$d_0'(x, y) \leq \mathrm{cr}(G_0 + xy) \leq d_0^*(x, y).$$

The proof of this result is given in Sect. 3.

If G_0 is a cubic graph, then for every planar embedding of G_0, $d'(x, y) = d^*(x, y)$. Therefore, $d_0'(x, y) = d_0^*(x, y)$, and Theorem 2 implies Theorem 1.

Theorem 2 is also the main ingredient to improve the approximation factor in the algorithm of Hliněný and Salazar [6]; see Corollary 3.

A key idea in our results is to show that $d_0^*(x, y)$ (respectively $d_0'(x, y)$) is closely related to the maximum number of edge-disjoint (respectively vertex-disjoint) cycles that separate x and y. The notion of the separation has to be understood in a certain strong sense that is introduced in Sect. 2. This result yields a dual expression for d_0^* (respectively d_0') and is used to show that $d_0^*(x, y)$ is closely related to the crossing number of $G_0 + xy$, while $d_0'(x, y)$ is in the same way related to the minor crossing number, $\mathrm{mcr}(G_0 + xy)$, a version of the crossing number that works well with minors; see Bokal et al. [2].

Finally, we show in Sect. 5 that computing the crossing number of weighted near-planar graphs is NP-hard. Our reduction uses weights that are not polynomially bounded, and therefore it does not imply NP-hardness for unweighted graphs.

Intuition. To understand the difficulty in computing the crossing number of a near-planar graph, let us consider the graph shown in Fig. 1 (taken from [13]), where the subgraph inside each of the "darker" triangles is a sufficiently dense triangulation that requires many crossings when crossed by an arc. By drawing the vertex x in the outside, we see that this graph is near-planar. The drawing in Fig. 1 shows that its crossing number is at most 11, but it is also clear that $d^*(x, y)$ can be made as large as we want.

This construction can be generalized such that a similar redrawing as made there for x is necessary also for y (in order to bring these two vertices "close together"). At the first sight this seems like the only possibility which may happen – to "flip" a part of the graph containing x and to "flip" a part containing y.

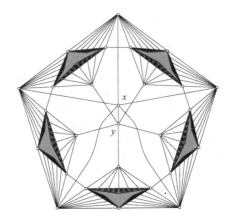

Fig. 1. The graph Q_k

And maybe some repetition of such changes may be needed. If this would be the only possibility of making the crossing number smaller than the one coming from the planar drawing of G_0, this would most likely give rise to a polynomial time algorithm for computing the crossing number of near-planar graphs. However, the authors can construct examples, in which additional complications arise.

Despite these examples and despite our NP-hardness result for the weighted case, the following question may still have a positive answer:

Problem 1. Is there a polynomial time algorithm which would determine the crossing number of $G_0 + xy$ if G_0 is an unweighted 3-connected planar graph?

2   Planar Separations and the Dual Distance

Let G_0 be a planar graph, x, y distinct vertices of G_0, and let Q be a subgraph of $G_0 - x - y$. We say that Q *planarly separates* vertices x and y if for every embedding of G_0 in the plane, x and y lie in the interiors of distinct faces of the induced embedding of Q.

Let Q be a subgraph of G. A Q-*bridge* in G is a subgraph of G that is either an edge not in Q but with both ends in Q (and its ends also belong to the bridge), or a connected component of $G - V(Q)$ together with all edges (and their endvertices in Q) which have one end in this component and the other end in Q. Let B be an Q-bridge. Vertices of $B \cap Q$ are *vertices of attachment* of B (shortly *attachments*).

Let C be a cycle in $G_0 - x - y$. Let B_x and B_y be the C-bridges in G_0 containing x and y, respectively. Two C-bridges B and B' are said to *overlap* if either (i) C contains four vertices a, a', b, b' in this order such that a and b are attachments of B and a', b' are attachments of B', or (ii) B and B' have (at least) three vertices of attachment in common. We define the *overlap graph* $O(G_0, C)$ of C-bridges (see [12]) as the graph whose vertices are the bridges of C, and two vertices are adjacent if the two bridges overlap on C. Since G_0 is planar, the overlap graph is bipartite. Distinct C-bridges are *weakly overlapping* if they are in the same connected component of $O(G_0, C)$, and in that component they belong to distinct bipartite classes. The following result follows easily from the definitions.

Lemma 1. *A cycle $C \subseteq G_0 - x - y$ planarly separates x and y if and only if B_x and B_y are weakly overlapping C-bridges.*

Tutte [17] characterized when $G_0 + xy$ is non-planar, i.e., when $\mathrm{cr}(G_0 + xy) \geq 1$ by proving

Theorem 3 (Tutte [17]). *Let x, y be vertices of a planar graph G_0. Then $G_0 + xy$ is non-planar if and only if $G_0 - x - y$ contains a cycle C such that the C-bridges of G containing x and y, respectively, are overlapping.*

Let us observe that $G_0 + xy$ is non-planar if and only if $G_0 - x - y$ planarly separates x and y. Therefore, the next lemma is closely related to Theorem 3.

Lemma 2. *If $Q \subseteq G_0 - x - y$ planarly separates x and y, then there is a cycle $C \subseteq Q$ that planarly separates x and y.*

The proof of this lemma is not hard but slightly technical, and we defer it to the full version of this paper.

For a plane graph G_0, a sequence $Q_1, \ldots, Q_k$ of edge-disjoint cycles of G_0 is *nested* if for $i = 1, \ldots, k-1$, all edges of the cycle Q_{i+1} lie in the exterior of Q_i.

Lemma 3. *Suppose that C and D are edge-disjoint cycles that planarly separate vertices x and y. Then there exist nested cycles $C_1, C_2 \subseteq C \cup D$ that planarly separate x and y.*

Again, the proof is deferred for the full version of the paper.

Lemma 4. *Let G_0 be a plane graph. If $Q_1, \ldots, Q_k$ are edge-disjoint cycles of G_0 that planarly separate vertices x and y of G_0, then there are nested edge-disjoint cycles $Q'_1, \ldots, Q'_k$ such that $\cup_{i=1}^{k} E(Q'_i) \subseteq \cup_{i=1}^{k} E(Q_i)$ and such that $Q'_1, \ldots, Q'_k$ planarly separate x and y.*

Proof. The proof follows rather easily by applying Lemma 3 consecutively on pairs of cycles Q_i, Q_j. One has to make sure that after finitely many steps we get a collection of nested cycles. This is done as follows. First we apply the lemma in such a way that one of the cycles in the family has none of the edges of the other $k-1$ cycles in its interior. After this is done, we repeat the process with the remaining $k-1$ cycles. $\square$

After this preparation, we are ready to discuss a dual expression for the dual distance, both for the 3-connected and for the general case.

Theorem 4. *Let G_0 be a planar graph and $x, y \in V(G)$. If r is an integer, then the following statements are equivalent:*

(a) $r \leq d_0^*(x, y)$.

(b) *There exists a family of r edge-disjoint cycles $Q_1, \ldots, Q_r$ that planarly separate x and y.*

(c) *There exists a family of r nested cycles $Q_1, \ldots, Q_r$ that planarly separate x and y.*

Equivalence of (b) and (c) follows from Lemma 4. It is also clear from the definitions (cf. Lemma 1) that (b) implies (a). The proof of the reverse implication that (a) yields (b) is by induction and also gives an efficient algorithm for finding $d_0^*(x, y)$ nested cycles planarly separating x and y. Let us observe that for 3-connected graphs, the maximum number of nested cycles can be determined by a simple "greedy" process.

Corollary 1. *The value of $d_0^*(x, y)$ is equal to the maximum number of edge-disjoint cycles that planarly separate x and y.*

The above dual expression for $d_0^*(x, y)$ is a min-max relation which offers an extension to the weighted case. Suppose that the edges of $G_0 + xy$ are weighted and that all weights are positive integers. Then we can replace each edge $e \neq xy$ by w_e parallel edges (each of weight 1). Let $\tilde{G}_0$ be the resulting unweighted graph. It is easy to argue that $d_0^*(G_0, x, y)$ is equal to $d_0^*(\tilde{G}_0, x, y) \cdot w_{xy}$. By Corollary 1, this value can be interpreted as the maximum number of edge-disjoint cycles planarly separating x and y in $\tilde{G}_0$.

3   Facial Distance

In this section we shall prove Theorem 2. First, we need a dual expression for $d'(x, y)$ which can be viewed as a surface version of Menger's Theorem.

Proposition 2. *Let G_0 be a plane graph and $x, y \in V(G_0)$ where y lies on the boundary of the exterior face. Let r be the maximum number of vertex-disjoint cycles, $Q_1, \ldots, Q_r$, contained in $G_0 - x - y$, such that for $i = 1, \ldots, r$, $x \in \text{int}(Q_i)$ and $y \in \text{ext}(Q_i)$. Then $d'(x, y) = r$.*

Proof. Since every (x, y)-arc intersects every Q_i, we conclude that $d'(x, y) \geq r$. The converse inequality is proved by induction on $d'(x, y)$. There is nothing to show if $d'(x, y) = 0$. Let F be the subgraph of G_0 containing all vertices and edges that are cofacial with x. Then F contains a cycle Q such that $x \in \text{int}(Q)$ and $y \in \text{ext}(Q)$. Delete all vertices and edges of F except x, and let G_1 be the resulting plane graph. It is easy to see that $d'_{G_1}(x, y) = d'_{G_0}(x, y) - 1$. By the induction hypothesis, G_1 has $d'_{G_0}(x, y) - 1$ disjoint cycles that contain x in their interior and y in the exterior. By adding Q to this family, we get $d'(x, y)$ such cycles. This shows that $d'(x, y) \leq r$.   □

The cycles $Q_1, \ldots, Q_r$ in Proposition 2 all contain x in their interior and y in their exterior. Therefore, they behave essentially like cycles on a cylinder that separate the two boundary components of the cylinder. Hence they are nested cycles separating x and y.

The main result of this section, Theorem 2, involves the minimum facial distance taken over all embeddings of G_0 in the plane. If G_0 is 3-connected, then $d'(x, y)$ is the same for every embedding of G_0, and Proposition 2 yields a dual expression for the facial distance. For general graphs, we need a similar concept as used in the previous section.

Let G_0 be a graph and $x, y \in V(G_0)$. Then we define $\rho(x, y, G_0)$ as the largest integer r for which there exists a collection of r vertex-disjoint cycles $Q_1, \ldots, Q_r$ in $G - x - y$ such that for every $i = 1, \ldots, r$, x and y belong to distinct weakly overlapping bridges of Q_i. It is convenient to realize that it may be required that the bridges containing x and y indeed overlap (not only weakly overlap), so we get an extension of Tutte's Theorem 3.

Lemma 5. *Let* $r = \rho(x, y, G_0)$. *Then there exists a collection of* r *vertex-disjoint cycles* $Q_1, \ldots, Q_r$ *in* $G_0 - x - y$ *such that for every* $i = 1, \ldots, r$, x *and* y *belong to distinct overlapping bridges of* Q_i.

Proof. For $i = 1, \ldots, r$, let B_x^i (resp. B_y^i) be the Q_i-bridge in G_0 containing x (resp. y). Note that every other cycle Q_j ($j \neq i$) is contained either in B_x^i or in B_y^i. Therefore we can define a linear order $\prec$ on $\{Q_1, \ldots, Q_r\}$ by setting $Q_i \prec Q_j$ if and only if $Q_j \subseteq B_y^i$. By adjusting indices, we may assume that $Q_1 \prec Q_2 \prec \cdots \prec Q_r$.

The proof method used in particular by Tutte in [17] is to change each cycle Q_i by rerouting it through the Q_i-bridges distinct from B_x^i and B_y^i in such a way that the two bridges with respect to the new cycle still weakly overlap, but contain more vertices. The actual goal is to minimize the number t of edges that are neither on the cycle nor in one of these two bridges. If B_x^i and B_y^i do not overlap but are weakly overlapping, it is possible to decrease t. It follows that after a series of changes, that do not affect any of the other cycles, the "big" bridges B_x^i and B_y^i overlap. We refer to [7] and to [11] for an algorithmic treatment showing that these changes can be made in linear time. □

The following lemma, whose proof is deferred to the full paper, is the analogue of Theorem 4.

Lemma 6. $d_0'(x, y) = \rho(x, y, G_0)$.

We are ready for the proof of Theorem 2.

Proof. (of Theorem 2). It has been shown before that $\mathbf{cr}(G_0 + xy) \leq d_0^*(x, y)$. The heart of the proof is to show that $d_0'(x, y)$ is a lower bound on $\mathbf{cr}(G_0 + xy)$. Let $r = d_0'(x, y)$. Lemmas 5 and 6 show that there are r vertex-disjoint $Q_1, \ldots, Q_r$ such that for every $i = 1, \ldots, r$, x and y belong to distinct overlapping bridges of Q_i. Let us denote these overlapping Q_i-bridges B_x^i and B_y^i as we did above. To simplify the notation in the sequel, we define $Q_0 = \{x\}$ and $Q_{r+1} = \{y\}$. Since B_x^i and B_y^i overlap, one of the following cases occurs:

(i) There are paths $P_1^+, P_2^+ \subseteq B_y^i$ joining Q_i with Q_{i+1}, and there are paths $P_1^-, P_2^- \subseteq B_x^i$ joining Q_i with Q_{i-1} such that the ends of these pairs of paths on Q_i interlace.

(ii) When the bridges B_x^i and B_y^i have precisely three vertices of attachment, they may overlap only because their attachments a, b, c on Q_i coincide. In that case, we have paths P_1^+, P_2^+, P_3^+ in B_y^i (resp. paths P_1^-, P_2^-, P_3^- in B_x^i) joining a, b, c with Q_{i+1} (resp. Q_{i-1}).

If Case (i) occurs, let S^i be the union of the paths P_1^- and P_2^- and let R^i be the union of the paths P_1^+ and P_2^+. If Case (ii) occurs, we define S^i and R^i similarly, as the union of the three paths in (ii) certifying the overlapping.

Suppose that we have a clean drawing of $G_0 + xy$ in the plane. If two cycles Q_i and Q_{i+1} intersect, then they make at least two crossings, and we declare one of them to be a crossing of type i, and the other one a crossing of type $i+1$. If two edges of the same cycle Q_i cross, we declare that crossing to be of type i as well. If an edge $e \notin E(Q_1 \cup \cdots \cup Q_r) \cup R^i \cup S^i \cup R^{i-1} \cup S^{i+1}$ (including the possibility that $e = xy$) crosses an edge of Q_i, we also declare the crossing to be of type i. Finally, if two edges, $e \in E(Q_{i-1} \cup S^i)$ and $f \in E(Q_{i+1} \cup R^i)$ cross, we also say that the crossing is of type i. Observe that by this definition, none of the crossings is of two different types (but for some of the crossings, the type may not have been specified).

Our goal is to show that for every $i = 1, \ldots, r$, there is a crossing of type i. This will show that there are at least r crossings, so the theorem holds.

Suppose, reductio ad absurdum, that there is no crossing of type i ($1 \leq i \leq r$). Then Q_i does not cross itself and both x and y are in the interior of Q_i (say) since the edge xy does not cross Q_i. Moreover, Q_i is not crossed by any of the other cycles Q_j. Suppose now that Q_{i-1} and Q_{i+1} are both inside Q_i (or both outside). Then it is easy to see that a crossing of type i occurs between an edge $e \in E(Q_{i-1} \cup S^i)$ and an edge $f \in E(Q_{i+1} \cup R^i)$. This shows that one of Q_{i-1} and Q_{i+1} is inside, while the other one is outside Q_i. We may assume that Q_{i+1} is inside and Q_{i-1} is outside Q_i. There is a path from Q_{i-1} to x that is disjoint from $V(Q_i)$ and does not use edges in S^i or in R^{i-1}. This path must clearly cross Q_i, and yields a crossing of type i. This contradiction completes the proof. □

As a corollary we get a generalization of Riskin's Theorem 1.

Corollary 2. *If the graph $G_0 - x - y$ has maximum degree 3, then $\mathrm{cr}(G_0 + xy) = d_0'(x, y) = d_0^*(x, y)$. In particular, the crossing number of $G_0 + xy$ is computable in linear time.*

Another corollary is an approximation formula for the crossing number of near-planar graphs if the maximum degree is bounded.

Corollary 3. *If the graph $G_0 - x - y$ has maximum degree Δ, then $d_0'(x, y) \leq \mathrm{cr}(G_0 + xy) \leq \frac{\Delta}{2} d_0'(x, y)$.*

Proof. Observe that $d_0^*(x, y) \leq \frac{\Delta}{2} d_0'(x, y)$ because there are at most $\frac{\Delta}{2}$ edge-disjoint cycles through any vertex and $d_0^*(x, y)$ is defined by a collection of $d_0^*(x, y)$ nested cycles (c.f. Theorem 4). □

Corollary 3 is an improvement of a theorem of Hliněný and Salazar [6] who proved analogous result with the factor Δ instead of $\Delta/2$.

A graph G is said to be d-apex if G has a vertex v of degree at most d such that $G - v$ is planar. Let us observe that every near-planar graph is essentially 2-apex (subdivide the "non-planar" edge).

Problem 2. Is there a result similar to Corollary 2 for 3-apex cubic graphs?

4   The Minor Crossing Number and d'

Structural graph theory based on the Robertson and Seymour theory of graph minors gives powerful results in relation to topological realizations of graphs. However, it does not work well with crossing numbers. To overcome this deficiency, Bokal et al. [2] introduced a related notion of the *minor crossing number*, $\mathrm{mcr}(G)$, which is defined as the minimum of $\mathrm{cr}(H)$ taken over all graphs H that contain G as a minor.

It is easy to see that $\mathrm{mcr}(G_0 + xy) \leq d_0'(x, y)$. However, a proof along similar lines as the proof of Theorem 2 shows even more intimate relationship.

Theorem 5. $\mathrm{mcr}(G_0 + xy) = d_0'(x, y)$.

5   NP-Hardness of $\mathrm{wcr}(\cdot)$ for Near-Planar Graphs

Consider the following decision problem:

> WEIGHTED CROSSING NUMBER
> Input: G, k, where G is an edge-weighted graph and $k > 0$.
> Question: Is $\mathrm{wcr}(G) \leq k$?

This problem is NP-complete because it generalizes the problem CROSSING NUMBER , which is NP-complete [3]. We will see that this problem remains NP-complete when restricted to near-planar graphs. We will use the notation $[n] = \{1, \ldots, n\}$.

Let $a_1, \ldots, a_n$ be natural numbers, and let $S = \sum_{i \in [n]} a_i$. We define the edge-weighted graph $G(a_1, \ldots, a_n)$ as follows (Fig. 2):

- its vertices are $u_1, \ldots, u_n$ and $v_1, \ldots, v_n$;
- there is a Hamiltonian cycle $Q = u_1 u_2 \cdots u_n v_1 v_2 \cdots v_n u_1$, each edge with weight S^2;
- there are edges $e_i = u_i v_i$ with weight a_i for each $i \in [n]$;

It is easy to note that $G(a_1, \ldots, a_n) - u_1 v_n$ planar, and hence $G(a_1, \ldots, a_n)$ is near-planar. For any subset of indices $I \subseteq [n]$, let $\sigma_I := \sum_{i \in I} a_i$. Consider a clean drawing $\mathcal{D}_0$ of G such that $\mathrm{wcr}(G) = \mathrm{wcr}(\mathcal{D}_0)$. It is easy to see that no edge of Q participates in a crossing, and therefore each edge e_i is contained either in the interior or in the exterior of the simple closed curve defined by Q. Using that all the edges in the interior (or the exterior) of Q must cross each other, we can show the following property.

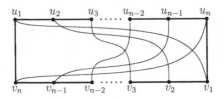

Fig. 2. The graph $G(a_1, \ldots, a_n)$ with the cycle Q bolder

Lemma 7. *It holds that*

$$2 \cdot \mathrm{wcr}(G(a_1, \ldots, a_n)) = \min_{I \subseteq [n]} \left\{ (\sigma_I)^2 + (\sigma_{[n] \setminus I})^2 \right\} - \sum_{i \in [n]} a_i^2.$$

Lemma 8. *It holds*

$$\mathrm{wcr}(G(a_1, \ldots, a_n)) = S^2/4 - \sum_{i \in [n]} a_i^2/2$$

if and only if there exists $I \subset [n]$ such that $\sigma_I = \sigma_{[n] \setminus I} = S/2$.

Proof. Note that

$$\min_{I \subseteq [n]} \left\{ (\sigma_I)^2 + (\sigma_{[n] \setminus I})^2 \right\} \geq \min\{A^2 + B^2 \mid A + B = S, \ A \geq 0, \ B \geq 0\} = S^2/2,$$

and there is equality if and only if there is some $I \subset [n]$ such that $\sigma_I = \sigma_{[n] \setminus I} = S/2$. The result then follows from Lemma 7.   □

Theorem 6. *The problem* WEIGHTED CROSSING NUMBER *is NP-complete for near-planar graphs.*

Proof. A standard planarizing argument shows that the problem WEIGHTED CROSSING NUMBER is in NP. To show NP-hardness, consider the following NP-complete problem [4].

> PARTITION
> Input: natural numbers $a_1, \ldots, a_n$.
> Question: is there $I \subset [n]$ such that $\sum_{i \in I} a_i = \sum_{i \in [n] \setminus I} a_i$?

Consider the function ϕ that maps the input $a_1, \ldots, a_n$ for PARTITION into the input

$$G(a_1, \ldots, a_n), \ S^2/4 - \sum_{i \in [n]} a_i^2/2$$

for WEIGHTED CROSSING NUMBER . Clearly, ϕ can be computed in polynomial time. Because of Lemma 8 both problems have the same answer. Therefore we have a polynomial time reduction from PARTITION to WEIGHTED CROSSING NUMBER that only uses near-planar graphs.   □

References

1. Bhatt, S.N., Leighton, F.T.: A framework for solving VLSI graph layout problems. J. Comput. System Sci. 28(2), 300–343 (1984)
2. Bokal, D., Fijavž, G., Mohar, B.: The minor crossing number. SIAM J. Discret. Math. 20(2), 344–356 (2006)
3. Garey, M.R., Johnson, D.S.: Crossing number is NP-complete. SIAM J. Alg. Discr. Meth. 4, 312–316 (1983)
4. Garey, M.R., Johnson, D.S.: Computers and Intractability: A Guide to the Theory of NP-Completeness. W. H. Freeman & Co., New York (1979)
5. Gutwenger, C., Mutzel, P., Weiskircher, R.: Inserting an edge into a planar graph. Algorithmica 41, 289–308 (2005)
6. Hliněný, P., Salazar, G.: On the crossing number of almost planar graphs. In: Kaufmann, M., Wagner, D. (eds.) GD 2006. LNCS, vol. 4372, pp. 162–173. Springer, Heidelberg (2007)
7. Juvan, M., Marinček, J., Mohar, B.: Elimination of local bridges. Math. Slovaca 47, 85–92 (1997)
8. Leighton, F.T.: Complexity issues in VLSI. MIT Press, MA (1983)
9. Leighton, F.T.: New lower bound techniques for vlsi. Math. Systems Theory 17, 47–70 (1984)
10. Liebers, A.: Planarizing graphs—a survey and annotated bibliography. J. Graph Algorithms Appl. 5, 74pp. (2001)
11. Mishra, B., Tarjan, R.E.: A linear-time algorithm for finding an ambitus. Algorithmica 7(5&6), 521–554 (1992)
12. Mohar, B., Thomassen, C.: Graphs on Surfaces. Johns Hopkins University Press, Baltimore (2001)
13. Mohar, B.: On the crossing number of almost planar graphs. Informatica 30, 301–303 (2006)
14. Riskin, A.: The crossing number of a cubic plane polyhedral map plus an edge. Studia Sci. Math. Hungar. 31, 405–413 (1996)
15. Shahrokhi, F., Sýkora, O., Székely, L.A., Vrt'o, I.: Crossing numbers: bounds and applications. In: Barany, I., Böröczky, K. (eds.) Intuitive geometry (Budapest, 1995). Bolyai Society Mathematical Studies, vol. 6, pp. 179–206. Akademia Kiado (1997)
16. Székely, L.A.: A successful concept for measuring non-planarity of graphs: The crossing number. Discrete Math. 276, 331–352 (2004)
17. Tutte, W.T.: Separation of vertices by a circuit. Discrete Math. 12, 173–184 (1975)
18. Vrt'o, I.: Crossing number of graphs: A bibliography, ftp://ftp.ifi.savba.sk/pub/imrich/crobib.pdf

Cubic Graphs Have Bounded Slope Parameter⋆

Balázs Keszegh[1,5], János Pach[2,4,5], Dömötör Pálvölgyi[3,4], and Géza Tóth[5]

[1] Central European University, Budapest
[2] City College, CUNY, New York
[3] Eötvös University, Budapest
[4] Ecole Polytechnique Fédérale de Lausanne
[5] A. Rényi Institute of Mathematics, Budapest

Abstract. We show that every finite connected graph G with maximum degree *three* and with at least one vertex of degree smaller than *three* has a straight-line drawing in the plane satisfying the following conditions. No three vertices are collinear, and a pair of vertices form an edge in G if and only if the segment connecting them is parallel to one of the sides of a previously fixed regular pentagon. It is also proved that every finite graph with maximum degree *three* permits a straight-line drawing with the above properties using only at most *seven* different edge slopes.

1 Introduction

A *drawing* of a graph G is a representation of its vertices by distinct points in the plane and the edges by continuous arcs connecting the corresponding endpoints, not passing through any other point corresponding to a vertex. In a *straight-line drawing* [8], the edges are represented by (possibly crossing) segments. If it leads to no confusion, we make no notational or terminological distinction between the vertices (edges) of G and the points (arcs) representing them.

There are several widely known parameters of graphs measuring how far G is from being planar. For instance, the *thickness* of G is the smallest number of its planar subgraphs whose union is G [14]. The *geometric thickness* of G is the smallest number of *crossing-free* subgraphs of a straight-line drawing of G, whose union is G [11]. The *slope number* of G is the minimum number of distinct edge slopes in a straight-line drawing of G [16]. It follows directly from the definitions that the thickness of any graph is at most as large as its geometric thickness, which, in turn, cannot exceed its slope number. For many interesting results about these parameters, consult [3,6,4,5,7,9,12,15].

The *slope parameter* of a graph was defined by Ambrus, Barát, and P. Hajnal [1], as follows. By abusing the usual terminology, we say that the *slope* of a line ℓ in the xy-plane is the smallest angle $\alpha \in [0, \pi)$ such that ℓ can be rotated into a position parallel to the x-axis by a clockwise turn through α. Given a set of points P in the plane and a set of slopes Σ, define $G(P, \Sigma)$ as the graph on the vertex set P, in which two vertices

---

⋆ Research supported by grants from NSF, NSA, PSC-CUNY and the Hungarian Research Foundation OTKA.

I.G. Tollis and M. Patrignani (Eds.): GD 2008, LNCS 5417, pp. 50–60, 2009.

$p, q \in P$ are connected by an edge if and only if the slope of the line pq belongs to Σ. The *slope parameter* $s(G)$ of G is the size of the smallest set of slopes Σ such that G is isomorphic to $G(P, \Sigma)$ for a suitable set of points P in the plane. This definition was motivated by the fact that all connections (edges) in an electrical circuit (graph) G can be easily realized by the overlay of $s(G)$ finely striped electrically conductive layers.

The slope parameter, $s(G)$, is closely related to the three other graph parameters mentioned before. For instance, for triangle-free graphs, $s(G)$ is at least as large as the slope number of G, the largest of the three quantities above. On the other hand, it sharply differs from them in the sense that the slope parameter of a complete graph on n vertices is *one*, while the thickness, the geometric thickness, and the slope number of K_n tend to infinity as $n \to \infty$. Jamison [10] proved that the slope number of K_n is n.

Any graph G of maximum degree *two* splits into vertex disjoint cycles, paths, and possibly isolated vertices. Hence, for such graphs we have $s(G) \leq 3$. In contrast, as was shown by Barát *et al.* [2], for any $d \geq 5$, there exist graphs of maximum degree d, whose slope parameters are arbitrarily large.

A graph is said to be *cubic* if the degree of each of its vertices is at most *three*. A cubic graph is *subcubic* if each of its connected components has a vertex of degree smaller than *three*.

The aim of this note is to prove

Theorem 1. *Every cubic graph has slope parameter at most* seven.

We will refer to the angles $i\pi/5$, $0 \leq i \leq 4$, as the *five basic slopes*. In Sect. 2, we prove the following statement, which constitutes the first step of the proof of Theorem 1.

Theorem 2. *Every subcubic graph has slope parameter at most* five. *Moreover, this can be realized by a straight-line drawing such that no* three *vertices are on a line and each edge has one of the five basic slopes.*

Using the fact that in the drawing guaranteed by Theorem 2 no *three* vertices are collinear, we can also conclude that the slope *number* of every subcubic graph is at most *five*. In [12], however, it was shown that this number is at most *four* and for cubic graphs it is at most *five*. This was improved for connected cubic graphs in [13] to *four*.

2   Proof of Theorem 2

The proof is by induction on the number of vertices of the graph. Clearly, the statement holds for graphs with fewer than *three* vertices. Let n be fixed and suppose that we have already established the statement for graphs with fewer than n vertices. Let G be a subcubic graph of n vertices. We can assume that G is connected, otherwise we can draw each of its connected components separately and translate the resulting drawings through suitable vectors.

To obtain a drawing of G, we have to find proper locations for its vertices. At each inductive step, we start with a drawing of a subgraph of G satisfying the conditions and extend it by adding a vertex. At a given stage of the procedure, for any vertex v that has already been added, consider the (basic) slopes of all edges adjacent to v that have already been drawn, and let $sl(v)$ denote the set of integers $0 \leq i < 5 \mod 5$

for which $i\pi/5$ is such a slope. That is, at the beginning $sl(v)$ is undefined, then it gets defined, and later it may change (expand). Analogously, for any edge uv of G, denote by $sl(uv)$ the integer $0 \leq i < 5 \mod 5$ for which the slope of uv is $i\pi/5$.

Case 1: *G has a vertex of degree one.* Assume without loss of generality, that v is a vertex of degree *one*, and let w denote its only neighbor. Deleting v from G, the degree of w in the resulting graph G' is at most *two*. Therefore, by the induction hypothesis, G' has a drawing meeting the requirements. As w has degree at most *two*, there is a basic slope σ such that no other vertex of G' lies on the line ℓ of slope σ that passes through w. Draw all *five* lines of basic slopes through each vertex of G'. These lines intersect ℓ in finitely many points. We can place v at any other point of ℓ, to obtain a proper drawing of G.

From now on, assume that G has no vertex of degree one.

Case 2: *G has no cycle that passes through a vertex of degree two.* Since G is subcubic, it contains a vertex w of degree *two* such that G is the union of two graphs, G_1 and G_2, having only vertex w in common. Both G_1 and G_2 are subcubic and have fewer than n vertices, so by the induction hypothesis both of them have a drawing satisfying the conditions. Translate the drawing of G_2 so that the points representing w in the two drawings coincide. Since w has degree *one* in both G_1 and G_2, by a possible rotation of G_2 about w through an angle that is a multiple of $\pi/5$, we can achieve that the two edges adjacent to w are not parallel. By scaling G_2 from w, if necessary, we can also achieve that the slope of no segment between a vertex of $G_1 \setminus w$ and a vertex of $G_2 \setminus w$ is a basic slope. Thus, the resulting drawing of G meets the requirements.

Case 3: *G has a cycle passing through a vertex of degree two.* If G itself is a cycle, we can easily draw it. If it is not the case, let C be a *shortest* cycle which contains a vertex of degree two. Let $u_0, u_1, \ldots, u_k$ denote the vertices of C, in this order, such that u_0 has degree *two* and u_1 has degree *three*. The indices are understood mod $k + 1$, that is, for instance, $u_{k+1} = u_0$. It follows from the minimality of C that u_i and u_j are not connected by an edge of G whenever $|i - j| > 1$.

Since $G \setminus C$ is subcubic, by assumption, it permits a straight-line drawing satisfying the conditions. Each u_i has at most *one* neighbor in $G \setminus C$. Denote this neighbor by t_i, if it exists. For every i for which t_i exists, we place u_i on a line passing through t_i. We place the u_i's one by one, "very far" from $G \setminus C$, starting with u_1. Finally, we arrive at u_0, which has no neighbor in $G \setminus C$, so that it can be placed at the intersection of two lines of basic slope, through u_1 and u_k, respectively. We have to argue that our method does not create "unnecessary" edges, that is, we never place two independent vertices in such a way that the slope of the segment connecting them is a basic slope. In what follows, we make this argument precise.

We determine the locations of $u_0, u_1, \ldots, u_k$ by using the below described PROCE-DURE(G, C, u_0, u_1, x), where G is our subcubic graph, C is the shortest cycle passing through a vertex of degree *two*, u_0 is such a vertex, u_1 is a neighbor of u_0 on C, whose degree is *three*, and x is a real parameter. Note that PROCEDURE(G, C, u_0, u_1, x) is a *nondeterministic* algorithm, as we have more than one choice at certain steps. (However, it is very easy to make it deterministic.)

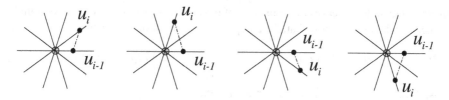

Fig. 1. The four possible locations of u_i

PROCEDURE(G, C, u_0, u_1, x)

- STEP 0. Since $G \setminus C$ is subcubic, it has a representation with the *five* basic slopes. Take such a representation, scaled and translated in such a way that t_1 (which exists since the degree of u_1 is three) is at the origin, and all other vertices are within unit distance from it.

  For any i, $2 \leq i \leq k$, for which u_i does not have a neighbor in $G \setminus C$, let t_i be any unoccupied point closer to the origin than 1, such that the slope of none of the lines connecting t_i to $t_1, t_2, \ldots t_{i-1}$ or to any other already embedded point of $G \setminus C$ is a basic slope.

  For any point p and for any i mod 5, let $\ell_i(p)$ denote the line with ith basic slope, $i\pi/5$, passing through p. Let ℓ_i stand for $\ell_i(O)$, where O denotes the origin. We will place $u_1, \ldots, u_k$ recursively, so that u_j is placed on $\ell_i(t_j)$, for a suitable i. Once the position of u_j has already been fixed on some $\ell_i(t_j)$, define $ind(u_j)$, the *index* of u_j, to be i. (Again, the indices are taken mod 5. Thus, for example, $|i - i'| \geq 2$ is equivalent to saying that $i \neq i'$ and $i \neq i' \pm 1 \mod 5$.) Start with u_1. The degree of t_1 in $G \setminus C$ is at most *two*, so that at the beginning the set $sl(t_1)$ (defined in the first paragraph of this section) has at most *two* elements. Let $l \notin sl(t_1)$. Direct the line $\ell_l(t_1)$ arbitrarily, and place u_1 on it at distance x from t_1 in the positive direction. (According to this rule, if $x < 0$, then u_1 is placed on $\ell_l(t_1)$ at distance $|x|$ from t_1 in the *negative* direction.)

  Suppose that $u_1, u_2, \ldots, u_{i-1}$ have been already placed and that u_{i-1} lies on the line $\ell_l(t_{i-1})$, that is, we have $ind(u_{i-1}) = l$.

- STEP i. We place u_i at one of the following four locations (see Fig. 1):
  (1) the intersection of $\ell_{l+1}(t_i)$ and $\ell_{l+2}(u_{i-1})$;
  (2) the intersection of $\ell_{l+2}(t_i)$ and $\ell_{l+3}(u_{i-1})$;
  (3) the intersection of $\ell_{l-1}(t_i)$ and $\ell_{l-2}(u_{i-1})$;
  (4) the intersection of $\ell_{l-2}(t_i)$ and $\ell_{l-3}(u_{i-1})$.
  Choose from the above four possibilities so that the edge $u_i t_i$ is not parallel to any other edge already drawn and adjacent to t_i, i.e., before adding the edge $u_i t_i$ to the drawing, $sl(t_i)$ did not include $sl(u_i t_i)$.

It follows directly from (1)–(4) that the edge $u_i u_{i-1}$ is not parallel to any other edge already drawn and adjacent to u_{i-1}. That is, before adding the edge $u_i u_{i-1}$ to the drawing, we had $sl(u_i u_{i-1}) \notin sl(u_{i-1})$. Avoiding for $u_i t_i$ the slopes of the edges already incident to t_i, leaves available two of the choices (1), (2), (3), (4). Some simple geometric calculations show that, for any possible location of u_i, we have

$$1.6\overline{Ou_{i-1}}-4 < 2\cos\left(\frac{\pi}{5}\right)\overline{Ou_{i-1}}-4 < \overline{Ou_i} < 2\cos\left(\frac{\pi}{5}\right)\overline{Ou_{i-1}}+4 < 1.7\overline{Ou_{i-1}}+4.$$

Thus, if $|x| \geq 50$, then we obtain by induction that

$$1.5\overline{Ou_{i-1}} < \overline{Ou_i}. \tag{1}$$

Here, we used that $x - 1 < \overline{Ou_1}$ and that, by the induction hypothesis, $\overline{Ou_j}$ is strictly increasing for $j < i$, therefore, we have $x - 1 < \overline{Ou_{i-1}}$.

We have to verify that the above procedure does not produce "unnecessary" edges, that is, the following statement is true.

Claim 1. *Suppose that* $|x| \geq 50$.
   (i) *The slope of* $u_i u_j$ *is not a basic slope, for any* $j < i - 1$.
   (ii) *The slope of* $u_i v$ *is not a basic slope, for any* $v \in V(G \setminus C)$.

Proof. (i) Suppose that the slope of $u_i u_j$ is a basic slope for some $j < i - 1$. By repeated application of inequality (1), we obtain that $\overline{Ou_i} > 1.5^{i-j}\overline{Ou_j} > 2\overline{Ou_j}$. On the other hand, if $u_i u_j$ has a basic slope, then easy geometric calculations show that $\overline{Ou_i} < 2\cos\left(\frac{\pi}{5}\right)\overline{Ou_j} + 4 < 2\overline{Ou_j}$, a contradiction.

(ii) Suppose for simplicity that $t_i u_i$ has slope 0, i.e., it is horizontal. By the construction, no vertex v of $G \setminus C$ determines a horizontal segment with t_i, but all of them are within distance 2 from t_i. As $\overline{Ou_i} > x - 1$, segment vu_i is almost, but not exactly horizontal. That is, we have $0 < |\angle t_i u_i v| < \pi/5$, contradiction. □

Suppose that STEP 0, STEP 1, ..., STEP k have already been completed. It remains to determine the position of u_0. We need some preparation.

Claim 2. *There exist two integers* $0 \leq l, l' < 5$ *with* $|l - l'| \geq 2 \mod 5$ *such that starting the* PROCEDURE *with* $ind(u_1) = l$ *and with* $ind(u_1) = l'$, *we can continue so that* $ind(u_2)$ *is the same.*

Proof. Suppose that the degrees of t_1 and t_2 in $G \setminus C$ are *two*, that is, there are two forbidden lines for both u_1 and u_2. In the other cases, when the degree of t_1 or the degree of t_2 is less than *two*, or when $t_1 = t_2$, the proof is similar, but simpler. We can place u_1 on $\ell_l(t_1)$ for any $l \notin sl(t_1)$. Therefore, we have three choices, two of which, $\ell_\alpha(t_1)$ and $\ell_\beta(t_1)$, are not consecutive, so that $|\alpha - \beta| \geq 2$.

The vertex u_2 cannot be placed on $\ell_m(t_2)$ for any $m \in sl(t_2)$, so there are three possible lines for u_2: $\ell_x(t_2)$, $\ell_y(t_2)$, $\ell_z(t_2)$, say. For any fixed location of u_1, we can place u_2 on at least two of the lines $\ell_x(t_2)$, $\ell_y(t_2)$, and $\ell_z(t_2)$. Therefore, at least one of them, $\ell_x(t_2)$, say, can be used for both locations of u_1. □

Claim 3. *We can place the vertices* $u_1, u_2, \ldots, u_k$ *using the* PROCEDURE *so that* $|ind(u_1) - ind(u_k)| \geq 2 \mod 5$.

Proof. By Claim 2, there are two placements of the vertices of $C \setminus \{u_0, u_k\}$, denoted by $u_1, u_2, \ldots, u_{k-1}$ and by $u_1', u_2', \ldots, u_{k-1}'$ such that $|ind(u_1) - ind(u_1')| \geq 2 \mod 5$, and $ind(u_i) = ind(u_i')$ for all $i \geq 2$. That is, we can start placing the vertices on

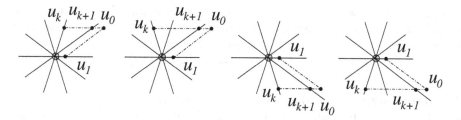

Fig. 2. The four possible locations of u_0

two nonneighboring lines so that from the second step of the PROCEDURE we use the same lines. We show that we can place u_k such that u_1 and u_k, or u'_1 and u_k are on nonneighboring lines. Having placed u_{k-1} (or u'_{k-1}), we have four choices for $ind(u_k)$. Two of them can be ruled out by the condition $ind(u_k) \notin sl(t_k)$. We still have two choices. Since u_1 and u'_1 are on nonneighboring lines, there is only one line which is neighboring of both of them. Therefore, we still have at least one choice for $ind(u_k)$ such that $|ind(u_1) - ind(u_k)| \geq 2$ or $|ind(u'_1) - ind(u_k)| \geq 2$. □

- STEP $k+1$. Let $i = ind(u_1)$, $j = ind(u_k)$, and assume, by Claim 3, that $|i-j| \geq 2$ mod 5. Consider the lines $\ell_{i-1}(u_1)$ and $\ell_{i+1}(u_1)$. One of them, $\ell_{i+1}(u_1)$, say, does not separate the vertices of $G \setminus C$ from u_k, the other one does. Place u_0 at the intersection of $\ell_{i+1}(u_1)$ and $\ell_i(u_k)$.

Claim 4. *Suppose that $|x| \geq 50$.*
  (i) *The slope of $u_0 u_j$ is not a basic slope, for any $1 < j < k$.*
  (ii) *The slope of $u_0 v$ is not a basic slope, for any $v \in V(G \setminus C)$.*

Proof. (i) Denote by u_{k+1} the intersection of $\ell_{i+1}(O)$ and $\ell_i(u_k)$. Suppose that the slope of $u_0 u_j$ is a basic slope for some $1 < j < k$. As in the proof of Claim 1, by repeated application of inequality 1, we obtain that $\overline{Ou_{k+1}} > 1.5^{k+1-j}\overline{Ou_j} > 2\overline{Ou_j}$. On the other hand, by an easy geometric argument, if the slope of $u_0 u_j$ is a basic slope, then $\overline{Ou_{k+1}} < 2\cos\left(\frac{\pi}{5}\right)\overline{Ou_j} + 4 < 2\overline{Ou_j}$, a contradiction, provided that $|x| \geq 50$.

(ii) For any vertex $v \in G \setminus C$, the slope of the segment $u_0 v$ is strictly between $i\pi/5$ and $(i+1)\pi/5$, therefore, it is not a basic slope. See Fig. 2. This concludes the proof of the claim and hence Theorem 2. □

3   Proof of Theorem 1

First we note that if G is connected, then Theorem 1 is an easy corollary to Theorem 2. Indeed, delete any vertex, and then put it back using two extra directions. If G is not connected, the only problem that may arise is that these extra directions can differ for different components. We will define a family of drawings for each component of G, depending on a parameter ε, and then choose the values of these parameters in such a way that the extra directions will coincide.

Suppose that G is a cubic graph. If a connected component is not 3-regular then, by Theorem 2, it can be drawn using the *five* basic slopes. If a connected component is a complete graph K_4 on *four* vertices, then it can also be drawn using the basic slopes. For the sake of simplicity, suppose that we do not have such components, ie. each connected component $G^1, \ldots, G^m$ of G is 3-regular and none of them is isomorphic to K_4.

First we concentrate on G^1. Let C be a shortest cycle in G^1. We distinguish two cases.

Case 1: C *is not a triangle.* Denote by $u_0, \ldots, u_k$ the vertices of C, and let t_0 be the neighbor of u_0 not belonging to C. Delete the edge $u_0 t_0$, and let $\bar{G}$ be the resulting graph.

Case 2: C *is a triangle.* Then every vertex of C has precisely *one* neighbor that does not belong to C. If all these neighbors coincide, then G^1 is a complete graph on *four* vertices, contradicting our assumption. So one vertex of C, u_0, say, has a neighbor t_0 which does not belong to C and which is not adjacent to the other two vertices, u_1 and u_2, of C. Delete the edge $u_0 t_0$, and let $\bar{G}$ be the resulting graph.

Observe that in both cases, u_k and t_0 are not connected in G^1. Indeed, suppose for a contradiction that they are connected. In the first case, G^1 would contain the triangle $u_0 u_k t_0$, contradicting the minimality of C. In the second case, the choice of u_0 would be violated.

There will be exactly two edges with extra directions, $u_0 u_1$ and $u_0 t_0$. The slope of $u_0 u_1$ will be very close to a basic slope and the slope of $u_0 t_0$ will be decided at the end, but we will show that almost any choice will do.

For any real x and $\varepsilon > 0$, define MODIFIEDPROCEDURE$(\bar{G}, C, u_0, u_1, x, \varepsilon)$, as follows. Let STEPS $0, 1, \ldots, k$ be identical to the corresponding STEPS of PROCEDURE$(\bar{G}, C, u_0, u_1, x)$.

- STEP $k + 1$. If there is a segment, determined by the vertices of $G \setminus C$, of slope $i\pi/5 + \varepsilon$ or $i\pi/5 - \varepsilon$, for any $0 \leq i < 5$, then STOP. In this case, we say that ε is *1-bad* for $\bar{G}$.

   Otherwise, when ε is *1-good*, let $i = ind(u_1)$ and $j = ind(u_k)$. We can assume that $|i - j| \geq 2 \mod 5$. Consider the lines $\ell_{i-1}(u_1)$ and $\ell_{i+1}(u_1)$. One of them does not separate the vertices of $G \setminus C$ from u_k, the other one does.

   If $\ell_{i-1}(u_1)$ separates $G \setminus C$ from u_k, then place u_0 at the intersection of $\ell_{i+1}(u_1)$ and the line through u_k with slope $i\pi/5 + \varepsilon$. If $\ell_{i+1}(u_1)$ separates $G \setminus C$ from u_k, then place u_0 at the intersection of $\ell_{i-1}(u_1)$ and the line through u_k with slope $i\pi/5 - \varepsilon$.

Since STEPS $0, \ldots, k$ are identical in MODIFIEDPROCEDURE$(\bar{G}, C, u_0, u_1, x, \varepsilon)$ and in PROCEDURE$(\bar{G}, C, u_0, u_1, x)$, the Claims 1, 2, and 3 also hold for the MODIFIEDPROCEDURE.

Moreover, it is easy to see that the analogue of Claim 4 also holds with an identical proof, provided that ε is sufficiently small: $0 < \varepsilon < 1/100$.

Claim 4'. *Suppose that* $|x| \geq 50$ *and* $0 < \varepsilon < 1/100$.
(i) *The slope of* $u_0 u_j$ *is not a basic slope, for any* $1 < j < k$.
(ii) *The slope of* $u_0 v$ *is not a basic slope, for any* $v \in V(\bar{G} \setminus C)$.   □

Perform MODIFIEDPROCEDURE$(\bar{G}, C, u_0, u_1, x, \varepsilon)$ for a fixed ε, and observe how the drawing changes as x varies. For any vertex u_i of C, let $u_i(x)$ denote the position of

u_i, as a function of x. For every i, the function $u_i(x)$ is linear, that is, u_i moves along a line as x varies.

Claim 5. *If ε is 1-good, then with finitely many exceptions, for every value of x,* MODI-FIEDPROCEDURE$(\bar{G}, C, u_0, u_1, x, \varepsilon)$ *produces a proper drawing of $\bar{G}$.*

Proof. Claims 1, 2, 3, and 4' imply Claim 5 for $|x| \geq 50$. Let u and v be two vertices of $\bar{G}$. Since $u(x)$ and $v(x)$ are linear functions, their difference, $\boldsymbol{uv}(x)$, is also linear.

If uv is an edge of $\bar{G}$, then the direction of $\boldsymbol{uv}(x)$ is the same for all $|x| \geq 50$. Therefore, it is the same for all values of x, with the possible exception of one value, for which $\boldsymbol{uv}(x) = 0$ holds.

If uv is not an edge of $\bar{G}$, then the slope of $\boldsymbol{uv}(x)$ is not a basic slope for any $|x| \geq 50$. Therefore, with the exception of at most *five* values of x, the slope of $\boldsymbol{uv}(x)$ is never a basic slope, nor does $\boldsymbol{uv}(x) = 0$ hold. □

Take a closer look at the relative position of the endpoints of the missing edge, $u_0(x)$ and $t_0(x)$. Since $t_0 \in \bar{G} \setminus C$, $t_0 = t_0(x)$ is the same for all values of x. The position of $u_0 = u_0(x)$ is a linear function of x. Let ℓ be the line determined by the function $u_0(x)$. If ℓ passes through t_0, then we say that ε is *2-bad* for $\bar{G}$. If ε is 1-good and it is not 2-bad for $\bar{G}$, then we say that it is *2-good* for $\bar{G}$. If ε is 2-good, then by varying x we can achieve almost any slope for the edge $t_0 u_0$. This will turn out to be crucially important, because we want to attain that these slopes coincide in all components.

Claim 6. *Suppose that the values $0 < \varepsilon, \delta < 1/100$ are 1-good for $\bar{G}$. Then at least one of them is 2-good for $\bar{G}$.*

Proof. Suppose, for simplicity, that $ind(u_1) = 0$, $ind(u_k) = 2$, and that u_1 and u_k are in the right half-plane (of the vertical line through the origin). The other cases can be settled analogously. To distinguish between MODIFIEDPROCEDURE$(\bar{G}, C, u_0, u_1, x, \varepsilon)$ and MODIFIEDPROCEDURE$(\bar{G}, C, u_0, u_1, x, \delta)$, let $u_0^{\varepsilon}(x)$ denote the position of u_0 obtained by the first procedure and $u_0^{\delta}(x)$ its position obtained by the second. Let ℓ^{ε} and ℓ^{δ} denote the lines determined by the functions $u_0^{\varepsilon}(x)$ and $u_0^{\delta}(x)$. Suppose that x is very large. Since, by (1), we have $\overline{u_k(x)O} > 1.5\overline{u_1(x)O}$, both $u_0^{\varepsilon}(x)$ and $u_0^{\delta}(x)$ are above the line $\ell_{\pi/10}$. On the other hand, if $x < 0$ is very small (i.e., if $|x|$ is very big), both $u_0^{\varepsilon}(x)$ and $u_0^{\delta}(x)$ lie below the line $\ell_{\pi/10}$. It follows that the slopes of ℓ^{ε} and ℓ^{δ} are larger than $\pi/10$, but smaller than $\pi/5$.

Suppose that neither ε nor δ is 2-good. Then both ℓ^{ε} and ℓ^{δ} pass through t_0. That is, for a suitable value of x, we have $u_0^{\varepsilon}(x) = t_0$. We distinguish two cases.

Case 1: $u_0^{\varepsilon}(x) = t_0 = u_k(x)$. Then, as x varies, the line determined by $u_k(x)$ coincides with $\ell_2(t_0)$. Consequently, t_0 and u_k are connected in G^1, a contradiction.

Case 2: $u_0^{\varepsilon}(x) = t_0 \neq u_k(x)$. In order to get a contradiction, we try to determine the position of $u_0^{\delta}(x)$. Considering STEP $k + 1$ in both MODIFIEDPROCEDURE $(\bar{G}, C, u_0, u_1, x, \varepsilon)$ and in MODIFIEDPROCEDURE$(\bar{G}, C, u_0, u_1, x, \delta)$, we can conclude that $u_1(x)$ lies on $\ell_1(t_0)$, $u_0^{\delta}(x)$ lies on $\ell_1(u_1(x))$, therefore, $u_0^{\delta}(x)$ lies on $\ell_1(t_0)$. On the other hand, $u_0^{\delta}(x)$ lies on ℓ^{δ}, and, by assumption, ℓ^{δ} passes through t_0. However, we have shown that ℓ^{δ} and $\ell_1(t_0)$ have different slopes, therefore, $u_0^{\delta}(x)$ must be at their intersection point, so we have $u_0^{\delta}(x) = u_0^{\varepsilon}(x) = t_0$.

Considering again STEP $k + 1$ in MODIFIEDPROCEDURE$(\bar{G}, C, u_0, u_1, x, \varepsilon)$ and in MODIFIEDPROCEDURE$(\bar{G}, C, u_0, u_1, x, \delta)$, we can conclude that the point $u_0^\delta(x) = t_0 = u_0^\varepsilon(x)$ belongs to both $\ell_\varepsilon(u_k(x))$ and $\ell_\delta(u_k(x))$. This contradicts our assumption that $u_k(x)$ is different from $u_0^\delta(x) = t_0 = u_0^\varepsilon(x)$. $\qquad\square$

By Claim 5, for every $\varepsilon < 1/100$ and with finitely many exceptions for every value of x, MODIFIEDPROCEDURE$(\bar{G}, C, u_0, u_1, x, \varepsilon)$ produces a proper drawing of $\bar{G}$. When we want to add the edge $u_0 t_0$, the slope of $u_0(x) t_0$ may coincide with the slope of $u(x) u'(x)$, for some $u, u' \in \bar{G}$. The following statement guarantees that this does not happen "too often". We use $\alpha(u)$ to denote the *slope* of a vector u.

Claim 7. *Let $u(x)$ and $v(x)$: $R \to R^2$ be two linear functions, and let $\ell(u)$ and $\ell(v)$ denote the lines determined by $u(x)$ and $v(x)$. Suppose that for some $x_1 < x_2 < x_3$, the vectors u, v do not vanish and that their slopes coincide, that is, $\alpha(u(x_1)) = \alpha(v(x_1))$, $\alpha(u(x_2)) = \alpha(v(x_2))$, and $\alpha(u(x_3)) = \alpha(v(x_3))$. Then $\ell(u)$ and $\ell(v)$ must be parallel.*

Proof. If $\ell(u)$ passes through the origin, then for every value of x, $u(x)$ has the same slope. In particular, $\alpha(v(x_1)) = \alpha(v(x_2)) = \alpha(v(x_3))$. Therefore, $\ell(v)$ also passes through the origin and is parallel to $\ell(u)$. (In fact, we have $\ell(u) = \ell(v)$.) We can argue analogously if $\ell(u)$ passes through the origin. Thus, in what follows, we can assume that neither $\ell(u)$ nor $\ell(v)$ passes through the origin.

Suppose that $\alpha(u(x_1)) = \alpha(v(x_1))$, $\alpha(u(x_2)) = \alpha(v(x_2))$, and $\alpha(u(x_3)) = \alpha(v(x_3))$. For any x, define $w(x)$ as the intersection point of $\ell(v)$ and the line connecting the origin to $u(x)$, provided that they intersect. Clearly, $v(x) = w(x)$ for $x = x_1, x_2, x_3$, and $u(x)$ and $w(x)$ have the same slope for every x. The transformation $u(x) \to w(x)$ is a projective transformation from $\ell(u)$ to $\ell(v)$, therefore, it preserves the cross ratio of any four points. That is, for any x, we have

$$(u(x_1), u(x_2); u(x_3), u(x)) = (w(x_1), w(x_2); w(x_3), w(x)) .$$

Since both $u(x)$ and $v(x)$ are linear functions, we also have

$$(u(x_1), u(x_2); u(x_3), u(x)) = (v(x_1), v(x_2); v(x_3), v(x)) .$$

Hence, we can conclude that $v(x) = w(x)$ for all x. However, this is impossible, unless $\ell(u)$ and $\ell(v)$ are parallel. Indeed, suppose that $\ell(u)$ and $\ell(v)$ are not parallel, and set x in such a way that $u(x)$ is parallel to $\ell(v)$. Then $w(x)$ cannot have the same slope as $u(x)$, a contradiction. $\qquad\square$

Suppose that ε is 2-good and let us fix it. As above, let $u_0^\varepsilon(x)$ be the position of u_0 obtained by MODIFIEDPROCEDURE$(\bar{G}, C, u_0, u_1, x, \varepsilon)$, and let ℓ^ε be the line determined by $u_0^\varepsilon(x)$.

Suppose also that there exist two independent vertices of $\bar{G}$, $u, u' \neq u_0$, such that the line determined by $uu'(x)$ is parallel to ℓ^ε. Then we say that ε is *3-bad* for $\bar{G}$. If ε is 2-good and it is not 3-bad for $\bar{G}$, then we say that it is *3-good* for $\bar{G}$.

It is easy to see that, for any $0 < \varepsilon, \delta < 1/100$, ℓ^ε and ℓ^δ are not parallel, therefore, for any fixed u, u', there is at most one value of ε for which the line determined by $uu'(x)$ is parallel to ℓ^ε. Thus, with finitely many exceptions, all values $0 < \varepsilon < 1/100$ are 3-good.

Summarizing, we have obtained the following.

Claim 8. *Suppose that ε is 3-good for $\bar{G}$. With finitely many exceptions, for every value of x,* MODIFIEDPROCEDURE$(\bar{G}, C, u_0, u_1, x, \varepsilon)$ *gives a proper drawing of G^1.* $\square$

Now we are in a position to complete the proof of Theorem 1. Proceed with each of the components as described above for G^1. For any fixed i, let $u_0^i v_0^i$ be the edge deleted from G^i, and denote the resulting graphs by $\bar{G}^1, \ldots, \bar{G}^m$. Let $0 < \varepsilon < 1/100$ be fixed in such a way that ε is 3-good for all graphs $\bar{G}^1, \ldots, \bar{G}^m$. This can be achieved, in view of the fact that there are only finitely many values of ε which are not 3-good. Perform MODIFIEDPROCEDURE$(\bar{G}^i, C^i, u_0^i, u_1^i, x^i, \varepsilon)$. Now the line ℓ^i determined by all possible locations of u_0^i does not pass through t_0^i.

Note that when MODIFIEDPROCEDURE$(\bar{G}^i, C^i, u_0^i, u_1^i, x^i, \varepsilon)$ is executed, then apart from edges with basic slopes, we use an edge with slope $r\pi/5 \pm \varepsilon$, for some integer r mod 5. By using rotations through $\pi/5$ and a reflection, if necessary, we can achieve that each component $\bar{G}^i$ is drawn using the basic slopes and one edge of slope ε.

It remains to set the values of x_i and draw the missing edges $u_0^i v_0^i$. Since the line ℓ^i determined by the possible locations of u_0^i does not pass through t_0^i, by varying the value of x^i, we can attain any slope for the missing edge $t_0^i u_0^i$, except for the slope of ℓ^i. By Claim 8, with finitely many exceptions, all values of x^i produce a proper drawing of G^i. Therefore, we can choose $x^1, x^2, \ldots, x^m$ so that all segments $t_0^i u_0^i$ have the same slope and every component G^i is properly drawn using the same seven slopes. Translating the resulting drawings through suitable vectors gives a proper drawing of G, this completes the proof of Theorem 1.

4   Concluding Remarks

In the proof of Theorem 1, the slopes we use depend on the graph G. However, the proof shows that one can simultaneously embed all cubic graphs using only *seven* fixed slopes.

It is unnecessary to use $|x| \geq 50$, in every step, we could pick any x, with finitely many exceptions.

It seems to be only a technical problem that we needed *two* extra directions in the proof of Theorem 1. We believe that *one* extra direction would suffice.

The most interesting problem that remains open is to decide whether the number of slopes needed for graphs of maximum degree *four* is bounded.

References

1. Ambrus, G., Barát, J., Hajnal, P.: The slope parameter of graphs. Acta Sci. Math.(Szeged) 72 (3–4), 875–889 (2006)
2. Barát, J., Matoušek, J., Wood, D.R.: Bounded-degree graphs have arbitrarily large geometric thickness. Electr. J. Combin. 13(1), R3, 14pp. (2006)
3. Dillencourt, M.B., Eppstein, D., Hirschberg, D.S.: Geometric thickness of complete graphs. J. Graph Algorithms Appl. 4(3), 5–17 (2000)
4. Dujmović, V., Suderman, M., Wood, D.R.: Graph drawings with few slopes. Comput. Geom. 38, 181–193 (2007)

5. Dujmović, V., Wood, D.R.: Graph treewidth and geometric thickness parameters. In: Healy, P., Nikolov, N.S. (eds.) GD 2005. LNCS, vol. 3843, pp. 129–140. Springer, Heidelberg (2006)

6. Duncan, C.A., Eppstein, D., Kobourov, S.G.: The geometric thickness of low degree graphs. In: SoCG 2004, pp. 340–346. ACM Press, New York (2004)

7. Eppstein, D.: Separating thickness from geometric thickness. In: Pach, J. (ed.) Towards a Theory of Geometric Graphs. Contemporary Math, vol. 342, pp. 75–86. AMS, Providence (2004)

8. Fáry, I.: On straight line representation of planar graphs. Acta Univ. Szeged. Sect. Sci. Math. 11, 229–233 (1948)

9. Hutchinson, J.P., Shermer, T.C., Vince, A.: On representations of some thickness-two graphs. Comput. Geom. 13, 161–171 (1999)

10. Jamison, R.E.: Few slopes without collinearity. Discrete Math. 60, 199–206 (1986)

11. Kainen, P.C.: Thickness and coarseness of graphs. Abh. Math. Sem. Univ. Hamburg 39, 88–95 (1973)

12. Keszegh, B., Pach, J., Pálvölgyi, D., Tóth, G.: Drawing cubic graphs with at most five slopes. In: Kaufmann, M., Wagner, D. (eds.) GD 2006. LNCS, vol. 4372, pp. 114–125. Springer, Heidelberg (2007)

13. Mukkamala, P., Szegedy, M.: Geometric representation of cubic graphs with four directions (manuscript, 2007)

14. Mutzel, P., Odenthal, T., Scharbrodt, M.: The thickness of graphs: a survey. Graphs Combin. 14, 59–73 (1998)

15. Pach, J., Pálvölgyi, D.: Bounded-degree graphs can have arbitrarily large slope numbers. Electr. J. Combin. 13(1), Note 1, 4pp. (2006)

16. Wade, G.A., Chu, J.-H.: Drawability of complete graphs using a minimal slope set. The Computer J. 37, 139–142 (1994)

Unimaximal Sequences of Pairs in Rectangle Visibility Drawing

Jan Štola

Department of Applied Mathematics, Charles University
Malostranské nám. 25, Prague, Czech Republic
Jan.Stola@mff.cuni.cz

Abstract. We study the existence of unimaximal subsequences in sequences of pairs of integers, e.g., the subsequences that have exactly one local maximum in each component of the subsequence. We show that every sequence of $\frac{1}{12}n^2(n^2-1)+1$ pairs has a unimaximal subsequence of length n. We prove that this bound is tight. We apply this result to the problem of the largest complete graph with a 3D rectangle visibility representation and improve the upper bound from 55 to 50.

1 Introduction

A 3D rectangle visibility drawing represents vertices by axis-aligned rectangles lying in planes parallel to the xy-plane. Edges correspond to the z-parallel visibility among these rectangles. This type of graph drawing was studied, for example, in [1,2,5,6,7,8].

We continue in the study of the maximum size of a complete graph with a 3D rectangle visibility representation. The representation of K_{22} given by Rote and Zelle (included in [8]) provides the best known lower bound. On the other hand, Bose et al. [2] showed that no complete graph with 103 or more vertices has such a representation. This result was then improved to 56 by Fekete et al. [1]. Their proof is based on the analysis of unimaximal subsequences in sequences of rectangle coordinates.

A sequence $x_1, x_2, \ldots$ of distinct integers is called *unimaximal* if it has exactly one local maximum, i. e., for all i, j, k with $i < j < k$ we have $x_j > \min\{x_i, x_k\}$. The following lemma (attributed by Chung [3] to V. Chvátal and J.M. Steele, among others) summarizes the most important properties of unimaximal sequences.

Lemma 1. *For all $n > 1$, in every sequence of $\binom{n}{2} + 1$ distinct integers, there exists a unimaximal subsequence of length n. On the other hand, there exists a sequence of $\binom{n}{2}$ distinct integers that has no unimaximal subsequence of length n.*

The notion of unimaximality can be generalized to sequences of pairs:

Definition 1. *A sequence $(x_1, y_1), (x_2, y_2), \ldots$ of pairs of integers is called uni-maximal if it is unimaximal in both components, i. e., if both sequences $x_1, x_2, \ldots$ and $y_1, y_2, \ldots$ are unimaximal.*

I.G. Tollis and M. Patrignani (Eds.): GD 2008, LNCS 5417, pp. 61–66, 2009.

If we apply the previous lemma twice on a sequence of pairs then we can see that every sequence of $\binom{\binom{n}{2}+1}{2} + 1 \approx \frac{1}{8}n^4$ pairs has a unimaximal subsequence of length n. In fact, the result of Fekete et al. [1] is based on this fact. We show in this paper that we can improve this bound to $\frac{1}{12}n^2(n^2 - 1) + 1$ if we consider both components of a sequence of pairs together. This result allows us to improve the upper bound on the size of the largest complete graph with a 3D rectangle visibility representation from 55 to 50.

2 Upper Bound

The definition of a unimaximal sequence requires distinct values in the sequence. Therefore both components of a unimaximal sequence of pairs must contain distinct values.[1] Hence we consider only sequences with this property in the sequel.

We show that every sufficiently long sequence of pairs contains a unimaximal subsequence of a given length. The following relations turn out to be useful in the analysis of this problem.

Definition 2. *Let* $(x_1, y_1), (x_2, y_2), \ldots$ *be a sequence of pairs of integers. We say that two pairs* $(x_i, y_i), (x_j, y_j), i < j$ *have a* $/\!\!/$*-relation if* $x_i < x_j$ *and* $y_i < y_j$*. The pairs have a* $\backslash\!\backslash$*-relation if* $x_i > x_j$ *and* $y_i > y_j$*.*

If both relations are forbidden then our problem becomes a simple consequence of the Erdős-Szekeres theorem [4].

Lemma 2. *If a sequence of* $(n - 1)^2 + 1$ *pairs of integers doesn't contain pairs with* $/\!\!/$*- and* $\backslash\!\backslash$*-relations then it has a unimaximal subsequence of length* n*.*

Proof. Let $((x_i, y_i))_i$ be a sequence of length $(n - 1)^2 + 1$. The sequence $(x_i)_i$ contains a monotone subsequence $(x_{i_j})_j$ of length n according to the Erdős-Szekeres theorem. The sequence $(y_{i_j})_j$ is monotone as well because the original sequence doesn't have pairs with $/\!\!/$- and $\backslash\!\backslash$-relations, e.g., if the sequence $(x_{i_j})_j$ is increasing then $(y_{i_j})_j$ is decreasing and vice versa.

Hence the subsequence $((x_{i_j}, y_{i_j}))_{j=1}^{n}$ is unimaximal.   □

Lemma 3 shows how the situation changes if only one relation is forbidden.

Lemma 3. *If a sequence of* $f_n = \frac{1}{6}(n - 1)n(2n - 1) + 1$ *pairs of integers doesn't contain pairs with a* $\backslash\!\backslash$*-relation then it has a unimaximal subsequence of length* n*.*

Proof. The lemma holds for $n = 1$. Let's suppose that it holds for $n = k \in \mathbb{N}$ and let $P = ((x_i, y_i))_{i=1}^{f_k + 1}$ be a sequence that doesn't contain pairs with a $\backslash\!\backslash$-relation. Let S be the set of pairs (x, y) such that P contains a unimaximal subsequence

---

[1] Both components $(x_i)_i$ and $(y_i)_i$ of $((x_i, y_i))_i$ must contain distinct values, but it may happen that $x_i = y_j$.

of length k starting at (x, y). We know that every sequence of length f_k contains at least one such a subsequence. Therefore $|S| \geq f_{k+1} - f_k + 1 = k^2 + 1$.

If there are two pairs $(x_i, y_i), (x_j, y_j), i < j$ in S that have a $/\!/$-relation then we can prepend (x_i, y_i) to the unimaximal subsequence of length k starting at (x_j, y_j) and obtain a unimaximal subsequence of length $k + 1$.

On the other hand, if there are no pairs in S that have a $/\!/$-relation then S contains a unimaximal subsequence of length $k + 1$ according to the previous lemma. Hence the lemma holds also for $n = k + 1$. □

The idea of the previous proof can be reused to analyze sequences with both relations allowed.

Theorem 1. *For all $n \in \mathbb{N}$, in every sequence of $g_n = \frac{1}{12}n^2(n^2 - 1) + 1$ pairs of integers, there exists a unimaximal subsequence of length n.*

Proof. We proceed in the same way as in the previous proof. The theorem holds for $n = 1$. Let's suppose that it holds for $n = k \in \mathbb{N}$ and let $P = ((x_i, y_i))_i$ be a sequence of length g_{k+1}. Let E be the set of pairs (x, y) such that P contains a unimaximal subsequence of length k ending at (x, y). We know that every sequence of length g_k contains at least one such a subsequence. Therefore $|E| \geq g_{k+1} - g_k + 1 = f_{k+1}$.

If there are two pairs $(x_i, y_i), (x_j, y_j), i < j$ in E that have a $\backslash\!\backslash$-relation then we can append (x_j, y_j) to the unimaximal subsequence of length k ending at (x_i, y_i) and obtain a unimaximal subsequence of length $k + 1$.

On the other hand, if there are no pairs in E that have a $\backslash\!\backslash$-relation then E contains a unimaximal subsequence of length $k + 1$ according to the previous lemma. Hence the theorem holds also for $n = k + 1$. □

3   Lower Bound

This section shows that the bounds derived in the previous section are tight.

Lemma 4. *For all $n > 1$ there exists a sequence P_n of $(n-1)^2$ pairs of integers that*

- *doesn't contain pairs with $/\!/$- and $\backslash\!\backslash$-relations,*
- *has no unimaximal subsequence of length n.*

Proof. According to the Erdős-Szekeres theorem there exists a sequence $(x_i)_{i=1}^{(n-1)^2}$ that doesn't contain a monotone subsequence of length n. The sequence $P_n = ((x_i, -x_i))_{i=1}^{(n-1)^2}$ clearly doesn't contain pairs with $/\!/$- and $\backslash\!\backslash$-relations.

A unimaximal subsequence of P_n (or any other sequence that doesn't contain pairs with $/\!/$- and $\backslash\!\backslash$-relations) must be monotone in both components. Therefore P_n cannot have a unimaximal subsequence of length n because otherwise $(x_i)_i$ would contain a monotone subsequence of this length. □

Let $P = ((x_i, y_i))_i$ be a sequence of pairs of integers and $m \in \mathbb{N}$. We denote the sequence $((x_i + m, y_i + m))_i$ by $P + m$ in the sequel.

Lemma 5. *For all $n > 1$ there exists a sequence Q_n of $\frac{1}{6}(n-1)n(2n-1)$ pairs of integers that*

- *doesn't contain pairs with a $\diagdown\!\!\!\diagdown$-relation,*
- *has no unimaximal subsequence of length n.*

Proof. Let $P_i, i = 2, \ldots, n$ be the sequences from the previous lemma. Let $P_i' = P_i + m_i$. The shifts m_i are selected such that for all $n \geq i > j \geq 2$ the pairs from P_i' have to pairs in P_j' $\diagup\!\!\!\diagup$-relations. Finally, let Q_n be a concatenation of the sequences $P_n', \ldots, P_2'$.

The length of Q_n is $\sum_{i=2}^{n}(i-1)^2 = \frac{1}{6}(n-1)n(2n-1)$.

Q_n doesn't contain a $\diagdown\!\!\!\diagdown$-relation because this relation is not present among pairs from the individual subsequences P_i' and there are $\diagup\!\!\!\diagup$-relations among pairs from the different subsequences.

Let U be a unimaximal subsequence of Q_n and k be the minimal index such that U contains a pair $(\bar{x}, \bar{y})$ from P_k'. Each pair from P_l', $l > k$ has a $\diagup\!\!\!\diagup$-relation to $(\bar{x}, \bar{y})$. If (x_i, y_i) and (x_j, y_j), $i < j$ are two pairs from a fixed P_l', $l > k$ then they cannot be both in U because the triple $(x_i, y_i), (x_j, y_j), (\bar{x}, \bar{y})$ is unimaximal only if (x_i, y_i) has a $\diagup\!\!\!\diagup$-relation to (x_j, y_j), but this cannot happen due to the definition of P_l'.

Therefore U contains at most one pair from each P_l', $l > k$ and at most $k-1$ pairs from P_k' (P_k' has no unimaximal subsequence of length k). Hence $|U| \leq (n-k) + (k-1) = n-1$ and Q_n has no unimaximal subsequence of length n. $\qquad\square$

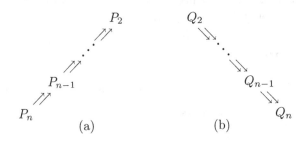

Fig. 1. Construction of (a) Q_n and (b) R_n

Lemmas 4 and 5 provide the lower bounds that match the upper bounds given by Lemmas 2 and 3. Finally, the following theorem shows that the bound in Theorem 1 is tight as well.

Theorem 2. *For all $n > 1$ there exists a sequence R_n of $\frac{1}{12}n^2(n^2-1)$ pairs of integers that has no unimaximal subsequence of length n.*

Proof. The proof is very similar to the proof of the previous lemma.

Let $Q_i, i = 2, \ldots, n$ be the sequences from the previous lemma. Let $Q_i' = Q_i + m_i$. The shifts m_i are selected such that for all $2 \leq i < j \leq n$ the pairs from Q'_i have to pairs in Q'_j $\searrow$-relations. Finally, let R_n be a concatenation of the sequences $Q'_2, \ldots, Q'_n$.

The length of R_n is $\sum_{i=2}^{n} \frac{1}{6}(i-1)i(2i-1) = \frac{1}{12}n^2(n^2-1)$.

Let U be a unimaximal subsequence of R_n and k be the minimal index such that U contains a pair $(\overline{x}, \overline{y})$ from Q'_k. $(\overline{x}, \overline{y})$ has a $\searrow$-relation to each pair from Q'_l, $l > k$. If (x_i, y_i) and (x_j, y_j), $i < j$ are two pairs from a fixed Q'_l, $l > k$ then they cannot be both in U because the triple $(\overline{x}, \overline{y})$, (x_i, y_i), (x_j, y_j) is unimaximal only if (x_i, y_i) has a $\searrow$-relation to (x_j, y_j), but this cannot happen due to the definition of Q'_l.

Therefore U contains at most one pair from each Q'_l, $l > k$ and at most $k-1$ pairs from Q'_k (Q'_k has no unimaximal subsequence of length k). Hence $|U| \le (n-k) + (k-1) = n-1$ and R_n has no unimaximal subsequence of length n. □

4   Application in 3D Rectangle Visibility Graphs

Fekete et al. [1] showed that every 3D rectangle visibility representation can be described using integer 4-tuples that denote perpendicular distances of sides of individual rectangles to the origin. They also proved the following lemma.

Lemma 6. *In a representation of K_5 by five rectangles $((e_i, n_i, w_i, s_i))_{i=1}^{5}$, it is impossible that both sequences $(n_i)_{i=1}^{5}$ and $(s_i)_{i=1}^{5}$ are unimaximal.*

Lemma 6 and Theorem 1 allow us to improve the best known upper bound on the size of the largest complete graph with a 3D rectangle visibility representation.

Theorem 3. *No complete graph K_n has a 3D rectangle visibility representation for $n \ge 51$.*

Proof. Let's assume we have a representation of K_n with $n \ge 51$ rectangles (e_i, n_i, w_i, s_i). Theorem 1 implies that the sequence $((n_i, s_i))_{i=1}^{51}$ has a unimaximal subsequence $(n'_i, s'_i)_i$ of length 5. Remove the rectangles not associated with the subsequence. The five remaining rectangles represent K_5, but this contradicts the previous lemma because both sequences $(n'_i)_{i=1}^{5}$ and $(s'_i)_{i=1}^{5}$ are unimaximal. □

5   Conclusion

We show that every sequence of $\frac{1}{12}n^2(n^2-1) + 1$ pairs of integers has a unimaximal subsequence of length n. On the other hand, there are sequences of $\frac{1}{12}n^2(n^2-1)$ pairs that do not contain such a sequence.

The analysis of unimaximal sequences of pairs allows us to improve the best known upper bound on the size of the largest complete graph with a 3D rectangle visibility representation from 55 to 50. The original bound by Fekete el al. [1] is also based on the study of unimaximal subsequences in the sequences of rectangle coordinates but they consider each coordinate independently. It remains an open problem how to analyze all four coordinates together to obtain a better bound.

References

1. Fekete, S.P., Houle, M.E., Whitesides, S.: New results on a visibility representation of graphs in 3D. In: Brandenburg, F.J. (ed.) GD 1995. LNCS, vol. 1027, pp. 234–241. Springer, Heidelberg (1996)
2. Bose, P., Everett, H., Fekete, S.P., Lubiw, A., Meijer, H., Romanik, K., Shermer, T., Whitesides, S.: On a visibility representation for graphs in three dimensions. In: Di Battista, G., Eades, P., de Fraysseix, H., Rosenstiehl, P., Tamassia, R. (eds.) GD 1993, pp. 38–39 (1993)
3. Chung, F.R.K.: On unimodal subsequences. J. Combinatorial Theory 29(A), 267–279 (1980)
4. Erdős, P., Szekeres, G.: A combinatorial problem in geometry. Compositio Math. 2, 463–470 (1935)
5. Štola, J.: Colorability in orthogonal graph drawing. In: Hong, S.-H., Nishizeki, T., Quan, W. (eds.) GD 2007. LNCS, vol. 4875, pp. 327–338. Springer, Heidelberg (2008)
6. Alt, H., Godau, M., Whitesides, S.: Universal 3-dimensional visibility representations for graphs. In: Brandenburg, F.J. (ed.) GD 1995. LNCS, vol. 1027, pp. 8–19. Springer, Heidelberg (1996)
7. Romanik, K.: Directed VR-Representable Graphs have Unbounded Dimension. In: Tamassia, R., Tollis, I(Y.) G. (eds.) GD 1994. LNCS, vol. 894, pp. 177–181. Springer, Heidelberg (1995)
8. Fekete, S.P., Meijer, H.: Rectangle and box visibility graphs in 3D. Int. J. Comput. Geom. Appl. 97, 1–28 (1997)

Visibility Representations of Four-Connected Plane Graphs with Near Optimal Heights

Chieh-Yu Chen[1], Ya-Fei Hung[2], and Hsueh-I Lu[1,2,*]

[1] Department of Computer Science and Information Engineering
National Taiwan University
[2] Graduate Institute of Networking and Multimedia
National Taiwan University
1 Roosevelt Road, Section 4, Taipei 106, Taiwan, ROC
f94922054@ntu.edu.tw, r94944014@ntu.edu.tw, hil@csie.ntu.edu.tw
http://www.csie.ntu.edu.tw/~hil

Abstract. A *visibility representation* of a graph G is to represent the nodes of G with non-overlapping horizontal line segments such that the line segments representing any two distinct adjacent nodes are vertically visible to each other. If G is a plane graph, i.e., a planar graph equipped with a planar embedding, a visibility representation of G has the additional requirement of reflecting the given planar embedding of G. For the case that G is an n-node four-connected plane graph, we give an $O(n)$-time algorithm to produce a visibility representation of G with height at most $\lceil \frac{n}{2} \rceil + 2 \left\lceil \sqrt{\frac{n-2}{2}} \right\rceil$. To ensure that the first-order term of the upper bound is optimal, we also show an n-node four-connected plane graph G, for infinite number of n, whose visibility representations require heights at least $\frac{n}{2}$.

1 Introduction

Unless clearly specified otherwise, all graphs in the present article are simple, i.e., having no self-loops and multiple edges. A *visibility representation* of a planar graph represents the nodes of the graph by non-overlapping horizontal line segments such that, for any nodes u and v adjacent in the graph, the line segments representing u and v are vertically visible to each other. Observe that if G_1 is a subgraph of G_2 on the save node set, then any visibility representation of G_2 is also a visibility representation of G_1. Therefore, we may assume without loss of generality that the input graph is maximally planar. Let G be an n-node plane triangulation, i.e., a maximally planar graph equipped with a planar embedding. A visibility representation of G has an additional requirement of reflecting the given planar embedding of G. Figure 1(b), for instance, is a visibility representation of the four-connected plane graph shown in Fig. 1(a). Under the conventional restriction of placing the endpoints of horizontal line segments on the integral grid points, any visibility representation of G requires width no more than $3n - 7$ and height no more than $n - 1$. Otten and van Wijk [7] gave the first known algorithm for constructing a visibility representation for any G. Rosenstiehl and Tarjan [8] and Tamassia

---

* Corresponding author. This author also holds a joint appointment in the Graduate Institute of Biomedical Electronics and Bioinformatics, National Taiwan University. Research supported in part by NSC grant 96-2221-E-002-033.

I.G. Tollis and M. Patrignani (Eds.): GD 2008, LNCS 5417, pp. 67–77, 2009.

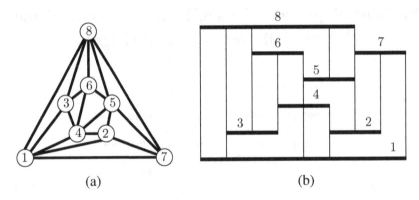

Fig. 1. (a) A four-connected plane triangulation G. (b) A visibility representation of G.

and Tollis [9] independently gave algorithms to compute a visibility representation of G with height at most $2n - 5$. Their work initiated a decade of competition on minimizing the width and height of the output visibility representation. All these algorithms run in linear time. In particular, the results of Fan, Lin, Lu, and Yen [2] and Zhang and He [16] are optimal in that the upper bounds differ from the best known lower bounds by very small constants.

The present article focuses on four-connected plane G. The $O(n)$-time algorithm of Kant and He [5] provides the optimal upper bound $n - 1$ on the width. The best previously known upper bound on the height, ensured by the $O(n)$-time algorithm of Zhang and He [12], is $\left\lceil \frac{3n}{4} \right\rceil$. In the present article, we obtain the following result with an improved upper bound on the required height.

Theorem 1. *For any n-node four-connected plane graph G, it takes $O(n)$ time to construct a visibility representation of G with height at most $\left\lceil \frac{n}{2} \right\rceil + 2 \left\lceil \sqrt{\frac{n-2}{2}} \right\rceil$.*

Table 1 compares our upper bound with previous results. All algorithms shown in Table 1 run in $O(n)$ time. Our algorithm follows the approach of Zhang and He [10, 15–17], originating from Rosenstiehl and Tarjan [8] and Tamassia and Tollis [9], that reduces the problem of computing a visibility representation for G with small height to finding an appropriate st-ordering of G. To find such an st-ordering of G, we resort to three linear-time obtainable node orderings:

- four-canonical orderings of four-connected plane graphs (Kant and He [5]),
- consistent orderings of ladder graphs (Zhang and He [15–17]), and
- post-orderings of canonical ordering spanning trees (He, Kao, and Lu [3]).

Our result is near optimal in that we can construct an n-node four-connected plane graph, for infinite number of n, whose visibility representations require heights at least $\left\lceil \frac{n}{2} \right\rceil$. That is, the first-order term of our upper bound is optimal.

The remainder of the paper is organized as follows. Section 2 gives the preliminaries. Section 3 describes and analyzes our algorithm. Section 4 ensures that the first-order term of our upper bound on height is optimal. Section 5 concludes the paper.

Table 1. Previous upper bounds and our result for any n-node plane graph G

	general G		four-connected G	
	width	height	width	height
Otten and van Wijk [7]	$3n - 7$	$n - 1$		
Rosenstiehl and Tarjan [8], Tamassia and Tollis [9]	$2n - 5$			
Kant [4]	$\lfloor \frac{3n-6}{2} \rfloor$			
Kant and He [5]			$n - 1$	
Lin, Lu, and Sun [6]	$\lfloor \frac{22n-24}{15} \rfloor$			
Zhang and He [10]		$\lceil \frac{15n}{16} \rceil$		
Zhang and He [14]		$\lfloor \frac{5n}{6} \rfloor$		
Zhang and He [11, 13]	$\lfloor \frac{13n-24}{9} \rfloor$			
Zhang and He [12]				$\lceil \frac{3n}{4} \rceil$
Zhang and He [15, 17]	$\frac{4n}{3} + 2\lceil \sqrt{n} \rceil$	$\frac{2n}{3} + 2\lceil \sqrt{\frac{n}{2}} \rceil$		
Zhang and He [16]		$\frac{2n}{3} + O(1)$		
Fan, Lin, Lu, and Yen [2]	$\lfloor \frac{4n}{3} \rfloor - 2$			
This paper				$\lceil \frac{n}{2} \rceil + 2\lceil \sqrt{\frac{n-2}{2}} \rceil$

2   Preliminaries

2.1   Ordering and st-Ordering

Let G be an n-node plane graph. An *ordering* of G is a one-to-one mapping σ from the nodes of G to $\{1, 2, \ldots, n\}$. A path of G is σ-*increasing* if $\sigma(u) < \sigma(v)$ holds for any nodes u and v such that u precedes v in the path. Let $length(G, \sigma)$ denote the maximum of the lengths of all σ-increasing paths in G. For instance, if G and σ are as shown in Fig. 1(a), then one can verify that $(1, 2, 5, 6, 8)$ is a σ-increasing path with maximum length. Therefore, $length(G, \sigma) = 4$.

Let s and t be two distinct external nodes of G. An st-*ordering* [1] of G is an ordering σ of G such that

- $\sigma(s) = 1$, $\sigma(t) = n$, and
- each node v of G other than s and t has neighbors u and w in G with $\sigma(u) < \sigma(v) < \sigma(w)$.

An example is shown in Fig. 1(a): the node labels form an st-ordering for the graph.

The following lemma reduces the problem of minimizing the height of visibility representation of G to that of finding an st-ordering σ of G with minimum $length(G, \sigma)$.

Lemma 1 (See [2, 8–10, 15, 17]). *If G admits an st-ordering σ for two distinct external nodes s and t of G, then it takes $O(n)$ time to obtain a visibility representation of G with height exactly length(G, σ).*

For instance, if G and σ are as shown in Fig. 1(a), then a visibility representation for G with height at most $length(G, \sigma) = 4$, as shown in Fig. 1(b), can be found in linear time.

2.2   Four-Canonical Ordering

Let G be an n-node four-connected plane triangulation. Let v_1, v_2, and v_n be the external nodes of G in counterclockwise order. Since G is a four-connected plane triangulation, G has exactly one internal node adjacent to both v_2 and v_n. Let v_{n-1} be the internal node adjacent to v_2 and v_n in G. A *four-canonical ordering* [5] of G is an ordering ϕ in G such that

- $\phi(v_1) = 1$, $\phi(v_2) = 2$, $\phi(v_{n-1}) = n - 1$, $\phi(v_n) = n$, and
- each node v of G other than v_1, v_2, v_{n-1} and v_n has neighbors u, u', w and w' in G with $\phi(u') < \phi(u) < \phi(v) < \phi(w) < \phi(w')$.

An example is shown in Fig. 2(a): the node labels form a four-canonical ordering of the four-connected plane triangulation.

Lemma 2 (Kant and He [5]). *It takes $O(n)$ time to compute a four-canonical ordering for any n-node G.*

2.3   Consistent Ordering of Ladder Graph

Let L be an $\lceil \frac{n}{2} \rceil$-node path. Let R be an $\lfloor \frac{n}{2} \rfloor$-node path. Let X consist of edges with one endpoint in L and the other endpoint in R. Let (L, R, X) denote the n-node graph $L \cup R \cup X$. We say that (L, R, X) is a *ladder graph* [15, 17] if $L \cup R \cup X$ is outerplanar. A ladder graph is shown in Fig. 3(a).

   An ordering σ of ladder graph (L, R, X) is *consistent* [15, 17] with respect to an outerplanar embedding $\mathcal{E}$ of (L, R, X) if L (respectively, R) forms a σ-increasing path in clockwise (respectively, counterclockwise) order according to $\mathcal{E}$. See Fig. 3(a) for an example: The node labels form a consistent ordering of the ladder graph with respect to the displayed outerplanar embedding.

Lemma 3 (He and Zhang [15, 17]). *Let (L, R, X) be an n-node ladder graph. It takes $O(n)$ time to compute a consistent ordering σ of (L, R, X) with respect to any given outerplanar embedding of (L, R, X) such that $length((L, R, X), \sigma) \leq \lceil \frac{n}{2} \rceil + 2 \lceil \sqrt{\frac{n}{2}} \rceil - 1$.*

For technical reason, we need a consistent ordering with additional properties, as stated in the next lemma, which is also illustrated by Fig. 3(a).

Lemma 4. *Let (L, R, X) be an n-node ladder graph. It takes $O(n)$ time to compute a consistent ordering σ of (L, R, X) with respect to any given outerplanar embedding $\mathcal{E}$ of (L, R, X) such that*

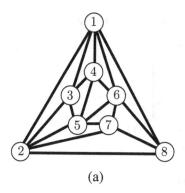

(a)

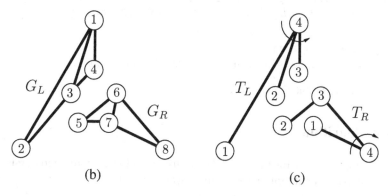

(b)   (c)

Fig. 2. (a) A four-canonical ordering ϕ of the four-connected plane triangulation G. (b) G_L is the subgraph induced by the nodes v with $1 \leq \phi(v) \leq 4$ and G_R is the subgraph induced by the nodes v with $5 \leq \phi(v) \leq 8$. (c) The counterclockwise post-ordering ψ_L of T_L and the clockwise post-ordering ψ_R of T_R.

- $\sigma(\ell_1) = 1$, $\sigma(r_1) = 2$, *and*
- $length((L, R, X), \sigma) \leq \left\lceil \frac{n}{2} \right\rceil + 2 \left\lceil \sqrt{\frac{n-2}{2}} \right\rceil$,

where ℓ_1 (respectively, r_1) is the first (respectively, last) node of L (respectively, R) in clockwise order around the external boundary of (L, R, X) with respect to $\mathcal{E}$.

Proof. Let $L' = L \setminus \{\ell_1\}$. Let $R' = R \setminus \{r_1\}$. Let $X' = X \setminus \{\ell_1, r_1\}$. Clearly, (L', R', X') is a ladder graph of $n - 2$ nodes. Let σ' be the consistent ordering of (L', R', X') with respect to $\mathcal{E}$ ensured by Lemma 3. We have

$$length((L', R', X'), \sigma') \leq \left\lceil \frac{n}{2} \right\rceil + 2 \left\lceil \sqrt{\frac{n-2}{2}} \right\rceil - 2.$$

Let σ be the ordering of (L, R, X) such that

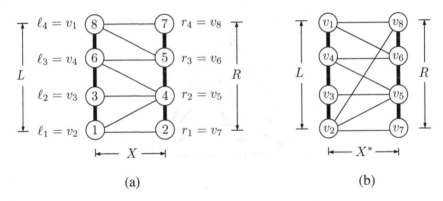

Fig. 3. (a) A consistent ordering of a ladder graph (L, R, X) with respect to the displayed outerplanar embedding. (b) $H^* = L \cup R \cup X^*$, where $X^* = X \cup \{(v_2, v_8)\}$.

- $\sigma(\ell_1) = 1$, $\sigma(r_1) = 2$, and
- $\sigma(u) = \sigma'(u) + 2$ holds for each node u other than ℓ_1 and r_1.

One can easily verify that the lemma holds.   □

3   Our Algorithm

Let G be the input n-node four-connected plane triangulation. According to Lemma 1, it suffices to describe our algorithm for computing an st-ordering σ for G in the following four steps.

3.1   Step 1

Let ϕ be a four-canonical ordering of G ensured by Lemma 2.

- Let G_L be the subgraph of G induced by the nodes v with $1 \leq \phi(v) \leq \lceil \frac{n}{2} \rceil$.
- Let G_R be the subgraph of G induced by the nodes v with $\lceil \frac{n}{2} \rceil < \phi(v) \leq n$.

Figure 2(b) illustrates this step, which runs in $O(n)$ time. Observe that each edge of G not in $G_L \cup G_R$ has one endpoint on the external boundary of G_L and the other endpoint on the external boundary of G_R.

3.2   Step 2

For each $i = 1, 2, \ldots, n$, let v_i denote the node of G with $\phi(v_i) = i$. It follows from the definition of ϕ that v_1, v_2, and v_n are the external nodes of G.

- For each $i = 2, 3, \ldots, \lceil \frac{n}{2} \rceil$, let $\pi(i)$ be the index j with $j < i$ such that v_j is the first neighbor of v_i in G_L in counterclockwise order around v_i. Let T_L be the spanning tree of G_L rooted at v_1 such that each $v_{\pi(i)}$ is the parent of v_i in T_L. Let ψ_L be the counterclockwise post-ordering of T_L.

– For each $i = \lceil \frac{n}{2} \rceil + 1, \lceil \frac{n}{2} \rceil + 2, \ldots, n - 1$, let $\pi(i)$ be the index j with $j > i$ such that v_j is the first neighbor of v_i in G_R in clockwise order around v_i. Let T_R be the spanning tree of G_R rooted at v_n such that each $v_{\pi(i)}$ is the parent of v_i in T_R. Let ψ_R be the clockwise post-ordering of T_R.

Figure 2(c) illustrates this step, which runs in $O(n)$ time. As a matter of fact, T_L is the canonical ordering spanning tree of G_L with respect to ϕ, as defined by He, Kao, and Lu [3].

Lemma 5. $\psi_L(v_2) = 1$, $\psi_L(v_1) = \lceil \frac{n}{2} \rceil$, $\psi_R(v_{n-1}) = 1$, and $\psi_R(v_n) = \lfloor \frac{n}{2} \rfloor$.

Proof. Since ϕ is a four-canonical ordering of G, if (v_2, v_i) with $i \geq 3$ is an edge of G_L, then v_i has to have a neighbor v_k with $2 \neq k < i$ in G_L. Observe that v_2 is the node immediately succeeding v_1 in counterclockwise order around the external boundary of G_L. One can verify that v_2 cannot be the first neighbor of v_i in G_L in counterclockwise order around v_i. That is, we have $\pi(i) \neq 2$. Since v_2 cannot be the parent of v_i in T_L, v_2 has to be a leaf of T_L. By the relative position between v_2 and v_1, it is clear that v_2 is the first node in the counterclockwise post-ordering of T_L, i.e., $\psi_L(v_2) = 1$.

One can prove $\psi_R(v_{n-1}) = 1$ analogously, where v_n (respectively, v_{n-1}, ψ_R, T_R, and G_R) plays the role of v_1 (respectively, v_2, ψ_L, T_L, and G_L). Since v_1 is the root of T_L and ψ_L is a post-ordering of T_L, we have $\psi_L(v_1) = \lceil \frac{n}{2} \rceil$. Since v_n is the root of T_R and ψ_R is a post-ordering of T_R, we have $\psi_R(v_n) = \lfloor \frac{n}{2} \rfloor$. □

3.3   Step 3

Let L, R, and X be defined as follows.

– Let L be the path $(\ell_1, \ell_2, \ldots, \ell_{\lceil n/2 \rceil})$, where ℓ_i is the node of G_L with $\psi_L(\ell_i) = i$.
– Let R be the path $(r_1, r_2, \ldots, r_{\lfloor n/2 \rfloor})$, where r_i is the node of G_R with $\psi_R(r_i) = i$.
– Let $X = X^* \setminus \{(v_2, v_n)\}$, where X^* consists of the edges of G with one endpoint in L and the other endpoint in R.

Figure 3(a) illustrates Lemma 5 and this step, which runs in $O(n)$ time. Figure 3(b) shows the corresponding $L \cup R \cup X^*$.

Lemma 6. (L, R, X) *is an n-node ladder graph.*

Proof. Consider any edge (ℓ_i, r_j) of X. By definition of ϕ, ℓ_i has to be on the external boundary of G_L and r_j has to be on the external boundary of G_R. By definition of T_L, ℓ_i is either a leaf of T_L or on the rightmost path of T_L. By definition of ψ_L, if $\ell_{i_1}, \ell_{i_2}, \ldots, \ell_{i_p}$ with $i_1 = 1$ are the nodes on the external boundary of G_L in counterclockwise order, then $i_1 < i_2 < \cdots < i_p$. Similarly, by definition of T_R, r_j is either a leaf of T_R or on the leftmost path of T_R. By definition of ψ_R, if $r_{j_1}, r_{j_2}, \ldots, r_{j_q}$ with $j_1 = 1$ are the nodes on the external boundary of G_R in clockwise order, then $j_1 < j_2 < \cdots < j_q$. Since G is a plane graph and the edges of X do not cross one another in G, the edges of X do not cross one another in (L, R, X). Therefore, (L, R, X) is outerplanar. □

3.4   Step 4

Let $H = (L, R, X)$. Lemma 6 ensures that H is an n-node ladder graph. Consider the outerplanar embedding $\mathcal{E}$ of H such that

$$\ell_1, \ell_2, \ldots, \ell_{\lceil n/2 \rceil}, r_{\lfloor n/2 \rfloor}, r_{\lfloor n/2 \rfloor - 1}, \ldots, r_1$$

are the nodes in clockwise order around the external boundary of H. Let the output σ of our algorithm be the consistent ordering of H with respect to $\mathcal{E}$ ensured by Lemma 4. Figure 3(a) illustrates this step, which also runs in $O(n)$ time.

Lemma 7. *The $O(n)$-time obtainable σ is an st-ordering of G with $\sigma(v_2) = 1$ and $\max(\sigma(v_1), \sigma(v_n)) = n$.*

Proof. We first show that ψ_L is an st-ordering of G_L. Let i be an index with $2 \leq i < \lceil \frac{n}{2} \rceil$. Let k be the index such that ℓ_k is the parent of ℓ_i in T_L. Since ψ_L is a post-ordering of T_L, we know that ℓ_k is a neighbor of ℓ_i in G_L with $i < k$. Let j be the index such that ℓ_j is the neighbor of ℓ_i in G_L immediately succeeding ℓ_k in counterclockwise order around ℓ_i. Recall that ℓ_k is the first neighbor of ℓ_i in G_L with $\phi(\ell_k) < \phi(\ell_i)$ in counterclockwise order around ℓ_i. Since ϕ is a four-canonical ordering of G, we also have $\phi(\ell_j) < \phi(\ell_i)$. Since ψ_L is the counterclockwise post-ordering of T_L, we have $\psi(\ell_j) < \psi(\ell_i)$, i.e., $j < i$. Since ℓ_j and ℓ_k are two neighbors of ℓ_i in G_L with $j < i < k$, we know that ψ_L is an st-ordering of G_L. It can be proved analogously that ψ_R is an st-ordering of G_R.

Since σ is a consistent ordering of H with respect to $\mathcal{E}$, we know that $1 \leq i < j \leq \lceil \frac{n}{2} \rceil$ implies $\sigma(\ell_i) < \sigma(\ell_j)$ and $1 \leq i < j \leq \lfloor \frac{n}{2} \rfloor$ implies $\sigma(r_i) < \sigma(r_j)$. We have the following observations.

- Since ψ_L is an st-ordering of G_L, for each $i = 1, \ldots, \lceil \frac{n}{2} \rceil - 1$, ℓ_i has a neighbor ℓ_k in G_L with $i < k$. Since G_L is a subgraph of G, ℓ_k is a neighbor of ℓ_i in G with $\sigma(\ell_i) < \sigma(\ell_k)$.
- Since ψ_L is an st-ordering of G_L, for each $i = 2, \ldots, \lceil \frac{n}{2} \rceil$, ℓ_i has a neighbor ℓ_j in G_L with $j < i$. Since G_L is a subgraph of G, we know that ℓ_j is a neighbor of ℓ_i in G with $\sigma(\ell_j) < \sigma(\ell_i)$.
- Since ψ_R is an st-ordering of G_R, for each $i = 1, \ldots, \lfloor \frac{n}{2} \rfloor - 1$, r_i has a neighbor r_k in G_R with $i < k$. Since G_R is a subgraph of G, we know that r_k is a neighbor of r_i in G with $\sigma(r_i) < \sigma(r_k)$.
- Since ψ_R is an st-ordering of G_R, for each $i = 2, \ldots, \lfloor \frac{n}{2} \rfloor$, r_i has a neighbor r_j in G_R with $j < i$. Since G_R is a subgraph of G, we know that r_j is a neighbor of r_i in G with $\sigma(r_j) < \sigma(r_i)$.

According to the above observations, it suffices to ensure that edges (ℓ_1, r_1) and $(\ell_{\lceil n/2 \rceil}, r_{\lfloor n/2 \rfloor})$ belong to G. By Lemma 5, $\ell_1 = v_2$, $r_1 = v_{n-1}$, $\ell_{\lceil n/2 \rceil} = v_1$, and $r_{\lfloor n/2 \rfloor} = v_n$. Since v_1 and v_n are external nodes of the plane triangulation G, we know that $(\ell_{\lceil n/2 \rceil}, r_{\lfloor n/2 \rfloor}) = (v_1, v_n)$ is an edge of G. By definition of four-canonical ordering ϕ, we know that v_{n-1} is adjacent to v_2. Therefore, $(\ell_1, r_1) = (v_2, v_{n-1})$ is an edge of G.   □

Figure 1(a) shows the resulting st-ordering σ of G computed by our algorithm.

3.5   Proving Theorem 1

Proof. Note that v_1, v_2, and v_n are the external nodes of G. By Lemmas 1 and 7, it suffices to ensure

$$length(G, \sigma) \leq \left\lceil \frac{n}{2} \right\rceil + 2 \left\lceil \sqrt{\frac{n-2}{2}} \right\rceil. \tag{1}$$

By Step 4 and Lemmas 4 and 6, we have

$$length(H, \sigma) \leq \left\lceil \frac{n}{2} \right\rceil + 2 \left\lceil \sqrt{\frac{n-2}{2}} \right\rceil. \tag{2}$$

Let $H^* = L \cup R \cup X^*$. That is, $H^* = H \cup \{(v_2, v_n)\}$, as illustrated by Fig. 3(a) and 3(b). By definition of σ and Lemma 5, we have $\sigma(v_2) = 1$ and $\sigma(v_n) \geq \max_j \sigma(r_j)$. Therefore, any σ-increasing path of H^* containing edge (v_2, v_n) contains exactly one node of R, i.e., v_n, and thus has length at most $\left\lceil \frac{n}{2} \right\rceil$. It follows from Inequality (2) that

$$length(H^*, \sigma) \leq \left\lceil \frac{n}{2} \right\rceil + 2 \left\lceil \sqrt{\frac{n-2}{2}} \right\rceil. \tag{3}$$

To prove Inequality (1), it remains to show that if P is a σ-increasing path of G, then there is a σ-increasing path Q of H^* such that the length of Q is no less than that of P. For each edge (u, v) of P with $\sigma(u) < \sigma(v)$, let $Q(u, v)$ be the σ-increasing path of H^* defined as follows.

- If $u = \ell_i$ and $v = r_j$, then let $Q(u, v) = (u, v)$, which is a σ-increasing path of X^*.
- If $u = r_i$ and $v = \ell_j$, then let $Q(u, v) = (u, v)$, which is a σ-increasing path of X^*.
- If $u = \ell_i$ and $v = \ell_j$, then by $\sigma(\ell_i) < \sigma(\ell_j)$ we know $\psi_L(\ell_i) < \psi_L(\ell_j)$ and thus $i < j$. Let $Q(u, v) = (\ell_i, \ell_{i+1}, \ldots, \ell_j)$. Since σ is a consistent ordering of H with respect to $\mathcal{E}$, $Q(u, v)$ is a σ-increasing path of L.
- If $u = r_i$ and $v = r_j$, then by $\sigma(r_i) < \sigma(r_j)$ we know $\psi_R(r_i) < \psi_R(r_j)$ and thus $i < j$. Let $Q(u, v) = (r_i, r_{i+1}, \ldots, r_j)$. Since σ is a consistent ordering of H with respect to $\mathcal{E}$, $Q(u, v)$ is a σ-increasing path of R.

Let Q be the union of $Q(u, v)$ for all edges (u, v) of P. Since each $Q(u, v)$ is a σ-increasing path of H^*, so is Q. The length of Q is no less than that of P. That is, we have

$$length(G, \sigma) \leq length(H^*, \sigma). \tag{4}$$

Since Inequality (1) is immediate from Inequalities (3) and (4), the lemma is proved. □

4   A Lower Bound

Let plane graph N_k be defined recursively as follows.

- Let N_1 be the four-node internally triangulated plane graph with four external nodes.

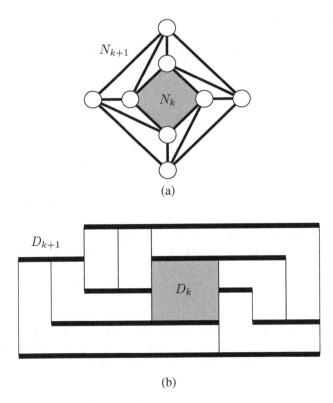

Fig. 4. (a) A four-connected plane graph N_{k+1} and its relation with N_k. (b) A visibility representation D_{k+1} of N_{k+1} and its relation with D_k.

- Let N_{k+1} be obtained from N_k by adding four nodes and twelve edges in the way as shown in Fig. 4(a).

One can easily verify that each N_k with $k \geq 1$ is indeed four-connected. The following lemma ensures that the the upper bound provided by Theorem 1 has an optimal first-order term.

Lemma 8. *All visibility representations of N_k have heights at least $2k$.*

Proof. We prove the lemma by induction on k. The lemma holds trivially for $k = 1$. Assume for a contradiction that N_{k+1} admits a visibility representation D_{k+1} with height no more than $2k + 1$. Let D_k be obtained from D_{k+1} by deleting all the horizontal segments representing those four external nodes of N_{k+1}. Since D_{k+1} has to reflect the planar embedding of N_{k+1}, D_k is a visibility representation of N_k. Since the external nodes of N_k are internal in N_{k+1}, the horizontal segments of D_{k+1} representing the external nodes of N_{k+1} have to wrap D_k completely. That is, D_{k+1} must have a horizontal segment above D_k and a horizontal segment below D_k. Therefore, the height of D_{k+1} is at least two more than that of D_k. It follows that the height of D_k is at most $2k - 1$, contradicting the inductive hypothesis. Since N_{k+1} cannot admit a visibility representation with height less than $2k + 2$, the lemma is proved. □

5   Concluding Remarks

It would be of interest to close the $\Theta(\sqrt{n})$ gap between the upper and lower bounds on the required height for the visibility representation of any n-node four-connected plane graph. We conjecture that the $\Theta(\sqrt{n})$ term in our upper bound can be reduced to $O(1)$.

References

1. Even, S., Tarjan, R.E.: Computing an st-numbering. Theoretical Computer Science 2(3), 339–344 (1976)
2. Fan, J.H., Lin, C.C., Lu, H.I., Yen, H.C.: Width-optimal visibility representations of plane graphs. In: Tokuyama, T. (ed.) ISAAC 2007. LNCS, vol. 4835, pp. 160–171. Springer, Heidelberg (2007)
3. He, X., Kao, M.Y., Lu, H.I.: Linear-time succinct encodings of planar graphs via canonical orderings. SIAM Journal on Discrete Mathematics 12(3), 317–325 (1999)
4. Kant, G.: A more compact visibility representation. International Journal Computational Geometry and Applications 7(3), 197–210 (1997)
5. Kant, G., He, X.: Regular edge labeling of 4-connected plane graphs and its applications in graph drawing problems. Theoretical Computer Science 172, 175–193 (1997)
6. Lin, C.C., Lu, H.I., Sun, I.F.: Improved compact visibility representation of planar graph via Schnyder's realizer. SIAM Journal on Discrete Mathematics 18(1), 19–29 (2004)
7. Otten, R.H.J.M., van Wijk, J.G.: Graph representations in interactive layout design. In: Proceedings of the IEEE International Symposium on Circuits and Systems, pp. 914–918 (1978)
8. Rosenstiehl, P., Tarjan, R.E.: Rectilinear planar layouts and bipolar orientations of planar graphs. Discrete and Computational Geometry 1, 343–353 (1986)
9. Tamassia, R., Tollis, I.G.: A unified approach to visibility representations of planar graphs. Discrete and Computational Geometry 1, 321–341 (1986)
10. Zhang, H., He, X.: Compact visibility representation and straight-line grid embedding of plane graphs. In: Dehne, F., Sack, J.-R., Smid, M. (eds.) WADS 2003. LNCS, vol. 2748, pp. 493–504. Springer, Heidelberg (2003)
11. Zhang, H., He, X.: On visibility representation of plane graphs. In: Diekert, V., Habib, M. (eds.) STACS 2004. LNCS, vol. 2996, pp. 477–488. Springer, Heidelberg (2004)
12. Zhang, H., He, X.: Canonical ordering trees and their applications in graph drawing. Discrete and Computational Geometry 33, 321–344 (2005)
13. Zhang, H., He, X.: Improved visibility representation of plane graphs. Computational Geometry 30(1), 29–39 (2005)
14. Zhang, H., He, X.: New theoretical bounds of visibility representation of plane graphs. In: Pach, J. (ed.) GD 2004. LNCS, vol. 3383, pp. 425–430. Springer, Heidelberg (2005)
15. Zhang, H., He, X.: Nearly optimal visibility representations of plane graphs. In: Bugliesi, M., Preneel, B., Sassone, V., Wegener, I. (eds.) ICALP 2006. LNCS, vol. 4051, pp. 407–418. Springer, Heidelberg (2006)
16. Zhang, H., He, X.: Optimal st-orientations for plane triangulations. In: Kao, M.-Y., Li, X.-Y. (eds.) AAIM 2007. LNCS, vol. 4508, pp. 296–305. Springer, Heidelberg (2007)
17. Zhang, H., He, X.: Nearly optimal visibility representations of plane triangulations. SIAM Journal on Discrete Mathematics 22(4), 1364–1380 (2008)

The Topology of Bendless Three-Dimensional Orthogonal Graph Drawing

David Eppstein

Computer Science Department, University of California, Irvine
eppstein@uci.edu

Abstract. We define an *xyz* graph to be a spatial embedding of a 3-regular graph such that the edges at each vertex are mutually perpendicular and no three points lie on an axis-parallel line. We describe an equivalence between *xyz* graphs and 3-face-colored polyhedral maps, under which bipartiteness of the graph is equivalent to orientability of the map. We show that planar graphs are *xyz* graphs if and only if they are bipartite, cubic, and three-connected. It is NP-complete to recognize *xyz* graphs, but we show how to do this in time $O(n2^{n/2})$.

1 Introduction

Consider a point set V in $\mathbb{R}^3$ (such as the vertices of an axis-aligned cube) with the property that every axis-parallel line in $\mathbb{R}^3$ contains either zero or two points of V. V forms the vertices of a cubic (that is, 3-regular) graph, in which each vertex v is connected to the other points that lie on the three axis-parallel lines through v. We call such a graph an *xyz graph*. Figure 1 depicts three examples.

Fig. 1. Three *xyz* graphs

In contrast to past work on three-dimensional orthogonal drawing with bends [2,3,6, 8,9,15,18,19,20], an *xyz* graph provides a simple form of bendless three-dimensional orthogonal drawing. In *xyz* graphs, edges may cross, but edge crossings may be distinguished visually from vertices by whether the edges stop or pass through them.

In three-dimensional layout of parallel processing intercommunication networks [5], *xyz* graphs provide a layout in which all connected pairs of processors have an open line of sight between each other. As we show, even-dimensional cube-connected-cycles networks, highly regular graphs used in parallel processing [16], have *xyz* graph layouts.

I.G. Tollis and M. Patrignani (Eds.): GD 2008, LNCS 5417, pp. 78–89, 2009.

These graphs also have an unexpected connection to topological graph theory and graph coloring: any xyz graph corresponds to a three-coloring of the faces of an embedding of a graph on a 2-manifold. Such face-colored embeddings arise naturally from the GEM (graph-embedded map) representation of manifold embeddings of graphs [4].

In this paper, we prove an equivalence between xyz graphs and certain 3-face-colored cell complexes, which we call xyz surfaces. As we show, an xyz graph is bipartite if and only if the corresponding xyz surface is orientable. We show that it is NP-complete to recognize xyz graphs, and we show how to find xyz graph embeddings in time $O(n2^{n/2})$; however, planar xyz graphs may be recognized in linear time. Due to space considerations we omit many results, details and proofs; we invite readers to find these in the longer version of this paper at http://arxiv.org/abs/0709.4087.

2   Topology of xyz Graphs

If C is a collection of cycles in an undirected graph G, we may define a *cell complex* with a point for each vertex, a line segment for each edge, and a disk for each cycle. For instance, if G is the graph of a cube, and C is the set of four-cycles in G, the resulting cell complex consists of the vertices, edges, and facets of a geometric cube. However, complexes may be defined independently of any spatial embedding. If the following conditions are satisfied, the cell complex is a 2-*manifold* (without boundary) or *map*:

1. Each edge of G belongs to exactly two cycles of C.
2. At each vertex v of G, one can reach any incident edge from any other incident edge by a chain of edge-face-edge steps in which each edge and face is incident to v.

Cubic graphs automatically satisfy the second condition. The complex defined from G and C is an *embedding* of G onto a manifold, and the cycles of C are its *faces*. We define an *xyz surface* to be an embedding of a cubic graph G with the following properties:

1. Any two faces intersect in either a single edge of G or the empty set.
2. The faces of C can be assigned three colors such that no two faces sharing an edge have the same color.

An embedded graph satisfying the first property is called *polyhedral* [14]. Polyhedral embeddings of non-cubic graphs may include faces that intersect in a single vertex, but

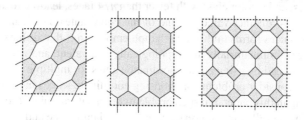

Fig. 2. Three xyz surfaces, each with the topology of the torus. In each case, the torus is depicted as cut and unrolled into a rectangle; the corresponding topological surface is formed by gluing opposite pairs of rectangle edges.

this cannot happen in a cubic graph. Craft and White [7] study a similar 3-coloring condition on orientable cubic maps without the polyhedral condition.

Theorem 1. *G is an xyz graph if and only if G can be embedded as an xyz surface.*

Proof. Let G be an xyz graph, and let C consist of the cycles in G that lie in an axis-parallel plane. Each edge of G belongs to two such cycles, so, C forms an embedding of G onto a manifold. The cycles of C can be colored according to the coordinate planes they are parallel to. The cycles of G in any single coordinate plane are disjoint, so if two cycles intersect, the intersection must lie on the axis-parallel line formed by the intersection of the two planes containing the cycles, and consists of the edge of G that lies on that same line, fulfilling the requirements of an xyz surface.

Conversely, suppose that G is embedded as an xyz surface, with cycle set C. Let X, Y, and Z be the three color classes of C, and let the faces in C be numbered $f_0, f_1, \ldots$. Each vertex v in G is incident to exactly three faces: f_i in X, f_j in Y, and f_k in Z for some i, j, k. We assign v the three-dimensional coordinates (i, j, k). If two vertices u and v are adjacent, they share the two coordinates determined by the two faces containing edge uv, and lie on an axis-parallel line of the embedding of G into $\mathbb{R}^3$. If two vertices are not adjacent, they can lie on at most one face of C and therefore have at most one coordinate in common. Thus, the three axis-parallel lines through each embedded vertex v each contain only v and one of its neighbors so the embedding forms an xyz graph. $\square$

The three xyz graphs in Fig. 1 correspond to xyz surfaces that are (left to right) a projective plane resembling the Roman surface, a spherical map combinatorially equivalent to a polyhedron with three hexagonal facets and six quadrilaterals, and an embedding of the Pappus graph on a torus. Figure 2 depicts three xyz surfaces, all tori. The leftmost is the Pappus graph again, the middle surface has 12 faces, 24 vertices, and 36 edges, and the right surface is a torus embedding of the 64-vertex four-dimensional cube-connected cycles network.

Theorem 1 can be used to embed any xyz graph into an $\frac{n}{4} \times \frac{n}{4} \times \frac{n}{4}$ grid: Each face of an xyz graph must have even length, at least four, because it alternates between edges parallel to two coordinate axes. Thus, any color class of an xyz surface coloring has at most $n/4$ faces: each vertex belongs to one face of that color, but each face contains at least four vertices. Each face provides a value for one of the coordinates in the grid embedding, so the number of distinct values for each coordinate is at most $n/4$. However this bound is tight only for the cube: any other xyz surface has a face with more than four vertices, and a color class with fewer than $n/4$ faces, leading to an embedding with fewer than $n/4$ distinct values in one of the coordinates. For many graphs, permuting the coordinates forms multiple xyz graph embeddings that differ geometrically, although they are combinatorially and topologically equivalent, and smaller grids may sometimes be obtained by using equal coordinate values for multiple faces of the same color. We do not consider problems of choosing coordinate values in order to improve the graph drawing in this paper, but such problems are a natural subject for future work.

As we show in the full version, every xyz graph is triangle-free and 3-vertex-connected. We conclude this section with an interesting connection between bipartiteness and topology. An *orientation* of a map can be described as a choice of cyclic order on each face of the map such that the two face cycles shared by any edge pass through it in opposite directions. A surface is *orientable* if graphs embedded on it may be oriented; the sphere and torus are orientable, while the projective plane is not.

Theorem 2. *Let G be a graph embedded onto an xyz surface. Then G is bipartite if and only if the surface is orientable.*

We omit the proof. 2-manifolds may be classified by their orientability and their *Euler characteristic* $|V| - |E| + |C|$, so by Theorem 2 one may determine the topology of any *xyz* surface by counting faces and testing bipartiteness.

3 Algorithms for *xyz* Embedding

As we now show, there exist efficient algorithms to determine whether an embedded surface is an *xyz* surface, or whether a partition of the edges of a graph into three perfect matchings can be used as the three parallel classes of edges in an *xyz* graph. However, it is not so easy to find an *xyz* graph representation for an initially unlabeled graph.

Theorem 3. *Let G be a connected undirected n-vertex graph, and let C be a collection of cycles in G. Then in time $O(n)$ we may determine whether C is the set of cycles of an xyz surface embedding of G, and if so construct an xyz graph representation of G.*

Proof. We first check that G is cubic and that C covers each edge of G twice. Next, we assign arbitrary index numbers to the cycles in C. Each edge has an associated pair of index numbers, which we order lexicographically. We sort the edges of G according to this lexicographic ordering by two passes of bucket sorting and verify that each consecutive pair of edges in the sorted order has a different pair of faces.

To test 3-colorability of the cycles in C, we store a set of the available colors for each cycle (initially, all three colors for each cycle) and a list L of cycles that have only one remaining color. When we color a cycle we remove that color from the available colors of all cycles that share an edge with it, and update L whenever that removal causes an adjacent cycle to have only one remaining available color. We begin by choosing arbitrarily two cycles that share an edge, and assigning arbitrarily two different colors to those two cycles. Then, while L remains nonempty, we remove a cycle from L, and assign it the one color that is available to it.

If this process terminates with a 3-coloring of all faces in C, we have found an *xyz* surface representation for G. Conversely, suppose that G has an *xyz* surface representation: we argue that this process will necessarily find a correct 3-coloring of all faces. To show this, permute the colors of the representation if necessary so that they match the colors chosen for the two faces at the start of the algorithm. Every color choice subsequent to that is forced, so the algorithm can neither choose an incorrect color for a face nor eliminate the correct color for any face; the only way it could fail to 3-color all faces would be to terminate with L empty before coloring all faces. But if f is any face of C, let p be any path connecting a vertex of the shared edge of the first two colored faces with any vertex of f. At any stage in the algorithm until f has been colored, let v be the vertex of p that is closest along the path to the first two colored faces, and that is incident to an uncolored face f'; then the two differently-colored neighboring faces of f' at v would force f' to belong to L. Thus, L cannot be empty until f is colored, and the algorithm cannot until all faces are colored.   □

Corollary 1. *Let G be a connected undirected n-vertex graph, and let E_1, E_2, E_3 be a partition of the edges of G into three matchings. Then in time $O(n)$ we may determine whether there is an xyz graph representation of G in which E_i is the set of edges parallel to the ith coordinate axis.*

Proof. For each pair E_i and E_j, $E_i \cup E_j$ is a disjoint union of cycles; we let C be the set of cycles formed in this way for all three pairs of matchings, and apply Theorem 3.   □

Lemma 1. *Let G be a biconnected cubic graph. Then there are at most $2^{(n-2)/2}$ partitions of the edges of G into three perfect matchings, and these partitions may be listed in time $O(2^{n/2})$.*

Proof. We compute an st-numbering of G [12]; that is, an ordering of the vertices of G in which each vertex, except for the ones at the start and the end of the sequence, has a neighbor that occurs earlier in the sequence and a neighbor that occurs later in the sequence. We define a *split vertex* to be one with one previous neighbor and two later neighbors, and a *merge vertex* to be one with two previous neighbors and one later neighbor. If there are k split vertices there would be $3 + 2k + (n - k - 2)$ edges, as the first vertex in the st-numbering is the earlier endpoint of three edges, the split vertices are each the earlier endpoint of two edges, the $n - k - 2$ merge vertices are each the earlier endpoint of only one edge, and the final vertex in the st-numbering is the earlier endpoint of no edges. Observing that the graph has $3n/2$ edges total and solving for k, we find that there must be exactly $(n-2)/2$ split vertices.

To list all partitions, we then perform a backtracking algorithm in which we assign the edges to partitions in order by their earlier endpoints in the st-numbering; once we make an assignment for an edge e we recursively list all partitions for edges occurring later in this ordering before backtracking and trying an alternative assignment for e (if an alternative exists). If this backtracking process ever reaches a contradictory state in which no possible assignment is available from an edge, it backtracks without recursing.

At the initial vertex of the st-numbering, the backtracking algorithm has no choices to make: it can partition the incident edges into three disjoint subsets in only one way. At the final vertex, there is again no choice to make, because all incident edges must already have been partitioned. And at each merge vertex, there is no choice to make, because there are two incident edges which must already have been placed into two sets of the partition, and the third incident edge can only go in the third set of the partition. Thus, the only branch points of this backtracking algorithm are the split vertices, at which the two edges for which the vertex is the earlier endpoint must be assigned to the two remaining partition sets, in either of two different ways.

Since the algorithm makes a binary choice at each of $(n-2)/2$ levels of its recursion, its total time is $O(2^{n/2})$. The number of partitions listed is at most the number of leaves in a binary tree of height $(n-2)/2$, which is $2^{(n-2)/2}$.   □

Greg Kuperberg (personal communication) has pointed out that the prisms over $n/2$-gons form biconnected cubic graphs with $\Omega(2^{n/2})$ partitions into three perfect matchings, showing that this bound is tight to within a constant factor.

Theorem 4. *We can test whether a given unlabeled graph is an xyz graph, and if so find an xyz graph representation of it, in time $O(n2^{n/2})$.*

Proof. We list all partitions into matchings using Lemma 1, and test whether any of them can be used to define an *xyz* graph representation using Corollary 1.   □

An implementation of our algorithms for listing all partitions of a cubic graph into perfect matchings and for testing whether a given graph is an *xyz* graph is available online at http://www.ics.uci.edu/~eppstein/PADS/xyzGraph.py.

4   Cayley and Symmetric Graphs

A *Cayley graph* is a graph having as its vertices the members of a finite group, and its edges determined by a subset of *generators* for that group; there is an edge from *g* to *gs* whenever *g* is a group element and *s* is one of the chosen generators. For instance, the cube-connected cycles network CCC_n, of importance in parallel processing [16], is a Cayley graph for the group of operations on *n*-bit binary words generated by single-bit rotations of the word and flips of the first bit of the word [1]. The cube-connected cycles of order three cannot be an *xyz* graph, as it is not triangle-free, but we have already seen (Fig. 2, right) that the cube-connected cycles of order four is an *xyz* graph.

Theorem 5. *Let n be any even number greater than or equal to four. Then the cube-connected cycles network CCC_n is an xyz graph.*

We omit the proof. We have not determined whether the cube-connected cycles of odd order greater than three may be an *xyz* graph.

Another important cubic Cayley graph is that of the *symmetric group* of permutations on four elements, generated by transpositions of adjacent elements. This graph forms the skeleton *permutohedron*, the convex hull of the 24 permutations of $(1,2,3,4)$ in the three-dimensional subspace $x+y+z+w = 10$ of $\mathbb{R}^4$ [13]. Moving each permutation to the position of its inverse causes the edges to fall into three parallel classes, and if we then transform the drawing affinely so that these three classes are perpendicular, the result is an *xyz* graph. Figure 3 shows the permutohedron, the resulting *xyz* drawing, and another *xyz* drawing in which we have permuted the coordinate values manually to reduce the number of crossings. A different Cayley graph for the same symmetric group, generated by the permutations $(12)(3)(4)$, $(13)(2)(4)$, and $(14)(2)(3)$, is the 24-vertex symmetric graph shown in Fig. 5. Higher dimensional permutohedra have too many edges per vertex to be xyz graphs, but a different Cayley graph for the symmetric group S_n, generated by a permutation that swaps the first two elements and another permutation that rotates all but the first element, is an *xyz* graph whenever *n* is an odd number greater than three. For $n = 5$ this graph forms the skeleton of a uniform polyhedron, the *truncated dodecadodecahedron* (Fig. 4), which has 30 square faces, 12 decagonal faces, and 12 star-shaped faces with ten vertices per face, interpenetrating each other to form a complex surface. The 3-face-coloring by which the truncated dodecadodecahedron can be recognized as an xyz surface coincides with the partition of its faces into different shapes.

Next, consider the points (x,y,z) in the $k \times k \times k$ grid for which $x+y+z$ is 0 or 1 (mod *k*). They form an *xyz* graph that is *symmetric*: that is, its symmetries act transitively on incident vertex-edge pairs. For $k = 3$ this produces the Pappus graph. The graph

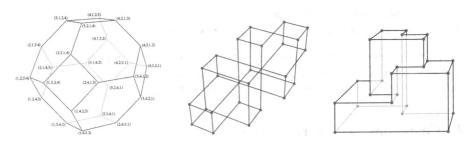

Fig. 3. The permutohedron (left) and two *xyz* drawings of the corresponding Cayley graph (center and right). The center drawing is formed by connecting pairs of permutohedron vertices that differ by swapping consecutive coordinates and affinely transforming so that the edges are perpendicular; the right drawing permutes the values of each coordinate to realize the Cayley graph as the skeleton of an orthogonal polyhedron.

F_{32} shown on the left of Fig. 5 is the *Dyck graph*, a 32-vertex symmetric cubic graph embedded by the same construction with $k = 4$. (Here F_n refers to the unique n-vertex cubic symmetric graph as listed in the Foster census [17].) Visible near the equatorial plane of the Dyck graph drawing are a number of six-vertex cycles that are not faces of the corresponding *xyz* surface (they use edges parallel to all three coordinate axes, while the surface faces are restricted to axis-parallel planes); this pattern persists for larger k, and if one analogously forms an infinite *xyz* graph from the points in a three-dimensional grid with coordinates summing to 0 or 1, the result is isomorphic as a graph to the hexagonal tiling of the plane [11].

A different construction for cubic symmetric *xyz* graphs is possible, based on the infinite tiling of the plane by regular hexagons. Three-color the hexagons of this tiling, choose a rhombus with angles of $\pi/3$ and $2\pi/3$, having its vertices at the centers of tiles that are all the same color, and form a torus by gluing opposite sides of this rhombus together. The result, as shown in Fig. 5, center, is an *xyz* surface. The graph embedded on this surface is symmetric, because we can transform any incident vertex-edge pair into any other such pair by a combination of translations and rotations by an angle of $\pi/3$. When $n = 18q^2$ for some q, one can form an n-vertex symmetric graph using both of the constructions above, either by forming a torus from a rhombus containing $n/2$ hexagons, with sides parallel to the edges of the hexagonal tiling, or by using the points congruent to 0 or 1 in a $3q \times 3q \times 3q$ grid. Both graphs formed in this way are isomorphic, but (except for $k = 1$) the *xyz* graph embeddings resulting from these constructions are inequivalent: the *xyz* surface resulting from the $k \times k \times k$ grid has fewer faces with more vertices per face. For instance the 72-vertex cubic symmetric graph F_{72} forms an *xyz* surface with 18 12-vertex faces (a $6 \times 6 \times 6$ grid) or with 36 6-vertex faces (a rhombus containing 36 hexagons).

Figure 5, right, shows another cubic symmetric graph, F_{40}, that does not fit into either of these constructions. F_{40} is the double cover of the regular dodecahedron; that is, it is the bipartite graph formed by making two copies of each dodecahedron vertex, colored black and white, and connecting the white copy of each vertex to the black copy of each of its neighbors. Its *xyz* graph representation has faces of three types: two decagons formed as the double covers of a pair of opposite dodecahedron faces, two

Fig. 4. The truncated dodecadodecahedron, from Wikimedia Commons, originally uploaded to Wikipedia by Tom Ruen in October 2005 and created using Robert Webb's Great Stella software (http://www.software3d.com/Stella.html). The vertices and edges of this shape form a Cayley graph for the symmetric group S_5, with generators $(12)(3)(4)(5)$ and $(1)(2345)$; the faces in the figure are 3-colored, giving an xyz surface representation of the graph.

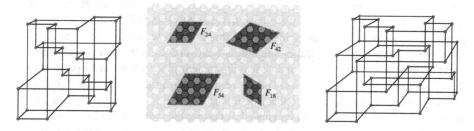

Fig. 5. Left and right: the Dyck graph F_{32} and the double cover of the dodecahedron F_{40}, two cubic symmetric graphs drawn as xyz graphs. Center: Construction of cubic xyz surfaces as toric quotients of the three-colored hexagonal tiling.

more decagons formed from the double cover of the equator between those two faces, and ten octagons formed as the boundary of a pair of adjacent dodecahedron faces that lie on opposite sides of the equator. There are six ways of choosing two opposite faces from which the decagons are formed, and once that choice is made there remain two ways of choosing the octagons to form an xyz surface, so F_{40}, viewed as a labeled graph, has 12 combinatorially distinct xyz surface representations.

We applied our implementation of an xyz graph embedding algorithm to the Foster census of symmetric cubic graphs [17] and did not find any other xyz graphs of this type on 56 or fewer vertices.

5   Planar and Nonplanar Graphs

We may exactly characterize the planar xyz graphs.

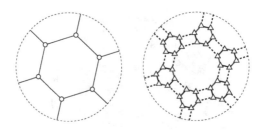

Fig. 6. An embedding of $K_{3,3}$ in the projective plane with one hexagonal face and three quadrilateral faces (left) and a GEM representation of the embedding (right). From [10].

Theorem 6. *Let G be a planar graph. Then G is an xyz graph if and only if G is bipartite, cubic, and 3-connected. If it is an xyz graph it has a unique representation as an xyz surface, up to permutation of the face colors of the surface.*

Corollary 2. *We may test in linear time whether a planar graph G is an xyz graph.*

The *xyz* graph formed from the points (x, y, z) in the $k \times k \times k$ grid for which $x + y + z$ is 0 or 1 (mod k) has $2k^2$ vertices and $3k^2$ edges but only $3k$ faces (one per axis-aligned plane) so its Euler characteristic is $3k - k^2$. If G and G' are *xyz* graphs, with designated vertices v and v', we may form the *connected sum* of G and G' by aligning the two graphs in $\mathbb{R}^3$ so that v and v' coincide (and so that no pairs of vertices, one from G and one from G', lie on an axis-parallel line unless both vertices in the pair are adjacent to v and v') and then by removing v and v', leaving in their place a non-vertex point where the lines through three edges cross. The 14-vertex planar graph in the center of Fig. 1 can be viewed in this way as a connected sum of two cubes. In terms of *xyz* surfaces, the connected sum operation can be viewed as cutting the two surfaces by a small disk surrounding each of v and v', and gluing the three faces surrounding this hole on one surface to the faces of corresponding colors surrounding the hole on the other surface, to form a handle connecting the two surfaces. By forming connected sums of tori and projective planes (the *xyz* graphs on the left and right of Fig. 1 respectively), we may form *xyz* surfaces of any topological type.

An alternative construction allows arbitrary surfaces to be represented as *xyz* surfaces: the *graph encoded map* (GEM, Fig. 6) [4, 10]. Let G be any graph embedded on a 2-manifold in such a way that each face of the embedding is a topological disk bounded by a simple cycle of G. A *flag* of this embedding is a triple of a vertex, edge, and face that are all incident to each other, and the graph encoded map M of this embedding is a 3-edge-colored cubic graph, having a vertex for each flag of the embedding of G. Two vertices of M are adjacent if the corresponding two flags differ only in a vertex, differ only in an edge, or differ only in a face; the edge coloring of M determines which type of difference each edge of M represents. M itself can be embedded on the same surface, with a $2k$-cycle for each vertex of degree k in G or each face in G that is surrounded by k edges, and a 4-cycle for each edge of G. These cycles form an *xyz* surface, in which the color of a face in the GEM is determined by whether it represents a vertex, face, or edge in G, so M is an *xyz* graph.

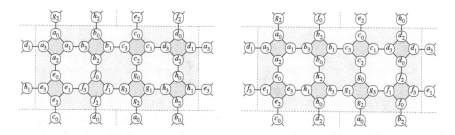

Fig. 7. A graph that can be embedded on the torus as an xyz surface in two different ways

The 32-vertex graph shown in Fig. 7 has two different torus embeddings, showing that the uniqueness of xyz surface representations for planar graphs does not directly generalize to other surfaces. The colored region of the figure shows a rectangle that can be glued to itself in a brick wall pattern to form a torus; vertices are repeated outside the colored rectangle to show the graph edges that cross the glued rectangle boundaries. This ambiguously-embeddable graph plays a key role in our NP-completeness proof in Sect. 6.

6 Complexity of xyz Graph Recognition

We show that recognizing xyz graphs is NP-complete, via a reduction from graph 3-colorability, using pieces of surfaces to represent the vertices to be 3-colored and the edges that connect them; these pieces are linked together using connected sum operations. The edge gadget is based on two copies of the graph of Fig. 7 connected to each other and the vertex gadgets by narrow tubes. We represent the choice of a color for a vertex by the choice of which coordinate axis to make parallel to certain edges of the vertex gadget. We omit the details for lack of space.

Theorem 7. *It is NP-complete, given an undirected graph G, to determine whether G can be represented as an xyz graph.*

7 Conclusions

We have studied examples, algorithms, topology, and complexity of xyz graph drawing. Our investigation opens up several avenues for further research:

– In our construction of an xyz graph from an xyz surface, we may permute the coordinate values associated with each face, giving drawings with different appearances for a single xyz surface representation (Fig. 3). How difficult is it, given an xyz surface, to find a permutation of coordinate values that minimizes the number of crossings?
– In some cases it may be possible to reduce the volume of the grid into which an xyz graph is embedded by allowing multiple faces of an xyz surface to share the same coordinate value. How difficult is it to find the minimum volume xyz graph drawing of a given xyz surface?

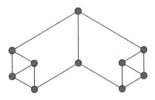

Fig. 8. A point set such that lines in three parallel families each contain zero or exactly two points, and the cubic graph derived from it. This graph is not an *xyz* graph, as it contains triangles.

- The rightmost drawing of the permutohedron in Fig. 3 shows it as the boundary of an orthogonal polyhedron (perhaps suitable for the design of a building). How difficult is it to determine whether such a representation exists for a given bipartite 3-connected cubic planar graph?
- Our reduction from graph coloring to *xyz* graph recognition produces graphs of high genus, but recognizing *xyz* graphs is easy in the case of genus zero (planar graphs). Is there an efficient algorithm for recognizing *xyz* graphs of bounded genus?
- For Cayley graphs with one self-inverse and one non-self-inverse generator, the difficulty in finding *xyz* graph representations is linked to the need to independently orient each cycle formed by the non-self-inverse generator. However, as these graphs are highly symmetric, it seems natural to hope that these cycles may be oriented in a symmetric way that avoids the need for testing all orientations of all cycles. Is there a Cayley graph that may be represented as an *xyz* graph only by orienting its cycles asymmetrically?
- Kuperberg's example of the prism shows that our algorithm for testing *xyz* graph representability using all partitions of the graph into three matchings cannot be improved, unless we avoid some partitions. However, for the prism itself, there are many partitions that can safely be avoided: for an *xyz* graph representation, we cannot use any partition into three matchings that uses three different orientations in a single quadrilateral. One can also devise similar conditions that restrict the matchings in hexagons and other short cycles of a given graph. Can one take advantage of these forbidden configurations to eliminate some partitions into matchings earlier in the algorithm and reduce its running time?
- In our discussion of graphs represented by the points with coordinates summing to 0 or 1 in a $k \times k \times k$ grid, we briefly referred to a similar construction of an infinite *xyz* graph in an infinite three-dimensional grid, isomorphic to the hexagonal tiling of the plane, a graph treated in more detail in another paper [11]. To what extent can the correspondence between *xyz* graphs and *xyz* surfaces be generalized to infinite graphs? What is the most appropriate way of handling the infinite chains of edges parallel to a single coordinate plane that can arise in the infinite case?
- If a planar point set intersects any line parallel to the sides of an equilateral triangle in either zero or two points, we may define a cubic graph from it analogously to the three-dimensional definition of *xyz* graphs; any *xyz* graph has a planar projection of this type. However, these planar three-orientation graphs are more general than *xyz* graphs; Fig. 8 shows a graph of this type that is not an *xyz* graph. To what extent may our theory be extended to these graphs?

Acknowledgements

We thank Ed Pegg, Jr., Tomo Pisanski, Frank Ruskey, Tom Tucker, Arthur White, and the anonymous reviewers for Graph Drawing 2008 for helpful comments. Work supported in part by NSF grant 0830403. Except as noted, all figures in this paper are by the author; all figures remain the copyright of their creators and are used by permission.

References

1. Annexstein, F., Baumslag, M., Rosenberg, A.L.: Group action graphs and parallel architectures. SIAM J. Comput. 19(3), 544–569 (1990)
2. Biedl, T., Shermer, T.C., Whitesides, S., Wismath, S.K.: Bounds for orthogonal 3-D graph drawing. J. Graph Alg. Appl. 3(4), 63–79 (1999)
3. Biedl, T., Thiele, T., Wood, D.R.: Three-dimensional orthogonal graph drawing with optimal volume. Algorithmica 44(3), 233–255 (2006)
4. Bonnington, C.P., Little, C.H.C.: The Foundations of Topological Graph Theory. Springer, Heidelberg (1995)
5. Calamoneri, T., Massini, A.: Optimal three-dimensional layout of interconnection networks. Theor. Comput. Sci. 255(1-2), 263–279 (2001)
6. Closson, M., Gartshore, S., Johansen, J.R., Wismath, S.K.: Fully dynamic 3-dimensional orthogonal graph drawing. J. Graph Alg. Appl. 5(2), 1–34 (2001)
7. Craft, D.L., White, A.T.: 3-maps. Discrete Math. (2008)
8. Eades, P., Stirk, C., Whitesides, S.: The techniques of Komolgorov and Bardzin for three-dimensional orthogonal graph drawings. Inf. Proc. Lett. 60(2), 97–103 (1996)
9. Eades, P., Symvonis, A., Whitesides, S.: Two algorithms for three dimensional orthogonal graph drawing. In: North, S.C. (ed.) GD 1996. LNCS, vol. 1190, pp. 139–154. Springer, Heidelberg (1997)
10. Eppstein, D.: Dynamic generators of topologically embedded graphs. In: Proc. 14th Symp. Discrete Algorithms, pp. 599–608. ACM and SIAM (January 2003)
11. Eppstein, D.: Isometric diamond subgraphs. In: Proc. 16th Int. Symp. Graph Drawing (2008)
12. Even, S., Tarjan, R.E.: Computing an st-numbering. Theor. Comput. Sci. 2(3), 339–344 (1976)
13. Gaiha, P., Gupta, S.K.: Adjacent vertices on a permutohedron. SIAM J. Appl. Math. 32(2), 323–327 (1977)
14. Kochol, M.: 3-regular non 3-edge-colorable graphs with polyhedral embeddings in orientable surfaces. In: Proc. 16th Int. Symp. Graph Drawing (2008)
15. Papakostas, A., Tollis, I.G.: Algorithms for incremental orthogonal graph drawing in three dimensions. J. Graph Alg. Appl. 3(4), 81–115 (1999)
16. Preparata, F.P., Vuillemin, J.: The cube-connected cycles: a versatile network for parallel computation. Commun. ACM 24(5), 300–309 (1981)
17. Royle, G., Conder, M., McKay, B., Dobscanyi, P.: Cubic symmetric graphs (The Foster Census). Web page (2001),
    http://people.csse.uwa.edu.au/gordon/remote/foster/
18. Wood, D.R.: An algorithm for three-dimensional orthogonal graph drawing. In: Whitesides, S.H. (ed.) GD 1998. LNCS, vol. 1547, pp. 332–346. Springer, Heidelberg (1999)
19. Wood, D.R.: Bounded degree book embeddings and three-dimensional orthogonal graph drawing. In: Mutzel, P., Jünger, M., Leipert, S. (eds.) GD 2001. LNCS, vol. 2265, pp. 312–327. Springer, Heidelberg (2002)
20. Wood, D.R.: Optimal three-dimensional orthogonal graph drawing in the general position model. Theor. Comput. Sci. 299(1-3), 151–178 (2003)

Rapid Multipole Graph Drawing on the GPU

Apeksha Godiyal[1], Jared Hoberock[1], Michael Garland[2], and John C. Hart[1]

[1] University of Illinois
{godiyal2,hoberock,jch}@illinois.edu
[2] NVIDIA Corp.,
mgarland@nvidia.com

Abstract. As graphics processors become powerful, ubiquitous and easier to program, they have also become more amenable to general purpose high-performance computing, including the computationally expensive task of drawing large graphs. This paper describes a new parallel analysis of the multipole method of graph drawing to support its efficient GPU implementation. We use a variation of the Fast Multipole Method to estimate the long distance repulsive forces in force directed layout. We support these multipole computations efficiently with a k-d tree constructed and traversed on the GPU. The algorithm achieves impressive speedup over previous CPU and GPU methods, drawing graphs with hundreds of thousands of vertices within a few seconds via CUDA on an NVIDIA GeForce 8800 GTX.

1 Introduction

Automatic graph layout algorithms convert the topology of vertex adjacency into the geometry of vertex position. These layouts usually represent vertices as points or icons in two or three dimensions connected by edges represented by lines or arcs. Automatic graph drawing has many important applications in information visualization, software engineering, database, web design, networking, VLSI circuit design, social network analysis, cartography, bioinformatics and the organization of visual interfaces for many other domains [4]. Growth in information technology and data processing has increased the size and complexity of graph datasets, posing the problem of drawing large graphs with millions of nodes that demand the consideration of new scalable parallel approaches.

Classical force directed algorithms [7, 9, 12, 22] layout graphs of hundreds of vertices, but run in $O(|V|^2 + |E|)$ time and do not scale well for larger graphs. Approximate force directed techniques [13, 14, 18, 20, 32] perform better, using a multilevel approach based on a graph hierarchy, where smaller coarser graph levels guide the initial drawing of progressively larger, finer levels of the graph hierarchy. The class of algorithms based on linear algebra [21, 23] are even faster. They perform best on grid-like regular graphs but can condense features on other graph types (e.g. with many biconnected components) [19, 21, 23].

These state-of-the-art algorithms for straight line graph drawing can still run too slow on modern graphs, e.g. six minutes for a graph of 143,437 nodes [18]. Other approaches work efficiently but with uneven layout quality across graph

I.G. Tollis and M. Patrignani (Eds.): GD 2008, LNCS 5417, pp. 90–101, 2009.
© Springer-Verlag Berlin Heidelberg 2009

type, e.g. extremely fast ACE[23] and HDE[21] methods work best only on quasi-grids. To address both limitations, this paper reworks the general-graph quality of approximated force directed layout into a form that can be efficiently processed on the GPU to layout hundreds of thousands of nodes within a few seconds. Our GPU implementation of the fast multipole multilevel method (FM^3) is more than 20× faster than the latest reported CPU version [18].

We parallelize a potential field based multilevel algorithm that uses only multipole expansions (no local expansions) to approximate long distance forces. This combines Barnes-Hut [3] and fast multipole methods (FMM) [16]. The FMM approach has proven error bounds and better asymptotic complexity, whereas Barnes-Hut is popular due to its simplicity and a low associated constant factor of implementation [15]. Their hybrid enjoys good error bounds and an $O(|V|\log|V| + E)$ time complexity with low constant factor, and yeilds high quality layouts that represent both local and global structures well, even for graphs deemed challenging [19].

The modern graphics processing unit (GPU) was initially designed for raster-based videogame graphics, but its marked improvement in performance and programmability has generated considerable interest in it as a high-performance computing platform [27, 29]. However, GPU programming remains challenging, and its performance relies on the ability to decompose a task into concurrent identical data-parallel instruction threads with limited support for stacks or recursion, and managing their access patterns to the various kinds of memory (shared, local, CPU, etc.). The contributions of this paper are the systems-level design and deployment of an efficient manycore graph drawing algorithm and to show that the acceleration of multipole-based layout justifies the challenges posed by the GPU's architecture and programming.

The main challenge of FMM processing on a single-instruction multiple-data (SIMD) processor (such as a GPU) is managing a shared spatial hierarchy. The k-d tree has been a popular choice for particle simulation [8, 2] as its size complexity is distribution independent [31], but does not map easily to the GPU's SIMD programming model. We combine the CPU and GPU to construct the tree, using the GPU for fast median selection so the CPU can construct a balanced k-d tree with $O(\log N)$ depth that keeps force calculation within $O(N \log N)$. We traverse the structure entirely on the GPU, using an efficient "stackless" k-d tree representation, where each node has a pair of pointers, one pointing to the first child and the other to the next node (in pre-order traversal order). Each processor of a data-parallel SIMD processor can efficiently traverse such a hierarchy by simply following one of two pointers [6, 10].

2   Related Work

The Fast Multipole Multilevel Method (FM^3) produces pleasing layouts in the general case and is relatively fast [18]. It combines a multilevel spatial partitioning with a multipole approximation of all pairs repulsive forces, specifically Greengard's FMM algorithm [16]. Our new GPU version uses only the multipole expansion coefficients and not the local expansion coefficients to approximating repulsive forces. We show that these multipole expansion coefficients alone are sufficient to produce high quality layout and the added complexity of working with local expansion coefficients is unnecessary. Our GPU implementation is $20\times - 60\times$ faster than the preveious CPU implementations of FM^3. Another improvement over the previous CPU FM^3 implementation [18] is that we use a k-d tree instead of quad tree for force calculations, motivated by GPU architecture as elaborated in Sec. 4.1.

Our implementation is more than 30% faster than a previous GPU multilevel force directed graph layout method [11]. That method approximated the all-pairs repulsive force with a center of gravity multipole acceptance criteria, which when compared to FM^3 has a larger aggregate error that can even become unbounded for unstructured distributions [28]. Our approach's time complexity, $O(|V|\log|V| + |E|)$, improves their's, $O(|V|^{1.5} + |E|)$.

Others have implemented general-purpose FMM on the GPU [30, 17]. Their approaches differ from ours as they include all FMM steps, most of which are unnecessary for graph drawing. Our approach utilizes the k-d tree which outperforms their quadtree, and we focus specifically on the issue of GPU tree construction.

3   Algorithm

Multilevel layout methods significantly reduce running times by converging to the optimal layout in fewer iterations [18, 23, 20, 14, 13, 32]. This approach recursively coarsens an input graph G^0 to produce a series of smaller graphs $G^1 \ldots G^k$, until the size of the coarsened graph falls below a threshold. An initial layout is first computed iteratively for the coarsest graph G^k. The converged vertex positions of a level i graph G^i are used as the initial vertex positions of the next finer level $i - 1$ graph G^{i-1}, which should relax into a converged state after a few iterations. This continues until the layout for the finest graph (the input), G^0, is obtained.

We use the multilevel method shown in Algorithm 1. The *ComputeLayout* step is the most expensive with runtime complexity of $O(|V|\log|V| + |E|)$, and is accelerated by the GPU. The remaining functions are linear $O(|V|)$ and computed on the CPU.

3.1   Coarsening

The function *CoarsenGraph* coarsens by maximal independent set (MIS) filtration, which has the advantage of being simple, efficient and produces a filtration controlled by the geometry of the graph [14, 13]. The vertex subset $S \subset V$ is an independent set of a graph $G = (V, E)$ if no two elements of S are connected by an edge. A maximal independent set filtration of G is a family of sets $V = V^0 \supset V^1 \supset \ldots \supset V^k \supset \emptyset$, such that each V^i is an independent set of V^{i-1}.

Calculating optimal independent sets is a NP-Complete problem, though an efficient 2-approximation exists. An independent set S of a set V can be computed by repeatedly deleting a vertex $v \in V$ and adding it to S and removing all vertices adjacent to v from V, until V is empty. The set S is the desired independent set.

Algorithm 1. Overall Algorithm

Input: $G = (V, E)$ with random initial placements
Output: $G = (V', E)$ with final placements
initialization;
graph $G^0 \longleftarrow G$;
$threshold \longleftarrow 50$;
$i \longleftarrow 0$;
while $|V^i| \geq threshold$ **do**
   | graph $G^{i+1} \longleftarrow CoarsenGraph(G^i)$;
   | $i \longleftarrow i + 1$
end
while $i \geq 0$ **do**
   | ComputeLayout(G^i) ; /* **via the GPU**   */
   | **if** $i \geq 1$ **then**
   | InterpolateInitialPositions(G^{i-1})
   | **end**
   | $i \longleftarrow i - 1$
end
return G^0

3.2 Interpolation

The function *InterpolateInitialPositions* derives the starting positions of vertices in G^i from the positions of vertices in the converged layout of G^{i+1}, using a relaxation method [11]. Each vertex $v \in V^i$ is initially placed at the position of its parent vertex $v' \in V^{i+1}$. Then several iterations (we used a maximum of 50) of a form of graph Laplacian move each vertex to an average of its current position, p_i, and that of its neighbors N_i,

$$p_i = \frac{1}{2} \left(p_i + \frac{1}{\deg(i)} \sum_{j \in N_i} p_j \right). \tag{1}$$

3.3 Force Calculation

For each graph G^i, the function *ComputeLayout* iteratively calculates and applies forces until it converges. The coarsest graph G^k typically requires 300 iterations, but this number decays rapidly for finer graphs and in most cases the finest graph G^0 needs zero iterations to converge. The pseudocode for one iteration is given in Algorithm 2.

3.4 Force Model

As in the force directed algorithm [12], we assume that the vertices of a graph $G(V, E)$ are charged particles that repel each other with an inverse-square law, and the edges are springs that contract with a non-physical but effective force [18]

Algorithm 2. Force Calculation Algorithm

Input: $G = (V, E)$ with initial placements
Output: $G = (V', E)$ with final placements
$kdTree \longleftarrow constructKDTree(V)$
Spawn $|V|$ threads on the GPU ; /* **Thread i calculates force on** v_i */
foreach *thread i* **do**
　　$force \longleftarrow calculateRepulsion(v_i, kdTree)$
　　$force \longleftarrow force + calculateAttraction(v_i, E)$
　　Send calculated force values to CPU in an array
end
; /* **Done on the CPU to avoid global synchronization on the GPU** */
forall v_i **do**
　　moveVertex(v_i, force)
end
return G

$$F = d^2 \log(d/d') \tag{2}$$

where d and d' are the actual and desired lengths of the edge.

3.5   Multipole Calculation

The most expensive step in force directed graph drawing is the all-pair repulsive force calculations. Although the force calculations may be quite complex in the near-field (when two vertices are very close to each other), force calculations are well-behaved in the far-field. In particular, if a vertex is sufficiently far from a set of charges, we may compute the aggregate effect of the charges on that vertex, and need not resort to computing every interaction. Greengard [16] first demonstrated how potential field based approximations can be used to find the far-field forces using quad trees. The idea is to construct a tree based spatial partition of particles and then evaluate multipole expansions using this tree.

Theorem 1. *(Multipole Expansion)Suppose that m charges of strengths q_i are located at points z_i, for $i = 1 \ldots m$, with center z_0 and $|z_i - z_0| < r$. Then for any $z \in C$ with $|z - z_0| > r$, the potential $\Phi(z)$ induced by the charges is given by*

$$\Phi(z) = Q \log(z - z_0) + \sum_{k=1}^{\infty} \frac{a_k}{(z - z_0)^k} \tag{3}$$

where $Q = \sum q_i$ and $a_k = \sum -q_i(z_i - z_0)^k/k$. As force is the negative of the gradient of the potential, the force that acts on a particle of unit charge at position z is given by $(\text{Re}(\Phi'(z)), -\text{Im}(\Phi'(z)))$.

Instead of summing up an infinite series for (3), only a constant number p of terms are calculated. The resulting truncated Laurent series is called *p-term multipole expansion*. We choose $p = 4$ as it is sufficient to keep the error of the approximation less than 10^{-2} [18].

As the k-d tree is constructed, the coefficients of this multipole expansion are calculated and stored for each node using (3). The center of a k-d tree node is the geometric center of the rectangular region it represents, and the radius used is the radius of a circle circumscribing this rectangular region. Each node in the k-d tree thus maintains a collection of charges (vertices of the graph) lying in its rectangular regions. Let $G(V, E)$ be a graph and K be the k-d tree of the vertices of G. Let n be a node of K with center z_0 and radius r. Let $\{v_i, v_2, \ldots, v_k\}$ be the set of vertices of graph G that are contained in k-d tree node n. To calculate the approximate repulsive force on each vertex $v \in V$ located at z, K is traversed from the root node. At a node n, if the distance between z_0 and z is greater than r, then the approximate repulsive force between v and vertices $v_i\{i = 1, \ldots, k\}$ are calculated using (3). Otherwise, if n is an internal node, the process is repeated for its children, and if n is a leaf node, the exact repulsive forces are calculated.

4   GPU Implementation

4.1   Processing the K-D Tree

Unlike the more traditional quadtree used in n-body simulation, we used a k-d tree [5]. Aluru *et al.*[1] has shown that the running time of adaptive FMM using quad tree [16] depends on the particle distribution and cannot be bounded in number of particles. In order to remedy this and guarantee $O(|V|log|V|)$ running time complexity, [18] uses complicated tree thinning and balancing techniques. These techniques do not translate into efficient GPU implementation because of the lack of recursion (no unbounded stack) and dynamic memory allocation. Since the k-d tree is a density decomposition tree and not a spatial decomposition tree, it does not suffer from distribution dependent running time [31].

The CUDA GPU programming model has a complex memory hierarchy and one has to keep in mind multiple factors to achieve good performance [26]. The k-d tree is traversed by all of the GPU threads and all the threads need the vertex position data for near field and attractive force calculations. Thus these data structures are passed to the GPU in texture memory, which is cached yielding higher bandwidth from k-d tree node locality. In our implementation, the k-d tree is constructed for the first four iterations and then for every twentieth iteration, because it changes only slightly in each later iteration and these changes do not significantly impact force calculations.

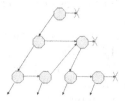

Fig. 1. A "stackless" k-d tree pre-threaded with first child (blue) and next neighbor (red) pointers

Traversal. Stackless traversal of the k-d tree on the GPU is achieved by a structure shown in Fig. 1 Each node of the tree has two pointers. The blue (success) pointer indicates its first child whereas the red (failure) pointer points to its next neighboring node. This tree threading allows the streaming SIMD GPU processing to parse a hierarchical data structure efficiently [6, 10]. The data parallel SIMD architecture of the GPU requires that when control flow reaches a condition, if some processors follow one side of the condition and the rest of the processors follow the other side of the condition, then all of the processors need to evaluate both sides of the condition, zeroing out the result of the side not used by each processor. Tree threading allows the processors instead to simply follow one of two pointers, replacing conditional control flow with data indirection which is fully supported by the GPU.

Construction. A k-d tree is constructed recursively. Each node of a k-d tree divides the set of vertices it represents V, into two equal sets by splitting along a chosen dimension. (In our implementation, the splitting dimension alternates between the two axes.) This bisection is achieved by a radix selection algorithm [24] whose worst case time complexity is $O(|V|)$. The process of finding the median and splitting the set of vertices is applied recursively until a node has less than threshold number of vertices (four, in our implementation). The multipole expansion coefficients from (3) of each node are calculated as the k-d tree is constructed. This median splitting approach generates a balanced k-d tree in $O(|V| \log |V|)$ time.

The radix selection algorithm is faster on the GPU for arrays of large size. In our configuration, the crossover array size, for which the GPU radix selection is faster than a well tuned CPU implementation, is 50,000, and we use the CPU for smaller arrays. We implemented radix selection using efficient GPU scan primitives [29] (which have also been used for GPU radix sort [25]).

4.2 Radix Selection with Prefix Scan

Radix select is the selection analog of the radix sort algorithm. It is recursive and selects the key (vertex coordinate in our case) whose rank is m, from an array $A[1 \ldots n]$ of n keys. The array is split at position s, into two sub-arrays based on the most significant bit: $A[1 \ldots s]$ contains all keys with 0 as the most significant bit, and $A[s + 1 \ldots n]$ contains all keys with 1 as the most significant bit. Then the next significant bit is considered. This goes on recursively until the key with rank m is found.

To carry out the split at each level of recursion in parallel, each thread needs to copy a different input key $A[i]$ to the split array. The address of each key $A[i]$, is the number of keys in $A[1 \ldots i - 1]$ whose most significant bit is 0. The array of these counts is called the prefix sum of A, denoted here as $B[1 \ldots n]$ such that $B[i] = \sum_{j<i} A[j]$. We compute this prefix sum on the GPU using an efficient O(n) CUDA prefix scan implementation [29]. This work-efficient scan of n elements requires two passes over the array: *reduce* and *down-sweep*. Each

requires $\log(n)$ parallel steps. The amount of work is cut in half at each step, resulting in an overall work complexity of $O(n)$.

4.3 Compressed Sparse Row Representation

We use a compressed sparse row (CSR) format, essentially a sparse matrix data structure [29], for representing the edges of the graph in GPU texture memory. It avoids conditional statements and thus makes the implementation fast. Let i be a vertex of graph G such that i has k edges $(i, j_1), (i, j_2)...(i, j_k)$. Then the graphs adjacency list is represented by 2 arrays:

1. Edge-value: For each vertex i, this array stores vertices $\{j_1, j_2...j_k\}$ i.e. the adjacency list of i.
2. Edge-index: Edge-Index[i-1] and Edge-Index[i] store the beginning and ending of the adjacency list of vertex i.

For each vertex i, a GPU processing thread uses this CSR representation to calculate the attractive forces due to its incident edges. This parallel computation is not perfectly load-balanced as the work done by each thread depends on the degree of the vertex it is handling. Processing the edges instead of the vertices would rectify this, but would require either atomic operations for adding up all the forces on a single vertex, or a prefix sum to add up the forces calculated by different threads, and neither option is very efficient.

The *edge-value* array is accessed frequently by each thread, and so is placed in the cached texture memory of the GPU. The *edge-index* array is accessed only twice per thread with negligible gain from caching, and so is placed in plain read-write GPU memory.

5 Results

The algorithm was tested on a single core 2.21 GHz AMD Athlon(tm) 64 Processor running Windows XP, with an NVIDIA GeForce 8800 GTX card programmed via the CUDA (Compute Unified Device Architecture) programming model, compiled by a C compiler with language extensions [26]. Both CPU and GPU implementations used single precision floating point.

The algorithm was tested on a variety of graphs extensively used in graph drawing research to support comparisons [18, 19, 33]. Figure 2 shows selected layouts and their associated run times. The layouts of all the tested artificial and real-world graphs resemble those produced by FM^3 [18]. Like FM^3, our algorithm is able to display the regularity of six-ary trees, the symmetry of spider and flower graphs and the global structure of snowflake graphs.

Figure 3 shows for various graphs the speedup our implementation achieves over FM^3 and over the GFDL force directed layout GPU implementation [11]. It shows our implementation to be $1.3\times - 4\times$ faster than GFDL and $20\times - 60\times$ faster than CPU implementation of FM^3. Figure 4 demonstrates the scalability of our GPU implementation. Its running time is largely a factor of graph size, though dependent

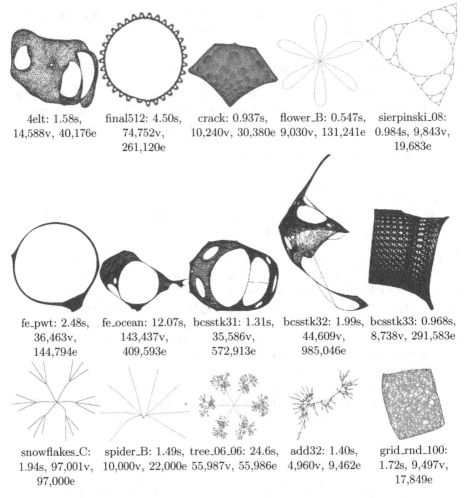

Fig. 2. Layouts of various graphs computed with out approach, indicated by name, running time (in seconds), followed by the numbers of vertices and edges

on the number of iterations needed to resolve vertex placement at each level of the graph hierarchy. Thus the large 6-ary tree required significantly more iterations (by a factor of five) to reach a planar embedding than did the others.

We recorded the running time of the major parts of the algorithm for both the CPU and the GPU implementations. Table 1 shows the result for a few graphs. The CPU implementation spends on an average nearly 85.5% of CPU cycles in calculating the forces and this step is clearly the performance bottleneck. The GPU implementation reduces the time spent in calculating forces by 7-40 times (depending upon the size of the graph). One disadvantage of the GPU implementation is that lots of cycles are wasted in copying data back and forth between the GPU and the CPU. GPU implementation spends 18%-25% of the

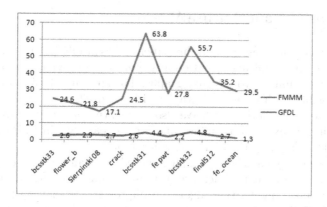

Fig. 3. Speedup factors over GPU force directed layout (GFDL) and Fast Multilevel Multipole Method (FMMM). The graphs are in increasing order of graph size.

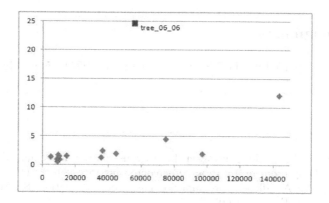

Fig. 4. Running time vs. graph size for GPU accelerated FM³ layout

Table 1. Running time (in seconds) comparing total and component run times on CPU (numerator) v. GPU (denominator)

Graph	$|V|$	$|E|$	Total	Coarsening	Data Trans.	Tree Const.	Force Calc.
bcsstk33	8,738	291,583	1.63 / 0.968	0.0 / 0.0	0.032 / 0.141	0.095 / 0.096	1.48 / 0.242
4elt	14,588	40,176	7.23 / 1.58	0.0 / 0.0	0.172 / 0.375	0.516 / 0.375	5.92 / 0.672
crack	10,240	30,380	3.51 / 0.937	0.0 / 0.0	0.080 / 0.172	0.456 / 0.203	2.81 / 0.449
final512	74,752	261,120	81.55 / 4.50	0.25 / 0.25	0.260 / 0.828	3.39 / 1.49	73.8 / 1.932
fe_ocean	143,437	409,593	90.9 / 12.07	4.1 / 4.1	1.30 / 1.50	5.20 / 3.89	83.0 / 2.48

running time in data movement as compared to 2%-3% time spent by the CPU implementation on the same. Time for constructing the k-d tree is nearly same in the CPU and GPU implementations, for graphs with less than 50,000 vertices. For larger graphs, k-d tree construction is more than 30% faster on the GPU.

6   Conclusions and Future Work

The parallel algorithm described in this paper makes graph drawing significantly faster without compromising layout quality, improving previous fast implementations that were limited to grid-like graphs. The speedup obtained shows that it is now possible to draw general graphs with hundreds of thousands of nodes within a few seconds via the GPU. We also showed that for the purpose of graph drawing multipole expansions suffice, and local expansions in FMM should be best avoided due to their the high constant factor.

The optimized layout of each graph required the hand tuning of a number of parameters, as automatic inference of these optimal parameters remains an open research problem. Further algorithm improvements may be possible. Increasing CPU-GPU bandwidth may lower the 50,000-node limit where the GPU outpaced the CPU on median finding, and further load balancing may improve force calculation.

Acknowledgments

This work is supported by the NSF under the grant #0534485, and by NVIDIA Corp.

References

[1] Aluru, S., Prabhu, G.M., Gustafson, J.: Truly distribution-independent algorithms for the n-body problem. In: Proc. Supercomputing, pp. 420–428 (1994)

[2] Appel, A.W.: An efficient program for many-body simulation. SIAM J. Sci. & Stat. Comp. 6(1), 85–103 (1985)

[3] Barnes, J., Hut, P.: A hierarchical o(n log n) force-calculation algorithm. Nature 324(6096), 446–449 (1986)

[4] Batini, C.: Applications of graph drawing to software engineering (abstract). SIGACT News 24(1), 57 (1993)

[5] Bentley, J.L.: Multidimensional binary search trees used for associative searching. CACM 18(9), 509–517 (1975)

[6] Carr, N.A., Hoberock, J., Crane, K., Hart, J.C.: Fast gpu ray tracing of dynamic meshes using geometry images. In: Proc. Graphics Interface, pp. 203–209 (2006)

[7] Davidson, R., Harel, D.: Drawing graphs nicely using simulated annealing. ACM Trans. Graph. 15(4), 301–331 (1996)

[8] Dikaiakos, M.D., Stadel, J.: A performance study of cosmological simulations on message-passing and shared-memory multiprocessors. In: Intl. Conf. on Supercomputing, pp. 94–101 (1996)

[9] Eades, P.A.: A heuristic for graph drawing. Congressus Numerantium 42, 149–160 (1984)

[10] Foley, T., Sugerman, J.: Kd-tree acceleration structures for a GPU raytracer. In: Proc. Graphics Hardware, pp. 15–22 (2005)

[11] Frishman, Y., Tal, M.-A.: Multi-level graph layout on the gpu. IEEE Trans. Vis. Comp. Graph. 13(6), 1310–1319 (2007)

[12] Fruchterman, T.M.J., Reingold, E.M.: Graph drawing by force-directed placement. Software - Practice and Experience 21(11), 1129–1164 (1991)
[13] Gajer, P., Goodrich, M.T., Kobourov, S.G.: A multi-dimensional approach to force-directed layouts of large graphs. Comput. Geom. Theory Appl. 29(1), 3–18 (2004)
[14] Gajer, P., Kobourov, S.G.: GRIP: Graph dRawing with intelligent placement. In: Marks, J. (ed.) GD 2000. LNCS, vol. 1984, pp. 222–228. Springer, Heidelberg (2001)
[15] Grama, A.Y., Kumar, V., Sameh, A.: Scalable parallel formulations of the Barnes-Hut method for n-body simulations. In: Proc. Supercomputing, pp. 439–448 (1994)
[16] Greengard, L.F.: The rapid evaluation of potential fields in particle systems. Ph.D. thesis, Yale, New Haven, CT, USA (1987)
[17] Gumerov, N.A., Duraiswami, R.: Fast multipole methods on graphics processors. J. Comp. Physics 227, 8290–8313 (2008)
[18] Hachul, S., Jünger, M.: Large-graph layout with the fast multipole multilevel method. Tech. rep., Zentrum für Angewandte Informatik Köln (December 2005)
[19] Hachul, S., Junger, M.: An experimental comparison of fast algorithms for drawing general large graphs. In: Healy, P., Nikolov, N.S. (eds.) GD 2005. LNCS, vol. 3843, pp. 235–250. Springer, Heidelberg (2006)
[20] Harel, D., Koren, Y.: A fast multi-scale method for drawing large graphs. In: Marks, J. (ed.) GD 2000. LNCS, vol. 1984, pp. 183–196. Springer, Heidelberg (2001)
[21] Harel, D., Koren, Y.: Graph drawing by high dimensional embedding. In: Goodrich, M.T., Kobourov, S.G. (eds.) GD 2002. LNCS, vol. 2528. Springer, Heidelberg (2002)
[22] Kamada, T., Kawai, S.: An algorithm for drawing general undirected graphs. Inf. Process. Lett. 31(1), 7–15 (1989)
[23] Koren, Y., Carmel, L., Harel, D.: ACE: a fast multiscale eigenvectors computation for drawing huge graphs (2001)
[24] Mahmoud, H.M.: Sorting: A Distribution Theory, chap. High Qulaity Ambient Occlusion. Wiley Interscience, Hoboken (2000)
[25] NVIDIA: CUDA data parallel primitives library, http://www.gpgpu.org/developer/cudpp/
[26] NVIDIA: CUDA programming guide (2007), http://developer.nvidia.com/object/cuda.html
[27] Pharr, M., Fernando, R.: GPU Gems 2: Programming Techniques for High-Performance Graphics and General-Purpose Computation. Addison-Wesley, Reading (2005)
[28] Sarin, V.: Analyzing the error bounds of multipole-based treecodes. In: Proc. Supercomputing, p. 19 (1998)
[29] Sengupta, S., Harris, M., Zhang, Y., Owens, J.D.: Scan primitives for gpu computing. In: Proc. Graphics Hardware, August 2007, pp. 97–106 (2007)
[30] Stock, M.J., Gharakhani, A.: Toward efficient gpu-accelerated n-body simulations. In: 46th AIAA Aerospace Sciences Meeting & Exhibit (2008)
[31] Uhlmann, J.K.: Enhancing multidimensional tree structures by using a bi-linear decomposition. Natl. Tech. Info. Svc. ADA229756 (1990)
[32] Walshaw, C.: A multilevel algorithm for force-directed graph drawing. In: Marks, J. (ed.) GD 2000. LNCS, vol. 1984, pp. 171–182. Springer, Heidelberg (2001)
[33] Walshaw, C.: Graph collection (2007), staffweb.cms.gre.ac.uk/~wc06/partition/

Clustered Planarity: Clusters with Few Outgoing Edges

Vít Jelínek*, Ondřej Suchý*, Marek Tesař, and Tomáš Vyskočil*

Department of Applied Mathematics
Charles University
Malostranské nám. 25, 118 00 Praha, Czech Republic
{jelinek,suchy,tesulo,tiger}@kam.mff.cuni.cz

Abstract. We present a linear algorithm for c-planarity testing of clustered graphs, in which every cluster has at most four outgoing edges.

1 Introduction

Clustered planarity is one of the challenges of contemporary Graph Drawing. It arises naturally when we want to draw the graph with further constraints on embedding of the vertices. This includes for example visualizing a computer network with the computers of the same department, faculty and institution being grouped together. Another application is in designing an integrated circuit with the connectors of each components being close to each other and the logical parts of the circuit being grouped together. There are many other applications including visualizations of process interaction, social networks etc.

The concept of the clustered graph—a graph equipped with a system of subsets of vertices (called clusters), that can be recursive— was first introduced by Feng et al. in [7]. In the same paper they also proved that clustered planarity (shortly c-planarity) can be tested in polynomial time for c-connected clustered graphs (where each cluster induces a connected subgraph of the underlying graph). This was later improved by Dahlhaus [4] to a linear time algorithm. The paper [7] also contains a useful characterization of the c-planar graphs: Graph is c-planar if and only if there is a set of edges (usually called a saturator) that can be added to this graph to obtain a c-connected c-planar clustered graph.

Since then many algorithms for testing the c-planarity were based on searching for a saturator. These include an $O(n^2)$-time algorithm for "almost" c-connected clustered graphs by Gutwenger et al. in [9,10]. An efficient algorithm for clusters with cyclic structure on a cycle was developed in [3]. The case of disjoint clusters on an embedded graph with small faces was recently addressed in [5]. Very similar result was at the same time independently published by Jelínková et al. [12]. The paper [12] also contains an $O(n^3)$-time algorithm for clusters of size at most three on a rib-Eulerian graph. This is an Eulerian graph that is obtained from a constant size 3-connected graph by multiplying and then subdividing edges.

---

* Supported by grant 201/05/H014 of the Czech Science Foundation.

I.G. Tollis and M. Patrignani (Eds.): GD 2008, LNCS 5417, pp. 102–113, 2009.

Another approach is to mimic the original proof of Feng et al. [7] where the behavior of the connected clusters is described by special trees. In this way a slight generalization to extrovert clustered graphs was given by Goodrich et al. [8]. In an extrovert clustered graph the parent cluster of any disconnected cluster is connected and every component of any disconnected cluster is incident to an edge which leads outside of its parent cluster.

We should also mention that every c-planar graph can be drawn by straight lines with clusters represented by convex polygons [6]. Another interesting contribution is the characterization of completely connected clustered graphs (where each subgraph induced by a cluster and its complement are connected) [1]: A completely connected clustered graph is c-planar if and only if the underlying graph is planar. More results on c-planarity can be found in [2]. Despite the number of results the complexity of testing the c-planarity for general instances remains open.

In this paper we focus on the situation where the number of outgoing edges of each cluster is small. We notice that in this case the behavior of the clusters can be simulated by special graphs, no matter whether the subgraph induced by the cluster is connected or not. We use these ideas to develop a linear time algorithm to test such graphs for c-planarity. As far as we know this is the first algorithm that can be used in the cases where the underlying graph is not connected at all or has very few edges in total. In particular we prove the following theorem:

Theorem 1. *Clustered planarity can be decided in linear time for instances, where each cluster has at most 4 outgoing edges.*

Section 2 is devoted to the basic definitions. We also show there that if there is a cluster with no outgoing edges, then the instance could be split into an instance formed by the subclusters of the cluster and one formed by the rest. In Section 3 we show how to replace the clusters by special graphs with the same behavior and prove that this does not affect the c-planarity. The algorithm is described in Section 4, together with the proofs of the correctness and the running time. In Section 5 we show that the approach cannot be generalized this way to the case of five or more outgoing edges.

2   Preliminaries

Let S_r denote the set of all permutations of the set $\{1, 2, \ldots, r\}$. A permutation $\pi \in S_r$ is represented by r-tuple $(\pi(1) \ldots \pi(r))$.

Regarding the graph notations, we follow the standard notation on finite loopless graphs. A graph is an ordered pair $G = (V, E)$, where V is the set of vertices and E is the set of edges i.e. pairs of vertices. We simply write uv instead of $\{u, v\}$ for edges. If $U \subseteq V$, then $G[U]$ is the induced subgraph of G on vertices U and $G \setminus U = G[V \setminus U]$. Let n denote the number of vertices $|V|$ of the graph G.

A *cluster set* on the graph $G = (V, E)$ is a set $\mathcal{C} \subseteq \mathcal{P}(V(G))$ such that for all $C, D \in \mathcal{C}$, either C and D are disjoint or they are in inclusion; the pair $(G, \mathcal{C})$ is called a *clustered graph*. The elements of $\mathcal{C}$ are called *clusters*. A *clustered planar embedding* of $(G,\mathcal{C})$ is a planar embedding *emb* of G together with a mapping emb_c that assigns to every cluster $C \in \mathcal{C}$ a planar region $emb_c(C)$ whose boundary is a closed Jordan curve and such that

- for each vertex $v \in V$ and every cluster $C \in \mathcal{C}$, it holds that $emb(v) \in emb_c(C)$ if and only if $v \in C$,
- for every two clusters C and D, the regions $emb_c(C)$ and $emb_c(D)$ are disjoint (in inclusion) if and only if C and D are disjoint (in inclusion, respectively), and
- for every edge $e \in E$ and every cluster $C \in \mathcal{C}$ the curve $emb(e)$ crosses the boundary of $emb_c(C)$ at most once.

A clustered graph is called *clustered planar* (shortly *c-planar*) if it allows a clustered planar embedding.

The following observation is a trivial consequence of the definition:

Remark 1. A pair $(G, \emptyset)$ is c-planar if and only if the graph G is planar.

We say that $C \in \mathcal{C}$ is a *cluster of the bottom-most level* if there is no $C' \in \mathcal{C}$ such that $C' \subset C$.

An edge $e = uv$ is an *outgoing edge* of a cluster C if $u \in C, v \in V \setminus C$ or vice versa.[1] Let $r(C) = |\{e = uv | e \in E, u \in C, v \in V \setminus C\}|$ denote the number of outgoing edges of a cluster C. If the cluster is clear from context we will just use notation r instead of $r(C)$.

Lemma 1. *If C has no outgoing edges then $(G,\mathcal{C})$ is c-planar if and only if $(G \setminus C, \mathcal{C}_1)$ and $(G[C], \mathcal{C}_2)$ are c-planar, where $\mathcal{C}_1 = \{A \setminus C | A \in \mathcal{C}, A \neq C, A \supset C\} \cup \{A | A \in \mathcal{C}, A \cap C = \emptyset\}$ and $\mathcal{C}_2 = \{B | B \in \mathcal{C}, B \neq C, B \subset C\}$.*

Proof. The direction from left to right is easy, we just omit from the embedding the parts that are no longer necessary.

So suppose that we have a c-planar embedding emb_1 of $(G \setminus C, \mathcal{C}_1)$ and a c-planar embedding emb_2 of $(G[C], \mathcal{C}_2)$. Take an arbitrary point x in the plane, such that for all clusters $A \in \mathcal{C}_1$ the following holds: x lies inside the region $(emb_1)_c(A)$ if and only if $C \subseteq A$. Suppose that there is neither vertex nor edge of $G \setminus C$ nor border of a cluster of $\mathcal{C}_1$ in distance less than ϵ from x in emb_1. Now shrink the embedding emb_2 so that it fits into the $\frac{\epsilon}{2}$-disc centered in x. Then take this disc as the embedding of C.

It is easy to check that we obtain a c-planar embedding of $(G,\mathcal{C})$, since the embeddings emb_1 and emb_2 cross neither each other nor the embedding of C, the inclusions of the clusters are preserved and the embedding of the cluster C contains exactly the embedding of the vertices, edges and clusters it should contain. $\square$

---

[1] Such an edge is called *edge incident with C* in [1,3,7,9] and *extrovert* edge in [8].

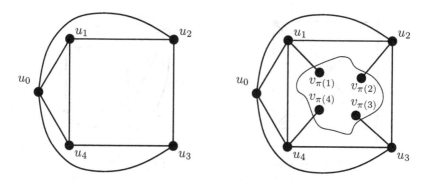

Fig. 1. The test graph T and the graph T_C^π from Definition 1

3   Replacement of Clusters by Graphs

Through this section we suppose, that we have some fixed cluster $C \in \mathcal{C}$ of bottom-most level, that has at most 4 outgoing edges. Having Lemma 1 in hand we assume that $1 \le r = r(C) \le 4$.

We denote the outgoing edges by $\{e_1, \ldots, e_r\}$. We also suppose that $e_i = v_i w_i$ for all i, where $v_i \in C$ and $w_i \in V \setminus C$ (maybe $w_i = w_j$ or $v_i = v_j$ for some $i \ne j$).

We denote by T the following *test graph* $T = (\{u_0, u_1, u_2, u_3, u_4\}, \{u_0 u_1, u_0 u_2, u_0 u_3, u_0 u_4, u_1 u_2, u_2 u_3, u_3 u_4, u_4 u_1\})$ (see Fig. 1).

Definition 1. *We say that the cluster C admits a permutation $\pi \in S_r$ if and only if the graph T_C^π created from $T \cup G[C]$ by adding edges $u_i v_{\pi(i)}, 1 \le i \le r$ is planar.*

Lemma 2. *If the cluster C admits the permutation $\pi \in S_r$ then there exists a planar embedding of the graph T_C^π such that the vertices of C are embedded inside and the vertex u_0 outside the cycle $u_1, \ldots, u_4, u_1$ of T. Moreover we can prescribe this cycle to be oriented clockwise in the embedding.*

Proof. First we take some planar embedding of the graph T_C^π. Now we take the edges incident with u_0 in the clockwise order $u_0 u_{i_1}, u_0 u_{i_2}, u_0 u_{i_3}, u_0 u_{i_4}$. For every $u_0 u_i$ and $u_0 u_j$ two consecutive of them (either $\{i, j\} = \{i_k, i_{k+1}\}$ for some $k = 1, 2$ or 3 or $\{i, j\} = \{i_1, i_4\}$) we can draw a new curve from u_i to u_j along the curve $u_i u_0$ and then $u_0 u_j$ so that it does not cross any other edge and area surrounded by the curves $u_i u_0$, $u_0 u_j$ and the new curve contains no vertex (see Fig. 2).

Suppose for a contradiction that some of the newly drawn curves connects two non-adjacent vertices, for example u_1 and u_3 (the case of u_2 and u_4 being similar). Since the new curves connect u_1 to at most one of the vertices u_2 and u_4 and we drew two curves from each u_i, we also connected u_2 and u_4. But this means that the newly drawn curves together with the original edges form

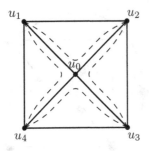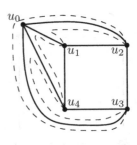

Fig. 2. Situation from the proof of Lemma 2

a planar embedding of K_5, which is a contradiction. So we know that all the curves that we drew newly connect two already adjacent vertices of the cycle.

Now we take these newly drawn curves as the embedding of the edges of the cycle. Then there is just u_0 inside the cycle and it remains to change the outer face to one of the newly obtained empty triangles, such that the vertex u_0 will be on the boundary of the outer face.

If the cycle is embedded in wrong direction, then we take the axis symmetry of the embedding.   □

Lemma 3. *If (G, C) is c-planar, then C admits some permutation.*

Proof. We suppose that (G, C) is c-planar and we fix a planar embedding *emb*. Let f be the boundary of $emb_C(C)$ (so f is a closed Jordan curve). Now we can start in an arbitrary point of this curve and move along this curve in the clockwise direction and we cross the edges $e_1, e_2, \ldots, e_r$ in some order $e_{i_1}, e_{i_2}, \ldots, e_{i_r}$. Denote the crossing points as $P_1, P_2, \ldots, P_r$ (in the same order). If $r < 4$ then we can choose new points $P_{r+1}, \ldots, P_4$ in such a way, that we meet the points $P_1, \ldots, P_4$ in this order when we move along the curve f in the clockwise direction and all these points are distinct.

Now we consider the planar embedding *emb'* of $G[C]$ which corresponds to the embedding *emb* of the graph G, place new vertices $u_1, \ldots, u_4$ to the points $P_1, \ldots, P_4$ and a vertex u_0 outside of the region bounded by the curve f. Clearly we can add edges $(u_1, v_{i_1}), \ldots, (u_r, v_{i_r})$ and embed these edges on curves which corresponded to edges $e_1, \ldots, e_r$ inside of the region $emb_C(C)$ and we can also add edges $(u_1, u_2), (u_2, u_3), (u_3, u_4)$ and (u_4, u_1) and embed them on the curve f in such a way that these edges may intersect only in vertices u_1, u_2, u_3 or u_4. It is clear that we can add edges $(u_0, u_1), \ldots, (u_0, u_4)$ and embed them in such a way that these edges will be outside of the region bounded by f and every two edges will cross only in the vertex u_0.

This way we obtain a planar embedding of the graph T_C^π where $\pi = (a_{i_1} \ldots a_{i_r})$. Thus C admits the permutation π.   □

Lemma 4. *If the cluster C admits a permutation $\pi = (a_1 a_2 \ldots a_r)$ then it also admits permutations $(a_r a_1 \ldots a_{r-1})$ and $(a_r a_{r-1} \ldots a_1)$.*

Proof. We obtain the planar embedding of $T_C^\delta, \delta = (a_r a_1 \ldots a_{r-1})$ from the planar embedding of T_C^π simply by relabeling the vertices such that u_1 becomes u_2, u_2 becomes u_3, u_3 becomes u_4 and u_4 becomes u_1 and if $r < 4$ then it is necessary to replace the edge $v_{a_r} u_{r+1}$ by a new edge $v_{a_r} u_1$ which goes along the edges $v_{a_r} u_{r+1}, u_{r+1} u_{r+2}, \ldots, u_4 u_1$ such that it doesn't cross any other edge.

For $r \geq 3$ the second part can be done similarly – it is enough to relabel such that u_1 becomes u_3 and u_3 becomes u_1 and if $r = 4$ then we use the first part to achieve permutation $(a_r a_{r-1} \ldots a_1)$. For $r < 3$ the first part also proves the second part.   □

We can now define a relation $\sim'$ on the permutations from the set S_r by $(a_1 a_2 \ldots a_r) \sim' (a_r a_1 \ldots a_{r-1})$ and $(a_1 a_2 \ldots a_r) \sim' (a_r a_{r-1} \ldots a_1)$. If we take $\sim$ to be the transitive closure of $\sim'$, then it is easy to show that $\sim$ is also reflexive and symetric. Thus $\sim$ is an equivalence. We will sometimes call the equivalence classes of this equivalence *circular permutations* The sets S_1, S_2, S_3 have just one equivalence class under $\sim$ while the set S_4 is partitioned into following three equivalence classes (they can be distinguished by the number that is "opposite" to the number 1):

$$S_4^2 = \{(1324), (3241), (2413), (4132), (4231), (1423), (3142), (2314)\},$$

$$S_4^3 = \{(1234), (2341), (3412), (4123), (4321), (1432), (2143), (3214)\},$$

$$S_4^4 = \{(1243), (2431), (4312), (3124), (3421), (1342), (2134), (4213)\}.$$

Definition 2. *We define the* corresponding graph for cluster C *as follows (see Fig 3).*

1. *If $r \leq 3$ and C admits some permutation, then the corresponding graph for C is R_r.*
2. *If there is a labeling of the outgoing edges such that C admits permutations from S_4^2, S_4^3, S_4^4 then the corresponding graph for C with this labeling is R_4^{234}.*
3. *If there is a labeling of the outgoing edges such that C admits a permutation from S_4^2 and from S_4^3, but no permutation from S_4^4 then the corresponding graph for C with this labeling is R_4^{23}.*
4. *If there is a labeling of the outgoing edges such that C admits a permutation from S_4^2, but no permutation from $S_4^3 \cup S_4^4$ then the corresponding graph for C with this labeling is R_4^2.*

Clearly, if $r \leq 3$ then the cluster C has unique corresponding graph. Since the sets S_4^2, S_4^3, and S_4^4 form a decomposition of S_4, from Lemma 4 we know that the cluster C admits all permutations from some non-empty combination of sets S_4^2, S_4^3, and S_4^4.

If the cluster C admits just permutations from the set S_4^i then by relabeling of edge e_2 by e_i and incident vertices v_2 by v_i and w_2 by w_i (if $i = 2$ we don't need to do it) we get labeling of the cluster C which admits only permutations from the set S_4^2. So the cluster C has unique corresponding graph.

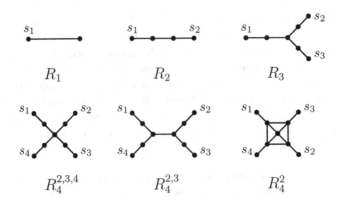

Fig. 3. The graphs $R_1, R_2, R_3, R_4^{234}, R_4^{23}$ and R_4^2

If the cluster C admits just permutations from two distinct sets S_4^i and S_4^j then we make similar relabeling of outgoing edges and incident vertices such that resulting relabeling makes the cluster C admit just permutations from the sets S_4^2 and S_4^3 and the cluster C has unique corresponding graph.

As a consequence we get the following corollary.

Corollary 1. *If C admits a permutation then there is a labeling of outgoing edges of C such that C has a corresponding graph with this labeling.*

For the rest of the paper we will use this new labeling.

Definition 3. *Let C be a cluster of the bottom-most level with outgoing edges $e_1, \ldots, e_r$ where $1 \leq r \leq 4$, $e_i = v_i w_i$ for all i, where $v_i \in C$ and $w_i \in V \backslash C$. Let R be a corresponding graph to the cluster C in this labeling. Then a replacement of cluster C by a corresponding graph R in (G, C) is a clustered graph (G', C') such that G' is created from $(G \setminus C) \cup R$ by unification of $w_1, \ldots, w_r$ with $s_1, \ldots, s_r$ (respectively) and C' is created from $C \setminus \{C\}$ by replacing every $C' \supseteq C$ by $(C' \setminus C) \cup (V(R) \setminus \{s_1, \ldots, s_r\})$.*

Proposition 1. *Let (G', C') be the replacement of cluster C by a corresponding graph R. Then (G, C) is c-planar if and only if (G', C') is c-planar.*

Proof. (" $\Rightarrow$ ":) We suppose that (G, C) is c-planar and we fix some planar embedding emb. Without loss of generality we can suppose that $emb_C(C)$ is a disc (because this region is homeomorphic to a disc). Suppose that the edges $e_1, \ldots, e_r$ cross the boundary of $emb_C(C)$ in (clockwise) order $e_{i_1}, \ldots, e_{i_r}$ and without loss of generality $i_1 = 1$.

If $r < 4$ then we simply remove cluster C with edges $e_1, \ldots, e_r$ and draw the graph R_r corresponding to C in a such way, that we identify vertex s_i with w_i for all $i \in \{1, \ldots, r\}$ and all other vertices of R_r draw inside $emb_C(C)$ in such a way, that edges of R_r don't cross any other edge of original graph nor other edge of R_r. This is clearly possible, it is enough to draw the edges outside the

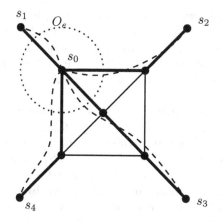

Fig. 4. Situation from the proof of Proposition 1(part "⇐")

disc $emb_C(C)$ along the deleted edges $e_1, \ldots, e_r$ and inside $emb_C(C)$ we can draw edges (or parts of edges) as noncrossing segments. This embedding of G' shows that (G', C') is c-planar.

If the corresponding graph for C is R_4^{234} then we can construct a c-planar embedding of C' in the same way as for $r < 4$.

If the corresponding graph for C is R_4^2 then the ordered set (i_1, i_2, i_3, i_4) must be equal to $(1, 3, 2, 4)$ or $(1, 4, 2, 3)$ because C admits only permutations from S_4^2 (otherwise we could find a permutation $\pi \notin S_4^2$ such that T_C^π is planar which is a contradiction). Now we delete the cluster C and add the graph R_4^2 in such a way that all the vertices of R_4^2 will be inside the disc $emb_C(C)$ and we identify vertices s_i with w_i for all $i \in \{1, \ldots, 4\}$ and any edge of R_4^2 will not cross any original edge nor any new edge of R_4^2. This is also clearly possible, it is enough to draw the edges outside the disc $emb_C(C)$ along the deleted edges $e_1, \ldots, e_4$ and inside $emb_C(C)$ we can draw the edges (or parts of the edges) as noncrossing segments. This embedding of G' shows that (G', C') is c-planar.

If the corresponding graph for cluster C is R_4^{23} then we continue similarly as in the previous cases. The ordered set (i_1, i_2, i_3, i_4) must be equal to $(1, 3, 2, 4)$, $(1, 4, 2, 3)$, $(1, 2, 3, 4)$ or $(1, 4, 3, 2)$ so again it is easy to replace the vertices and the edges of C by the graph R_4^{23} by identifying the vertices s_i with w_i for all $i \in \{1, \ldots, 4\}$ which proves that (G', C') is c-planar again.

("⇐":) Suppose we have a c-planar embedding of (G', C'). Moreover suppose that in the case $R = R_4^2$ there is nothing embedded in any interior face of R. This can be easily achieved in a similar way as in the proof of Lemma 2. We take an arbitrary spanning tree of the graph R and let s_0 denote its arbitrary vertex different from $s_1, \ldots, s_r$. Now draw the r curves connecting s_0 to $s_1, s_2, \ldots, s_r$ along the unique paths connecting the vertices in the tree, so that they do not cross each other nor anything in the embedding, except possibly for the edges of R. Then remove the original edges of R.

Now take some ϵ such that there are no edges, vertices nor clusters embedded in distance less than ϵ from s_0, except for the curves incident with s_0. Denote by O_ϵ the circle of radius ϵ with center s_0. Suppose P_i is the last intersection of the curve s_0s_i with O_ϵ. We can assume, that ϵ is so small, that if we label these curves clockwise $s_0s_{i_1}, s_0s_{i_2}, \ldots, s_0s_{i_r}$ as they leave s_0, then $P_{i_1}, P_{i_2}, \ldots, P_{i_r}$ are the points P_i in the clockwise order along O_ϵ. (We can assume, that each curve in the embedding is formed by finitely many straight line segments and circular arcs.)

By case analysis we show, that C admits the permutation $\pi = (i_1i_2 \ldots i_r)$. This is clear if $R = R_1, R_2, R_3$ or R_4^{234}. The graph $R_4^2 \backslash \{s_1, \ldots, s_4\}$ is 3-connected so the order of the edges is given in this case (up to the equivalence $\sim$) and the permutation π is in S_4^2. If $R = R_4^{23}$ and $\pi \in S_4^4$, then by connecting the neigbouring edges we obtain a planar embedding of $K_{3,3}$ — a contradiction.

So we take the planar embedding of T_C^π guaranteed by Lemma 2 and remove the vertex u_0. We can take a homeomorphic copy of this embedding of $T_C^\pi \backslash \{u_0\}$, in which the cycle u_1, u_2, u_3, u_4, u_1 coincides with a circle O_ϵ and the vertices $u_1, u_2, \ldots, u_r$ are embedded at the points $P_{i_1}, P_{i_2}, \ldots, P_{i_r}$, respectively. We replace the interior of O_ϵ by such an embedding.

We are ready to describe an embedding of (G, C). For every i the concatenation of the curve $v_{\pi(i)}u_i = P_{\pi(i)}$ and $P_{\pi(i)}s_{\pi(i)}$ forms an embedding of the edge $v_{\pi(i)}w_{\pi(i)}$ that crosses no other edge of G' or $G[C]$. Moreover, it crosses the boundary of each cluster of C' at most once, since there were no cluster boundaries inside O_ϵ, curve $P_{\pi(i)}s_{\pi(i)}$ was drawn along some edges of R and among them only the one incident with $s_{\pi(i)}$ could cross some cluster boundary and also at most once, because we started with a c-planar embedding of (G', C'). It remains to take O_ϵ as the boundary of the cluster C. It only crosses the edges w_iv_i. Furthermore, since curve s_iP_i (recall that $s_i = w_i$) lies completely outside O_ϵ (except for P_i), while P_iv_i lies completely inside O_ϵ (except for P_i), O_ϵ crosses the edge w_iv_i exactly once (in the point P_i). There are no other crossings, since they would have to be in the original c-planar embedding of (G', C') too. □

4   The Algorithm

The algorithm is described in Fig. 5.

Proposition 2. *The algorithm correctly decides c-planarity for instances, where each cluster has at most 4 outgoing edges.*

Proof. We first prove by the mathematical induction that for every $0 \leq i \leq |C|$, the pair (G_i, C_i) is defined and c-planar if and only if (G, C) is c-planar. This is certainly true for $i = 0$. Now suppose that this is true for every $i' < i$ and let us prove it for i.

In the case $r(C) = 0$ we have two possibilities. Either $G_{i-1}[C]$ is not planar, then also G is not planar and (G_{i-1}, C_{i-1}) is definitely not c-planar. Then the algorithm correctly rejects (and G_j, C_j is not defined for $j \geq i$). Or $G_{i-1}[C]$ is planar and by Lemma 1 and Remark 1 pair (G_{i-1}, C_{i-1}) is c-planar if and only

---

Input: Graph G and cluster set $\mathcal{C}$, where each cluster has at most 4 outgoing edges.
Task: Accept (G, C) if and only if (G, C) is clustered planar.

1. Set $G_0 := G, \mathcal{C}_0 := \mathcal{C}$.
2. For $i := 1$ to $|\mathcal{C}|$ do:
   (a) Let C be some cluster on the bottom-most level in $\mathcal{C}_{i-1}$.
   (b) If $r(C) = 0$ then
      i. If $G_{i-1}[C]$ is planar then set
         $G_i := G_{i-1} \setminus C$
         $\mathcal{C}_i := \{A \setminus C | A \in \mathcal{C}_{i-1} \setminus \{C\}, A \supseteq C\} \cup \{A | A \in \mathcal{C}_{i-1}, A \not\supseteq C \}$
      ii. else REJECT.
   (c) else
      i. For each permutation $\pi \in S_{r(C)}$ test whether C admits π (whether T_C^π is planar)
      ii. If C admits no permutation, then REJECT.
      iii. Let $(G_i, \mathcal{C}_i)$ be the replacement of cluster C by the corresponding graph in $(G_{i-1}, \mathcal{C}_{i-1})$.
3. If $G_{|\mathcal{C}|}$ is planar then ACCEPT, otherwise REJECT $(\mathcal{C}_{|\mathcal{C}|} = \emptyset)$.

Fig. 5. An overview of the algorithm

if $(G_i, \mathcal{C}_i)$ is, since $\{B | B \in \mathcal{C} \setminus \{C\}, B \subseteq C\}$ is empty (C is on the bottom-most level).

Now consider the case $1 \leq r(C) \leq 4$. If C admits no permutation, then by Lemma 3 the pair $(G_{i-1}, \mathcal{C}_{i-1})$ is not c-planar and the algorithm correctly rejects (and does not define $G_j, \mathcal{C}_j$ for $j \geq i$). Otherwise C has a corresponding graph by Corollary 1 and from the Proposition 1 we know that $(G_{i-1}, \mathcal{C}_{i-1})$ is c-planar if and only if $(G_i, \mathcal{C}_i)$ is c-planar.

Since $|\mathcal{C}_i| = |\mathcal{C}_{i-1}| - 1$ whenever defined, we have $|\mathcal{C}_{|\mathcal{C}|}| = 0$ and thus $\mathcal{C}_{|\mathcal{C}|} = \emptyset$ if $\mathcal{C}_{|\mathcal{C}|}$ is defined. But then $(G_{|\mathcal{C}|}, \mathcal{C}_{|\mathcal{C}|}) = (G_{|\mathcal{C}|}, \emptyset)$ is c-planar if and only if $G_{|\mathcal{C}|}$ is planar due to Remark 1, which completes the proof. □

Proposition 3. *The algorithm works in time $O(n)$.*

Proof. The cycle is executed at most $|\mathcal{C}|$ times, in each time we delete one cluster or reject. When we omit a planarity testing, complexity of each step of cycle in the algorithm is bounded by constant. We add constant number of vertices and if we have a suitable representation of clusters (for example tree representation) we can find cluster on the bottom-most level in constant time too. And then for these operations we need $|\mathcal{C}|$ in complexity time. The algorithm touches each vertex at most three times, when we add, test, and remove it. For vertices which we added later we paid before, by constant in each iteration. And for the original vertices we need extra n for planarity testing. Each vertex from the original graph we touch only once, because if we touch it we remove it or reject whole graph. Since $|\mathcal{C}|$ is bounded by $O(n)$, the complexity of our algorithm is $O(n)$. □

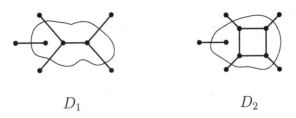

$$D_1 \qquad\qquad\qquad D_2$$

Fig. 6. Two clusters with 5 outgoing edges that cannot be represented by any connected graph

5   The Limits of the Approach

Let us consider clusters with more than 4 outgoing edges. Definition 1, Lemmas 2 and 4 easily generalize to this case as well as Lemma 3. The problem with the generalization is that there are disconnected clusters with 5 outgoing edges that admit a combination of permutations which cannot be represented by a connected graph. In particular it can be shown that the two clusters from Fig. 6 have this property.

Let us try to formalize the result. Consider a graph R that is supposed to be corresponding to some cluster C. Hence it has some distinguished vertices $s_1, \ldots, s_r$ of degree 1 that are supposed to be identified with the vertices of $G \setminus C$ when the cluster C is replaced by R in a graph G. Let $R' = V(R) \setminus \{s_1, \ldots, s_r\}$. We say that the graph R admits a permutation π, if the cluster R' of the clustered graph $(R, \{R'\})$ admits a permutation π.

Proposition 4. *There is no connected graph that admits the same number of permutations as the cluster D_1 from Fig. 6.*

Proof. We will count the circular permutations. In total there are 12 circular permutations on 5 elements, each representing 10 (standard) permutations. Observe first that D_1 admits 8 circular permutations. Now assume for a contradiction that there is a connected graph R (with distinguished vertices $s_1, \ldots, s_5$) that admits also 8 circular permutations.

We observe that whenever we take a subgraph Q of the graph R, $s_1, \ldots, s_5 \in V(Q)$, then the graph Q admits at least the same number of permutations, since we can just ommit the unnecessary parts from the appropriate embedding. Now consider a subtree T of R with leaves of T being exactly the vertices $s_1, \ldots, s_5$. It is clear that R has such a subgraph since R is connected.

Since T has 5 leaves, it has at most 3 vertices of degree at least 3 — either it has 3 vertices of degree 3, or one of degree 3 and one of degree 4, or just one of degree 5. It is not hard to check that in the first two cases T admits 4 and 6 circular permutations, respectively. Thus in this cases R cannot admit 8 circular permutations. The tree with just one vertex of degree 5 (among the vertices of degree at least 3) admits all 12 circular permutations. Thus we know that T must be some subdivision of $K_{1,5}$.

If R contains no path connecting two different branches of T, then clearly R admits the same permutations as T i.e. all 12 circular permutations. On the other hand, if R contains a path between two branches of T, then there is another tree T', subgraph of R, that has one vertex of degree 4 and one vertex of degree 3. But this means that R admits at most 6 permutations — a contradiction.   □

References

1. Cornelsen, S., Wagner, D.: Completely connected clustered graphs. Journal of Discrete Algorithms 4(2), 313–323 (2006)
2. Cortese, P.F., Di Battista, G.: Clustered planarity. In: ACM SoCG 2005, pp. 32–34 (2005)
3. Cortese, P.F., Di Battista, G., Patrignani, M., Pizzonia, M.: Clustering cycles into cycles of clusters. Journal of Graph Algorithms and Applications 9(3), 391–413 (2005); In: Pach, J. (ed.) GD 2004. LNCS, vol. 3383, pp. 391–413. Springer, Heidelberg (2005)
4. Dahlhaus, E.: A linear time algorithm to recognize clustered planar graphs and its parallelization. In: Lucchesi, C.L., Moura, A.V. (eds.) LATIN 1998. LNCS, vol. 1380, pp. 239–248. Springer, Heidelberg (1998)
5. Di Battista, G., Frati, F.: Efficient C-planarity testing for embedded flat clustered graphs with small faces. In: Hong, S.-H., Nishizeki, T., Quan, W. (eds.) GD 2007. LNCS, vol. 4875, pp. 291–302. Springer, Heidelberg (2008)
6. Eades, P., Feng, Q.W., Lin, X., Nagamochi, H.: Straight-line drawing algorithms for hierarchical graphs and clustered Graphs. Algorithmica 44, 1–32 (2006)
7. Feng, Q.W., Cohen, R.F., Eades, P.: Planarity for clustered graphs. In: Spirakis, P.G. (ed.) ESA 1995. LNCS, vol. 979, pp. 213–226. Springer, Heidelberg (1995)
8. Goodrich, M.T., Lueker, G.S., Sun, J.Z.: C-planarity of extrovert clustered graphs. In: Healy, P., Nikolov, N.S. (eds.) GD 2005. LNCS, vol. 3843, pp. 211–222. Springer, Heidelberg (2006)
9. Gutwenger, C., Jünger, M., Leipert, S., Mutzel, P., Percan, M., Weiskircher, R.: Advances in c-planarity testing of clustered graphs. In: Goodrich, M.T., Kobourov, S.G. (eds.) GD 2002. LNCS, vol. 2528, pp. 220–235. Springer, Heidelberg (2002)
10. Gutwenger, C., Jünger, M., Leipert, S., Mutzel, P., Percan, M., Weiskircher, R.: Subgraph induced planar connectivity augmentation. In: Bodlaender, H.L. (ed.) WG 2003. LNCS, vol. 2880, pp. 261–272. Springer, Heidelberg (2003)
11. Hopcroft, J., Tarjan, R.E.: Efficient planarity testing. J. ACM 21(4), 549–568 (1974)
12. Jelínková, E., Kára, J., Kratochvíl, J., Pergel, M., Suchý, O., Vyskočil, T.: Clustered planarity: Small clusters in eulerian graphs. In: Hong, S.-H., Nishizeki, T., Quan, W. (eds.) GD 2007. LNCS, vol. 4875, pp. 303–314. Springer, Heidelberg (2008)

Computing Maximum C-Planar Subgraphs

Markus Chimani, Carsten Gutwenger, Mathias Jansen,
Karsten Klein, and Petra Mutzel

Technische Universität Dortmund, Germany
{markus.chimani,carsten.gutwenger,mathias.jansen,
karsten.klein,petra.mutzel}@cs.tu-dortmund.de

Abstract. Deciding c-planarity for a given clustered graph $C = (G, T)$ is one of the most challenging problems in current graph drawing research. Though it is yet unknown if this problem is solvable in polynomial time, latest research focused on algorithmic approaches for special classes of clustered graphs. In this paper, we introduce an approach to solve the *general* problem using integer linear programming (ILP) techniques. We give an ILP formulation that also includes the natural generalization of c-planarity testing—the *maximum c-planar subgraph problem*—and solve this ILP with a branch-and-cut algorithm. Our computational results show that this approach is already successful for many clustered graphs of small to medium sizes and thus can be the foundation of a practically efficient algorithm that integrates further sophisticated ILP techniques.

1 Introduction

Drawing clustered graphs is a prevalent problem in practical applications of graph drawing, e.g., to group nodes into departments, as well as in graph theory, since the occurring graph theoretical problems are in particular challenging, even in simplified special cases. A clustered graph $C = (G, T)$ is formally defined as a graph $G = (V, E)$ together with a rooted tree T, the *inclusion tree* of C, where the leaves of T are the vertices of G. Each node ν of T represents a *cluster* of the vertices $V(\nu)$ of G that are leaves of the subtree rooted at ν.

In a drawing of a clustered graph, the clusters themselves are drawn as simple regions, e.g., rectangles, and special aesthetic criteria on the drawing need to be met to guarantee readability. In particular, we call a drawing *c-planar*, if there are neither edge–edge nor edge–region crossings and the drawing of a cluster ν is contained in the interior of the region of a cluster μ if and only if μ lies on the path from ν to the root of T. A *c-planar* clustered graph is a clustered graph for which a c-planar drawing exists.

Though c-planarity has been intensively studied in the past years, the complexity of deciding c-planarity is still unknown. Instead of considering the general problem, latest research focused on special classes of clustered graphs. Besides the well known results by Feng et al. [8] and Dahlhaus [4] for *c-connected* clustered graphs, i.e., clustered graphs where the vertices $V(\mu)$ of each cluster μ induce a connected graph, various classes of non-c-connected clustered graphs

I.G. Tollis and M. Patrignani (Eds.): GD 2008, LNCS 5417, pp. 114–120, 2009.

have been studied [10,3,9,6,11]. In contrast to this, we tackle the general c-planarity problem in this paper by presenting the foundation of an ILP-based approach consisting of an ILP formulation and a branch-and-cut algorithm.

In order to draw not necessarily c-planar clustered graphs, Di Battista et al. [5] adapted the topology-shape-metrics approach to clustered graphs and described a planarization-based method for crossing minimization. This method first computes a c-planar subgraph C', and then reinserts the deleted edges successively into a c-planar embedding of C', so that only a small number of crossings is produced. Our ILP approach also solves the first problem of this c-planarization approach, i.e., finding a c-planar subgraph of maximum size:

Definition 1 (Maximum C-planar Subgraph Problem (MCPSP)). *Given a clustered graph $C = (G = (V, E), T)$ find a c-planar clustered graph $C' = (G' = (V, E'), T)$ with $E' \subseteq E$ such that E' has maximum cardinality.*

Obviously, MCPSP is NP-hard, since the maximum planar subgraph problem is already NP-hard. This paper is organized as follows. Section 2 presents our ILP formulation for MCPSP and a branch-and-cut algorithm for solving the ILP; an experimental evaluation of this algorithm is given in Sect. 3.

2   ILP and Branch-and-Cut

In the following, let $C = (G = (V, E), T)$ be the given clustered graph with edge set E. For a cluster ν in C let $E(\nu)$ denote the edge set induced by the vertices $V(\nu)$ in cluster ν, and let $E(\bar{\nu})$ denote the edge set induced by the vertices in $V(\bar{\nu}) = V \setminus V(\nu)$.

We say a c-connected clustered graph is *completely connected* if, for each non-root cluster ν, the subgraph by $V(\bar{\nu})$ is connected. For our formulation we need the following result by Cornelsen and Wagner [2]:

Theorem 1. *A clustered graph is c-planar if and only if it is a subgraph of a c-planar completely connected clustered graph. A completely connected clustered graph $C = (G, T)$ is c-planar if and only if its underlying graph G is planar.*

Our central concept for the formulation then is to (a) augment the given clustered graph such that it becomes completely connected, and (b) to ensure that the resulting graph, disregarding the cluster structure, is planar:

Corollary 1. C^* *is a maximum c-planar subgraph of C if and only if it is the largest subgraph with the property that there exists a completely connected clustered graph C' such that (a) C^* is its subgraph and (b) the underlying graph of C' is planar. If $C^* = C$, C is c-planar.*

In the following we will hence concentrate on finding such a completely connected *solution graph $C' = (G' = (V, E'), T)$.*

2.1 The ILP Formulation

We define the set F as the complement of E, i.e., F are the potential edges for the augmentation. This allows us to introduce our two variables

$$x_e, y_f \in \{0, 1\} \quad \forall e \in E, f \in F \tag{1}$$

which are 1 if the corresponding edge is contained in the solution graph, and 0 otherwise. Then we can write the objective function as

$$\max \sum_{e \in E} x_e - \varepsilon \sum_{f \in F} y_f. \tag{2}$$

We want to maximize the number of original edges in the solution and use as few augmenting edges as possible. In order for the latter criterion to not interfere with the main optimization goal, we restrict its influence by the introduction of $\varepsilon := \frac{0.1}{3n}$; due to Euler's formula this guarantees that the second term in (2) does not grow larger that 0.1.

We have two sets of constraints: the first set guarantees that the solution graph C' is completely connected; the second set ensures planarity of G'.

Connectivity Constraints. A *cut set* $W|A$ with $W \subseteq V$ and $A \subseteq E$ in the graph $G = (V, E)$ is defined as the set of edges in A that are incident to exactly one vertex of W. A graph is connected if and only if the cardinality of $W|E$ is at least 1 for any $\emptyset \neq W \subset V$. We define the *connectivity constraints* as:

$$\sum_{e \in W|E(\xi)} x_e + \sum_{f \in W|F(\xi)} y_f \geq 1 \quad \forall \nu \in T, \forall \xi \in \{\nu, \bar{\nu}\}, \forall \emptyset \neq W \subseteq V(\xi) \setminus \{w_\xi\} \tag{3}$$

While the case $\xi \in T$ only guarantees c-connectivity, the additional constraints with $\xi \notin T$ are necessary to ensure complete connectivity. We use $W \subseteq V(\xi) \setminus \{w_\xi\}$ for some fixed $w_\xi \in V(\xi)$ instead of $W \subset V(\xi)$ to avoid redundancy.

Kuratowski Constraints. In order to guarantee that the solution graph is planar we use Kuratowski constraints as introduced for the maximum planar subgraph problem [12]. These constraints are based on Kuratowski's theorem [14] which states that a graph is planar if and only if it does not contain a subdivision of K_5 or $K_{3,3}$. We call these subdivisions *Kuratowski subdivisions*, and represent them by their edge sets. Let $\mathcal{K}$ be the set of all Kuratowski subdivisions in $(V, E \cup F)$. For any $K \in \mathcal{K}$, the solution graph will not contain all edges of K, as this would contradict its planarity. We hence formulate the Kuratowski constraints as

$$\sum_{e \in K} x_e + \sum_{e \in K} y_e \leq |K| - 1 \quad \forall K \in \mathcal{K}. \tag{4}$$

Theorem 2. *The ILP*

$$\left\{ \max \sum_{e \in E} x_e - \varepsilon \sum_{e \in F} y_e, \text{ subject to (1), (3), and (4)} \right\}$$

solves the maximum c-planar subgraph problem. If $x_e = 1$ for all $e \in E$, the given clustered graph is c-planar.

2.2   Branch-and-Cut

Both constraint sets contain an exponential number of constraints and hence it is not applicable to generate all constraints in advance. We solve the ILP within a branch-and-cut framework: we start with a small subset of constraints, drop the integrality constraints, and apply cutting-plane algorithms to add additional constraints as required. The problem of identifying such cuts after obtaining a fractional solution of the partial LP-relaxation is called *separation problem*.

Separation. Separating the connectivity constraints can be done in polynomial time by computing minimum cuts on the graph, using the fractional solution as edge capacities. On the other hand, there are no known polynomial algorithms for the Kuratowski constraint separation, and we have to resort to a heuristic routine, similar to the ones described in [12]: we round the fractional solution to an integer solution, which we can interpret as our *support graph S*, and search for Kuratowski subdivisions in S. For any such subdivision K we can test whether the current fractional solution violates the constraint induced by K.

Traditional planarity test algorithms can extract a single Kuratowski subdivision per run; in our experiments we use the extended test algorithm presented in [1] which extracts multiple different subdivisions in linear time. Note that we separate all cut constraints before separating any Kuratowski constraints.

Branching and Primal Heuristic. If we have a fractional solution, but cannot find any violated constraints, we have to resort to branching. In such cases good LP-based heuristics become crucial, to prune nodes early in the branch-and-bound tree. Our heuristic works as follows: We start by computing a spanning tree recursively for each cluster in a bottom-up scheme on T, using the fractional solution as negative weights. Merging all these minimum spanning trees, we obtain a c-connected and c-planar spanning tree R. We sort the remaining edges based on their fractional values, and iteratively try to add them to R in decreasing order. This can be done in polynomial time, since planarity testing of a c-connected clustered graph is polynomial. We obtain a maximal c-connected, c-planar subgraph R that implies a c-planar subgraph of C.

3   Computational Experiments and Discussion

We report on the results of our experimental evaluation. The main intention of this short study is to point out the feasibility of our approach, without giving attention to speed-up techniques like strong preprocessing and heuristics, column generation, etc. We implemented our approach within the Open Graph Drawing Framework (www.ogdf.net) using the branch-and-cut framework ABACUS [13] with CPLEX 9.0 as LP-solver. The experiments were run on a 2.33GHz Intel Xeon with 2GB RAM per process and a time limit of 30 minutes per instance.

In addition to solving the MCPSP, we also experimented with a variant were only c-planarity is tested; in this case no maximum c-planar subgraph needs to be computed and subproblems are pruned as soon as their dual bound proves that an original edge would have to be deleted.

	# inst.	c-plan.	c-con.	compl. con.	Clusters min	avg	max	Vertices max	Edges max
Planar graphs	1815	1494	25	2	3	4	9	29	30
Non-planar graphs	116	0	3	0	3	5.2	9	26	30

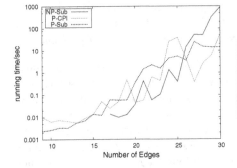

	Running time (sec) min	avg	95%	max
P-Sub	0.01	4.9	9.6	1460.9
NP-Sub	0.01	40.9	18.7	1456.5
P-CPl	0.01	4.3	6.1	249.9

Fig. 1. (top) Properties of the benchmark instances. **(bottom)** Average runtime performance of the branch-and-cut algorithm. P-Sub and NP-Sub are running times for solving the MCPSP on the (non-)planar graphs, respectively. P-CPl denotes the runtime for the c-planarity test on the instances with underlying planar graphs. 95% denotes the 95%-percentile.

Benchmark Set. We created a benchmark set based on the Rome graphs [7] by generating cluster hierarchies on top of each graph of the library. The library contains planar and non-planar graphs; key properties are shown in Table 1(top).

We create a cluster structure by randomly picking vertices in a cluster ν, starting with the root cluster, and after each pick, a random decision is made if a new cluster is generated with the vertices picked so far, up to a maximum number of 9 clusters. We restrict the maximum cluster tree depth to two levels (in addition to the root cluster), the number of edges to 30, and divide the created clustered graphs into two groups depending on the planarity of the underlying graph. The benchmark set can be found at `ls11-www.cs.uni-dortmund.de/people/klein/clusterbenchmarks08.zip`.

Results and Discussion. Figure 1(bottom-left) shows the resulting running times required by our approach, relative to the graph size; the table on the bottom right summarizes the runtime performance of the instances, depending on the planarity of the underlying graph. We see that restricting the computation to pure c-planarity testing by pruning leads to decreases in the overall average computation time, but does not necessarily need to speed up the computation for each instance, because subproblems containing the maximum c-planar subgraph may be pruned, which extends the search in the branch tree.

Our main observation is that the performance on most of the test graphs is promising: only 2 non-planar and 17 planar graphs could not be solved within the time limit; the 95%-percentile shows that long running cases are extremely rare. The average running time of the c-connected clustered graphs is below 0.02 seconds, indicating that the ILP performs well on this polynomial time solvable class. We therefore conjecture that the ILP may be useful as a tool when developing c-planarity tests for special graph classes, as the ILP may give hints on the classes' hardness.

Conclusion and Future Work. We introduced the Maximum C-planar Subgraph Problem and presented an ILP formulation together with a branch-and-cut approach to solve it to optimality. Our brief experimental evaluation showed the general feasibility of the concept. We believe that our branch-and-cut approach can be improved to also cope with harder instances, which is part of our future work, especially by using stronger heuristics and preprocessing to reduce the search space, as well as pricing instead of adding all possible variables in advance. Encouraged by the results on the c-connected graphs, we also plan to perform a closer investigation of the behavior of our branch-and-cut approach with regard to other polynomial time solvable classes of clustered graphs.

References

1. Chimani, M., Mutzel, P., Schmidt, J.M.: Efficient extraction of multiple kuratowski subdivisions. In: Hong, S.-H., Nishizeki, T., Quan, W. (eds.) GD 2007. LNCS, vol. 4875, pp. 159–170. Springer, Heidelberg (2008)
2. Cornelsen, S., Wagner, D.: Completely connected clustered graphs. J. Discrete Algorithms 4(2), 313–323 (2006)
3. Cortese, P.F., Di Battista, G., Patrignani, M., Pizzonia, M.: Clustering cycles into cycles of clusters. In: Pach, J. (ed.) GD 2004. LNCS, vol. 3383, pp. 100–110. Springer, Heidelberg (2005)
4. Dahlhaus, E.: A linear time algorithm to recognize clustered planar graphs and its parallelization. In: Lucchesi, C.L., Moura, A.V. (eds.) LATIN 1998. LNCS, vol. 1380, pp. 239–248. Springer, Heidelberg (1998)
5. Di Battista, G., Didimo, W., Marcandalli, A.: Planarization of clustered graphs. In: Mutzel, P., Jünger, M., Leipert, S. (eds.) GD 2001. LNCS, vol. 2265, pp. 60–74. Springer, Heidelberg (2002)
6. Di Battista, G., Frati, F.: Efficient c-planarity testing for embedded flat clustered graphs with small faces. In: Hong, S.-H., Nishizeki, T., Quan, W. (eds.) GD 2007. LNCS, vol. 4875, pp. 291–302. Springer, Heidelberg (2008)
7. Di Battista, G., Garg, A., Liotta, G., Tamassia, R., Tassinari, E., Vargiu, F.: An experimental comparison of four graph drawing algorithms. Comput. Geom. Theory Appl. 7(5-6), 303–325 (1997)
8. Feng, Q.W., Cohen, R.F., Eades, P.: Planarity for clustered graphs. In: Spirakis, P.G. (ed.) ESA 1995. LNCS, vol. 979, pp. 213–226. Springer, Heidelberg (1995)
9. Goodrich, M.T., Lueker, G.S., Sun, J.Z.: C-planarity of extrovert clustered graphs. In: Healy, P., Nikolov, N.S. (eds.) GD 2005. LNCS, vol. 3843, pp. 211–222. Springer, Heidelberg (2006)
10. Gutwenger, C., Jünger, M., Leipert, S., Mutzel, P., Percan, M., Weiskircher, R.: Advances in c-planarity testing of clustered graphs. In: Goodrich, M.T., Kobourov, S.G. (eds.) GD 2002. LNCS, vol. 2528, pp. 220–235. Springer, Heidelberg (2002)

11. Jelínková, E., Kára, J., Kratochvíl, J., Pergel, M., Suchý, O., Vyskocil, T.: Clustered planarity: Small clusters in eulerian graphs. In: Hong, S.-H., Nishizeki, T., Quan, W. (eds.) GD 2007. LNCS, vol. 4875, pp. 303–314. Springer, Heidelberg (2008)
12. Jünger, M., Mutzel, P.: Maximum planar subgraphs and nice embeddings: Practical layout tools. Algorithmica 16(1), 33–59 (1996)
13. Jünger, M., Thienel, S.: The ABACUS system for branch-and-cut-and-price algorithms in integer programming and combinatorial optimization. Software: Practice and Experience 30, 1325–1352 (2000)
14. Kuratowski, K.: Sur le problème des courbes gauches en topologie. Fundamenta Mathematicae 15, 271–283 (1930)

Clustered Planarity: Embedded Clustered Graphs with Two-Component Clusters

(Extended Abstract)

Vít Jelínek[1,*], Eva Jelínková[1], Jan Kratochvíl[1,2], and Bernard Lidický[1]

[1] Department of Applied Mathematics[**]
[2] Institute for Theoretical Computer Science[***]
Charles University
Malostranské nám. 25, 118 00 Praha, Czech Republic
{jelinek,eva,honza,bernard}@kam.mff.cuni.cz

Abstract. We present a polynomial-time algorithm for c-planarity testing of clustered graphs with fixed plane embedding and such that every cluster induces a subgraph with at most two connected components.

1 Introduction

Clustered planarity (or shortly, c-planarity) has recently become an intensively studied topic in the area of graph and network visualization. In many situations one needs to visualize a complicated inner structure of graphs and networks. Clustered graphs provide a possible model of such a visualization, and as such they find applications in many practical problems, e.g., management information systems, social networks or VLSI design tools [5]. However, from the theoretical point of view, the computational complexity of deciding c-planarity is still an open problem and it is regarded as one of the challenges of contemporary graph drawing.

A *clustered graph* is a pair (G, C), where $G = (V, E)$ is a graph and C is a family of subsets of V (called *clusters*), with the property that each two clusters are either disjoint or in inclusion. We always assume that the vertex set V is in C, and we call it *the root cluster*. We say that a clustered graph (G, C) is *clustered-planar* (or shortly *c-planar*), if the graph G has a planar drawing such that we may assign to every cluster $X \in C$ a compact simply connected region of the plane which contains precisely the vertices of X and whose boundary crosses every edge of G at most once (see Sect. 2 for the precise definition).

It is well known that planar graphs can be recognized in polynomial (even linear) time. For c-planarity, determining the time-complexity of the decision problem remains open; only partial results are known. If every cluster of (G, C) induces a connected subgraph of G, then the c-planarity of (G, C) can be tested in

---

[*] Supported by the grant 201/05/H014 of the Czech Science Foundation.
[**] Supported by project MSM0021620838 of the Czech Ministry of Education.
[***] Supported by grant 1M0545 of the Czech Ministry of Education.

I.G. Tollis and M. Patrignani (Eds.): GD 2008, LNCS 5417, pp. 121–132, 2009.
© Springer-Verlag Berlin Heidelberg 2009

linear time by an algorithm of Dahlhaus [3], which improves upon a polynomial algorithm of Feng et al. [5]. Several generalizations of this result are known: c-planarity testing is polynomial for clustered graphs in which all disconnected clusters form a single chain in the cluster hierarchy [7], for clustered graphs in which for every disconnected cluster X, the parent cluster and all the sibling clusters of X are connected [7], and for clustered graphs where every disconnected cluster X has connected parent cluster, with the additional assumption that each component of X is adjacent to a vertex not belonging to the parent of X [6].

Another approach to c-planarity testing is to consider *flat clustered graphs*, which are clustered graphs in which all non-root clusters are disjoint. Even in this restricted setting, the complexity of c-planarity testing is unknown. However, polynomial-time algorithms exist for special types of flat clustered graphs, e.g., if the underlying graph is a cycle and the clusters are arranged in a cycle [2], if the underlying graph is a cycle and the clusters are arranged into an embedded plane graph [1], or if the underlying graph is a cycle and the clusters contain at most three vertices [9]. Even for these very restricted settings, the algorithms are quite non-trivial.

Suppose an embedding of the underlying graph is fixed. Does the c-planarity testing become easier? This question was already addressed in [4], who provide a linear algorithm for flat clustered graphs with a prescribed embedding in which all faces have size at most five.

In this paper, we also deal with clustered graphs $(G, \mathcal{C})$, for which the embedding of G is fixed. In this setting, we obtain a polynomial algorithm for c-planarity of clustered graphs in which each cluster induces a subgraph with at most two connected components.

Theorem 1. *There is a polynomial time algorithm for deciding c-planarity of a clustered graph $(G, \mathcal{C})$, where G is a plane graph and every cluster of $\mathcal{C}$ induces a subgraph of G with at most two connected components.*

In this extended abstract, we present a simplified version of the algorithm which assumes that the cluster hierarchy is flat. We also omit some of the proofs.

2   Preliminaries

We follow standard terminology on finite simple loopless plane graphs. A *plane graph* is an ordered pair $G = (V, E)$, where V is a finite set of points in the plane (called *vertices*) and E is a set of Jordan arcs (called *edges*), such that every edge connects two distinct vertices of G and avoids any other vertex, every pair of vertices is connected by at most one edge, and no two edges intersect, except in a possible common endpoint.

If $G = (V, E)$ is a plane graph and $X \subseteq V$ is a set of vertices, we let $\overline{X}$ denote the set $V \setminus X$ and we let $G[X]$ denote the subgraph of G induced by X.

Two plane graphs $G = (V, E)$ and $G' = (V', E')$ are *isomorphic* if there is a continuous bijection f of the plane with continuous inverse such that $V' = \{f(v) : v \in V\}$ and $E' = \{f[e] : e \in E\}$ (where $f[e]$ is the set $\{f(x) : x \in e\}$).

The algorithm we will present in this paper expects a representation of a plane graph as part of its input. Since the algorithm does not need to make a distinction between isomorphic plane graphs, we may represent a plane graph G by a data structure which identifies G uniquely up to isomorphism. We may identify the isomorphism class of G by specifying, for every vertex of G, the cyclic order of edges and faces incident to v, and by specifying the outer face of G. The isomorphism class of a plane graph can be thus represented by a data structure whose size is polynomial in $|V|$.

Let $G = (V, E)$ be a plane graph. A *cluster set* on G is a set $\mathcal{C} \subseteq \mathcal{P}(V(G))$ such that for all $X, Y \in \mathcal{C}$, either X and Y are disjoint or they are in inclusion; the pair $(G, \mathcal{C})$ is called a *plane clustered graph*. The elements of $\mathcal{C}$ are called *clusters*. We assume that the set $V(G)$ is always in $\mathcal{C}$, and we call it the *root cluster*. A cluster that does not contain any other cluster as a subset is called *minimal*.

Clusters are naturally ordered by inclusion. The set $V(G)$ is the maximum of this ordering. A cluster is called *connected* if it induces in G a connected subgraph and *disconnected* otherwise. A *component* of a cluster $X \in \mathcal{C}$ is a maximal set $X_1 \subseteq X$ such that $G[X_1]$ is a connected subgraph of $G[X]$.

We say that a plane clustered graph $(G, \mathcal{C})$ is *connected* (or *2-connected*, or *disconnected*) if the graph G is connected (or 2-connected, or disconnected). Let us remark that some earlier papers use the term 'connected clustered graph' to denote a clustered graph in which every cluster is connected; we break with this convention for the sake of consistency of our definitions.

In this paper, we consider clustered graphs $(G, \mathcal{C})$ in which every disconnected cluster in $\mathcal{C}$ has exactly two components. We will call such a pair $(G, \mathcal{C})$ a *2-component clustered graph*.

For a plane clustered graph $(G, \mathcal{C})$, a *clustered planar embedding* is a mapping emb_c that assigns to every cluster $X \in \mathcal{C}$ a compact simply connected planar region $emb_c(X)$ (called *the cluster region of X*) whose boundary $\gamma(X)$ is a closed Jordan curve (called *the cluster boundary of X*), such that

- for each vertex $v \in V$ and each cluster $X \in \mathcal{C}$, v is in $emb_c(X)$ if and only if $v \in X$,
- for each cluster $X \in \mathcal{C}$, the cluster boundary $\gamma(X)$ does not contain any vertex from V,
- for every two clusters X and Y, the regions $emb_c(X)$ and $emb_c(Y)$ are disjoint (in inclusion) if and only if X and Y are disjoint (in inclusion, respectively), and
- for every edge $e \in E$ and every cluster $X \in \mathcal{C}$, the edge e crosses the cluster boundary of X at most once.

A plane clustered graph is called *clustered planar* (shortly *c-planar*) if it allows a clustered planar embedding.

When testing c-planarity, we adopt the approach first used in [5] of adding extra edges to the underlying graph in order to make each cluster connected.

Definition 1. *Let* $(G, \mathcal{C})$ *be a plane clustered graph. Let* c *be a cycle in* G *whose vertices all belong to a cluster* $X \in \mathcal{C}$. *We say that* c *is a* hole *of the cluster* X, *if the interior region of* c *contains a vertex not belonging to* X.

Clearly, a plane clustered graph with a hole is not c-planar. On the other hand, it is known [5] that a plane clustered graph without holes whose clusters are all connected is c-planar. For a given plane clustered graph $(G, \mathcal{C})$ the existence of a hole can be determined in polynomial time [5].

Definition 2. *Let* G *be a plane graph. A* candidate edge *of* G *is a simple curve* $e \notin E$ *such that* $(V, E \cup \{e\})$ *is a plane graph. A* candidate set *is a set* S *of candidate edges of* G *such that* $(V, E \cup S)$ *is a plane graph. We use the notation* $G \cup e$ *and* $G \cup S$ *as a shorthand for* $(V, E \cup \{e\})$ *and* $(V, E \cup S)$ *respectively.*

We say that two candidate edges e *and* e' *are* isomorphic *if* $G \cup e$ *and* $G \cup e'$ *are isomorphic plane graphs.*

Note that a pair of vertices u, v of a plane graph G may be connected by two distinct non-isomorphic candidate edges. On the other hand, it is not hard to see that a plane graph on n vertices has at most $O(n^2)$ non-isomorphic candidate edges.

The following theorem reduces c-planarity testing to searching for a specific set of candidate edges. It was proved in an equivalent version by Feng et al. [5].

Theorem 2. *A plane clustered graph* $(G, \mathcal{C})$ *is c-planar if and only if there exists a candidate set* S *with the following properties:*

1. $(G \cup S, \mathcal{C})$ *has no hole,*
2. every cluster X *of* $\mathcal{C}$ *induces a connected subgraph in* $G \cup S$.

A set S of candidate edges satisfying the above conditions is called a *saturator*[1]. A set S that satisfies the first condition will be called a *partial saturator*. We say that a candidate edge e *saturates* a cluster X, if e connects a pair of vertices belonging to different components of X. A saturator S is *minimal* if no proper subset of S is a saturator. Note that every candidate edge from a minimal saturator S saturates a cluster from $\mathcal{C}$. Moreover, if X is a cluster with two components that does not contain any disconnected subcluster, then a minimal saturator S has exactly one candidate edge saturating X.

Definition 3. *If* e *is a candidate edge of a plane clustered graph* $(G, \mathcal{C})$ *such that* $(G, \mathcal{C})$ *is c-planar if and only if* $(G \cup e, \mathcal{C})$ *is c-planar, then the edge* e *is called* harmless. *Similarly, a candidate set* S *is* harmless *provided* $(G, \mathcal{C})$ *is c-planar if and only if* $(G \cup S, \mathcal{C})$ *is c-planar.*

Note that if $(G, \mathcal{C})$ is a c-planar clustered graph, then a candidate set is harmless if and only if it is a subset of a saturator of $(G, \mathcal{C})$. On the other hand, if $(G, \mathcal{C})$ is not c-planar, then any candidate set is harmless.

Let us now present several simple but useful lemmas, whose proofs are omitted due to space constraints.

---

[1] Note that this definition of saturator differs slightly from that of some other papers— here, candidate edges are already embedded.

Lemma 1. *Let $(G, \mathcal{C})$ be a plane clustered graph without holes, let $X \in \mathcal{C}$ be a cluster which is minimal and connected. Then $(G, \mathcal{C})$ is c-planar if and only if $(G, \mathcal{C} \setminus \{X\})$ is c-planar.*

The next lemma shows that c-planarity testing of 2-component graphs can be reduced to c-planarity testing of 2-component connected plane clustered graphs.

Lemma 2. *If there is a polynomial time algorithm for deciding c-planarity for connected 2-component plane clustered graphs, then there is a polynomial time algorithm for deciding c-planarity for arbitrary 2-component plane clustered graphs.*

The following lemma allows us to reduce c-planarity testing of a connected graph to an equivalent instance of c-planarity where the underlying graph is 2-connected.

Lemma 3. *Let $(G, \mathcal{C})$ be a connected plane clustered graph with at least three vertices which is not 2-connected. There is a polynomial-time transformation which constructs a plane clustered graph $(G', \mathcal{C}')$ such that G' is connected, G' has fewer components of 2-connectivity than G, $(G', \mathcal{C}')$ is c-planar if and only if $(G, \mathcal{C})$ is c-planar, and there is a bijection f between $\mathcal{C}$ and $\mathcal{C}'$ such that for every cluster $X \in \mathcal{C}$, the graph $G[X]$ has the same number of components as the graph $G'[f(X)]$.*

Thanks to Lemma 3, a connected 2-component plane c-planarity instance $(G, \mathcal{C})$ can be polynomially transformed into an equivalent 2-connected 2-component instance $(G', \mathcal{C}')$. To achieve this, we simply perform repeatedly the transformation described in Lemma 3, until the resulting graph has only one 2-connected component.

Combining Lemma 2 and Lemma 3, we see that to decide the c-planarity of 2-component plane graphs, it is sufficient to provide an algorithm that decides c-planarity of 2-connected 2-component plane graph. This is an important technical simplification, because in a 2-connected plane graph, the boundary of every face is a cycle, and a candidate edge in every inner face is uniquely determined (up to isomorphism) by its end-vertices and the face where it should be drawn.

Unfortunately, if F is the outer face of G, a pair of vertices of F may still be connected by two non-isomorphic candidate edges belonging to F (see Fig. 1). To avoid this technical nuisance, we will restrict the set of candidate edges. Let $(G, \mathcal{C})$ be a 2-connected plane clustered graph, let $f \in E(G)$ be an edge which connects a pair of vertices $u, v \in V(G)$, with the following properties:

- f appears on the boundary of the outer face of G,
- every non-root cluster contains at most one of the two vertices u, v.

Such an edge f exists, otherwise the boundary of the outer face would be a hole of a non-root cluster. We say that a candidate edge e of G is *properly drawn* if f is on the boundary of the outer face of $G \cup e$. Note that every candidate edge in an inner face of G is properly drawn, while a pair of non-adjacent vertices on the boundary of the outer face may be connected by two non-isomorphic candidate

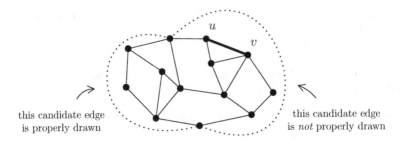

Fig. 1. Two candidate edges connecting the same pair of vertices in the outer face

edges, exactly one of which is properly drawn. Thus, a properly drawn candidate edge is uniquely determined (up to isomorphism) by its pair of endpoints and the face where it should be embedded.

It can be shown that if a 2-connected plane clustered graph is c-planar, then it has a saturator that only contains properly drawn candidate edges.

3   The Algorithm

In this section, we present our algorithm deciding the c-planarity of 2-component plane clustered graphs. As mentioned in the introduction, we will only deal with the restricted setting of *flat* clustered graph, i.e., the clustered graphs where all the non-root clusters are minimal.

Our aim is to find a polynomial algorithm deciding the c-planarity of plane 2-connected 2-component flat clustered graph $(G, \mathcal{C})$.

To achieve this, we will present a polynomial-time procedure FIND-EDGE which, when presented with a 2-component 2-connected hole-free plane clustered graph $(G, \mathcal{C})$ as an input, will either determine that $(G, \mathcal{C})$ is not c-planar, or it will output a harmless candidate edge e that saturates a cluster $X \in \mathcal{C}$. Observe that such a candidate edge e cannot create a hole in $G \cup e$, because both its endpoints belong to different components of X by assumption, and there is no other non-root cluster containing the endpoints of e. This is the main reason why the flat clustered graphs are much easier to deal with than general clustered graphs.

If the procedure FIND-EDGE outputs a harmless candidate edge e, it does not necessarily mean that $(G, \mathcal{C})$ is c-planar. However, since e is harmless, we know that $(G, \mathcal{C})$ is c-planar if and only if $(G \cup e, \mathcal{C})$ is c-planar. We may then call FIND-EDGE again on the input $(G \cup e, \mathcal{C})$, to determine that $(G \cup e, \mathcal{C})$ (and hence also $(G, \mathcal{C})$) is not c-planar, or to find another harmless edge. Since every candidate edge output by the FIND-EDGE procedure saturates a cluster from $\mathcal{C}$, after at most $|\mathcal{C}|$ invocations of FIND-EDGE we will either obtain a saturator of $(G, \mathcal{C})$ or determine that $(G, \mathcal{C})$ is not c-planar.

The FIND-EDGE algorithm maintains a set P of *permitted edges*. In the beginning, the set P is initialized to contain all the properly drawn candidate edges that saturate a cluster from $\mathcal{C}$. In the first phase of the algorithm, called *the pruning phase*, the algorithm iteratively removes some candidate edges from P, using a set of *pruning rules*, which will be described in Subsection 3.1. The pruning rules guarantee that if $(G, \mathcal{C})$ has a saturator, then it also has a saturator which is a subset of P.

When the set P cannot be further pruned, the algorithm performs the following *triviality checks*, described in detail in Subsection 3.2:

- if there a disconnected cluster that cannot be saturated by any of the permitted edges, then $(G, \mathcal{C})$ is not c-planar,
- if there is a disconnected cluster saturated by a unique permitted edge $e \in P$, then e is harmless,
- if there is a permitted edge e that does not cross any other permitted edge, then e is harmless.

If any of the above conditions is satisfied, the algorithm outputs the corresponding solution and stops. Otherwise, it distinguishes two cases:

1. If there is a disconnected cluster $X \in C$ and a face F of G such that every permitted edge saturating X appears in the face F, then the algorithm performs a subroutine LOCATE-IN-FACE, which will output a harmless permitted edge inside F and stop. This subroutine, together with a brief sketch of its proof, is presented in Subsection 3.3.
2. If the previous case does not apply, it can be shown that any permitted edge is harmless. The algorithm then performs a subroutine called OUTPUT-ANYTHING which outputs an arbitrary permitted edge and stops. The proof of its correctness is sketched in Subsection 3.4.

Before we describe the main parts of the algorithm in greater detail, we need some more terminology.

Let G be a 2-connected plane graph. Let a, b, c, d be a quadruple of distinct vertices on the boundary of a face F of G. We say that the pair ab *crosses* the pair cd in F, if the four vertices appear on the boundary of F in the cyclic order $acbd$. If e and f are two candidate edges of a 2-connected clustered graph $(G, \mathcal{C})$, we say that e *crosses* f if the two candidate edges belong to the same face F of G and the endpoints of e cross with the endpoints of f. For two sets of vertices X and Y, we say that X *crosses* Y *in face* F, if there are vertices $a, b \in X$ and $c, d \in Y$ such that ab crosses cd in the face F.

Most of our arguments rely on the following basic properties of connected subgraphs of 2-connected plane graphs:

- If G is a 2-connected plane graph, and X and Y are disjoint sets of vertices such that $G[X]$ and $G[Y]$ are both connected, then X and Y do not cross in any face of G.
- Let G be a 2-connected plane graph. Let X, Y and Z be disjoint sets of vertices, each of them inducing a connected subgraph of G. Then G has at most two faces that contain vertices of all the three sets on their boundary.

The proof of these properties are omitted from this extended abstract.

3.1   The Pruning Phase

In the pruning phase, the algorithm FIND-EDGE iteratively restricts the set P of permitted candidate edges. In the beginning of the pruning phase, the set P is initialized to contain all the properly drawn candidate edges that saturate at least one cluster. Note that every permitted edge $e \in P$ saturates a unique cluster $X \in \mathcal{C}$, since we assume that $\mathcal{C}$ is flat. A permitted edge that saturates X will be called an X-*edge*.

If X is a minimal cluster, and if e and e' are two X-edges, we say that e and e' are *equivalent*, if for every permitted edge $f \in P$ that is not an X-edge, the edge f crosses e if and only if it crosses e'.

Throughout the pruning phase, the set P will satisfy the following three invariants.

- For each cluster X and each face F, all the X-edges that belong to F form a vertex-disjoint union of complete bipartite subgraphs; these complete bipartite subgraphs will be called X-*bundles* (or just *bundles*, if X is clear from the context). Two X-edges from different bundles do not cross (see Fig. 2).
- If X and Y are distinct clusters, then if an X-edge e crosses two Y-edges f and f', then f and f' belong to the same bundle.
- If $(G, \mathcal{C})$ is c-planar, then it has a saturator that is a subset of P.

In the beginning, when P contains all the properly drawn candidate edges that saturate some cluster from $\mathcal{C}$, the three invariants above are satisfied. In fact, if F is a face that contains at least one X-edge, then all the X-edges in F form a complete bipartite graph. Thus, each face has at most one X-bundle.

To prune the set P, we apply the following two rules.

- If, for a cluster X, there is a permitted edge that crosses all the X-edges, then remove from P each edge that crosses all the X-edges.
- Let $e = uv$ and $e' = u'v$ be two X-edges that belong to the same face F and that share a common vertex v. If e and e' are equivalent, remove from P all the X-edges in F incident to u'.

It can be proven that an arbitrary application of one of the rules above preserves all the invariants. The algorithm applies the pruning rules in arbitrary order, reducing the number of permitted edges in each step, until it reaches the situation when none of the rules is applicable. Let us remark that in the general (i.e., non-flat) situation, the pruning is slightly more complicated: there are four pruning rules instead of two, and the rules have assigned priorities which are taken into account when the algorithm selects which rule to apply.

Fig. 2. A face F with two bundles of X-edges

3.2   Triviality Checks

When there is no rule applicable to the set P of permitted edges, the pruning phase ends. The FIND-EDGE algorithm then proceeds with three types of triviality checks, described below.

First, the algorithm checks whether there is a cluster X that is not saturated by any permitted edge. If this is the case, the algorithm concludes that the clustered graph (G, C) is not c-planar and stops. This is a correct conclusion, since if (G, C) were c-planar, then by the last invariant there would have to be a saturator made of permitted edges, which is clearly impossible.

As the next triviality check, the algorithm tries to find a cluster X, such that the set P contains a single X-edge e. If such a cluster X is found, the algorithm outputs e as a harmless edge and stops. This is again a correct output, since by the last invariant, if G is c-planar, then it has a saturator S which is a subset of P. Necessarily, S contains the edge e. This implies that e is harmless.

In the last type of triviality check, the algorithm looks for a permitted edge e that does not cross any permitted edge belonging to a different cluster. If such an edge e is found, the algorithm outputs e as a harmless edge and stops. This is again easily seen to be a correct output.

If none of the triviality checks succeeds, the algorithm counts, for each cluster X, the number of faces of G that contain at least one X-edge. We will say that a cluster X is *one-faced* if all the X-edges belong to a single face of G, X is *two-faced* if all the X-edges appear in the union of two distinct faces, and X is *many-faced* otherwise.

If there is a one-faced cluster X whose permitted edges belong to a face F, then the algorithm performs a subroutine LOCATE-IN-FACE to find a harmless permitted edge in F. This subroutine is described in the next subsection.

If there is no one-faced cluster, it can be shown that all the clusters are two-faced, and that any permitted edge is harmless. The algorithm then outputs an arbitrary permitted edge and stops. The main arguments involved in proving the correctness of this step are sketched in Subsection 3.4.

3.3   LOCATE-IN-FACE

Assume that we are given a set P of permitted edges satisfying all the invariants described in Subsection 3.1. Assume furthermore than none of the pruning rules is applicable to P, and none of the triviality checks has succeeded.

For a face F, we say that a cluster X is an F-*cluster*, if all the X-edges belong to F. We say that a vertex of X is *active*, if it is incident to at least one X-edge.

Assume that F is a face with at least one F-cluster. Using our assumptions about P, we are able to deduce the following facts:

- If X is an F-cluster, and Y is a cluster that has a permitted edge which crosses a permitted edge of X, then Y is also an F-cluster.
- If X is an F-cluster with two components X_1 and X_2, then each component X_i has at most two active vertices. It follows that X has either four permitted

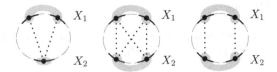

Fig. 3. Possible configurations of permitted edges of an F-cluster X

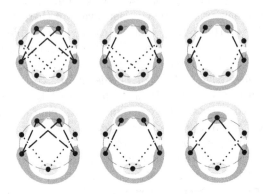

Fig. 4. Mutual positions of permitted edges of two crossing F-clusters

edges which all belong to a single bundle, or X has exactly two permitted edges (see Fig. 3; recall that due to the triviality checks, each cluster has at least two permitted edges).

Let X be an arbitrary F-cluster, let X_1 and X_2 be its two components. From the triviality checks, we know that every X-edge is crossed by a permitted edge of another cluster. Let $Y \neq X$ be a cluster whose permitted edge crosses an X-edge, and let Y_1 and Y_2 be its two components. Note that a set Y_i may not cross with the set X_j on the boundary of F, because these two sets induce connected subgraphs of G. Recall also, that no Y-edge may intersect all the X-edges (and vice versa), because it would have been pruned.

Putting all these facts together, we conclude that the mutual position of the X-edges and Y-edges corresponds to one of the situations depicted on Fig. 4.

Note that all the configurations of Fig. 4 exhibit a 'mirror symmetry'. To make this observation rigorous, we define a 'symmetry mapping' σ on the set of all the F-active vertices as follows: let X be an arbitrary F-cluster, with components X_1 and X_2. If a component X_i contains two active vertices x and x', then we define $\sigma(x) = x'$ and $\sigma(x') = x$. If X_i contains only one active vertex x, then we put $\sigma(x) = x$. We then extend the mapping σ to the set of X-edges in a natural way: for an X-edge e with endpoints x and y, we define $\sigma(e)$ to be the X-edge with endpoints $\sigma(x)$ and $\sigma(y)$.

The mapping σ has the following properties:

– For an F-cluster X and an X-edge e, $\sigma(e)$ is an X-edge different from e.

- If X and Y are F-clusters, an X-edge e crosses a Y-edge f if an only if $\sigma(e)$ crosses $\sigma(f)$.
- An X-edge e is harmless if and only if $\sigma(e)$ is harmless.

From these properties, it can be easily deduced that if an F-cluster X has only two permitted edges, then both these edges are harmless.

Furthermore, it is possible to show that if there is at least one F-cluster in a face F, then there is also an F-cluster that has only two permitted edges.

The procedure LOCATE-IN-FACE is then easy to describe: as an input, the procedure expects a face F for which there is at least one F-cluster. The procedure then finds an F-cluster X that has only two permitted edges, and outputs any X-edge as a harmless edge.

3.4 OUTPUT-ANYTHING

If, after the end of the pruning phase, each cluster has permitted edges in at least two distinct faces, and if none of the triviality checks is applicable, we can show that the set P of permitted edges has the following properties:

- For each cluster X, there are exactly two faces of G that contain the X-edges.
- All the X-edges that appear in the same face are equivalent.
- If X and Y are distinct clusters, and if an X-edge crosses a Y-edge, then all the X-edges and all the Y-edges appear in the same pair of faces, and every Y-edge crosses all the X-edges in its face.
- Let $S \subseteq P$ be a minimal saturator of permitted edges. For each edge $e \in S$ find an arbitrary permitted edge $\bar{e}$ that saturates the same cluster as e and appears in a different face than e. The set $\bar{S} = \{\bar{e} \colon e \in S\}$ is another minimal saturator of permitted edges.

From these properties, we may deduce that every permitted edge $e \in P$ is harmless. The procedure OUTPUT-ANYTHING simply outputs an arbitrary permitted edge and stops.

This completes the description of the simplified version of the FIND-EDGE algorithm. It is clear that the algorithm runs in polynomial time.

4 Concluding Remarks

We have shown that c-planarity of 2-component plane clustered graphs can be determined in polynomial time. This result raises several related open problems.

Problem 1. What is the complexity of the c-planarity problem for 2-component graphs $(G, \mathcal{C})$ if the embedding of G is not prescribed?

Problem 2. What is the complexity of deciding the c-planarity of clustered graphs with $O(1)$ components per cluster?

Problem 3. What if we relax the 2-component assumption by allowing the graph G to have arbitrarily many components, and only restricting the number of components of the non-root clusters?

References

1. Cortese, P.F., Di Battista, G., Patrignani, M., Pizzonia, M.: On embedding a cycle in a plane graph. In: Healy, P., Nikolov, N.S. (eds.) GD 2005. LNCS, vol. 3843, pp. 49–60. Springer, Heidelberg (2006)
2. Cortese, P.F., Di Battista, G., Patrignani, M., Pizzonia, M.: Clustering cycles into cycles of clusters. Journal of Graph Algorithms and Applications 9(3), 391–413 (2005); special issue In: Pach, J. (ed.) GD 2004. LNCS, vol. 3383, pp. 100–110. Springer, Heidelberg (2005)
3. Dahlhaus, E.: A linear time algorithm to recognize clustered planar graphs and its parallelization. In: Lucchesi, C.L., Moura, A.V. (eds.) LATIN 1998. LNCS, vol. 1380, pp. 239–248. Springer, Heidelberg (1998)
4. Di Battista, G., Frati, F.: Efficient C-planarity testing for embedded flat clustered graphs with small faces. In: Hong, S.-H., Nishizeki, T., Quan, W. (eds.) GD 2007. LNCS, vol. 4875, pp. 291–302. Springer, Heidelberg (2008)
5. Feng, Q.W., Cohen, R.F., Eades, P.: Planarity for clustered graphs. In: Spirakis, P.G. (ed.) ESA 1995. LNCS, vol. 979, pp. 213–226. Springer, Heidelberg (1995)
6. Goodrich, M.T., Lueker, G.S., Sun, J.Z.: C-planarity of extrovert clustered graphs. In: Healy, P., Nikolov, N.S. (eds.) GD 2005. LNCS, vol. 3843, pp. 211–222. Springer, Heidelberg (2006)
7. Gutwenger, C., Jünger, M., Leipert, S., Mutzel, P., Percan, M., Weiskircher, R.: Advances in c-planarity testing of clustered graphs. In: Goodrich, M.T., Kobourov, S.G. (eds.) GD 2002. LNCS, vol. 2528, pp. 220–235. Springer, Heidelberg (2002)
8. Gutwenger, C., Jünger, M., Leipert, S., Mutzel, P., Percan, M., Weiskircher, R.: Subgraph induced planar connectivity augmentation. In: Bodlaender, H.L. (ed.) WG 2003. LNCS, vol. 2880, pp. 261–272. Springer, Heidelberg (2003)
9. Jelínková, E., Kára, J., Kratochvíl, J., Pergel, M., Suchý, O., Vyskočil, T.: Clustered planarity: Small clusters in eulerian graphs. In: Hong, S.-H., Nishizeki, T., Quan, W. (eds.) GD 2007. LNCS, vol. 4875, pp. 303–314. Springer, Heidelberg (2008)

Visual Analysis of One-to-Many Matched Graphs*

Emilio Di Giacomo, Walter Didimo, Giuseppe Liotta, and Pietro Palladino

Dipartimento di Ingegneria Elettronica e dell'Informazione,
Università degli Studi di Perugia, Italy
{digiacomo,didimo,liotta,palladino}@diei.unipg.it

Abstract. Motivated by applications of social network analysis and of Web-search clustering engines, we describe an algorithm and a system for the display and the visual analysis of two graphs G_1 and G_2 such that each G_i is defined on a different data set with its own primary relationships and there are secondary relationships between the vertices of G_1 and those of G_2. Our main goal is to compute a drawing of G_1 and G_2 that makes clearly visible the relations between the two graphs by avoiding their crossings, and that also takes into account some other important aesthetic requirements like number of bends, area, and aspect ratio. Application examples and experiments on the system performances are also presented.

1 Introduction

The visual analysis of complex data sets is one of the most natural applications of graph drawing technologies (see, e.g., [2–4]). A typical application scenario consists of a set of data (nodes) and one or more relationships among these data (each relationship is a set of edges); therefore one is given one or more graphs on the same set of nodes. Both each graph must be visualized in a readable way and possible similarities among the different graphs must be easily detected by looking at the different drawings. This scenario has, for example, motivated a rich body of papers and systems about simultaneous graph embeddings and visualizations of evolving graphs (see, e.g., [7, 13–16]).

Recently, Collins and Carpendale [8] proposed a new research direction devoted to the visual comparison and analysis of heterogeneous data sets. The input consists of n sets of data $D_1, D_2, \ldots, D_n$, such that for each D_i a distinct set of *primary relationships* (i.e., a distinct graph) is defined; also, there are *secondary relationships* which model semantic connections between data belonging to different sets. The visualization consists of a set of n drawings (one for each graph) on top of which the edges that represent the secondary relationships are displayed. Collins and Carpendale present a system, called VISLINK, where each graph is drawn on a distinct plane and the secondary relationships are links between these planes (see Fig. 1(a) for a schematic illustration). The work by Collins and Carpendale extends a previous work by Schneiderman and Aris where multi-plane views with inter-plane edges are used to visualize different semantic substrates of a same graph [19](see Fig. 1(b) for an illustration).

---

* Research partially supported by the MIUR Project "MAINSTREAM: Algorithms for Massive Information Structures and Data Streams".

I.G. Tollis and M. Patrignani (Eds.): GD 2008, LNCS 5417, pp. 133–144, 2009.

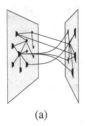 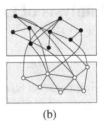

(a)   (b)

Fig. 1. Schematic illustrations of a visualization (a) adopted by VisLink, (b) using different semantic substrates of a same network. In both the visualizations the drawing on each plane has been computed without taking into account the relationships with the other. This may cause many crossings between inter-sets relationships.

Motivated by applications of social network analysis and of Web search clustering engines, we elaborate on the concepts by Collins and Carpendale by studying the following problem: We are given two graphs G_1 and G_2 and a function that defines a set of secondary relationships by mapping some of the vertices of G_1 to some other vertices of G_2; we aim at visually analyzing and interacting both with G_1, G_2 and with their secondary relationships. We observe that the systems described in [8, 19] follow the common approach of drawing each graph independently of each other. As a result, the secondary edges may be difficult to read as they can have many crossings. Our main goal is to design a system where the two drawings are computed by taking into account the edge-crossing minimization of the secondary edges. We focus on *one-to many relationships* between G_1 and G_2, i.e., vertices of G_1 are associated with disjoint subsets of vertices of G_2. The main contributions of the paper are the following:

- We introduce the concept of *one-to-many matched graphs* and define drawing conventions for these graphs in a *strong* and *non-strong* model. Both drawings require the secondary relationships between the graphs not to cross each other (Sect. 2).
- We describe a system that computes strong and non-strong one-to-many matched drawings of the input graphs by also taking into account the optimization of important aesthetic requirements. Furthermore, the system provides the user with several interaction functionalities that make it possible to analyze the drawings at different levels of details by collapsing/expanding clusters and by filtering information with the definition of node/edge thresholds (Sect. 3). Our drawing approach combines orthogonal drawings in the topology driven approach with circular drawing algorithms, and adopts an edge bundling technique to reduce the visual complexity introduced by some links.
- We show the effectiveness of the system by presenting application examples (Sect. 4), and an experimental study on the system performances (Sect. 5).

We finally remark that the problem of drawing two matched planar graphs G_1 and G_2 with one-to-one secondary relationships between them have been originally studied in [11], where it is required that the drawing of each G_i is planar and that the secondary edges are represented as non-intersecting horizontal segments.

2   One-to-Many Matched Graphs and Drawings

We assume familiarity with basic concepts of graph planarity and graph drawing [10]. If G is a graph, we denote by $\Gamma(G)$ a drawing of G. $\Gamma(G)$ is an *orthogonal drawing* if each edge is drawn as a chain of horizontal and vertical segments. A *bend* in $\Gamma(G)$ is a point of an edge shared by a horizontal and a vertical segment of the edge. A drawing $\Gamma(G)$ is a *circular drawing* if there is a circle passing through all vertices and each edge is drawn as a straight-line segment. In the following, if $G = (V, E)$ is a graph and $V' \subseteq V$ we denote by $G(V')$ the subgraph of G induced by the vertices of V'.

Let $G_1 = (V_1, E_1)$ and $G_2 = (V_2, E_2)$ be two distinct graphs. We say that $\langle G_1, G_2 \rangle$ is a pair of *one-to-many matched graphs* if: (i) Each vertex u of G_1 is associated with a subset $M(u) = \{v_1, v_2, \ldots, v_k\}$ of vertices of G_2, which we call the *cluster of u* in G_2; (ii) the set of clusters $\{M(u) \subseteq V_2 : u \in V_1\}$ is a partition of V_2, i.e., $\bigcup_{u \in V_1} M(u) = V_2$ and $\bigcap_{u \in V_1} M(u) = \emptyset$.

Let $\langle G_1, G_2 \rangle$ be a pair of one-to-many matched graphs, and let $\Gamma(G_1)$, $\Gamma(G_2)$ be drawings of G_1 and G_2, respectively. We say that $\langle \Gamma(G_1), \Gamma(G_2) \rangle$ is a *one-to-many matched drawing* if the following properties hold: (**P1**) The bounding boxes of $\Gamma(G_1)$ and $\Gamma(G_2)$ do not intersect. (**P2**) For each vertex u of G_1, cluster $M(u)$ in $\Gamma(G_2)$ is bounded by a rectangular region $R(u)$ such that: (i) $G(M(u))$ is completely contained in $R(u)$; (ii) each vertex $v \in V_2 \setminus M(u)$ is outside $R(u)$; (iii) each edge of G_2 intersects the boundary of $R(u)$ at most once. (**P3**) For each vertex u of G_1, there exists a simple curve $\ell(u)$ that connects the geometric shape p_u representing u in $\Gamma(G_1)$ to the boundary of $R(u)$ in $\Gamma(G_2)$, in such a way that $\bigcap_{u \in V_1} \ell(u) = \emptyset$.

In the paper, simple curves $\ell(u)$ are referred to as *matching connections*. Property (**P3**) guarantees that there is no intersection between distinct matching connections. A one-to-many matched drawing is said to be *strong* if the centers of the vertices of $\Gamma(G_1)$ have distinct y-coordinates and regions $R(u)$ are vertically ordered in $\Gamma(G_2)$ according to the positions of the corresponding vertices in $\Gamma(G_1)$. More formally, if $u_1, u_2 \in V_1$ and p_{u_1} is above p_{u_2} in $\Gamma(G_1)$, then $R(u_1)$ is completely above $R(u_2)$ in $\Gamma(G_2)$. In the paper, a one-to-many matched drawing that is not strong will be referred to as a *non-strong* one-to-many matched drawing. Figure 2 shows two examples of one-to-many matched drawings for the same pair of graphs. The one in Fig. 2(b) is a strong one-to-many matched drawing.

3   The System MOM

In this section we present a system for the display and the visual analysis of one-to-many matched drawings. We call our system MOM[1]. Let $\langle G_1, G_2 \rangle$ be a pair of one-to-many matched graphs to be visualized. MOM displays the drawing of G_1 to the left of the drawing of G_2, according to the following main criteria: (**C1**) It assumes that a drawing $\Gamma(G_1)$ is given as part of the input or that it can be computed using some classical graph drawing algorithm. (**C2**) It concentrates on the computation of $\Gamma(G_2)$, while trying to optimize a certain number of aesthetic criteria, other than guaranteeing that $\langle \Gamma(G_1), \Gamma(G_2) \rangle$ is a one-to-many matched drawing. (**C3**) Once $\Gamma(G_2)$ has

---

[1] MOM stands for **M**atched **O**ne-to-**M**any graphs.

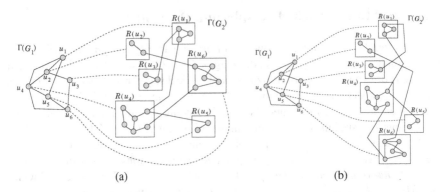

Fig. 2. (a) A (non-strong) one-to-many matched drawing of a pair of matched graphs. (b) A strong one-to-many matched drawing for the same pair of graphs.

been computed, it draws the matching connections and provides the user with a set of interaction functionalities for the visual analysis of the resulting drawing.

Criterion (**C1**) is motivated by several application scenarios that we had in mind during the design of the system. In these applications G_1 is often a graph whose entities represent geographic locations and therefore their position is either fixed or strongly constrained (examples are given in Sect. 4). About (**C2**), we focus on well recognized aesthetic criteria like number of crossings, number of bends, drawing area. Since the optimization of these criteria typically leads to an NP-hard problem, we propose some heuristics based on engineered versions of popular graph drawing algorithms, which are able to deal with the constraints of a one-to-many matched drawing. As an additional aesthetic criterion we require that $\langle \Gamma(G_1), \Gamma(G_2) \rangle$ is computed in such a way that the matching connections can be always drawn without intersecting the edges of G_2. When G_2 is a dense graph, $\Gamma(G_2)$ may have a high visual complexity, which makes it difficult to read the drawing at a whole, independently of the applied drawing strategy. This is the motivation for (**C3**).

3.1 Drawing Algorithm

Our drawing strategy for $\Gamma(G_2)$ combines different drawing conventions. We use orthogonal drawings for the layout of the rectangular regions $R(u)$ and their connections. Circular drawings are used to represent $G(M(u))$ inside $R(u)$. Finally, in order to simplify the visual complexity, we adopt a bundling operation for the edges connecting a vertex inside a region $R(u)$ to vertices outside $R(u)$; to avoid ambiguity, we use a "confluent-like" representation for these edges, as explained later. The algorithms used for the different drawing conventions have been engineered in order to deal with a certain number of constraints. In the following we describe in detail the steps performed by our drawing algorithm. We denote by V_i and E_i the set of vertices and edges of G_i, respectively ($i \in \{1, 2\}$).

Step 1: Planarization. The goal of this step is to compute a suitable planar embedding of the graph consisting of "cluster vertices" and their interconnections, possibly replacing edge crossings with dummy vertices. More precisely, let $u_1, u_2, \ldots, u_n$ be the vertices

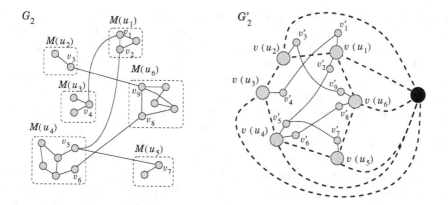

Fig. 3. (a) A graph G_2. (b) The graph G'_2 used in Step 1 plus the wheel gadget (black node and dashed bold edges) adopted to guarantee (**E1**); the wheel gadget is removed at the end of Step 1.

of G_1 in the top-to-bottom order[2] they appear in $\Gamma(G_1)$, and let G'_2 be the graph obtained from G_2 by collapsing each cluster $M(u_i)$ into a single vertex $v(u_i)$ $(1 \leq i \leq n)$, called a *cluster vertex*. In G'_2 edges connecting vertices in the same cluster $M(u)$ disappear, while an edge connecting a vertex in $M(u_i)$ to a vertex in $M(u_j)$ $(i \neq j)$ is transformed to a *corresponding* edge between $v(u_i)$ and $v(u_j)$. We aim at computing a planar embedding Ψ of G'_2 that satisfies the following two conditions: (**E1**) Cluster vertices $v(u_1), v(u_2), \ldots, v(u_n)$ appear counterclockwise in this order on the external face of Ψ; (**E2**) If $v \in M(u_i)$ in G_2 and if $e_1, \ldots, e_k$ are edges of G_2 incident to v, then the edges corresponding to $e_1, \ldots, e_k$ in G'_2 appear consecutively (not necessarily in this order) around $v(u_i)$ in Ψ. Condition (**E1**) will guarantee Property (**P3**), i.e., the possibility of routing the matching connections without crossings among them; it also avoids crossings between matching edges and the edges of G_2. Condition (**E2**) makes it possible to simplify the links between the outside and the inside of each region $R(u_i)$ in the final drawing and to bundle these links as it will be explained in Step 3. To force (**E2**) we further transform G'_2 by attaching to $v(u_i)$ a vertex v' for each vertex $v \in M(u_i)$ connected to vertices outside $M(u_i)$, and by replacing the edges $e_1, \ldots, e_k$ that are incident to v with corresponding edges $e'_1, \ldots, e'_k$ connected to v'. Vertex v' is called the *image* of v.

On G'_2 we apply a standard planarization algorithm based on first extracting a maximal planar subgraph and then on iteratively reinserting the discarded edges by computing shortest paths in the dual graph and by replacing edge crossings with dummy vertices [10]. To force (**E1**), we use a "wheel gadget" of uncrossable edges that will be removed at the end of the planarization phase. Figure 3 shows an example of a graph G'_2 and the wheel gadget used to guarantee (**E1**).

Notice that, quadratic and linear-time algorithms for planarity testing and edge reinsertion within the above described embedding constraints have been also proposed in [1, 17]. Our planarization phase takes $O(|E_2|(c + |V_2|) \log(c + |V_2|))$ time, where c is the number of edge crossings in the final embedding of G'_2.

---

[2] If u_i and u_j have the same y-coordinate, they are ordered from right to left.

Step 2: Orthogonalization and Compaction. Once a planar embedding Ψ of G_2' (with possible cross vertices) has been found, an orthogonal drawing of G_2' that preserves Ψ is computed. The basic idea is to use an orthogonal drawing algorithm that deals with arbitrary vertex degree and that allows for vertex size customization. Indeed, we want that $v(u_i)$ is drawn as a box big enough to host all vertices of $M(u_i)$. To this aim, the system uses the network flow based drawing algorithm described in [9], which represents a good heuristic both in terms of bend minimization and in terms of area drawing compaction. Denoted by $B(v(u_i))$ the box representing vertex $v(u_i)$, we draw $B(v(u_i))$ as a square of a certain size r_i. In the final drawing we place a circle of radius ρ_i inside $B(v(u_i))$ and equi-distribute along its perimeter the vertices of $M(u_i)$. To determine ρ_i, we fix a minimum distance δ we want to guarantee between any two vertices of $M(u_i)$ and we set $\rho_i = \delta \cdot |M(u_i)|/2\pi$. We choose r_i to be larger enough than ρ_i so that it is possible to route the edges connecting vertices inside $B(v(u_i))$ with the outside. Each square $B(v(u_i))$ will correspond to region $R(u_i)$ in the final drawing. Also, in order to guarantee the properties of a one-to-many matched drawing, we add a certain number of constraints as described below.

If one wants to compute a strong one-to-many matched drawing, then all vertices $v(u_1), v(u_2), \ldots, v(u_n)$ are temporarily connected in this order to form a simple cycle C that becomes the new boundary for the external face. Then the following angle and bend constraints on the vertices and edges of C are imposed: Each edge of C connecting $v(u_i)$ to $v(u_{i+1})$ $(1 \leq i \leq n - 1)$ is constrained to be straight-line in the drawing, while the edge of C connecting $v(u_n)$ to $v(u_1)$ is constrained to turn always in the left direction while moving from $v(u_n)$ to $v(u_1)$. Each angle formed at a vertex $v(u_i)$ on the external face is set to be of 180 degrees. These constraints guarantee that $v(u_1), v(u_2), \ldots, v(u_n)$ are encountered from top-to-bottom in the final drawing and that they are all visible from left. Once a drawing has been computed the edges of C are removed. If one wants to compute a (not necessarily strong) one-to-many matched drawing, then we still construct cycle C, but we only impose the constraint that the edges of C turn in the left direction or go straight while moving along C counter-clockwise. Finally, in order to correctly perform the next step (i.e., the edge bundling operation), we also require that for each image vertex v' attached to a vertex $v(u_i)$, there is no other edge incident to v' from the same direction of edge $(v(u_i), v')$.

All the orthogonalization constraints described above are translated into constraints on the flow network of the algorithm in [9]. The orthogonalization and compaction phases take $O((|V_1||V_2| + c)^2 \log(|V_1||V_2| + c))$ time, where c is still the number of cross vertices in the embedding Ψ.

Step 3: Edge Bundling. This step removes each image vertex v' and creates in its place a "confluent-like" structure for the edges incident to v'. Namely, let v be the vertex of the original graph that has v' as its image and let $M(u_i)$ be the cluster that contains v. Let $e_1', \ldots, e_k'$ be the edges incident to v' other than edge $(v', v(u_i))$. We want that v' is no longer present in the final drawing and that the edges $e_1', \ldots, e_k'$ are replaced by the edges $e_1, e_2, \ldots, e_k$ that were originally connected to v. To simplify the final drawing however, we bundle the edges $e_1, e_2, \ldots, e_k$ from v to v'; this edge bundle follows the drawing of e from the boundary of $R(u_i)$ to v' and then it divides in k branches at v' using splines, as shown in Fig. 4(a). It is important to remark that the

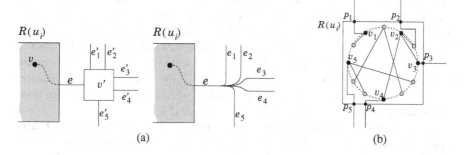

(a)   (b)

Fig. 4. (a) Illustration of Step 3. The image vertex v' is removed and its incident edges are replaced by a "confluent-like" structure. The dashed curve is the part of edge bundle that will be drawn in Step 4. (b) Illustration of Step 4. The black vertices inside $R(u_i)$ denote the vertices whose relative circular ordering is fixed according to their corresponding external connections.

edge bundling operation guarantees that for each vertex v inside a region $R(u_i)$ there will be at most one link (a bundle of edges) incident to v from the outside of $R(u_i)$. Since these links must be routed around the circular drawing representing $G(M(u_i))$, this property strongly simplifies the visual complexity introduced by these connections. The edge bundling step takes $O(|E_2|)$ time.

Step 4: Circular Drawing Computation. At the end of the previous step, we have a partial drawing of G_2 such that for each cluster vertex $v(u_i)$ there is a corresponding rectangular region $R(u_i)$ and some edges incident to the boundary of $R(u_i)$ at certain points $p_1, p_2, \ldots, p_k$. To complete the drawing of G_2 we construct a circular drawing for each $G(M(u_i))$, and then connect p_j to its corresponding vertex v_j of $M(u_i)$ ($1 \leq j \leq k$). See Fig. 4(b) for an illustration. In order to avoid crossings between links (p_j, v_j), we force the circular order of vertices v_j to be consistent with the circular order of points $p_1, p_2, \ldots, p_k$ around $R(u_i)$, i.e., if $p_1, p_2, \ldots, p_k$ occur clockwise in this order around $R(u_i)$ then we force $v_1, v_2, \ldots, v_k$ to occur clockwise in this order in the circular drawing. Conversely, all vertices of $M(u_i)$ distinct from v_j ($1 \leq j \leq k$) can be placed everywhere in the circular ordering (these vertices are not connected to vertices outside $R(u_i)$). In other words, if $V_{fix} = \{v_1, v_2, \ldots, v_k\}$ and $V_{free} = M(u_i) \setminus V_{fix}$, we want to find a "good" circular order for the vertices of $M(u_i)$ such that the relative order of the vertices of V_{fix} is fixed; our goal is the minimization of the number of edge crossings, which is however an NP-Hard problem [18]. To solve it, we designed a variation of the heuristic described by Baur and Brandes [5], which has been experimentally shown to produce better results in terms of crossing reduction than previous heuristics for computing circular drawings, and that has been successfully adopted for the layout of two-level networks that are similar to the clustered structure of G_2 [6]. We also recall that faster but less effective circular drawing algorithms in terms of edge crossings have been described in [20]. The heuristic by Baur and Brandes computes an ordering of the vertices on a straight line ℓ, assuming that all edges are drawn on the same half-plane determined by ℓ. In terms of edge crossings this model is equivalent to place the vertices on a circle and to draw the edges as straight-line segments. At the end of this placement greedy heuristic, a post-processing step, called *circular sifting* is applied to further reduce the number of edge crossings if possible. The idea is to iteratively swapping a vertex with its successor vertex in the linear order on ℓ and recording the change in crossing count; the vertex is then placed in the position that corresponds to its local optimal. Denoted by n and m the number of vertices and the number of edges of the input graph, respectively, the placement greedy heuristic can be performed in $O((n+m)\log n)$ time, while repositioning each vertex once in the circular sifting phase can be done in $O(nm)$ time (see [5]).

Our variation of the algorithm in [5] works as follows. The placement greedy heuristic performs analogously to the one of Baur and Brandes, but it assumes that the vertices of V_{fix} are already placed on ℓ in a preassigned order; therefore the placement decisions are restricted to the vertices of V_{free}. The circular sifting phase is modified so that swaps between vertices both belonging to V_{fix} are not allowed. Once the circular ordering of the vertices of $M(u_i)$ has been computed, the algorithm equi-distributes these vertices on a circle inside $R(u_i)$ and rotates this circle in order to reduce the total length of the connections (p_j, v_j) $(1 \leq j \leq k)$, which are routed as polygonal chains of vertical and horizontal segments. The circular drawing computation over all cluster vertices takes $O(|V_1|((|V_2| + |E_2|)\log|V_2| + |V_2||E_2|))$ time (recall that $|V_1|$ corresponds to the number of cluster vertices).

Step 5: Drawing of Matching Edges. This step is simply performed by routing the matching edges as polygonal chains from the location of a vertex u_i of $\Gamma(G_1)$ to the boundary of the corresponding region $R(u_i)$ in $\Gamma(G_2)$. Since the circular ordering of the regions on the external face of $\Gamma(G_2)$ is consistent with the top-down ordering of the corresponding vertices in $\Gamma(G_1)$, this can be done without crossing between matching edges. Also, in a strong one-to-many matched drawing, each matching edge can be routed with at most two bends.

Time Complexity. The next theorem summarizes the discussion about the drawing algorithm implemented in MOM. To simplify the time complexity of this algorithm, the statement of the theorem assumes that $|V_1|$ is bounded by a constant. This appears as a reasonable assumption if $|V_1| \ll |V_2|$.

Theorem 1. *Let $\langle G_1, G_2 \rangle$ be a pair of one-to-many matched graphs such that $G_1 = (V_1, E_1)$ and $G_2 = (V_2, E_2)$. Let $\Gamma(G_1)$ be any drawing of G_1. There exists a polynomial-time algorithm that computes a one-to-many matched drawing $\langle \Gamma(G_1), \Gamma(G_2) \rangle$ (either in the strong or in the non-strong model) with the additional property that the matching edges can be drawn without intersecting any vertex and edge of $\Gamma(G_2)$. Also, if $|V_1|$ is bounded by a constant, and denoted by N the number $N = |V_2| + c$, where c is the number of inter-cluster edge crossings in $\Gamma(G_2)$, then the time complexity of the drawing algorithm is: $O((|E_2|N + N^2)\log N)$.*

3.2 Interaction Functionalities

In order to facilitate the visual analysis of the computed one-to-many matched drawings, we equipped our system with a certain number of interaction functionalities, other than conventional zooming and translation primitives. We briefly describe them in the following.

Cluster Expansion/Contraction: By default, all cluster regions $R(u)$ in $\Gamma(G_2)$ are expanded, i.e., the whole subgraph inside each $R(u)$ is displayed by the system. In order to compact the drawing and/or to hide some details, the user can decide to contract a certain number of clusters by simply clicking on them. A cluster contraction redraws the cluster as a small box and hides its content. Every cluster can be expanded or contracted an infinite number of times without any restriction. After a cluster expansion/contraction, the drawing is automatically re-compacted by the system, but the orthogonal shape of the drawing remains unchanged, so to avoid that the user mental map is lost. Contracting clusters can be useful to get an overview of the inter-cluster relations before analyzing the intra-cluster ones.

Cluster Filtering: If the user is interested in focusing on some of the clusters, she can select them and hide the remaining clusters and their connections. After such an operation, the user can also decide to re-compact the remaining part of the drawing to save space if possible. When the drawing of $\Gamma(G_2)$ has many clusters and/or many inter-cluster links, the cluster filtering primitive can help to explore the graph structure portion by portion.

Edge Filtering: Our system allows the representation of edge weighted graphs. This means that a weight can be assigned to each edge of G_1 and of G_2. When a graph is too dense, the user can sparsify the links by setting an edge visibility threshold. All links having the weight below the given threshold are not shown by the system. Again, the drawing is re-compacted if required.

Edge/Vertex Highlighting: Moving the mouse over a certain vertex or cluster region, the user can decide to highlight all edges incident to that vertex or to that cluster region. A tooltip with information about the selected vertex is also displayed. This helps to get local information on the drawing. Furthermore, moving the mouse over an edge, a tooltip that displays the labels of its end-vertices is shown. This helps when just a portion of the selected edge fits in the current view.

4 Application Examples

One-To-Many matched graphs occur in several applications contexts. Here we briefly present an example on social network analysis. Another application example on Web search clustering engines is described in [12].

Our example focuses on the co-authorship network of the last Symposium on Graph Drawing, GD 2007. G_1 is the graph having European countries as vertices and edges between countries that cooperated in co-authoring some papers. Each edge has a weight equal to the number of papers resulting from the cooperation of the connected countries. The drawing $\Gamma(G_1)$ is a simple straight-line drawing, where each vertex is placed at a fixed location on a geographic map. Graph G_2 represents authors and their cooperations in the articles. Figure 5 shows a one-to-many matched drawing in the strong model. The drawing gives an overview of the network structure, which reveals the number of contributing authors for each country and a relevant level of cooperation among the different countries. Looking inside a country, it is possible to see its different sub-communities. For example, it is easy to recognize two sub-communities in Greece, in Italy, and in Czech Republic, several communities in Germany, and one big community

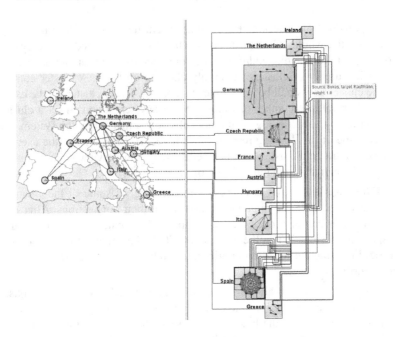

Fig. 5. A one-to-many matched drawing showing the European co-authorship network of GD 2007

in Spain. Selecting an author in a country, all her connections with other authors are highlighted by the system. In the figure, author "Kaufmann" inside Germany is selected, and the system highlights (in bold red color) his connections with other authors, three in Greece and one in Italy. Moving the mouse over one of the bold red edges, it is displayed a tooltip that reports the labels of its end-vertices. Figure 6 shows an example of edge and vertex filtering on the previous drawing, which makes it easier to focus on specific relationships. Namely, the edges of $\Gamma(G_1)$ has been filtered so that only those edges with a weight greater than 1 are shown. The vertices of $\Gamma(G_2)$ have been filtered in such a way that only the countries having some incident links in $\Gamma(G_1)$ are shown (i.e., Germany, Italy, and The Netherlands). Then, cluster Germany has been contracted to focus on the interplay between Italy and The Netherlands. After the vertex filtering and contraction operations, $\Gamma(G_2)$ is recomputed so to become more compact without destroying the user's mental map. In the figure, the connections of author "Meijer" are highlighted in bold red.

5   System Performances

We have tested our system in order to measure its performances. Our main goal was to measure the running time and some important aesthetic requirements, like number of crossings, number of bends, drawing area, and aspect ratio (width/height). We compared the algorithm for strong one-to-many matched drawings against the algorithm for non-strong one-to-many matched drawings, so to understand the trade-off between the

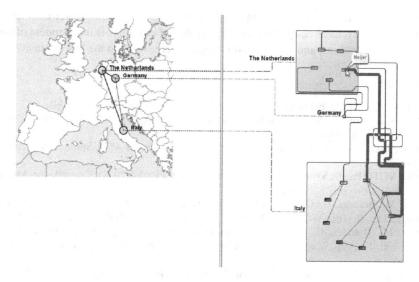

Fig. 6. The same one-to-many matched graphs of Fig. 5 after some edge and vertex filtering

results of the two algorithms. A strong drawing greatly helps in the readability of the matching between G_1 and G_2, but we expect that a strong drawing has worst values for some aesthetics (e.g., aspect ratio and number of bends) than for a non-strong drawing.

The focus is on the drawing of G_2, because we are assuming that a drawing $\Gamma(G_1)$ is given as part of the input or that it is computed with some classical drawing algorithm. For the experiments we used a test suite of instances for G_2, with given number of cluster vertices. We generated 240 graphs in total, 5 graphs for each *sample*. A sample is obtained by fixing number of vertices, number of clusters, and density (number of edges/number of vertices). The number of vertices is a value in the set $\{100, 400, 700, 1000\}$, the number of clusters is a value in $\{5, 10, 15, 20\}$, and the density is a value in $\{1, 1.5, 2\}$. Each graph was generated at random, by assuming that 10% of the edges are inter-cluster edges and that 90% of the edges are intra-cluster edges. The experiments have been executed under the Windows 2003 server OS, on an Intel Pentium IV with 3.0GHz and 2GB of RAM.

The charts of the experimental results are omitted for reasons of space and can be found in [12]. As for the running time, the computation of strong drawings is slightly slower than for non-strong drawings (in the average, it requires about 10% more). In general, both types of computations take a few seconds for graphs up to 400 vertices and low density values. Graphs with the highest density and 700 vertices are computed in a few minutes, while the computations may require up to 30 minutes for the hardest instances of our test suite, i.e., graphs with 1000 vertices and density 2. About the area and the aspect ratio, since in a strong one-to-many matched drawing every two cluster regions are constrained to stay one below the other, strong drawings have a worst aspect ratio but smaller area than non-strong drawings, which have aspect ratio close to 1. About the number of bends, strong drawings present in the average $11 - 12\%$

of bends more than non-strong drawings, which are caused by their greater number of constraints. Finally, as already observed, the number of crossings is independent of the two drawing algorithms, and as expected it rapidly increases with the graph density.

References

1. GDToolkit: Graph drawing toolkit, http://www.dia.uniroma3.it/~gdt/
2. IEEE Symposium on Information Visualization (InfoVis 2007), October 28-30. IEEE Computer Society, Sacramento (2007)
3. Hong, S.-H., Nishizeki, T., Quan, W. (eds.): GD 2007. LNCS, vol. 4875. Springer, Heidelberg (2008)
4. IEEE VGTC Pacific Visualization Symposium 2008 (PacificVis 2008), Kyoto, Japan, March 4-7. IEEE Computer Society, Los Alamitos (2008)
5. Baur, M., Brandes, U.: Crossing reduction in circular layouts. In: Hromkovič, J., Nagl, M., Westfechtel, B. (eds.) WG 2004. LNCS, vol. 3353, pp. 332–343. Springer, Heidelberg (2004)
6. Baur, M., Brandes, U.: Multi-circular layout of micro/Macro graphs. In: Hong, S.-H., Nishizeki, T., Quan, W. (eds.) GD 2007. LNCS, vol. 4875, pp. 255–267. Springer, Heidelberg (2008)
7. Braß, P., Cenek, E., Duncan, C.A., Efrat, A., Erten, C., Ismailescu, D., Kobourov, S.G., Lubiw, A., Mitchell, J.S.B.: On simultaneous planar graph embeddings. Comput. Geom. Theory and Appl. 36(2), 117–130 (2007)
8. Collins, C., Carpendale, M.S.T.: Vislink: Revealing relationships amongst visualizations. IEEE Trans. Vis. Comput. Graph. 13(6), 1192–1199 (2007)
9. Di Battista, G., Didimo, W., Patrignani, M., Pizzonia, M.: Orthogonal and quasi-upward drawings with vertices of prescribed size. In: Kratochvíl, J. (ed.) GD 1999. LNCS, vol. 1731, pp. 297–310. Springer, Heidelberg (1999)
10. Di Battista, G., Eades, P., Tamassia, R., Tollis, I.G.: Graph Drawing. Prentice-Hall, Englewood Cliffs (1999)
11. Di Giacomo, E., Didimo, W., van Kreveld, M., Liotta, G., Speckmann, B.: Matched drawings of planar graphs. In: Hong, S.-H., Nishizeki, T., Quan, W. (eds.) GD 2007. LNCS, vol. 4875, pp. 183–194. Springer, Heidelberg (2008)
12. Di Giacomo, E., Didimo, W., Palladino, P., Liotta, G.: Visual analysis of one-to-many matched graphs. Tech. rep, RT-003-08, DIEI - Università di Perugia, Italy (2008)
13. Di Giacomo, E., Liotta, G.: Simultaneous embedding of outerplanar graphs, paths, and cycles. Intern. Journ. of Comput. Geom. and Appl. 17(2), 139–160 (2007)
14. Erten, C., Harding, P.J., Kobourov, S.G., Wampler, K., Yee, G.V.: Graphael: Graph animations with evolving layouts. In: Liotta, G. (ed.) GD 2003. LNCS, vol. 2912, pp. 98–110. Springer, Heidelberg (2004)
15. Erten, C., Kobourov, S.G.: Simultaneous embedding of a planar graph and its dual on the grid. Theory Comput. Syst. 38(3), 313–327 (2005)
16. Erten, C., Kobourov, S.G., Le, V., Navabi, A.: Simultaneous graph drawing: Layout algorithms and visualization schemes. Journ. Graph Alg. and Appl. 9(1), 165–182 (2005)
17. Gutwenger, C., Klein, K., Mutzel, P.: Planarity testing and optimal edge insertion with embedding constraints. Journ. of Graph Alg. and Appl. 12(1), 73–95 (2008)
18. Masuda, S., Kashiwabara, T., Nakajima, K., Fujisawa, T.: On the NP-completeness of a computer network layout problem. In: 20th IEEE Int. Symposium on Circuits and Systems, pp. 292–295 (1987)
19. Shneiderman, B., Aris, A.: Network visualization by semantic substrates. IEEE Trans. Vis. Comput. Graph. 12(5), 733–740 (2006)
20. Six, J.M., Tollis, I.G.: A framework and algorithms for circular drawings of graphs. Journ. of Discr. Alg. 4(1), 25–50 (2006)

Topological Morphing of Planar Graphs*

Patrizio Angelini, Pier Francesco Cortese,
Giuseppe Di Battista, and Maurizio Patrignani

Università Roma Tre
{angelini,cortese,gdb,patrigna}@dia.uniroma3.it

Abstract. In this paper we study how two planar embeddings of the same biconnected graph can be morphed one into the other while minimizing the number of elementary changes.

1 Introduction

A useful feature of a graph drawing editor is the possibility of selecting a certain face of the drawing and of promoting it to be the external face (see, e.g., [9]). In order to preserve the mental map, the user would like that the editor executed such an operation by performing a few changes to the drawing.

The above operation is just an example of a topological feature that would be useful to have at disposal from an editor. More generally, it would be interesting to have an editor allowing the user to look at a drawing and to specify in some way, e.g. pointing at vertices or edges, a new embedding. Such an embedding could be even requested at a more abstract level, asking the editor to go to one with minimum depth, or with minimum radius, etc. Again, the editor should transform the current embedding into the new one smoothly, i.e. with the minimum number of changes.

A similar problem occurs when, keeping the topology unchanged, an editor has to geometrically morph a drawing into another one, specified in some way from the user. In this case the operations that the editor can perform are topology-preserving translations and scaling of objects. The user would like to see a geometric morphing with the minimum number of intermediate snapshots.

The existence of a geometric morphing between two drawings was addressed surprisingly long ago. Cairns proved in 1944 that between any two straight-line drawings of a triangulated planar graph there exists a morph in which any intermediate drawing is straight-line planar [7]. This was extended to general planar graphs by Thomassen in 1983 [19]. The first algorithms to find such morphings were proposed by Floater and Gotsman for triangulations [12] and by Gotsman and Surazhsky for general plane graphs [13]. While the search for a geometric morph between two given drawings of a planar graph with a polynomial number of steps and with a bounded size of the needed grid is still open, some recent studies address the problem for the special cases of orthogonal drawings [15,6] and arbitrary plane drawings [11].

---

* Work partially supported by MUR under Project "MAINSTREAM: Algoritmi per strutture informative di grandi dimensioni e data streams."

I.G. Tollis and M. Patrignani (Eds.): GD 2008, LNCS 5417, pp. 145–156, 2009.

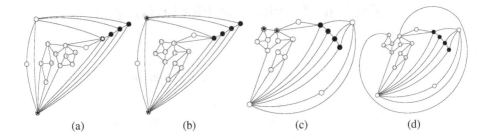

(a)	(b)	(c)	(d)

Fig. 1. A sequence of flips and skips transforming an embedding

We study the morphing between two drawings from the topological perspective and we call it *topological morphing*. There are many ways to state the problem, ranging from the family of graphs, to the operations that an editor can perform, their completeness, their ability to capture changes that are "natural" for the user, and to the metrics that distinguish a good from a bad morphing. This work starts from the following basic hypotheses. (i) We consider biconnected planar graphs, since such graphs are the building block of several graph drawing methodologies. (ii) We consider operations that move in one step entire blocks of the drawing, that are identified by some connectivity features. Namely, using a term that is common in planarity testing literature, we call *flip* the operation that allows to "flip" a component around its separation pair. Also, borrowing the term from the common rope skipping game played by children, we call *skip* the operation that allows to move the external face by "skipping" an entire component without modifying the combinatorial embedding. (iii) The metric is the number of performed operations. Namely, we have that a topological morphing is "good" if the editor performs it with a few flips and skips. Intuitively, the fewer operations are performed, the better the user preserves the mental map.

As an example, suppose that the graph is embedded as shown in Fig. 1.a and that the user would like to obtain the embedding in Fig. 1.d. A minimum sequence of operations that leads to Fig. 1.d consists of flipping the component separated by the starred vertices of Fig. 1.a, then by skipping the component separated by the starred vertices of Fig. 1.b, and finally by skipping the edge separated by the starred vertices of Fig. 1.c.

We present the following results. Let G be a biconnected planar graph and denote by $\langle \Gamma, f \rangle$ one of its combinatorial embeddings Γ with f as external face. Suppose that pair $\langle \Gamma_1, f_1 \rangle$ is the current topology and that $\langle \Gamma_2, f_2 \rangle$ represents a target topology chosen by the user. (1) In Sect. 2 we show that if both flips and skips are allowed the general problem of morphing $\langle \Gamma_1, f_1 \rangle$ into $\langle \Gamma_2, f_2 \rangle$ with the minimum number of flips and skips is NP-complete. Motivated by such a result we tackle several more restricted problems. (2) Suppose that $\Gamma_1 = \Gamma_2$ and that only skips are allowed. In Sect. 3 we give a linear time algorithm to move the external face from f_1 to f_2 with the minimum number of skips. (3) In Sect. 4 we show that the topological morphing problem can be efficiently solved if G does not have parallel triconnected components. (4) In Sect. 5 we show that the problem is fixed-parameter tractable. Basic definitions are in Sect. 2 while concluding remarks are in Sect. 6.

2 Basic Concepts

In this section we define the flip and skip operations and their properties. The proofs of the lemmas and theorems can be found in [1].

We assume familiarity with planarity and connectivity of graphs. A *planar drawing* of a graph is a mapping of its vertices to distinct points of the plane and of its edges to non-intersecting open Jordan curves between their end-points. A graph is *planar* if it has a planar drawing. A planar drawing partitions the plane into *faces* (topologically connected regions). The unbounded face is the *external face*. Two planar drawings of a graph G are *equivalent* if they determine the same circular ordering of the edges around each vertex. An equivalence class of planar drawings is a *combinatorial embedding* of G. A *planar embedding* is a pair $\langle \Gamma, f \rangle$, where Γ is a combinatorial embedding and f is the external face.

The *SPQR-tree* $\mathcal{T}$ of a biconnected graph G describes the arrangement of its triconnected components. We assume familiarity with SPQR-trees. For details see [10].

Let G be a biconnected planar graph and let $\mathcal{T}$ be the SPQR-tree of G. A planar embedding $\langle \Gamma, f \rangle$ of G can be represented by a labeling of $\mathcal{T}$. Namely, Γ is described by the combinatorial embedding of the skeleton of each node in $\mathcal{T}$, which can be succinctly represented by labeling each R-node with a Boolean value and each P-node with a circular ordering of its adjacent nodes, as described in [5].

In order to account for the external face f in the SPQR-tree $\mathcal{T}$, we introduce the following definitions. A node μ of $\mathcal{T}$ is an *allocation node* of a face f of Γ either if μ is a Q-node incident to f or if there exist no virtual edge e of $skel(\mu)$ such that $pertinent(e)$ contains all the edges of f. Observe that if μ is an allocation node of f, then there is exactly one face f_μ in $skel(\mu)$ such that all the pertinent graphs of its virtual edges contain at least one edge of f. Face f_μ is the *representative* of f in $skel(\mu)$. In the following, we will denote by f both a face of Γ and its representative face in the skeleton of one of its allocation nodes. We say that f *belongs* to all its allocation nodes. The set of all the allocation nodes of f are a subtree of $\mathcal{T}$, called the *allocation tree* of f. Figure 2.a shows examples of allocation trees.

Property 1. The allocation tree of a face f is the subtree of $\mathcal{T}$ whose leaves are the Q-nodes corresponding to the edges of f.

The external face of Γ can be provided by specifying its allocation tree in $\mathcal{T}$. The following lemma shows how adjacent nodes in $\mathcal{T}$ share exactly two faces.

Lemma 1. *Let μ_1 and μ_2 be two adjacent nodes of an SPQR-tree $\mathcal{T}$. There are exactly two faces f' and f'' of Γ that belong both to μ_1 and to μ_2. In $skel(\mu_1)$ ($skel(\mu_2)$) f' and f'' share edge $e(\mu_2)$ ($e(\mu_1)$). If μ_1 (μ_2) is not an S-node, then $e(\mu_2)$ ($e(\mu_1)$) is the only edge shared by f' and f'' in $skel(\mu_1)$ ($skel(\mu_2)$).*

Now we define the flip and skip operations and we show how they change the embedding of a planar graph. Let G be a planar graph, and let $\langle \Gamma, f \rangle$ be one of its embeddings. Let (u, v) be a split pair of G and let G_1 be a set of topologically contiguous maximal split components of G w.r.t. (u, v) such that G_1 does not contain all the edges of f. We define the *flip* operation on $\langle \Gamma, f \rangle$ with respect to G_1: $flip(\langle \Gamma, f \rangle, G_1) = \langle \Gamma', f' \rangle$

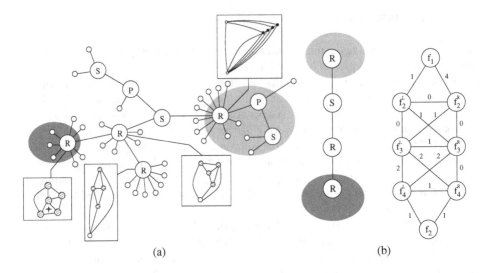

(a)   (b)

Fig. 2. SPQR-tree of the graph of Fig. 1. (a) The light gray ellipse circles the allocation tree of the external face of Fig. 1.a. The dark Gray ellipse circles the allocation tree of the external face of Fig. 1.d. (b) The skip path and the corresponding track graph of the two faces.

where Γ' is obtained from Γ by reversing the adjacency lists of all the vertices of G_1, but for u and v, and by reversing the order of the edges of G_1 in the adjacency lists of u and v. Face f' is determined as follows. If at least one of u and v is not in f, then $f' = f$. Otherwise, f' is the unique face of Γ' containing both the edges belonging to f and not belonging to G_1 and some edge of G_1 not belonging to f. As an example, see the flip applied to the embedding of Fig. 1.a that yields the embedding of Fig. 1.b.

We add the constraint that a flip operation cannot be performed if G_1 contains f because a flipping of the entire external structure of the graph around an internal component is undesirable from a comprehension point of view.

The following property describes three basic features of the flip operation and is trivial to prove.

Property 2. (a) $flip(flip(\langle \Gamma, f \rangle, G_1), G_1) = \langle \Gamma, f \rangle$. (b) If G_1 is a path, then $flip(\langle \Gamma, f \rangle, G_1) = \langle \Gamma, f \rangle$. (c) If $G - G_1$ is a path, then $flip(\langle \Gamma, f \rangle, G_1) = \langle \overline{\Gamma}, f \rangle$, where $\overline{\Gamma}$ is Γ with reversed adjacency lists.

Let $\langle \Gamma_1, f_1 \rangle$ be a planar embedding of G and Γ_2 be a "target" combinatorial embedding of G. It is easy to see that there always exists a sequence of flip operations that leads from $\langle \Gamma_1, f_1 \rangle$ to $\langle \Gamma_2, f_2 \rangle$ for some choice of f_2 in Γ_2. We denote by $\mathcal{F}(\langle \Gamma_1, f_1 \rangle, \langle \Gamma_2, f_2 \rangle)$ the minimum number of flips to obtain $\langle \Gamma_2, f_2 \rangle$ from $\langle \Gamma_1, f_2 \rangle$ for any f_2.

Now we define the skip operation, which provides the ability to modify the external face of an embedding. Let G be a planar graph, and let $\langle \Gamma, f_1 \rangle$ be one of its planar embeddings. Let (u, v) be a split pair of G incident to faces f_1 and f_2 in Γ. Skip is defined as follows: $skip(\langle \Gamma, f_1 \rangle, f_2) = \langle \Gamma, f_2 \rangle$. It is easy to see that there exists a sequence of skip operations that leads from $\langle \Gamma, f_1 \rangle$ to $\langle \Gamma, f_2 \rangle$ for any choice of f_2

in Γ. As an example, see the skip applied to the embedding of Fig. 1.b that yields the embedding of Fig. 1.c. We denote by $\mathcal{S}(\langle \Gamma, f_1 \rangle, \langle \Gamma, f_2 \rangle)$ the minimum number of skips to obtain $\langle \Gamma, f_2 \rangle$ from $\langle \Gamma, f_1 \rangle$.

Given two planar embeddings $\langle \Gamma_1, f_1 \rangle$ and $\langle \Gamma_2, f_2 \rangle$ of a graph G, one could ask which is the minimum number of flip and skip operations for obtaining $\langle \Gamma_2, f_2 \rangle$ from $\langle \Gamma_1, f_1 \rangle$. We denote by $\mathcal{FS}(\langle \Gamma_1, f_1 \rangle, \langle \Gamma_2, f_2 \rangle)$ such a number.

Property 3. $\mathcal{FS}(\langle \Gamma_1, f_1 \rangle, \langle \Gamma_2, f_2 \rangle) \leq \mathcal{F}(\langle \Gamma_1, f_1 \rangle, \langle \Gamma_2, f_3 \rangle) + \mathcal{S}(\langle \Gamma_2, f_3 \rangle, \langle \Gamma_2, f_2 \rangle).$

Lemma 2. *The values of* $\mathcal{F}(\langle \Gamma_1, f_1 \rangle, \langle \Gamma_2, f_2 \rangle), \mathcal{S}(\langle \Gamma_1, f_1 \rangle, \langle \Gamma_1, f_2 \rangle),$ *and* $\mathcal{FS}(\langle \Gamma_1, f_1 \rangle,$ $\langle \Gamma_2, f_2 \rangle)$ *are* $O(n)$, *where* n *is the number of vertices of* G.

Unfortunately, given a biconnected planar graph G and two of its planar embeddings $\langle \Gamma_1, f_1 \rangle$ and $\langle \Gamma_2, f_2 \rangle$, the problem of transforming $\langle \Gamma_1, f_1 \rangle$ into $\langle \Gamma_2, f_2 \rangle$ with the minimum number of flip/skip operations is NP-complete.

Theorem 1. *Let* G *be a biconnected planar graph and let* $\langle \Gamma_1, f_1 \rangle$ *and* $\langle \Gamma_2, f_2 \rangle$ *be two planar embeddings of* G. *Both computing* $\mathcal{FS}(\langle \Gamma_1, f_1 \rangle, \langle \Gamma_2, f_2 \rangle)$ *and computing* $\mathcal{F}(\langle \Gamma_1, f_1 \rangle, \langle \Gamma_2, f_2 \rangle)$ *is NP-complete.*

3   Linearity of the Case with Fixed Combinatorial Embedding

Let G be a biconnected planar graph, and let $\langle \Gamma, f_1 \rangle$ and $\langle \Gamma, f_2 \rangle$ be two planar embeddings of G. In this section, we show how to compute the value of $\mathcal{S}(\langle \Gamma, f_1 \rangle, \langle \Gamma, f_2 \rangle)$.

First, we need to introduce the following lemma whose proof is given in [1].

Lemma 3. *Let* G *be a biconnected planar graph and let* $\mathcal{T}$ *be the SPQR-tree of* G. *Let* $\langle \Gamma, f_1 \rangle$ *and* $\langle \Gamma, f_2 \rangle$ *be two planar embeddings of* G. *If there exists an R-node* μ *of* $\mathcal{T}$ *such that* $skel(\mu)$ *contains both* f_1 *and* f_2, *then* $\mathcal{S}(\langle \Gamma, f_1 \rangle, \langle \Gamma, f_2 \rangle)$ *is the length of the shortest path from* f_1 *to* f_2 *on the dual of* $skel(\mu)$.

Let $\mathcal{T}$ be the SPQR-tree of G and let $\mathcal{T}_1$ and $\mathcal{T}_2$ be the allocation trees of f_1 and f_2, respectively. The value of $\mathcal{S} = \mathcal{S}(\langle \Gamma, f_1 \rangle, \langle \Gamma, f_2 \rangle)$ can be easily computed when $\mathcal{T}_1 \cap \mathcal{T}_2 \neq \emptyset$. If this is true we have to tackle three cases: $\mathcal{T}_1 \cap \mathcal{T}_2 = \{\mu\}$, $\mathcal{T}_1 \cap \mathcal{T}_2 = \{\mu, \nu\}$, and $\mathcal{T}_1 \cap \mathcal{T}_2 = \{\mu_1, \mu_2, \ldots, \mu_k\}$. Conversely, the case $\mathcal{T}_1 \cap \mathcal{T}_2 = \emptyset$ is more complex.

Case $\mathcal{T}_1 \cap \mathcal{T}_2 = \{\mu\}$. In this case μ is the only node of $\mathcal{T}$ whose skeleton contains both f_1 and f_2. If μ is an S-node, then G is a cycle and so $\mathcal{S} = 1$. If μ is a P-node, since a skip operation can move the external face from f_1 to any face of $skel(\mu)$, we have that $\mathcal{S} = 1$. Finally, if μ is an R-node, by Lemma 3, $\mathcal{S}$ is the length of the shortest path on the dual of $skel(\mu)$ from f_1 to f_2.

Case $\mathcal{T}_1 \cap \mathcal{T}_2 = \{\mu, \nu\}$. Observe that, in this case, μ and ν are adjacent in $\mathcal{T}$, and hence they cannot be both P-nodes or both S-nodes. Also, by Lemma 1, f_1 and f_2 are adjacent in both $skel(\mu)$ and $skel(\nu)$. Hence, we have that $\mathcal{S} = 1$. Notice that, if one of the two nodes, say μ, is an S-node, then all edges in $skel(\mu)$ are real edges, but for $e(\nu)$.

Case $\mathcal{T}_1 \cap \mathcal{T}_2 = \{\mu_1, \mu_2, \ldots, \mu_k\}$, with $k \geq 3$. As this case is more involved, we treat it separately in the following lemma, the proof of which is left out of this extended abstract.

Lemma 4. *Let T_1 and T_2 be the allocation tree of two faces f_1 and f_2 of a graph G. If $T_1 \cap T_2 = T_3$, $T_3 = \{\mu_1, \mu_2, \ldots, \mu_k\}$, with $k \geq 3$, then T_3 is a star graph whose central node is an S-node.*

By Lemma 4 and since f_1 and f_2 belong to the same S-node, it follows that $S = 1$.

If $T_1 \cap T_2 = \emptyset$, the computation of S is not trivial; however, we provide a linear time algorithm, called SKIPONLY, to solve this problem. The algorithm is described below. We define *skip path $sp(f_1, f_2)$* in T the (unique) shortest path in T between a node of T_1 and a node of T_2 (see Fig. 2.b). Since a skip operation can only move the external face from $skel(\mu)$ to $skel(\nu)$, with μ adjacent to ν, the following Property holds.

Property 4. Any sequence of skip operations that moves the external face from f_1 to f_2 must traverse all the nodes of the skip path between T_1 and T_2.

In order to compute the sequence of skip operations to move the external face from f_1 to f_2 with S steps, we define a *weighted track graph* [3] $Track(f_1, f_2)$ (see Fig. 2.b). The nodes of $Track(f_1, f_2)$ are faces of the skeletons of the nodes in $sp(f_1, f_2)$. In particular, let $\{\mu_1, \ldots \mu_k\}$ be the nodes in $sp(f_1, f_2)$, where f_1 is the external face of $skel(\mu_1)$, while f_2 is a face of $skel(\mu_k)$. Faces f_1 and f_2 are nodes of $Track(f_1, f_2)$. For each node μ_i, $i = 2, \ldots, k$, $Track(f_1, f_2)$ contains two nodes, called $f_i^{\,l}$ and $f_i^{\,r}$, corresponding to the two faces of $skel(\mu_i)$ adjacent to the virtual edge representing μ_{i-1} in $skel(\mu_i)$. Notice that such faces also correspond to the two faces of $skel(\mu_{i-1})$ adjacent to the virtual edge representing μ_i in $skel(\mu_{i-1})$. Node f_1 belongs to level 1, nodes $f_i^{\,l}$ and $f_i^{\,r}$, for $i = 2, \ldots, k$, belong to level i, and node f_2 belongs to level $k+1$.

We insert in $Track(f_1, f_2)$ two types of edges, called *horizontal edges*, connecting nodes of the same level, and *vertical edges*, connecting nodes of adjacent levels. More precisely, horizontal edges are $(f_i^{\,r}, f_i^{\,l})$, for $i = 2, \ldots, k$, with weight 1, while vertical edges are, for $i = 2, \ldots, k-1$, $(f_i^{\,l}, f_{i+1}^{\,l})$, $(f_i^{\,l}, f_{i+1}^{\,r})$, $(f_i^{\,r}, f_{i+1}^{\,l})$, $(f_i^{\,r}, f_{i+1}^{\,r})$, and edges $(f_1, f_2^{\,l})$, $(f_1, f_2^{\,r})$, $(f_k^{\,l}, f_2)$, $(f_k^{\,r}, f_2)$.

Consider a vertical edge $(f_i^{s_i}, f_{i+1}^{s_{i+1}})$, with $s_i, s_{i+1} \in \{l, r\}$, spanning levels i and $i+1$. If μ_i is a P-node, then the weight is either 0 or 1, depending on the fact that virtual edges corresponding to μ_i and μ_{i+1} are consecutive or not in the circular ordering of the nodes. If μ_i is an S-node, then the weight is either 0 or 1, depending on whether $s_1 = s_2$ or not. Finally, if μ_i is an R-node the weight is the length of the shortest path on the dual of $skel(\mu_i)$ from f_i^* to f_{i+1}^*.

The weight of an edge (f', f'') in $Track(f_1, f_2)$ represents the number of skip operations needed to move the external face from f' to f''. Weight 1 assigned to an horizontal edge $(f_i^{\,r}, f_i^{\,l})$ represents the possibility to skip the virtual edge representing μ_i in $skel(\mu_{i-1})$.

Theorem 2. *Let G be a biconnected planar graph, and let $\langle \Gamma, f_1 \rangle$ and $\langle \Gamma, f_2 \rangle$ be two planar embeddings of G. If only skip operations are allowed, then there exists an algorithm to compute $S(\langle \Gamma, f_1 \rangle, \langle \Gamma, f_2 \rangle)$ in linear time.*

Proof sketch. Consider the shortest path $sp(f_1, f_2)$ on $Track(f_1, f_2)$ from f_1 to f_2 computed by Algorithm SKIPONLY. The proof is based on the fact that any sequence of skip operations leading from f_1 to f_2, by Property 4, must traverse all the levels of

$Track(f_1, f_2)$, and hence can not be shorter than the sequence identified by $sp(f_1, f_2)$. Regarding the computational complexity, since the needed operations on the SPQR-tree T and the sizes of involved structures are linear, it is possible to show that Algorithm SKIPONLY can be implemented to run in linear time [1].   □

4 Linearity of the Case without P-Nodes

In this section we show that if T does not contain P-nodes, the problem of computing $\mathcal{FS}(\langle \Gamma_1, f_1 \rangle, \langle \Gamma_2, f_2 \rangle)$ can be solved in linear time. For simplicity, the algorithm described in this section only considers a subset of the possible flip operations. Namely, given an S-node μ, although a legitimate flip operation may concern the split components of any split pair of μ, we only consider flip operations that concern split components of maximal split pairs of μ. Intuitively, this corresponds to flipping a single neighbor ν of μ or all the neighbors of μ with the exception of ν. At the end of the section we handle the general case.

In order to compute $\mathcal{FS}(\langle \Gamma_1, f_1 \rangle, \langle \Gamma_2, f_2 \rangle)$ when T does not contain P-nodes, we first assign a label in $\{\texttt{turned}, \texttt{unturned}\}$ to each node μ of T. Intuitively, the label of node μ indicates whether some transformation is needed on the skeleton of μ in order to obtain Γ_2 from Γ_1. If μ is a Q-node, then μ is labeled $\texttt{unturned}$. If μ is an R-node, μ is labeled $\texttt{unturned}$ if it has the same Boolean value in both the labellings representing Γ_1 and Γ_2, and $\texttt{turned}$ otherwise. Finally, if μ is an S-node, it is labeled $\texttt{unturned}$ ($\texttt{turned}$) if the majority of its adjacent R-nodes is $\texttt{unturned}$ ($\texttt{turned}$). In case of a tie, we give μ an arbitrary label, unless μ is an internal S-node of the skip path sp. In this case, we give μ a label that is different from one of its adjacent R-nodes in sp.

Second, we suitably extend the labeling from the nodes to the edges. An edge e incident to a Q-node is labeled $\texttt{unturned}$. Otherwise, e is labeled $\texttt{unturned}$ ($\texttt{turned}$) if its incident nodes have the same label (a different label). The number of $\texttt{turned}$ edges of T corresponds to the minimum number of flips to be performed on Γ_1 in order to obtain Γ_2, that is $\mathcal{F}(\langle \Gamma_1, f_1 \rangle, \langle \Gamma_2, f_2 \rangle)$. In particular, each $\texttt{turned}$ edge e identifies a split pair, which, since T has not P-nodes, identifies in its turn two split components G_1 and G_2. Any minimum sequence of flips that transforms Γ_1 into Γ_2 contains either $flip(\langle \Gamma', f' \rangle, G_1)$ or $flip(\langle \Gamma'', f'' \rangle, G_2)$, for some suitable Γ', Γ'', f', and f''.

A trivial case is when the intersection of the two allocation trees T_1 and T_2 of f_1 and f_2 is non-empty. In such a case, since f_1 and f_2 belong to the same skeleton, there is no flip that can help to reduce the number of skips, and the trivial algorithm that first performs all flips and then all skips uses $\mathcal{FS}(\langle \Gamma_1, f_1 \rangle, \langle \Gamma_2, f_2 \rangle)$ operations. Since, in general, a flip operation may modify the distance between two faces (and hence modify the number of needed skips), in order to compute $\mathcal{FS}(\langle \Gamma_1, f_1 \rangle, \langle \Gamma_2, f_2 \rangle)$ we have to consider the case in which flip and skip operations are allowed to be alternated.

We propose an algorithm, called NOPARALLEL, to compute $\mathcal{FS}(\langle \Gamma_1, f_1 \rangle, \langle \Gamma_2, f_2 \rangle)$ when $T_1 \cap T_2 = \emptyset$ and T does not contain P-nodes. Such an algorithm is similar to Algorithm SKIPONLY. The weights of the edges of graph $Track(f_1, f_2)$ are modified in order to take into account the possibility of performing some flip operations in advance in order to reduce the number of skip operations. Namely, consider two nodes μ_i and

μ_{i+1} of the skip path sp, which are adjacent through the `turned` edge e, and consider a skip operation on μ_{i+1}. Such a skip operation has the effect of transferring the external face from f^l_{i+1} to f^r_{i+1} or vice versa. The same effect is obtained by flipping μ_{i+1} with respect to μ_i. Therefore, we set to 0 the weight of the horizontal edge linking f^l_{i+1} to f^r_{i+1} in graph $Track(f_1, f_2)$ and call *shortcut* such an edge. Using a shortcut in the shortest path from f_1 to f_2 corresponds to performing a flip in advance and saving a skip operation.

The sequence of skip and flip operations that transform $\langle \Gamma_1, f_1 \rangle$ into $\langle \Gamma_2, f_2 \rangle$ is given by the edges of a suitably selected weighted shortest path p from f_1 to f_2 in graph $Track(f_1, f_2)$ as follows. First, perform the flip operations corresponding to the short-cuts that are traversed by p, while the external face is still f_1. Second, perform the skip operations corresponding to the non-shortcuts edges of p. Finally, perform the flip operations corresponding to all the other `turned` edges, while the external face is f_2.

Observe that graph $Track(f_1, f_2)$ may admit more than one weighted shortest path from f_1 to f_2. Suppose that the last node of the skip path sp is a `turned` (unturned) R-node μ. Also suppose that a weighted shortest path p_1 from f_1 to f_2 uses an even (odd) number of shortcuts. By performing the corresponding flip operations in advance, while f_1 is the external face, the embedding of node μ will be reversed an even (odd) number of times, i.e., μ will end up `turned`. Hence, in order to obtain Γ_2, according to Property 2, we would need to perform a final flip operation with respect to an edge belonging to f_2. In this case, by using an equal cost weighted shortest path p_2 from f_1 to f_2 that traverses an odd (even) number of horizontal edges whose weight is 0 we would save the last flip. Hence, we need to compute, for any intermediate node f of $Track(f_1, f_2)$, the two weighted shortest path from f_1 to f, if both exist, using an odd and even number of shortcuts. This computation can be performed in linear time and, since all other operations can be performed in linear time, the following theorem holds.

Theorem 3. *Let G be a biconnected planar graph, and let $\langle \Gamma_1, f_1 \rangle$ and $\langle \Gamma_2, f_2 \rangle$ be two planar embeddings of G. Let T be the SPQR-tree of G. If T does not contain P-nodes then $\mathcal{FS}(\langle \Gamma_1, f_1 \rangle, \langle \Gamma_2, f_2 \rangle)$ can be efficiently computed in linear time.*

Now we show how to modify Algorithm NOPARALLEL in order to handle the general case in which a flip operation may concern the split component of any split pair of an S-node μ. Intuitively, this corresponds to allow flipping with a single operation an arbitrary number of consecutive neighbors of μ. The general idea is to modify the SPQR-tree T of G, relaxing the constraint that S-nodes can not be adjacent. Namely, for any maximal sequence $\sigma_i = \nu_1, \nu_2, \ldots, \nu_k$ of consecutive R-nodes with the same label adjacent to μ, we add an S-node μ_i adjacent to μ and move σ_i from the adjacency list of μ to that of μ_i. The label of μ_i is the same as the one of σ_i. The label of μ is computed as for Algorithm NOPARALLEL.

5   Fixed Parameter Tractability of the General Case

Since transforming $\langle \Gamma_1, f_1 \rangle$ into $\langle \Gamma_2, f_2 \rangle$ is NP-complete when G is an arbitrary bi-connected planar graph, in this section we study the fixed parameter tractability of the problem when the structure of G is of limited complexity.

Let $\mathcal{T}$ be the SPQR-tree of a biconnected planar graph G and let $\langle \Gamma_1, f_1 \rangle$ and $\langle \Gamma_2, f_2 \rangle$ be two planar embeddings of G. We present an algorithm that computes $\mathcal{FS}(\langle \Gamma_1, f_1 \rangle, \langle \Gamma_2, f_2 \rangle)$ in $O(n^2 \times 2^{k+h})$ time, where k and h are two parameters that describe the arrangement of P-nodes in $\mathcal{T}$ and their relationships with S-nodes.

We first describe how to handle P-nodes, which are responsible for the NP-hardness of the general problem, with a fixed parameter tractability approach. Recall that the embedding of the skeleton of each P-node μ_P is described in the labeled SPQR-trees representing Γ_1 and Γ_2 by two circular sequences of virtual edges σ_1 and σ_2, respectively. As shown in [1], the problem of morphing with the minimum number of flips σ_1 into σ_2 is equivalent to the *sorting by reversal* problem (SBR), which has been proved to be NP-hard in both cases of linear and circular sequences [8,17]. In fact, sorting virtual edges is equivalent to sorting integer numbers, where a flip of l contiguous edges corresponds to a reversal of l contiguous elements of the sequence.

The fixed parameter approach is based on the fact that SBR problem can be solved in polynomial time, both in its linear and in its circular formulation, when each number has a sign and the reversal of l contiguous elements also changes their signs [14,18,16]. Indeed, when the virtual edges of a P-node correspond to components that have to be reordered and suitably "flipped", then the problem of morphing σ_1 into σ_2 can be modeled as an instance of signed SBR problem, hence admitting a polynomial time solution. For example, if all nodes adjacent to the P-node are R-nodes, then the problem of finding the minimum number of flips that sort them is polynomial. Unfortunately, some virtual edges, as for example those corresponding to paths, do not need to be flipped in a specific way. If k such virtual edges are present, we conventionally assign to them all combinations of signs, and apply 2^k times the signed SBR polynomial algorithm. In fact, there exists an assignment of signs that make it possible to find the minimum number of flips that order a mixed signed/unsigned sequence [2].

Let $\mathcal{T}$ be the SPQR-tree of G and let $\mathcal{T}_1$ and $\mathcal{T}_2$ be the allocation trees of f_1 and f_2, respectively. We concentrate on the case when $\mathcal{T}_1 \cap \mathcal{T}_2 = \emptyset$ that is the most complex.

In order to compute $\mathcal{FS}(\langle \Gamma_1, f_1 \rangle, \langle \Gamma_2, f_2 \rangle)$ each node of $\mathcal{T}$ is labeled as turned, unturned, or neutral. We order them based on their distance from sp. First, starting from the farthest ones, we label nodes that are not in sp with the strategy described below. Second, we label nodes of sp with a different strategy. Consider the current unlabeled node μ not in sp. Observe that μ has all labeled adjacent nodes with the exception of the node that links μ to sp. If μ is an R-node, then we label μ based on its embedding as described in Algorithm NoPARALLEL. If μ is a Q-node, we label μ neutral. If μ is an S-node, we assign μ the label of the majority of its non-neutral labeled adjacent nodes. In case of a tie, we label μ neutral. If μ is a P-node, denote by σ_1 and σ_2 the two circular sequences representing the embedding of μ in Γ_1 and Γ_2. While labeling μ, we also compute the flips that are needed to transform σ_1 into σ_2. Observe that, since the external face can not be internal to a subgraph that is flipped, σ_1 and σ_2 are actually linear sequences as far as flip operations are concerned. In particular, we denote by σ_1' and σ_2' the two linear sequences obtained from σ_1 and σ_2, respectively, by removing the virtual edge e corresponding to the node that links μ to sp and starting from the virtual edge following e in the sequence. Let k be the number of neutral elements of σ_1' and σ_2'. We assign all possible combinations of turned and unturned values to them, and compute 2^k times the linear signed SBR distance d from σ_1' to σ_2', and the analogous distance $\bar{d}$ from σ_1' to $\bar{\sigma}_2'$, where $\bar{\sigma}_2'$ is obtained from σ_2' by reversing the order and changing the signs. If $d < \bar{d}$ ($d > \bar{d}$, $d = \bar{d}$, respectively) we assign μ the label unturned (turned, neutral, respectively).

Now we describe how to assign labels to the elements of the skip path $sp = \mu_1, \mu_2,$ $\ldots, \mu_m$ from T_1 to T_2. Nodes in sp are never labeled neutral. If we have h P-nodes in sp, we consider for them all the combinations of the two possible values turned and unturned, and we repeat 2^h times the computation that follows. R-nodes and S-nodes of sp are labeled as described in Sect. 4. Analogously to Algorithm NoPARALLEL, we extend the labeling to the edges of T. In particular, an edge is labeled turned if it links a turned node to an unturned one and such nodes are not P-nodes, otherwise is labeled unturned. We construct a *weighted track graph* $Track(f_1, f_2)$ as in Algorithm NoPARALLEL where P-nodes were not present, and we describe how to set the weights of the edges exiting nodes f_i^l and f_i^r corresponding to a P-node μ_i of sp. All other weights are set as described in Sect. 4. Denote by σ_1 and σ_2 the two circular sequences representing the embedding of μ_i in Γ_1 and Γ_2. From σ_1 we obtain the linear sequence σ_1^l (σ_1^r) ending with (starting with, respectively) the virtual edge corresponding to μ_{i-1}. Intuitively, sequence σ_1^l (σ_1^r) corresponds to the configuration of the parallel component when the external face is f_i^l (f_i^r). Analogously, from σ_2 we obtain the linear sequence σ_2^l (σ_2^r) ending with (starting with, respectively) the virtual edge corresponding to μ_{i+1}. Our aim is to set the weight of each vertical edge (f_i^s, f_{i+1}^t), for $s, t \in \{l, r\}$, as the minimum number of operations needed to transform σ_1^s into σ_2^t. Observe that, when the external face is moved from f_i^s to another face f_i of $skel(\mu_i)$ in Γ_1, we obtain a new linear sequence σ_1^* with the same circular order as σ_1^s. Namely, σ_1^* is obtained from σ_1 by opening it between the two virtual edges adjacent to f. Hence, when computing the minimum number of operations needed to transform σ_1^s into σ_2^t, we have to consider the possibility to first transforming σ_1^s into another linear sequence σ_1^* with the same circular order, that can be done by performing one skip operation, and then transforming σ_1^* into σ_2^t with the minimum number of flips, that can be done by applying the signed SBR algorithm. In order to do this, observe that all nodes adjacent to μ_i in T are labeled as turned, unturned, or neutral. Let k be the number of nodes adjacent to μ_i and labeled neutral. As described above, we consider all possible assignments of turned and unturned values to such nodes, and we compute 2^k times the linear signed SBR distance from σ_1^* to σ_2^t. The weight of vertical edge (f_i^s, f_{i+1}^t) is the minimum of such $n_i \times 2^k$ values, where n_i is the number of nodes adjacent to μ_i in T. The weight of an horizontal edge for a P-node is 1.

The remaining part of the algorithm strictly follows the lines of Algorithm NoPAR-ALLEL. Namely, we compute the minimum weight path from f_1 to f_2 in $Track(f_1, f_2)$ and, based on such a path, we decide the sequence of skip and flip operations to be performed. Again, if $Track(f_1, f_2)$ admits more than one minimum weight path, we choose among such paths taking into account the number of shortcuts traversed, corresponding to flip operations that are convenient to be performed in advance.

Here we analyze the computational complexity of the algorithm. All the operations, except those involving P-nodes, can be performed in linear time. For each P-node μ_i not belonging to the skip path, the computation of the minimum number of flips that are needed to transform σ_1 into σ_2 can be performed in $O(n_i \times 2^k)$, where n_i is the number of neighbors of μ_i in $\mathcal{T}$. Observe that computing the minimum SBR distance can be done in linear time [4], while actually finding the sequence of operations that yield that minimum can be done in time $O(n^{\frac{3}{2}}\sqrt{log(n)})$ time [18]. Hence, when considering the 2^k assignments, we only compute the distance and then, when the optimal assignment has been found, we perform the algorithm for finding the actual sequence of flips. For each P-node μ belonging to sp, the computation of the minimum number of flips that are needed to transform σ_1^s into σ_2^s can be performed in $O(n_i^2 \times 2^k)$. Namely, we have to consider the 2^k assignments of signs to the k neutral neighbors of μ_i and the possibility to transform σ_1^s into σ_2^t by first moving the external face to each of the n_i faces of $skel(\mu_i)$ in Γ_1 and then performing the computation of the signed linear SBR distance in linear time. Since such a computation has to be performed for each of the 2^h assignments of labels to the h P-nodes of sp, the global computational complexity of the algorithm is $O(2^h \times \sum_{i=1}^{h}(n_i^2 \times 2^{k+h}))$, which is equal to $O(n^2 \times 2^{k+h})$, since the total number of neighbors of all the P-nodes is less or equal than the total number of edges of $\mathcal{T}$, that is $O(n)$. Based on the above discussion we have:

Theorem 4. *Let G be a biconnected planar graph, let $\langle \Gamma_1, f_1 \rangle$ and $\langle \Gamma_2, f_2 \rangle$ be two planar embeddings of G. Let $\mathcal{T}$ be the SPQR-tree of G, let k be the maximum number of neutral S-nodes adjacent to a P-node in $\mathcal{T}$, and let h be the number of P-nodes in the skip path $sp(f_1, f_2)$. If both flip and skip operations are allowed, then $\mathcal{FS}(\langle \Gamma_1, f_1 \rangle, \langle \Gamma_2, f_2 \rangle)$ can be computed in $O(n^2 \times 2^{k+h})$ time.*

6 Conclusions

Preserving the user mental map while coping with ever-changing information is a common goal of the Graph Drawing and the Information Visualization areas. The information represented, in fact, may change with respect to three different levels of abstraction: (i) structural changes may modify the graph that the user is inspecting; (ii) topological changes may affect the way the same graph is embedded on the plane; and (iii) drawing changes may map the same embedded graph to differently positioned graphic objects. A large body of literature has been devoted to structural changes, addressing the representation models and techniques in the so-called dynamic and on-line settings. Also, much research effort has been devoted to manage drawing changes, where the target is to preserve the mental map by morphing the picture while avoiding intersections and overlappings. On the contrary, to our knowledge, no attention at all has been devoted to topological changes, that is, changes of the embedding of a graph in the plane.

In this paper we addressed the topological morphing problem. Namely, the problem of morphing a topology into another one with a limited number of changes. This paper leaves many open problems. (1) Primitives. We considered two topological primitives, called flip and skip. It would be important to enrich such a set with other operations that can be considered "natural" for the user perception. (2) Connectivity. It is easy to extend the results presented in Sect. 3 to simply connected graphs. However, the other presented results are deeply related to biconnectivity. There is a lot of space here for further investigation. (3) We gave the same weight to the operations performed by the morphing. However, other metrics are possible. One could weight an operation as a non-decreasing function of the moved edges or of the thickness of the moved component.

As a final remark we underline how usually the Computational Biology field looks at Graph Drawing as a tool. In this paper it happened the opposite. In fact, Theorems 1 and 4 exploit Computational Biology results.

References

1. Angelini, P., Cortese, P.F., Di Battista, G., Patrignani, M.: Topological morphing of planar graphs. Tech. Report RT-DIA-134-2008, Dept. of Computer Sci., Univ. di Roma Tre (2008)
2. Auyeung, A., Abraham, A.: Estimating genome reversal distance by genetic algorithm. CoRR cs.AI/0405014 (2004)
3. Bachmaier, C., Brandenburg, F.J., Forster, M.: Track planarity testing and embedding. In: Van Emde Boas, P., Pokorný, J., Bieliková, M., Štuller, J. (eds.) SOFSEM 2004. LNCS, vol. 2932, pp. 3–17. Springer, Heidelberg (2004)
4. Bader, D.A., Moret, B.M.E., Yan, M.: A linear-time algorithm for computing inversion distance between signed permutations with an experimental study. Journal of Computational Biology 8(5), 483–491 (2001)
5. Bertolazzi, P., Di Battista, G., Didimo, W.: Computing orthogonal drawings with the minimum number of bends. IEEE Transactions on Computers 49(8), 826–840 (2000)
6. Biedl, T.C., Lubiw, A., Spriggs, M.J.: Morphing planar graphs while preserving edge directions. In: Healy, P., Nikolov, N.S. (eds.) GD 2005. LNCS, vol. 3843, pp. 13–24. Springer, Heidelberg (2006)
7. Cairns, S.S.: Deformations of plane rectilinear complexes. American Math. Monthly 51, 247–252 (1944)
8. Caprara, A.: Sorting by reversals is difficult. In: RECOMB 1997: Proceedings of the first annual international conference on Computational molecular biology, pp. 75–83. ACM Press, New York (1997)
9. de Fraysseix, H., Ossona de Mendez, P.: P.I.G.A.L.E - Public Implementation of a Graph Algorithm Library and Editor, sourceForge project page,
   http://sourceforge.net/projects/pigale
10. Di Battista, G., Tamassia, R.: On-line planarity testing. SIAM J. Comput. 25, 956–997 (1996)
11. Erten, C., Kobourov, S.G., Pitta, C.: Intersection-free morphing of planar graphs. In: Liotta, G. (ed.) GD 2003. LNCS, vol. 2912, pp. 320–331. Springer, Heidelberg (2004)
12. Floater, M., Gotsman, C.: How to morph tilings injectively. Journal of Computational and Applied Mathematics 101, 117–129 (1999)
13. Gotsman, C., Surazhsky, V.: Guaranteed intersection-free polygon morphing. Computers and Graphics 25, 67–75 (2001)
14. Kaplan, H., Shamir, R., Tarjan, R.E.: Faster and simpler algorithm for sorting signed permutations by reversals. In: SODA 1997, pp. 344–351 (1997)
15. Lubiw, A., Petrick, M., Spriggs, M.: Morphing orthogonal planar graph drawings. In: SODA 2006, pp. 222–230. ACM Press, New York (2006)
16. Meidanis, J., Walter, M., Dias, Z.: Reversal distance of sorting circular chromosomes. Tech. Report IC-00-23, Institute of Computing, Universidade Estadual de Campinas (2000)
17. Solomon, A., Sutcliffe, P., Lister, R.: Sorting circular permutations by reversal. In: Dehne, F., Sack, J.-R., Smid, M. (eds.) WADS 2003. LNCS, vol. 2748, pp. 319–328. Springer, Heidelberg (2003)
18. Tannier, E., Bergeron, A., Sagot, M.F.: Advances on sorting by reversals. Discrete Appl. Math. 155(6-7), 881–888 (2007)
19. Thomassen, C.: Deformations of plane graphs. Journal of Combinatorial Theory, Series B 34, 244–257 (1983)

An SPQR-Tree Approach to Decide Special Cases of Simultaneous Embedding with Fixed Edges

J. Joseph Fowler[1], Carsten Gutwenger[2], Michael Jünger[3,*],
Petra Mutzel[2], and Michael Schulz[3,*]

[1] Department of Computer Science, University of Arizona, USA
jfowler@cs.arizona.edu
[2] Department of Computer Science, Technische Universität Dortmund, Germany
{petra.mutzel,carsten.gutwenger}@cs.tu-dortmund.de
[3] Department of Computer Science, University of Cologne, Germany
{mjuenger,schulz}@informatik.uni-koeln.de

Abstract. We present a linear-time algorithm for solving the simultaneous embedding problem with fixed edges (**SEFE**) for a planar graph and a pseudoforest (a graph with at most one cycle) by reducing it to the following embedding problem: Given a planar graph G, a cycle C of G, and a partitioning of the remaining vertices of G, does there exist a planar embedding in which the induced subgraph on each vertex partite of $G \setminus C$ is contained entirely inside or outside C? For the latter problem, we present an algorithm that is based on SPQR-trees and has linear running time. We also show how we can employ SPQR-trees to decide **SEFE** for two planar graphs where one graph has at most two cycles and the intersection is a pseudoforest in linear time. These results give rise to our hope that our SPQR-tree approach might eventually lead to a polynomial-time algorithm for deciding the general **SEFE** problem for two planar graphs.

1 Introduction

Many practical graph drawing applications demand planar embeddings of a graph that yield additional constraints. One natural application is in obtaining simultaneous drawings of a set of related planar graphs. This is useful in the areas of bioinformatics, social sciences and software engineering. A single drawing can be insufficient in depicting complex interrelationships of different models of a system. Instead, multiple drawings may be required, each from a different perspective. The challenge is to preserve the "mental map" of the common structures in each layout so that the scientist can easily navigate between the different diagrams. To do this, common vertices and edges are placed and drawn equally in each drawing. This can be modeled via embedding constraints.

Various embedding constraints have already been studied in [2,5,6]; Gutwenger *et al.* [12] apply SPQR-trees to efficiently decide if a graph has a combinatorial embedding with respect to a set of hierarchical constraints modeling grouping and fixed orders of edges around a vertex. We instead address a problem that cannot be modeled by any

---

* Partially supported by the German Science Foundation (JU204/11-1).

I.G. Tollis and M. Patrignani (Eds.): GD 2008, LNCS 5417, pp. 157–168, 2009.

of the previous approaches. Given a planar graph G, a cycle $C \subset G$, and a partition P of all vertices of $G \setminus C$, we ask whether there is a planar embedding of G where all vertices $v \in p$ for some part $p \in P$ lie completely inside or outside C. We give an efficient decision algorithm using SPQR-trees that can be used to solve a simultaneous embedding problem.

Given a set of planar graphs $\{G_1, G_2, \ldots, G_n\}$ on the same vertex set, a *simultaneous embedding with fixed edges* (SEFE) of $\{G_i\}$ are planar drawings Γ_i of G_i, $i \in [1..n]$, such that all vertices and all edges belonging to two graphs G_i and G_j are drawn identically in the corresponding drawings Γ_i and Γ_j. SEFE and its variant of simultaneous geometric embedding (SGE) with planar straight-line drawings as well as the other variations of simultaneous embedding have become an important branch within the field of graph drawing. It is known that deciding SEFE is NP-complete for three graphs [11] while deciding SGE is NP-hard for two graphs [8]. The complexity of deciding SEFE for two graphs is still open.

Many approaches have been made to decide the problem for some classes of graph pairs [4,7,9,10]. Frati [10] showed that trees and planar graphs always have a SEFE. Fowler *et al.* [9] improved this result to show that forests, circular caterpillars (removal of all degree-1 vertices yields a cycle), K_4, and subgraphs of K_3-multiedges (an edge (x, y) with any number of edges with x or y as endpoints) are the only graphs to always have a SEFE with any planar graph. Their drawing algorithms are based upon using an optimal Euclidean shortest path algorithm [13]. We also apply this technique in our algorithms.

In this paper we examine the pairs of a planar graph G_1 with a pseudoforest G_2. A SEFE is not always guaranteed unless all non-cycle edges of G_2 are incident to the cycle, i.e., the pseudoforest happens to be a circular caterpillar. However, we show that SEFE for such pairs can be decided in polynomial time by presenting an efficient decision algorithm. We further discuss efficient decision algorithms for the case that G_2 contains two cycles and $G_1 \cap G_2$ is a pseudoforest. We think that our approach is promising in that it may eventually lead to a general polynomial time decision algorithm for testing SEFE of two graphs.

2   Preliminaries

Given some planar drawing Γ of a planar graph G, a cycle C in G forms a Jordan curve that splits the plane into two connected components. One is bounded by C and the other is unbounded as given by the Jordan curve theorem [14]. We say that some vertex $v \in G \setminus C$ lies in the *interior (exterior)* of C if it is mapped to a position in the bounded (unbounded) component.

A *combinatorial embedding* of a planar graph G is defined as a clockwise ordering of the incident edges for each vertex with respect to a crossing-free drawing of G in the Euclidean plane. A *planar embedding* is a combinatorial embedding together with a fixed *external face*.

A *block* is a maximal 2-connected subgraph of a graph G. If G is 2-connected, the *SPQR-tree* T of G represents its decomposition into 3-connected components comprising serial, parallel, and 3-connected structures [3]. The respective structure is given by a skeleton graph associated with each tree node which is either a cycle (S-node), a bundle of parallel edges (P-node), or a 3-connected simple graph (R-node); Q-nodes serve as representatives for the edges of G.

If G is 2-connected and planar, its SPQR-tree $\mathcal{T}$ represents all combinatorial embeddings of G. In particular, a combinatorial embedding of G uniquely defines a combinatorial embedding of each skeleton in $\mathcal{T}$, and fixing the combinatorial embedding of each skeleton uniquely defines a combinatorial embedding of G.

Given two planar graphs $G_1 = (V, E_1)$ and $G_2 = (V, E_2)$ on the same vertex set V, a *simultaneous embedding with fixed edges* (SEFE) consists of planar drawings Γ_i of $G_i, i \in [1, 2]$, such that each vertex is mapped to the same point in the plane for Γ_1 and Γ_2 and each edge in $G_1 \cap G_2$ is represented by the same simple curve in the plane for both drawings.

3   A Planar Graph, a Cycle, and a Partition

In this section, we consider the following graph embedding problem. Given a planar graph $G = (V, E)$, a cycle $C = (V_C, E_C) \subset G$, and a partition P of $V \setminus V_C$, decide whether G has a planar drawing such that all vertices of each part in P either lie completely inside or outside of C; see Algorithm 1.

The input partition P or the planar embeddings of the graph may force two vertices to be on the same side of the cycle (either both inside or both outside). We call this situation a *same-side constraint*. On the other hand, by examining all embeddings of the graph we may reveal that two vertices must be positioned on opposite sides of the cycle (one inside and one outside). We refer to this situation as an *opposite-side constraint*. The idea of the algorithm is to find all such constraints and then check whether all these constraints can be satisfied at once, i.e., whether a planar embedding with the required property exists.

The following algorithm uses an SPQR-tree $\mathcal{T}$ to examine all embeddings of the block of graph G containing the given cycle C. Each skeleton of a node of $\mathcal{T}$ may lead to constraints prohibiting some of the possible embeddings as discussed above. We use an auxiliary graph H containing all of the vertices of the original graph to maintain the occuring constraints. Same-side constraints are represented by green edges and opposite-side constraints by red edges.

We say that H is 2-colorable if its vertices can be colored with two colors, say red and green, in such a way that both endpoints of a green edge have the same color and both endpoints of a red edge have different color.

As cycles are 2-connected, the given cycle C is contained in a single block B of graph G. All other blocks are either completely inside or outside of C in all planar drawings of G. Hence, we get one same-side constraint for all vertices of each block $B' \neq B$. We can now assume to deal with a 2-connected graph G and its SPQR-tree $\mathcal{T}$ that represents all planar embeddings of G together with some cycle $C \subseteq G$. Let $\nu \in T$ be some node of the SPQR-tree, S be its skeleton and $e \in S$ be any skeleton edge. If the expansion graph of e includes any edge of C, we call e a *cycle edge*. We consider the different possibilities for ν in turn.

Algorithm 1. Deciding the embeddability of parts respecting a cycle

---

Input: Planar graph $G = (V, E)$, cycle $C = (V_C, E_C) \subseteq G$, partition P of $V \setminus V_C$
Output: Returns YES if and only if G has a planar embedding such that all induced
       subgraphs of each $p \in P$ lie on one side of C

Let $H = (V, \emptyset)$
for *all parts $p \in P$* **do**
   ⌊ Construct path in H with green edges of all vertices in p
Block B := Biconnected component of G containing C
for *all blocks $B' \neq B$* **do**
   ⌊ Construct path in H with green edges of all vertices in B'
Tree T := SPQR-tree of B
for *all nodes $\mu \in T$* **do**
   if *skeleton S of μ has at least two cycle edges* **then**
      Cycle C' := cycle consisting of all cycle edges in S
      if *μ is R-node* **then**
         Expand all non-cycle edges in S
         Construct path in H with green edges of all vertices inside C'
         Construct path in H with green edges of all vertices outside C'
         if *there exist vertex v in the interior of C' and vertex w in the exterior of C'*
         then
            ⌊ Add red edge to H between v and w

      if *μ is P-node* **then**
         for *all edges e in $S \setminus C'$* **do**
            Construct path in H with green edges of all vertices in the expansion
            ⌊ graph of e

if *H is 2-colorable* **then**
   ⌊ **return** YES
else
   ⌊ **return** NO

If S contains exactly one cycle edge e, then the edges belonging to the skeleton of all the other vertices must lie on the same side of C. When regarding the node of T belonging to e, all these vertices are contained in the expansion graph of a single edge that is not a cycle edge. Repeating this process, if necessary, we get a T-node that has more than one cycle edge but also has a single non-cycle edge containing all of the vertices from above. When dealing with this T-node, the necessary auxiliary graph augmentation to handle this same-side constraint is performed.

If S contains two or more cycle edges, then these cycle edges comprise a cycle in S. If S also contains non-cycle edges, ν is a P-node or an R-node.

1. In an S-node this can only occur if all edges of the skeleton are cycle edges. In this situation there is nothing to be done as this does not lead to any same-side constraints or opposite-side constraints.
2. Let ν be a P-node (see Fig. 1). All the vertices occurring in an expansion graph of any other edge in S are forced to be on one side of the cycle C.

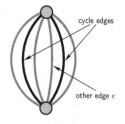

Fig. 1. Cycle edges in the skeleton of a P-node lead to same-side constraints: all vertices in the expansion graph of a non-cycle edge e are on the same side of the cycle

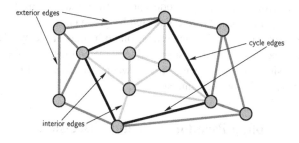

Fig. 2. Cycle edges in the skeleton of an R-node yield two same-side and one opposite-side constraints: All vertices in the expansion graphs of the interior component are on one side of the cycle while all vertices of the exterior component are on the other side

3. Having ν as an R-node (see Fig. 2) is the most involved. The skeleton S of ν is a 3-connected graph and has hence a unique embedding (besides mirroring and choosing the outer face). The cycle edges split S into two halves: the interior and the exterior components of S. All vertices belonging to all expansion graphs of edges of one side must be on one side of the cycle in the final embedding. Neither pair of vertices w_1 and w_2 being the interior and the exterior components, respectively, may end up on the same side of the cycle. Hence, we get two same-side constraints (between all vertices in the interior and exterior components, respectively) and one opposite-side constraint (the edges from the interior and the exterior components must be separated).

Theorem 1. *Algorithm 1 has a runtime of* $O(|V|)$ *and works correctly, i.e., it returns* YES *if and only if the input graph G has a planar embedding $\mathcal{E}$ such that for each $p \in P$ all vertices in p lie on one side of C in $\mathcal{E}$.*

Proof. Obviously, the first two for-loops including the construction of $\mathcal{T}$ require only $O(|V|)$ time, thus add only $O(|V|)$ green edges to H. The third loop iterates over all nodes $\mu \in \mathcal{T}$ and expands some non-cycle edges. Observe that—for all nodes μ—the expansion graphs of these non-cycle edges do not share any edge, and thus no vertex except for vertices on the cycle C. Therefore, the whole for-loop takes $O(|V|)$ time, and we add only $O(|V|)$ green and red edges to H. Since the size of H is linear in $|V|$, we can check if H is 2-colorable using breadth-first-search in $O(|V|)$ time.

We next show that the algorithm works correctly. First, assume that the algorithm returns NO. Then the constructed auxiliary graph H is not 2-colorable. This means that two vertices v and w in H are connected by two paths: one containing an odd number of red edges and one containing an even number. This implies that v and w must lie on the same side of C (due to the path with even number of red edges), as well as on opposite sides (path with odd number of red edges). Hence, G has no such embedding.

Next assume that the algorithm returns YES in which H is 2-colorable. We pick one of the two colors to lie in the interior of C and one to lie on the outside. The choice of embeddings for every P-node and every R-node implies an embedding $\mathcal{E}$ for G. For each such node in $\mathcal{T}$, we can choose an embedding that satisfies the given choice of interior and exterior of C. In each P-node, the vertices that belong to the expansion graph of one of the parallel edges are connected by green edges in H, thus, they lie on the same side in $\mathcal{E}$. In each R-node, the vertices on both sides of the cycle are connected by green edges, respectively, while a single edge between these sets forbids both parts to lie on the same side. Finally, green edges between the vertices of the input partition yield that these vertices lie on the same side of C. $\qquad\square$

4   A Planar Graph, a Pseudoforest, and a Decision

In this section, we apply Algorithm 1 to solve the following open problem in simultaneous embedding: Given a planar graph G_1 and a pseudoforest G_2, find an efficient algorithm to decide whether the pair $\{G_1, G_2\}$ has an SEFE; see Algorithm 2. For a few special cases of G_2 the situation becomes trivial as described by the next theorem.

Theorem 2 (Fowler et al. [9]). *Let G_1 be a planar graph and G_2 be a forest or a circular caterpillar. Then G_1 and G_2 have a* SEFE.

Next, we consider the more general case of a pseudoforest containing a cycle C in which not all non-cycle edges are incident to C. We see by the next theorem that the case is also trivial if C is not in the intersection of G_1 and G_2.

Theorem 3. *Let $G_1 = (V, E_1)$ be a planar graph and $G_2 = (V, E_2)$ be a pseudoforest with a cycle C. If C is not in $G_1 \cap G_2$, then the pair has a* SEFE.

Proof. Let edge $e \in C \setminus G_1$. Create a planar drawing of Γ_1 of G_1 in the plane using any suitable graph drawing algorithm (e.g. [1]). We construct a planar drawing Γ_2 of G_2 that, together with Γ_1, creates a SEFE of G_1 and G_2.

Draw all vertices and all edges of $G_1 \cap G_2$ in Γ_2 in the same way as in Γ_1 guaranteeing a simultaneous drawing. We still must draw all edges of $G_2 \setminus G_1$ without introducing any crossings in Γ_2. As e is not part of G_1, it has not been drawn in Γ_2 yet. We draw all edges of $G_2 \setminus G_1$ in Γ_2 one after another with e as the last edge. The order of the other edges can be chosen arbitrarily.

To do this we use an optimal Euclidean shortest path algorithm [13]. We apply the modification as done by Fowler *et al.* [9] in their drawing algorithms. A distance ε is always maintained between the shortest path and any line segment corresponding to previous part of Γ_2. This allows subsequent edges to be routed as need be in between any pair of non-incident edges that would otherwise be touching. Applying this algorithm adds at most $O(|V|)$ edge bends for each new edge (as new bends hide old bends as argued in [9]) so that the final complexity of the drawing is $O(|V|^2)$ giving an overall running time of $O(|V|^2 \log |V|)$.

As G_2 has only one cycle C and e is part of C, $G_2 \setminus \{e\}$ is a forest. Any drawing of any subgraph of $G_2 \setminus \{e\}$ has exactly one face. Hence, starting with the partial drawing of Γ_2 it is always possible to insert a route for the edges not yet drawn maintaining planarity. Even, in the last step, when edge e is inserted, the partial drawing of Γ_2 has exactly one face and thus, e can be safely inserted into Γ_2. Then Γ_2 is completed and $\{\Gamma_1, \Gamma_2\}$ is a SEFE of $\{G_1, G_2\}$. □

Due to Theorems 2 and 3 we assume G_2 to have exactly one cycle C in the intersection $G_1 \cap G_2$. By construction G_1 is planar. However, to ensure a SEFE of G_1 and G_2 we must embed G_1 in such a way that the cycle C does not separate any pair of vertices that are adjacent in G_2. On the other hand, as $G_2 \setminus C$ is a forest, this condition suffices to guarantee a SEFE of the pair $\{G_1, G_2\}$.

Theorem 4. *Let $G_1 = (V, E_1)$ be a planar graph and $G_2 = (V, E_2)$ be a pseudoforest each on n vertices with a cycle $C \subseteq G_1 \cap G_2$. G_1 and G_2 have a SEFE if and only if there exists a planar drawing of G_1 such that for all edges $e = \{v, w\} \in G_2 \setminus G_1$ either both v and w lie inside or both lie outside of C.*

Proof. Assume first that G_1 has a planar drawing Γ_1 with the described property. We create a planar drawing Γ_2 of G_2 that, together with Γ_1, yields a SEFE of G_1 and G_2. Draw all vertices and all edges of $G_1 \cap G_2$ in Γ_2 in the same way as in Γ_1. As $C \subseteq G_1 \cap G_2$, the cycle is now present in Γ_2. We draw all remaining edges of $G_2 \setminus G_1$ next by using the same approach of the proof of Theorem 3.

We start with the edges e that have one endpoint in the exterior of C in Γ_1. Due to the condition on Γ_1, both endpoints of e are in the exterior of C or one endpoint is on C. As we have just drawn C and all vertices in the same way as in Γ_1, this condition also holds for the partial drawing of Γ_2. As $G_2 \setminus G_1$ is a forest there is a way to route e without introducing crossings: Imagine C and its interior as one big vertex. The partial drawing Γ_2 then has exactly one face. This also holds for edges connecting the exterior of C with C itself. The same argument holds for all the edges in the interior of C as well as the edges connecting the interior with C. Hence, by construction we have a planar drawing Γ_2 of G_2 that, together with Γ_1, yields a SEFE of G_1 and G_2.

Now let G_1 be without a planar drawing with the described property. Assume G_1 and G_2 have a SEFE. By definition there exist planar drawings Γ_i of G_i, $i \in [1, 2]$, such that the intersection $G_1 \cap G_2$ is drawn in the same way in both Γ_1 and Γ_2. As G_1 has no planar drawing with the described property, there exists an edge $e = \{v, w\} \in G_2 \setminus C$ such that v lies in the interior of C and w lies in the exterior of C in Γ_1. As vertices v and w and cycle C are part of $G_1 \cap G_2$, the same condition holds for Γ_2. But this means that e cannot be routed in Γ_2 without introducing a crossing in Γ_2, which is a contradiction to our assumption. Hence, G_1 and G_2 have no SEFE. □

Algorithm 2. Deciding SEFE for planar graph and pseudoforest pair

---

Input: Planar graph G_1 and pseudoforest G_2.
Output: YES if and only if $\{G_1, G_2\}$ has a SEFE.

if G_2 *contains no cycle* **then**
  └ Return YES.
Cycle $C :=$ the only cycle of G_2
if $C \nsubseteq G_1$ **then**
  └ Return YES.
Partition $P = \{P_v \mid v \in G_1 \setminus C\} :=$ trivial partition of $G_1 \setminus C$
for *all edges* $\{v, w\}$ *of* $G_2 \setminus C$ **do**
  └ UNION P_v and P_w.
Run Algorithm 1 with input (G_1, C, P).
Return output of Algorithm 1.

We use the previously discussed results to create an efficient algorithm deciding the problem mentioned in the beginning of this section.

Theorem 5. *Algorithm 2 works correctly, i.e., it returns* YES *if and only if* $\{G_1, G_2\}$ *has a* SEFE. *Moreover, it has a linear runtime.*

Proof. Assume first that the algorithm returns YES, which is by one of three statements. The first returns YES if G_2 contains no cycle. But then Theorem 2 states that $\{G_1, G_2\}$ has a SEFE. The second statement returns YES if cycle C is not completely part of G_1. Theorem 3 guarantees that G_1 and G_2 have a SEFE in this situation. The last instruction is that the run of Algorithm 1 returns YES. Algorithm 1 checks whether graph G_1 can be embedded in the plane such that all partition sets of P lie completely inside or outside C. By the construction of P, this is equivalent to saying that both endpoints of every edge of $G_2 \setminus C$ lie both inside or both outside C. Then Theorem 4 yields a SEFE of G_1 and G_2.

Assume next, that the algorithm returns NO, which implies Algorithm 1 returned NO. Hence, G_1 has no planar drawing with the property of Theorem 4, which implies that G_1 and G_2 are without a SEFE.

The proposed runtime $O(|V|)$ follows directly from the complexity analysis of Algorithm 1.   □

5   A Planar Graph, a Path, and a Cyclic Edge Order

In this section, we consider two embedding problems with requirements on the cyclic order of some of the edges around a vertex x or two vertices x and y that can be used to decide some special SEFE problems in Section 6.

In the first problem, x and y are two distinct vertices connected by a path p. Let e_p and e_p' be the first and last edges on p incident to x and y, respectively, where $\{e_a, e_b\}$ and $\{e_a', e_b'\}$ are distinct edges also incident to x and y. We want to ensure that the order of these edges around x and y (amongst other possible incident edges) in a combinatorial embedding Γ of G is consistent with an embedding of a graph in which x and y are connected by the three edge-disjoint paths $p, p_a = e_a, \ldots, e'_a$, and $p_b = e_b, \ldots, e'_b$. This implies that either the cyclic order around x is e_p, e_a, e_b and around y is e'_p, e'_b, e'_a or both orders are reversed. It suffices to test only one possibility, since we can generate a combinatorial embedding with the reversed orders simply by mirroring the embedding.

Let $E_x = \{e_p, e_a, e_b\}$ and $E_y = \{e'_p, e'_a, e'_b\}$. We observe that—if not all edges in $E_x \cup E_y$ are in the same block—such a required combinatorial embedding always exists; in this case, x or y is a cut vertex. We can insert the embedding of one block B' into a face of an embedding of the other block B (mirroring the embedding of B' if necessary) so that the requirements on the embedding are met. On the other hand, if all the edges in $E_x \cup E_y$ are contained in a single block B, it is sufficient to test a few simple conditions in the SPQR-tree T of B. The necessary and sufficient conditions are given in the lemma below.

Lemma 1. *G has a combinatorial embedding Γ such that the cyclic order induced by Γ on E_x is e_p, e_a, e_b and the cyclic order induced on E_y is e'_p, e'_b, e'_a if and only if*

1. *there is no block B of G containing all edges in $E_x \cup E_y$; or*
2. *there is a block B containing $E_x \cup E_y$, and its SPQR-tree T has neither*
    (a) *a P-node whose skeleton contains three distinct edges e_1, e_2, e_3 such that e_p and e'_p are contained in the expansion graph of e_1, e_a and e'_b in the expansion graph of e_2, and e_b and e'_a in the expansion graph of e_3; nor*
    (b) *an R-node whose skeleton has a combinatorial embedding such that e_p, e_a, e_b are in the expansion graphs of three distinct skeleton edges $\tilde{e}_p, \tilde{e}_a, \tilde{e}_b$ in this cyclic order, and e'_p, e'_a, e'_b are in the expansion graphs of three distinct skeleton edges $\tilde{e}'_p, \tilde{e}'_a, \tilde{e}'_b$ in this cyclic order.*

These conditions can be checked in linear time, since constructing an SQPR-tree and determining for each edge $e \in E_x \cup E_y$ in the expansion graphs of which skeleton edges it is contained, requires only linear time, and there are only two combinatorial embeddings of each R-node's skeleton.

In the second embedding problem, we consider a planar graph G with a vertex x and four distinct edges e_a, e'_a, e_b, e'_b incident to x. We want to decide if there exists an embedding Γ of G that induces a cyclic order on these four edges in which e_a and e'_a (and thus also e_b and e'_b) are consecutive. The motivation for this problem is similar as for the first problem, where p is an empty path and thus x and y are identical. In this case, deciding if a feasible combinatorial embedding of G exists is even easier. We only need to consider only R-node skeletons containing x in which x is incident to at least four skeleton edges. This gives the following lemma whose conditions can be verified in linear time:

Lemma 2. *G has a combinatorial embedding Γ such that the cyclic order induced by Γ on $E_x = \{e_a, e'_a, e_b, e'_b\}$ is such that e_a and e'_a are consecutive, if and only if either*

1. *no block of G contains all edges in E_x; or*
2. *there is a block B containing all edges in E_x, and its SPQR-tree contains no R-node whose skeleton S contains x and the edges in E_x are in the expansion graphs of four distinct skeleton edges $\tilde{E}_x = \tilde{e}_a, \tilde{e}'_a, \tilde{e}_b, \tilde{e}'_b$ such that there exists a combinatorial embedding of S that induces a cyclic order on the edges in $\tilde{E}_x$ in which $\tilde{e}_a$ and $\tilde{e}'_a$ are not consecutive.*

6   Two Planar Graphs with Restrictions and a Decision

We now consider how this approach of using SPQR-trees might be extended to address more general decision problems for deciding whether a pair of graphs has a SEFE. We examine pairs of planar graphs G_1 and G_2 where we restrict both the number and the arrangement of cycles in G_2 and in $G_1 \cap G_2$.

$G_1 \cap G_2$ **is a forest:** We start with a more general version of Theorem 3 where we have a larger number of cycles in G_2 but still the intersection is a forest.

Theorem 6. *Let $G_1 = (V, E_1)$ be a planar graph and $G_2 = (V, E_2)$ be a planar graph where all cycles $C_i \subseteq G_2, i \in [1..k]$, are pairwise disjoint. If no C_i is contained in $G_1 \cap G_2$, then the pair $\{G_1, G_2\}$ has a SEFE.*

Proof (sketch). We adapt the proof of Theorem 3. When drawing G_2, remove one edge e_i from each $C_i \setminus G_1$ and draw the rest of G_2, which is a forest. Then insert one edge e_i after another in the same way as done with edge e in the proof of Theorem 3. As all cycles are disjoint and no further cycles exist, this method can be applied without introducing any crossings in the drawing of G_2.   □

Next, we discuss the case where G_2 contains exactly two cycles that either touch in exactly one point or share a common path. With the ideas developed in Section 5 we can handle this situation efficiently.

Theorem 7. *The SEFE decision problem for two planar graphs G_1 and G_2 where G_2 contains exactly two cycles and $G_1 \cap G_2$ is a forest can be decided in linear time.*

Proof (sketch). Let C_1 and C_2 be the two cycles of G_2. If $C_1 \cap C_2 = \emptyset$, the case is trivial as given by Theorem 6. As G_2 contains no more cycles, $C_1 \cap C_2$ is a path p with endpoints x and y; see Fig. 3. A planar embedding of G_1 can force the outgoing edges of x and y to have a specific order leading to the situation in Fig. 3(b) in G_2 that prevents a SEFE of G_1 and G_2. However, if G_1 has an embedding that allows the right cyclic order for both x and y as in Fig. 3(a), then a SEFE can be achieved. All other edges of G_2 can be drawn without introducing crossings as in the proof of Theorem 4. Lemma 1 gives a linear time check to determine whether G_1 has an embedding such

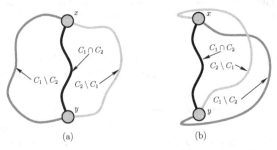

(a)   (b)

Fig. 3. The two cycles C_1 and C_2 drawn without and with crossings. The respective clockwise ordering of the edges incident to x and y differ.

Algorithm 3. Deciding SEFE for restricted planar graph pair

Input: Planar graphs G_1 and G_2 where G_2 contains exactly two cycles and $G_1 \cap G_2$ is a pseudoforest but not a forest.
Output: YES if and only if $\{G_1, G_2\}$ has a SEFE.

Cycle $C :=$ the only cycle of $G_1 \cap G_2$
Partition $P = \{P_v \mid v \in G_1 \setminus C\} :=$ trivial partition of $G_1 \setminus C$
for *all edges* $\{v, w\}$ *of* $G_2 \setminus C$ **do**
  $\llcorner$ UNION P_v and P_w.
Run Algorithm 1 with input (G_1, C, P).
Return output of Algorithm 1.

---

that the cyclic order for the three outgoing edges corresponds to the paths shown in Fig. 3. Lemma 2 handles the degenerate case for $x = y$, also determinable in linear time.   □

$G_1 \cap G_2$ **is a pseudoforest:** Assume now that both G_1 and G_2 are planar graphs in which G_2 contains exactly two cycles C_1 and C_2 of which only one, say C_1, is contained in $G_1 \cap G_2$. When removing one edge of $C_2 \setminus G_1$ we are in the situation described in Section 4. This correlation allows us to construct a new decision algorithm based on Algorithm 2. We start by generalizing Theorem 4, which we use as the key ingredient to Algorithm 3.

Theorem 8. *Let* $G_1 = (V, E_1)$ *be a planar graph and* $G_2 = (V, E_2)$ *be a planar graph with exactly two cycles* C_1 *and* C_2 *where* $C_1 \subseteq G_1 \cap G_2$ *and* $C_2 \nsubseteq G_1 \cap G_2$. G_1 *and* G_2 *have a* SEFE *if and only if there exists a planar drawing of* G_1 *such that for all edges* $e = \{v, w\} \in G_2 \setminus G_1$ *either both* v *and* w *lie inside or both lie outside of* C_1.

Theorem 8 can be proved by using Theorem 4 to determine whether $\{G_1, G_2 \setminus \{e\}\}$ has a SEFE. In an SEFE of this smaller pair, edge $e = \{v, w\}$ can be inserted if and only if both endpoints v and w lie on the same side of C_1.

It is easy to see that Algorithm 3 works correctly. We can imitate the proof of correctness of Algorithm 2 (see Theorem 5) where this time Theorem 8 plays the role of Theorem 4.

7   Concluding Remarks and Future Applications

We have shown how to use SPQR-trees in the context of simultaneous embedding with fixed edges by presenting several new decision algorithms for some classes of graph pairs. Clearly, much future works remains, but overall this approach of using SPQR-trees seems promising in potentially yielding a polynomial-time decision algorithm for deciding whether two graphs have a SEFE, if one exists.

References

1. de Fraysseix, H., Pach, J., Pollack, R.: How to draw a planar graph on a grid. Combinatorica 10(1), 41–51 (1990)
2. Di Battista, G., Didimo, W., Patrignani, M., Pizzonia, M.: Drawing database schemas. Software: Practice and Experience 32(11), 1065–1098 (2002)
3. Di Battista, G., Tamassia, R.: On-line planarity testing. SIAM Journal on Computing 25(5), 956–997 (1996)
4. Di Giacomo, E., Liotta, G.: A note on simultaneous embedding of planar graphs. In: EWCG 2005, pp. 207–210 (2005)
5. Dornheim, C.: Planar graphs with topological constraints. Journal on Graph Algorithms and Applications 6(1), 27–66 (2002)
6. Eiglsperger, M., Fößmeier, U., Kaufmann, M.: Orthogonal graph drawing with constraints. In: Proc. SODA 2000, pp. 3–11 (2000)
7. Erten, C., Kobourov, S.G.: Simultaneous embedding of planar graphs with few bends. In: Pach, J. (ed.) GD 2004. LNCS, vol. 3383, pp. 195–205. Springer, Heidelberg (2005)
8. Estrella-Balderrama, A., Gassner, E., Jünger, M., Percan, M., Schaefer, M., Schulz, M.: Simultaneous geometric graph embeddings. In: Hong, S.-H., Nishizeki, T., Quan, W. (eds.) GD 2007. LNCS, vol. 4875, pp. 280–290. Springer, Heidelberg (2008)
9. Fowler, J.J., Jünger, M., Kobourov, S.G., Schulz, M.: Characterizations of restricted pairs of planar graphs allowing simultaneous embedding with fixed edges. In: WG 2008 (to appear)
10. Frati, F.: Embedding graphs simultaneously with fixed edges. In: Kaufmann, M., Wagner, D. (eds.) GD 2006. LNCS, vol. 4372, pp. 108–113. Springer, Heidelberg (2007)
11. Gassner, E., Jünger, M., Percan, M., Schaefer, M., Schulz, M.: Simultaneous graph embeddings with fixed edges. In: Fomin, F.V. (ed.) WG 2006. LNCS, vol. 4271, pp. 325–335. Springer, Heidelberg (2006)
12. Gutwenger, C., Klein, K., Mutzel, P.: Planarity testing and optimal edge insertion with embedding constraints. In: Kaufmann, M., Wagner, D. (eds.) GD 2006. LNCS, vol. 4372, pp. 126–137. Springer, Heidelberg (2007)
13. Hershberger, J., Suri, S.: An optimal algorithm for Euclidean shortest paths in the plane. SIAM Journal on Computing 28(6), 2215–2256 (1999)
14. Veblen, O.: Theory on plane curves in non-metrical analysis situs. Transactions of the American Mathematical Society 6, 83–98 (1905)

Graph Simultaneous Embedding Tool, GraphSET*

Alejandro Estrella-Balderrama, J. Joseph Fowler, and Stephen G. Kobourov

Department of Computer Science, University of Arizona
{aestrell,jfowler,kobourov}@cs.arizona.edu

Abstract. Problems in simultaneous graph drawing involve the layout of several graphs on a shared vertex set. This paper describes a
Graph Simultaneous Embedding Tool, GraphSET, designed to allow the
investigation of a wide range of embedding problems. GraphSET can
be used in the study of several variants of simultaneous embedding including *simultaneous geometric embedding, simultaneous embedding with
fixed edges* and *colored simultaneous embedding* with the vertex set partitioned into color classes. The tool has two primary uses: (i) studying
theoretical problems in simultaneous graph drawing through the pro-
duction of examples and counterexamples and (ii) producing layouts of
given classes of graphs using built-in implementations of known algorithms. GraphSET along with movies illustrating its utility are available
at http://graphset.cs.arizona.edu.

1 Introduction

Drawing multiple graphs simultaneously is a problem motivated by its applications in bioinformatics, social sciences, and software engineering. The large
networks defined by multiple relationships make using a single layout impractical. Instead, such networks can be viewed from different perspectives according
to the particular structure, behavior, or scale of interest. When looking for common patterns and substructures among the heterogeneous representations of the
same data it is essential to preserve the "mental map" of the user. A natural way
to accomplish this is to have common vertices and edges laid out in a similar
manner throughout the various layouts.

Simultaneous embedding problems are difficult to solve and require extensive
manipulation of different instances in order to gain insight. A useful tool is one
that allows for the dynamic manipulation of common vertices while accounting
for how the edge crossings in each graph can change. In addition, having the
ability to visualize each graph separately or as a whole while simultaneously
manipulating each graph can allow one to solve complex problems. Finally, having built-in implementations of algorithms related to simultaneous embedding
can also aid in further research.

* This work was supported in part by NSF grants CCF-0545743 and ACR-0222920.

I.G. Tollis and M. Patrignani (Eds.): GD 2008, LNCS 5417, pp. 169–180, 2009.

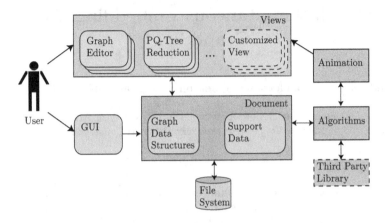

Fig. 1. An overview of the GraphSET system

Our *Graph Simultaneous Embedding Tool, GraphSET* meets the above goals, allowing the manipulation of up to eight graphs simultaneously with the capability of displaying each graph separately in its own window. This is an essential feature that has enabled us to solve several simultaneous embedding problems.

A related tool is the *Interactive Multi-User System for Simultaneous Graph Drawing* [15]. It only considers simultaneous geometric embedding of two graphs and the emphasis is on collaboration with the aid of the DiamondTouch device [3]. Another related tool that can be used to obtain simultaneous drawings of graphs using force-directed methods is described in [5].

2   System Architecture

Figure 1 gives a high-level overview of the system architecture of GraphSET. The user can introduce commands using the GUI (menus, dialog boxes, toolbar, etc.) or directly manipulating the view (Graph Editor). When the user makes modifications, they are done in the document (graph data structures, application settings, etc.) and those changes are reflected on every active view of the document. When the modifications are done from the view (such as moving a vertex) the document is modified and reflected back in all active views. The document can be loaded/saved in the file system. Algorithms are called from the document. Some algorithms (such as drawing or recognition) only reflect temporary modifications directly in the view (animation, for example).

Dashed boxes represent plugin components that include customized views (such as a 3D view we have used for studying 3D morphing). The other dashed box corresponds to third-party libraries that can be hooked into the algorithms module via a proxy. For example, we have proxies for LEDA [16] and OGDF (the Open Graph Drawing Framework available at http://www.ogdf.net). Algorithms from these libraries are called through these proxies. Overlapping boxes

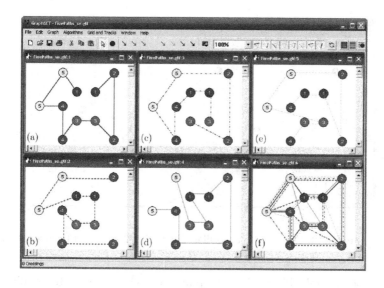

Fig. 2. Example of a simultaneous geometric embedding of five paths: blue path is solid in (a), red is dashed in (b), green is dash-dotted in (c), cyan is light-solid in (d) and yellow is light-dashed in (e). The SGE of all 5 paths is shown in (f).

represent several views of the same type in the document. This allows for the different graph views in which to work with a simultaneous embedding; see Fig. 2. Features like toggling the grid, snapping, and visibility of a given edge set are properties of the graph editor view allowing each to have individual settings.

3   Preliminaries

We begin with a few definitions to clarify the various problems of interest.

Two n-vertex graphs $G_1(V_1, E_1)$ and $G_2(V_2, E_2)$ have a *simultaneous embedding with mapping* if, given a bijection $f : V_1 \mapsto V_2$, each graph can be drawn in the plane $\mathbb{R}^2$ without crossings such that for all $v \in V_1$ and $f(v) \in V_2$, v and $f(v)$ are represented by the same point in their respective drawings. If f is not given, but this can be done for some bijection, then G_1 and G_2 are *simultaneously embeddable without mapping*. Unless indicated otherwise, a simultaneous embedding (SE) refers to one with mapping.

A *simultaneous geometric embedding* (SGE) consists of a simultaneous embedding in which only straight-line edges are used. *Simultaneous embedding with fixed edges* (SEFE) is less restricted since edges are drawn with simple curves and common, or *fixed edges*, use the same curve. Clearly, SGE ⊂ SEFE ⊂ SE.

The problem of *colored simultaneous embedding* (CSE) is a generalization of simultaneous embedding with mapping in which each V_i is strictly partitioned into k *colors* with respect to a k-coloring of a pointset P. Each vertex of a given color can be mapped to a point of the same color. When $k = n$ this is equivalent

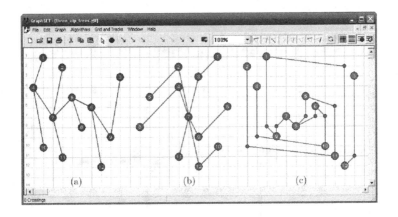

Fig. 3. Layouts of ULP trees: (a) caterpillar, (b) radius-2 star, and (c) degree-3 spider

to simultaneous embedding with mapping, and when $k = 1$, to simultaneous embedding without mapping. Figure 2 is an example of five 5-colored paths on ten vertices in which there are $5 \cdot 2^5 = 160$ possible mappings, one of which is shown.

Finding simultaneous embeddings with paths drawn monotonically uses a restricted form of planarity, called *level planarity*. Only one of the Cartesian coordinates is allowed to change when attempting to find a crossings-free drawing.

An undirected *level graph* $G(V, E, \phi)$ has a *labeling* $\phi : V \mapsto [1..k]$ assigning each vertex to one of k levels so that $\phi(u) \neq \phi(v)$ for every edge (u, v). This prevents any pair of adjacent vertices from being in the same level. In a *level drawing* all the vertices of the same level share the same y-coordinate, placed along a horizontal *track*, and each edge is drawn strictly y-monotone. If G can still be drawn planarly, then G is *level planar*, otherwise, G is *level non-planar*. Any level planar drawing with bends has one without bends [4]. Hence, adding edge bends does not affect the level planarity of a graph.

If G is level planar over all possible labelings, then G is *unlabeled level planar* (ULP). In [6], ULP trees were characterized as consisting of three classes of trees: (i) *caterpillars* (the removal of vertices that have degree-1 yields a path or an empty graph); (ii) *radius-2 stars* (any number of paths of length one or two that all share a common endpoint); and (iii) *degree-3 spiders* (three paths that share a common endpoint); see Fig. 3.

4   Applications

In this section, we describe several successful uses of GraphSET. First, we discuss how GraphSET has been used in working with ULP trees [6] and a related problem on colored trees. Second, we consider a pair of trees whose union is homeomorphic to complete graph K_n for $n > 3$ for which there is a pair without a SGE [10]. Third, we discuss how GraphSET has aided in verifying gadgets of

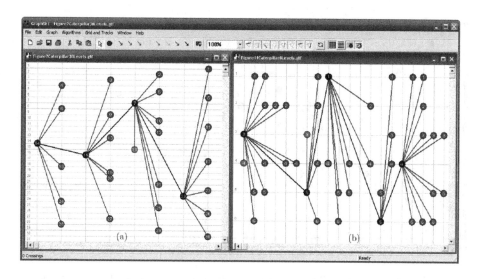

Fig. 4. Layouts of (a) a 25-level caterpillar and (b) a 6-level caterpillar

reductions used to show that deciding whether a graph pair has a SGE is NP-hard and whether a graph triple has a SEFE is NP-complete [7,9]. Finally, we show how GraphSET can be used to find CSE counterexamples, as in [2].

4.1 Unlabeled Level Planar Trees

When there are more vertices than levels, caterpillars are the only class of trees that remains ULP. This allows GraphSET to draw any caterpillar without crossings; see Fig. 4. When there is exactly one vertex per level, GraphSET can also provide level planar layouts of the other two classes of ULP trees; see Fig. 5.

GraphSET also implements the ULP recognition algorithms that highlight the ULP trees by their class. If the graph is not ULP, a subgraph homeomorphic to one of the forbidden ULP trees is highlighted as the user modifies the graph; see Fig. 6. GraphSET has been instrumental in determining correct and implementable algorithms for these purposes. Movies of the tool demonstrating all the ULP tree algorithms can be found at http://ulp.cs.arizona.edu.

4.2 Colored Level Planar Trees

Our tool has the feature of allowing the user to snap and lock vertices to tracks in order to investigate not only unlabeled level planar graphs but the planarity of multiple level graphs being simultaneously embedded. Tracks can be colored so that only vertices of that color can be snapped to that track.

As an example of this utility, we consider the open problem of whether a 3-colored tree-path pair always has a SGE. One approach is to attempt to layout the path monotonically. Here each colored track has one vertex of its color.

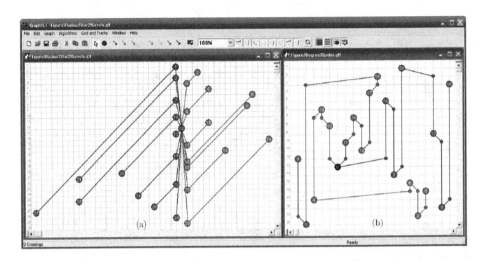

Fig. 5. Layouts of (a) a 30-level radius-2 star and (b) a 20-level degree-3 spider

The idea is to find an algorithm to swap vertices between tracks of the same color until the 3-colored tree becomes level planar. This is not always possible if the tracks are colored sequentially as in Fig. 7(a). However, if the tracks are colored randomly, then it may be possible to find a sequence of swaps in going from a level non-planar assignment as in Fig. 7(b) to a level planar one as in Fig. 7(c). Even in the worst case of sequentially colored tracks there may be relatively few interchanges of colored tracks needed so that a CSE then becomes possible. This would then correspond to paths consisting of relatively few monotonic segments that may have a SGE.

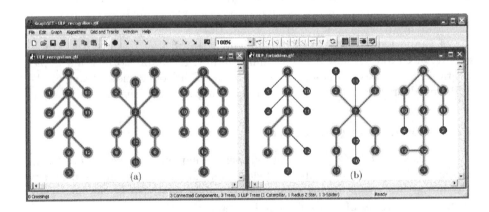

Fig. 6. ULP recognition algorithms highlighting a caterpillar, a radius-2 star, and a degree-3 spider (a) and the forbidden trees T_7, T_8, and T_9 (b)

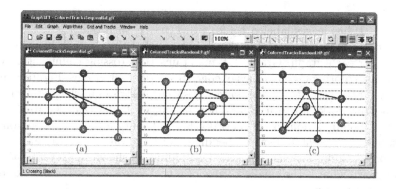

Fig. 7. This 3-colored tree (vertices 1-3 are blue, 4-7 red and 8-10 green) is level non-planar for sequentially colored tracks (tracks 1-3 are blue and solid, tracks 4-7 are red and dashed, tracks 8-10 are green and dash-dotted) as in (a), but may or may not be level planar for randomly colored tracks as in (b) and (c), respectively

4.3   Simultaneous Geometric Embedding of Pairs of Trees

In this section, we consider the simultaneous geometric embedding of two trees $T_1(V, E_1)$ and $T_2(V, E_2)$ on $n^2 - 2n + 2$ vertices whose union contains a subgraph homeomorphic to the complete graph K_n on n vertices for a given $n > 3$. Both T_1 and T_2 have a root vertex labeled '0' that is adjacent to the remaining $n - 1$ vertices of V labeled '1', '2', ..., '$n-1$'. In each tree, these $n - 1$ vertices have $n - 2$ leaves so that each non-leaf vertex has degree $n - 1$. Leaves are labeled i, j for $i, j \in [1..n - 1]$ and $i \neq j$. In T_1 the vertex labeled $i \in [1..n - 1]$ has leaves

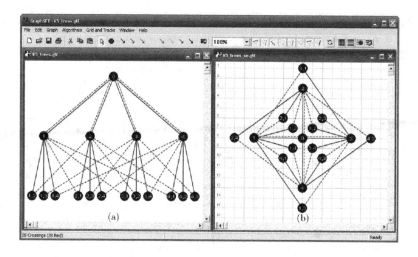

Fig. 8. A pair of trees whose union is homeomorphic to K_5 (a) in which one tree has red edges (dashed) and the other has blue edges (solid) with a SGE shown in (b)

labeled i, j for $j \in [1..n-1]$ such that $i \neq j$. Similarly, in T_2 the vertex labeled $j \in [1..n-1]$ has leaves labeled i, j for $i \in [1..n-1]$ such that $i \neq j$.

The tool is especially useful in this case given that the user can have different windows for each graph. GraphSET maintains the crossing count within each graph while ignoring the crossings of edges from different graphs. Figure 8 shows two trees for the case of $n = 5$ on 17 vertices that illustrates a schema to generate a layout that works up to $n = 6$. When $n > 6$, we found that the root vertex labeled '0' could no longer be centrally located, but rather had to be on the convex hull of the simultaneous embedding. For large values of n these tree pairs do not have a simultaneous geometric embedding, as shown by Geyer *et al.* [10]. It is unknown what is the smallest value of n that forces a crossing; for example, the case $n = 8$ is open.

4.4 Gadgets for Planar 3-SAT Reductions

GraphSET supports multiple edges with different colors. These edges may include bends and can be treated as a single edge (for fixed edges) or as different edges (for multi-graphs). An application of this is the manipulation of gadgets for Planar 3-SAT reductions.

In [9] Gassner *et al.* proved that SEFE is NP-Complete for three graphs. The proof is a reduction using clause gadgets and literal gadgets; see Fig. 9(a). There are two possible embeddings for each literal gadget and these embeddings correspond to true or false values in the matching literals. The argument is that a drawing of the clause without crossings is only possible if one of the literals is true. In the drawing this implies that we can only get rid of a crossing by flipping a literal gadget (changing the embedding of the gadget).

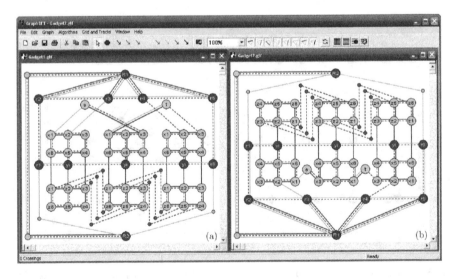

Fig. 9. Gadget for a clause with 3 literals (a) and a SEFE of the gadget in (b)

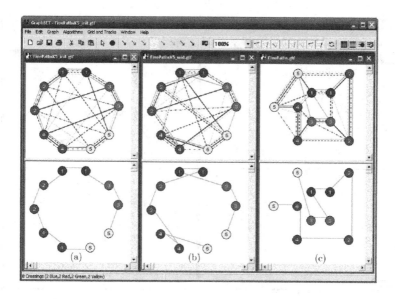

Fig. 10. Five colored paths (blue path is solid, red is dashed, green is dash-dotted, cyan is light-solid and yellow is light-dashed) without a **SEFE** in (a) and after some swaps among vertices of the same color in (b) have a **SGE** in (c). The split window shows the cyan path for each step.

GraphSET is useful in exhibiting problems in the gadget construction by finding initially less than obvious embeddings that may break the argument; see for example Fig. 9(b). With the aid of GraphSET the correct reduction was found [7]. The flipping/rotation and the cut/paste operations included in the tool are essential in constructing and manipulating these kinds of gadgets.

4.5   Colored Simultaneous Embeddings

GraphSET was used to build a counterexample of five 5-colored paths on five distinctly colored vertices without a **SGE** to show that there does not exist a universal pointset for 5-colored paths [2]. One open **CSE** problem is whether there exists four paths on four colors that do not always have a **SGE**. We illustrate the difficulty of this problem with a potential alternate counterexample of five 5-colored paths not using distinctly colored vertices with Figs. 2 and 10. Here the five 5-colored paths are on ten vertices in which each path has two vertices of the same color corresponding to its endpoints. As given in Fig. 10(a) a crossing will always occur regardless of the placement of vertices. This is due to the fact that each pair of vertices with the same color is connected by four edges of the other colors. This means that when each of these vertex pairs are contracted they form the example of five paths on five colors in [2].

However, vertices of the same color can exchange adjacencies. The tool lets one swap the adjacency lists between two vertices of the same color in one

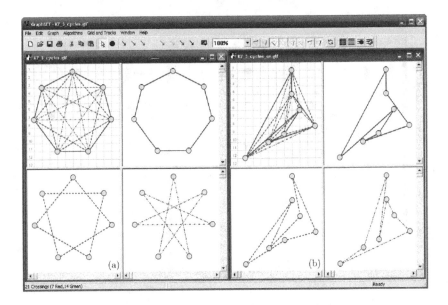

Fig. 11. Three cycles whose union forms a K_7 with no common edges in (a) and the corresponding SGE in (b)

of the graphs. A series of such swaps in the five graphs results in Fig. 10(b), which has the SGE in Fig. 10(c). While this is not the counterexample we are after, it illustrates the utility of GraphSET when attempting to construct such counterexamples.

Another open CSE problem in which GraphSET is very useful is shown in Fig. 11(a). One starts with an arrangement of three cycles whose union forms a K_7. In general, any odd prime p has a decomposition into $(p-1)/2$ cycles whose union forms a K_p in which each edge in the union is in exactly one cycle. This is of interest because the three 6-colored cycles whose union forms a $K_{3,3}$ without a SGE given in [2] are constructed so that each edge in the union belongs to two of the three paths. This forces one of the cycles to have a self crossing.

It is an open problem to find a set of cycles without any common edges that do not have a SGE. While this example for K_7 has a SGE shown in Fig. 11(b), this requires several small angles between pairs of incident edges along the same cycle. We conjecture for sufficiently large p that such a SGE no longer exists.

5   Implementation

GraphSET is a stand-alone Windows application written in C++ that can be downloaded from http://graphset.cs.arizona.edu, where the source code is also available. GraphSET can also run under Linux and MacOS using wine.

GraphSET contains other related algorithms for graph drawing as support for the previous applications. This includes implementation of the PQ-tree data

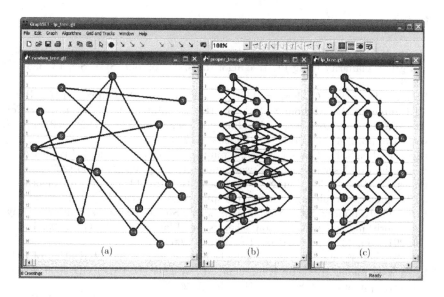

Fig. 12. A random labeled tree (a) made proper (b) with a level planar embedding (c)

structure and the planarity testing algorithm by Booth and Lueker [1] . The level planarity testing and embedding algorithms by Healy *et al.* [11,12] are also available. These algorithms require the graph to be proper, i.e., there are only edges between vertices in consecutive levels. When the graph is non-proper, GraphSET adds dummy vertices along edges; see Fig. 12. The runtime for these algorithms is $O(|V|^2)$ provided the graph is proper.

6   Conclusions and Future Work

We presented GraphSET, a tool that has been valuable in studying problems related to simultaneous embedding. We hope that other researchers interested in these problems will find this tool useful.

While currently GraphSET only includes the recognition and drawing algorithms for ULP trees, we plan to incorporate algorithms for all ULP graphs. We foresee using this tool in the research of minimal level non-planar (MLNP) patterns; the first step is to implement MLNP patterns recognition algorithms for trees [8]. We also plan to incorporate the faster $O(|V| \log |V|)$ level planarity testing and embedding algorithms by Jünger, Leipert and Mutzel [13,14].

Acknowledgments

We would like to thank Martin Harrigan for the explanations on the embedding part of the vertex-exchange algorithm, Markus Geyer in helping finding the 3-colored tree counterexample, and Michael Jünger and Sebastian Leipert for providing us with their level planarity testing code.

References

1. Booth, K., Lueker, G.: Testing for the consecutive ones property, interval graphs, and graph planarity using PQ-tree algorithms. J. Comput. Syst. Sci. 13, 335–379 (1976)
2. Brandes, U., Erten, C., Fowler, J.J., Frati, F., Geyer, M., Gutwenger, C., Hong, S., Kaufmann, M., Kobourov, S.G., Liotta, G., Mutzel, P., Symvonis, A.: Colored simultaneous geometric embeddings. In: Lin, G. (ed.) COCOON 2007. LNCS, vol. 4598, pp. 254–263. Springer, Heidelberg (2007)
3. Dietz, P., Leigh, D.: Diamondtouch: a multi-user touch technology. In: 14th ACM Symposium on User interface software and technology, pp. 219–226 (2001)
4. Eades, P., Feng, Q.-W., Lin, X., Nagamochi, H.: Straight-line drawing algorithms for hierarchical graphs and clustered graphs. Algorithmica 44(1), 1–32 (2006)
5. Erten, C., Kobourov, S.G., Navabia, A., Le, V.: Simultaneous graph drawing: Layout algorithms and visualization schemes. In: Liotta, G. (ed.) GD 2003. LNCS, vol. 2912, pp. 437–449. Springer, Heidelberg (2004)
6. Estrella-Balderrama, A., Fowler, J.J., Kobourov, S.G.: Characterization of unlabeled level planar trees. In: Kaufmann, M., Wagner, D. (eds.) GD 2006. LNCS, vol. 4372, pp. 367–369. Springer, Heidelberg (2007)
7. Estrella-Balderrama, A., Gassner, E., Jünger, M., Percan, M., Schaefer, M., Schulz, M.: Simultaneous geometric graph embeddings. In: Hong, S.-H., Nishizeki, T., Quan, W. (eds.) GD 2007. LNCS, vol. 4875, pp. 280–290. Springer, Heidelberg (2008)
8. Fowler, J.J., Kobourov, S.G.: Minimum level nonplanar patterns for trees. In: Hong, S.-H., Nishizeki, T., Quan, W. (eds.) GD 2007. LNCS, vol. 4875, pp. 69–75. Springer, Heidelberg (2008)
9. Gassner, E., Jünger, M., Percan, M., Schaefer, M., Schulz, M.: Simultaneous graph embeddings with fixed edges. In: Fomin, F.V. (ed.) WG 2006. LNCS, vol. 4271, pp. 325–335. Springer, Heidelberg (2006)
10. Geyer, M., Kaufmann, M., Vrťo, I.: Two trees which are self-intersecting when drawn simultaneously. In: Healy, P., Nikolov, N.S. (eds.) GD 2005. LNCS, vol. 3843, pp. 201–210. Springer, Heidelberg (2006)
11. Healy, P., Harrigan, M.: Practical level planarity testing and layout with embedding constraints. In: Hong, S.-H., Nishizeki, T., Quan, W. (eds.) GD 2007. LNCS, vol. 4875, pp. 62–68. Springer, Heidelberg (2008)
12. Healy, P., Kuusik, A.: The vertex-exchange graph: A new concept for multi-level crossing minimisation. In: Kratochvíl, J. (ed.) GD 1999. LNCS, vol. 1731, pp. 205–216. Springer, Heidelberg (1999)
13. Jünger, M., Leipert, S.: Level planar embedding in linear time. In: Kratochvíl, J. (ed.) GD 1999. LNCS, vol. 1731, pp. 72–81. Springer, Heidelberg (1999)
14. Jünger, M., Leipert, S., Mutzel, P.: Level planarity testing in linear time. In: Whitesides, S.H. (ed.) GD 1998. LNCS, vol. 1547, pp. 224–237. Springer, Heidelberg (1999)
15. Kobourov, S.G., Pitta, C.: An interactive multi-user system for simultaneous graph drawing. In: Pach, J. (ed.) GD 2004. LNCS, vol. 3383, pp. 492–501. Springer, Heidelberg (2005)
16. Mehlhorn, K., Näher, S.: The LEDA Platform of Combinatorial and Geometric Computing. Cambridge University Press, Cambridge (1999)

Hamiltonian Alternating Paths on Bicolored Double-Chains*

Josef Cibulka[1], Jan Kynčl[2], Viola Mészáros[3,4], Rudolf Stolař[1], and Pavel Valtr[2]

[1] Department of Applied Mathematics, Charles University, Faculty of Mathematics and Physics, Malostranské nám. 25, 118 00 Prague, Czech Republic
cibulka@kam.mff.cuni.cz, ruda@kam.mff.cuni.cz
[2] Department of Applied Mathematics and Institute for Theoretical Computer Science (ITI), Charles University, Faculty of Mathematics and Physics, Malostranské nám. 25, 118 00 Prague, Czech Republic
kyncl@kam.mff.cuni.cz
[3] Department of Applied Mathematics and Institute for Theoretical Computer Science, Charles University, Malostranské nám. 25, 118 00 Prague, Czech Republic
[4] Bolyai Institute, University of Szeged, Aradi vértanúk tere 1, 6720 Szeged, Hungary
viola@math.u-szeged.hu

Abstract. We find arbitrarily large finite sets S of points in general position in the plane with the following property. If the points of S are equitably 2-colored (i.e., the sizes of the two color classes differ by at most one), then there is a polygonal line consisting of straight-line segments with endpoints in S, which is Hamiltonian, non-crossing, and alternating (i.e., each point of S is visited exactly once, every two non-consecutive segments are disjoint, and every segment connects points of different colors).

We show that the above property holds for so-called double-chains with each of the two chains containing at least one fifth of all the points. Our proof is constructive and can be turned into a linear-time algorithm. On the other hand, we show that the above property does not hold for double-chains in which one of the chains contains at most $\approx 1/29$ of all the points.

1 Introduction

1.1 Previous Results

One of the basic problems in geometric graph theory is to decide if a given graph can be drawn on a given planar point set using pairwise non-crossing straight-line edges. In a more demanding version, the points and the vertices of the graph are colored and each vertex has to be placed in a point of the same color (see the survey [5] for further references). Interesting and non-trivial questions arise already if we want to embed a 2-colored path on a 2-colored point set. The authors of several papers have focused on embeddings of so-called alternating paths, which are paths with no monochromatic

---

* Work on this paper was supported by the project 1M0545 of the Ministry of Education of the Czech Republic. Work by Viola Mészáros was also partially supported by OTKA grant T049398 and by European project IST-FET AEOLUS.

I.G. Tollis and M. Patrignani (Eds.): GD 2008, LNCS 5417, pp. 181–192, 2009.
© Springer-Verlag Berlin Heidelberg 2009

edge. Since the colors on a 2-colored alternating path must alternate along the path, a 2-colored point set S may admit a Hamiltonian alternating path only if the coloring of S is equitable, i.e., the sizes of the color classes differ by at most one.

Let S be an equitably 2-colored set of points in general position in the plane. It is known that if the two color classes of S can be separated by a line then there is a non-crossing Hamiltonian alternating path on S [1]. The same result holds if one of the color classes is exactly the set of vertices of the convex hull [1]. Kaneko et al. [6] proved that any equitably 2-colored set S of at most 12 points or of 14 points admits a non-crossing Hamiltonian alternating path. On the other hand, Kaneko et al. [6] gave examples of equitably 2-colored sets S of n points admitting no non-crossing Hamiltonian alternating path for any $n > 12$, $n \neq 14$.

The above result on sets with color classes separated by a line easily implies that any equitably 2-colored set S of size n admits a non-crossing alternating path on at least $n/2$ points of S. It is an open problem if this lower bound can be improved to $n/2 + f(n)$, where $f(n)$ is unbounded (see also the book [3]). On the other hand, there are equitably 2-colored sets admitting no non-crossing alternating path of length more than $\approx 2n/3$ [2,7]. This upper bound is proved for certain colorings of sets in convex position. The above general lower bound $n/2$ can be slightly improved to $n/2 + \Omega(\sqrt{n/\log n})$ for sets in convex position [7].

In this paper we find arbitrarily large "universal" sets for which any equitable 2-coloring admits a non-crossing Hamiltonian alternating path. We prove the "universality" for so-called double-chains with each chain containing at least one fifth of all the points. Double-chains were first considered in [4].

1.2 Our Results

A *convex* or a *concave chain* is a finite set of points in the plane lying on the graph of a strictly convex or a strictly concave function, respectively. A *double-chain* (C_1, C_2) consists of a convex chain C_1 and a concave chain C_2 such that each point of C_2 lies strictly below every line determined by C_1 and similarly, each point of C_1 lies strictly above every line determined by C_2 (see Fig. 1). Note that we allow different sizes of the chains C_1 and C_2.

Let (C_1, C_2) be a double-chain, and let $p_1, p_2, \ldots, p_k \in C_1 \cup C_2$ be distinct points of $C_1 \cup C_2$. The polygonal line $p_1 p_2 \ldots p_k$ consisting of the $k-1$ straight-line segments $p_1 p_2, p_2 p_3, \ldots, p_{k-1} p_k$ is shortly called *the path* $p_1 p_2 \ldots p_k$. The path $p_1 p_2 \ldots p_k$ is

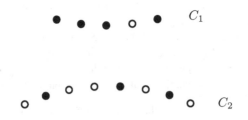

Fig. 1. An equitably 2-colored double-chain (C_1, C_2)

non-crossing if any two non-consecutive segments in it are disjoint. The path $p_1 p_2 \ldots p_k$ is *Hamiltonian (for the double-chain* (C_1, C_2)*)* if it visits all the points of $C_1 \cup C_2$ (i.e., $k = |C_1| + |C_2|$).

Suppose that the points of a double-chain (C_1, C_2) are colored by two colors. Then a path $p_1 p_2 \ldots p_k$ is *alternating* if the endpoints of each segment are colored by different colors. A path on $C_1 \cup C_2$ is a *good path* if it is non-crossing, Hamiltonian and alternating.

An *equitable* 2-*coloring* of a double-chain (C_1, C_2) is a coloring of $C_1 \cup C_2$ by two colors such that the sizes of the color classes differ by at most one. We use *black* and *white* as the colors in the colorings. Here is our main result:

Theorem 1. *Let* (C_1, C_2) *be a double-chain whose points are colored by an equitable* 2-*coloring, and let* $|C_i| \geq \frac{1}{5}(|C_1| + |C_2|)$ *for* $i = 1, 2$. *Then* (C_1, C_2) *has a good path. Moreover, a good path on* (C_1, C_2) *can be found in linear time.*

On the other hand, we show that double-chains with highly unbalanced sizes of chains do not admit a good path for some equitable 2-colorings:

Theorem 2. *Let* (C_1, C_2) *be a double-chain whose points are colored by an equitable* 2-*coloring, and let* C_1 *be periodic with the following period of length 16: 2 black, 4 white, 6 black and 4 white points. If* $|C_1| \geq 28(|C_2| + 1)$, *then* (C_1, C_2) *has no good path.*

2 Proof of Theorem 1

This section contains only the proof for double-chains with an even number of points. The proof for the odd number of points can be found in the Appendix.

The main idea of our proof is to cover the chains C_i by a special type of pairwise non-crossing paths, so called hedgehogs, and then to connect these hedgehogs into a good path by adding some edges between C_1 and C_2.

2.1 Notation Used in the Proof

For $i = 1, 2$, let b_i be the number of black points of C_i and let $w_i := |C_i| - b_i$ denote the number of white points of C_i.

Since the coloring is equitable, we may assume that $b_1 \geq w_1$ and $w_2 \geq b_2$. Then black is *the major color of* C_1 and *the minor color of* C_2, and white is *the major color of* C_2 and *the minor color of* C_1. Points in the major color, i.e., black points on C_1 and white points on C_2, are called *major points*. Points in the minor color are called *minor points*.

Points on each C_i are linearly ordered according to the x-coordinate. An *interval* of C_i is a sequence of consecutive points of C_i. An *inner point* of an interval I is any point of I which is neither the leftmost nor the rightmost point of I.

A *body* D is a non-empty interval of a chain C_i ($i = 1, 2$) such that all inner points of D are major. If the leftmost point of D is minor, then we call it a *head* of D. Otherwise

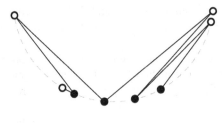

Fig. 2. A hedgehog in C_1

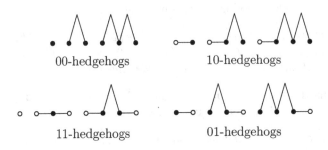

Fig. 3. Types of hedgehogs (sketch)

D has no head. If the rightmost point of D is minor, then we call it a *tail* of D. Otherwise D has no tail. If a body consists of just one minor point, this point is both the head and the tail.

Bodies are of the following four types. A 00-*body* is a body with no head and no tail. A 11-*body* is a body with both head and tail. The bodies of remaining two types have exactly one endpoint major and the other one minor. We will call the body a 10-*body* or a 01-*body* if the minor endpoint is a head or a tail, respectively.

Let D be a body on C_i. A *hedgehog (built on the body $D \subseteq C_i$)* is a non-crossing alternating path H with vertices in C_i satisfying the following three conditions: (1) H contains all points of D, (2) H contains no major points outside of D, (3) the endpoints of H are the first and the last point of D. A hedgehog built on an $\alpha\beta$-body is an $\alpha\beta$-*hedgehog* ($\alpha, \beta = 0, 1$). If a hedgehog H is built on a body D, then D is *the body of H* and the points of H that do not lie in D are *spines*. Note that each spine is a minor point. All possible types of hedgehogs can be seen on Fig. 3 (for better lucidity, we will draw hedgehogs with bodies on a horizontal line and spines indicated only by a "peak" from now on).

On each C_i, maximal intervals containing only major points are called *runs*. Clearly, runs form a partition of major points. For $i = 1, 2$, let r_i denote the number of runs in C_i.

2.2 Proof in the Even Case

Throughout this subsection, (C_1, C_2) denotes a double-chain with $|C_1| + |C_2|$ even. Since the coloring is equitable, we have $b_1 + b_2 = w_1 + w_2$. Set

$$\Delta := b_1 - w_1 = w_2 - b_2.$$

First we give a lemma characterizing collections of bodies on a chain C_i that are bodies of some pairwise non-crossing hedgehogs covering the whole chain C_i.

Lemma 3. *Let $i \in \{1, 2\}$. Let all major points of C_i be covered by a set $\mathcal{D}$ of pairwise disjoint bodies. Then the bodies of $\mathcal{D}$ are the bodies of some pairwise non-crossing hedgehogs covering the whole C_i if and only if $\Delta = d_{00} - d_{11}$, where $d_{\alpha\alpha}$ is the number of $\alpha\alpha$-bodies in $\mathcal{D}$.*

Proof. An $\alpha\beta$-hedgehog containing t major points contains $(t-1) + \alpha + \beta$ minor points. It follows that the equality $\Delta = d_{00} - d_{11}$ is necessary for the existence of a covering of C_i by disjoint hedgehogs built on the bodies of $\mathcal{D}$.

Suppose now that $\Delta = d_{00} - d_{11}$. Let F be the set of minor points on C_i that lie in no body of $\mathcal{D}$, and let M be the set of the mid-points of straight-line segments connecting pairs of consecutive major points lying in the same body. It is easily checked that $|F| = |M|$. Clearly $F \cup M$ is a convex or a concave chain. Now it is easy to prove that there is a non-crossing perfect matching formed by $|F| = |M|$ straight-line segments between F and M (for the proof, take any segment connecting a point of F with a neighboring point of M, remove the two points, and continue by induction); see Fig. 4.

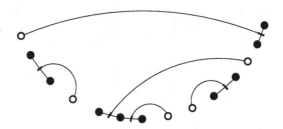

Fig. 4. A non-crossing matching of minor points and midpoints (in C_1)

If $f \in F$ is connected to a point $m \in M$ in the matching, then f will be a spine with edges going from it to those two major points that determined m. Obviously, these spines and edges define non-crossing hedgehogs with bodies in $\mathcal{D}$ and with all the required properties. $\qquad\square$

The following three lemmas and their proofs show how to construct a good path in some special cases.

Lemma 4. *If $\Delta \geq \max\{r_1, r_2\}$ then (C_1, C_2) has a good path.*

Proof. Let $i \in \{1, 2\}$. Since $r_i \leq \Delta \leq \max(b_i, w_i)$, the runs in C_i may be partitioned into Δ 00-bodies. By Lemma 3, these 00-bodies may be extended to pairwise non-crossing hedgehogs covering C_i. This gives us 2Δ hedgehogs on the double-chain. They may be connected into a good path by $2\Delta - 1$ edges between the chains in the way shown in Fig. 5. $\qquad\square$

Lemma 5. *If $r_1 = r_2$ then (C_1, C_2) has a good path.*

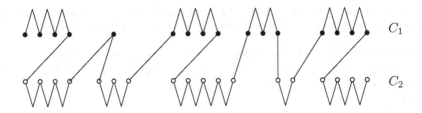

Fig. 5. 00-hedgehogs connected to a good path

Proof. Set $r := r_1 = r_2$. If $r \leq \Delta$ then we may apply Lemma 4. Thus, let $r > \Delta$.

Suppose first that $\Delta \geq 1$. We cover each run on each C_i by a single body whose type is as follows. On C_1 we take Δ 00-bodies followed by $(r - \Delta)$ 10-bodies. On C_2 we take (from left to right) $(\Delta - 1)$ 00-bodies, $(r - \Delta)$ 01-bodies, and one 00-body. By Lemma 3, the r bodies on each C_i can be extended to hedgehogs covering C_i. Altogether we obtain $2r$ hedgehogs. They can be connected to a good path by $2r - 1$ edges between C_1 and C_2 (see Fig. 6).

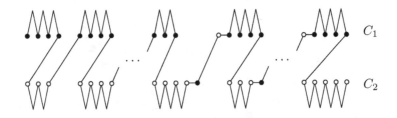

Fig. 6. A good path in the case $r_1 = r_2 > \Delta \geq 1$

Suppose now that $\Delta = 0$. We add one auxiliary major point on each C_i as follows. On C_1, the auxiliary point extends the leftmost run on the left. On C_2, the auxiliary point extends the rightmost run on the right. This does not change the number of runs and increases Δ to 1. Thus, we may proceed as above. The good path obtained has the two auxiliary points on its ends. We may remove the auxiliary points from the path, obtaining a good path for (C_1, C_2). □

A *singleton* $s \in C_i$ is an inner point of C_i ($i = 1, 2$) such that its two neighbors on C_i are colored differently from s.

Lemma 6. *Suppose that C_1 has no singletons and C_2 can be covered by $r_1 - 1$ pairwise disjoint hedgehogs. Then (C_1, C_2) has a good path.*

Proof. For simplicity of notation, set $r := r_1$. We denote the $r - 1$ hedgehogs on C_2 by $P_1, P_2, \ldots, P_{r-1}$ in the left-to-right order in which the bodies of these hedgehogs appear on C_2. For technical reasons, we enlarge the leftmost run of C_1 from the left by an auxiliary major point σ.

Our goal is to find r hedgehogs $H_1, H_2, \ldots, H_r$ on $C_1 \cup \{\sigma\}$ such that they may be connected with the hedgehogs $P_1, P_2, \ldots, P_{r-1}$ into a good path. For each $j = 1, \ldots, r$, the body of the hedgehog H_j will be denoted by D_j. For each $j = 1, \ldots, r$, D_j covers the j-th run of $C_1 \cup \{\sigma\}$ (in the left-to-right order). We now finish the definition of the bodies D_j by specifying for each D_j if it has a head and/or a tail. The body D_1 is without head. For $j > 1$, D_j has a head if and only if P_{j-1} has a tail. The last body D_r is without tail and $D_j, j < r$, has a tail if and only if P_j has a head.

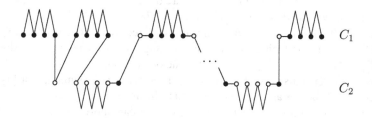

Fig. 7. A good path in the case of no singletons on C_1

It follows from Lemma 3 that we may add or remove some minor points on $C_1 \cup \{\sigma\}$ so that $D_1, \ldots, D_r$ can then be extended to pairwise non-crossing hedgehogs $H_1, \ldots, H_r$ covering the "new" C_1. More precisely, there is a double-chain (C_1', C_2) such that $D_1, \ldots, D_r$ can be extended to pairwise non-crossing hedgehogs $H_1, \ldots, H_r$ covering C_1', where either $C_1' = C_1 \cup \{\sigma\}$ or C_1' is obtained from $C_1 \cup \{\sigma\}$ by adding some minor (white) points on the left of $C_1 \cup \{\sigma\}$ (say) or C_1' is obtained from $C_1 \cup \{\sigma\}$ by removal of some minor (white) points lying in none of the bodies $D_1, \ldots, D_r$. Then the concatenation $H_1 P_1 H_2 P_2 \cdots H_{r-1} P_{r-1} H_r$ shown in Fig. 7 gives a good path on (C_1', C_2). This good path starts with the point σ. Removal of σ from it gives a good path P for the double-chain $(C_1' \setminus \{\sigma\}, C_2)$. The endpoints of P have different colors. Thus, P covers the same number of black and white points. Black points on P are the $\frac{|C_1| + |C_2|}{2}$ black points of (C_1, C_2). Thus, P covers exactly $|C_1| + |C_2|$ points. It follows that $|C_1' \setminus \{\sigma\}| = |C_1|$ and thus $C_1' \setminus \{\sigma\} = C_1$. The path P is a good path on the double-chain (C_1, C_2). $\square$

The following lemma will be used to find a covering needed in Lemma 6.

Lemma 7. *Suppose that $|C_i| \geq k$, $r_i \leq k$ and $\Delta \leq k$ for some $i \in \{1, 2\}$ and for some integer k. Then C_i can be covered by k pairwise disjoint hedgehogs.*

Proof. The idea of the proof is to start with the set $\mathcal{D}$ of $|C_i|$ bodies, each of them being a single point, and then gradually decrease the number of bodies in $\mathcal{D}$ by joining some of the bodies together. We see that $\Delta = d_{00} - d_{11}$, where $d_{\alpha\alpha}$ is the number of $\alpha\alpha$-bodies in $\mathcal{D}$. If we join two neighboring 00-bodies to one 00-body and withdraw a single-point 11-body from $\mathcal{D}$ (to let the minor point become a spine) at the same time, the difference between the number of 00-bodies and the number of 11-bodies remains the same and $|\mathcal{D}|$ decreases by two. We can reduce $|\mathcal{D}|$ by one while preserving the difference $d_{00} - d_{11}$ by joining a 00-body with a neighboring single-point 11-body into a 01- or a 10-body. Similarly we can join a 01- or a 10-body with a neighboring (from the proper side) single-point 11-body into a new 11-body to decrease $|\mathcal{D}|$ by one as well. When we are joining two 00-bodies, we choose the single-point 11-body to remove in such a way to keep as many single-point 11-bodies adjacent to 00-bodies as possible. This guarantees that we can use up to r_i of them for heads and tails.

We start with joining neighboring 00-bodies and we do this as long as $|\mathcal{D}| > k + 1$ and $d_{00} > r_i$. Note that by the assumption $\Delta \leq k$, we will have enough single-point 11-bodies to do that. When we end, one of the following conditions holds: $|\mathcal{D}| = k$, $|\mathcal{D}| = k + 1$ or $d_{00} = r_i$. In the first case we are done. If $|\mathcal{D}| = k + 1$, we just add one head or one tail (we can do this since $d_{00} + d_{11} = |\mathcal{D}| = k + 1 \geq d_{00} - d_{11} + 1$, which implies $d_{11} > 0$). If $d_{00} = r_i$, then each run is covered by just one 00-body. We need to add $|\mathcal{D}| - k$ heads and tails. We have enough single-point 11-bodies to do that since $d_{11} = |\mathcal{D}| - d_{00} = |\mathcal{D}| - r_i \geq |\mathcal{D}| - k$. On the other hand, $r_i - d_{11} = \Delta \geq 0$, so the number of heads and tails needed is at most r_i. Therefore, all the single-point 11-bodies are adjacent to 00-bodies and we can use them to form heads and tails.

In all cases we get a set $\mathcal{D}$ of k bodies. Now we can apply Lemma 3 to obtain k pairwise disjoint hedgehogs covering C_i.   □

By a *contraction* we mean removing a singleton with both its neighbors and putting a point of the color of its neighbors in its place instead. It is easy to verify that if there is a good path in the new double-chain obtained by this contraction, it can be expanded to a good path in the original double-chain.

Now we can prove our main theorem in the even case.

Proof of Theorem 1 (even case).   Without loss of generality we may assume that $r_1 \geq r_2$. In the case $r_1 = r_2$, we get a good path by Lemma 5. In the case $\Delta \geq r_1$, we get a good path by Lemma 4. Therefore, the only case left is $r_1 > r_2, r_1 > \Delta$.

If there is a singleton on C_1, we make a contraction of it. By this we decrease r_1 by one and both r_2 and Δ remain unchanged. If now $r_1 = r_2$ or $r_1 = \Delta$, we again get a good path, otherwise we keep making contractions until one of the previous cases appears or there are no more singletons to contract.

If there is no more singleton to contract on C_1 and still $r_1 > r_2$ and $r_1 > \Delta$, we try to cover C_2 by $r_1 - 1$ pairwise disjoint paths. Before the contractions, $|C_2| \geq \frac{|C_1|}{4}$ did hold and by the contractions we could just decrease $|C_1|$, therefore it still holds.

All the maximal intervals on the chain C_1 (with possible exception of the first and the last one) have now length at least two, which implies that $r_1 \leq \frac{|C_1|}{4} + 1$. Hence $|C_2| \geq \frac{|C_1|}{4} \geq r_1 - 1$, so we can create $r_1 - 1$ pairwise disjoint hedgehogs covering C_2 using Lemma 7. Then we apply Lemma 6 and expand the good path obtained by Lemma 6 to a good path on the original double-chain.

There is a straightforward linear-time algorithm for finding a good path on (C_1, C_2) based on the above proof.   □

3   Unbalanced Double-Chains with No Good Path

In this section we prove Theorem 2. Let (C_1, C_2) be a double-chain whose points are colored by an equitable 2-coloring, and let C_1 be periodic with the following period: 2

black, 4 white, 6 black and 4 white points. Let $|C_1| \geq 28(|C_2| + 1)$. We want to show that (C_1, C_2) has no good path.

Suppose on the contrary that (C_1, C_2) has a good path. Let $P_1, P_2, \ldots, P_t$ denote the maximal subpaths of the good path containing only points of C_1. Since between every two consecutive paths P_i, P_j in the good path there is at least one point of C_2, we have $t \leq |C_2| + 1$. In the following we think of C_1 as of a cyclic sequence of points on the circle. Note that we get more intervals in this way. Theorem 2 now directly follows from the following theorem.

Theorem 8. *Let C_1 be a set of points on a circle periodically 2-colored with the following period of length 16: 2 black, 4 white, 6 black and 4 white points. Suppose that all points of C_1 are covered by a set of t non-crossing alternating and pairwise disjoint paths $P_1, P_2, \ldots, P_t$. Then $t > |C_1|/28$.*

Proof. Each maximal interval spanned by a path P_i on the circle is called a *base*. Let $b(P_i)$ denote the number of bases of P_i. A path with one base only is called a *leaf*. We consider the following special types of edges in the paths. *Long edges* connect points that belong to different bases. *Short edges* connect consecutive points on C_1. Note that short edges cannot be adjacent to each other. A maximal subpath of a path P_i spanning two subintervals of two different bases and consisting of long edges only is called a *zig-zag*. A path is *separated* if all of its edges can be crossed by a line. Note that each zig-zag is a separated path. A maximal separated subpath of P_i that contains an endpoint of P_i and spans one interval only is a *rainbow*. We find all the zig-zags and rainbows in each P_i, $i = 1, 2, \ldots, t$. Note that two zig-zags, or a zig-zag and a rainbow, are either disjoint or share an endpoint. A *branch* is a maximal subpath of P_i that spans two intervals and is induced by a union of zig-zags.

For each path P_i that is not a leaf construct the following graph G_i. The vertices of G_i are the bases of P_i. We add an edge between two vertices for each branch that connects the corresponding bases. If G_i has a cycle (including the case of a "2-cycle"), then one of the corresponding branches consists of a single edge that lies on the convex hull of P_i. We delete such an edge from P_i and don't call it a branch anymore. By deleting a corresponding edge from each cycle of G_i we obtain a graph G_i', which is a spanning tree of G_i. The *branch graph* G' is a union of all graphs G_i'.

Let $\mathcal{L}$ denote the set of leaves and $\mathcal{B}$ the set of branches. Let $\mathcal{P} = \{P_1, P_2, \ldots, P_t\}$.

Observation 1. *The branch graph G' is a forest with components G_i'. Therefore,*

$$|\mathcal{B}| = \sum_{i, P_i \notin \mathcal{L}} (b(P_i) - 1).$$

The branches and rainbows in P_i do not necessarily cover all the points of P_i. Each point that is not covered is adjacent to a deleted long edge and to a short edge that connects this point to a branch or a rainbow. It follows that between two consecutive branches (and between a rainbow and the nearest branch) there are at most two uncovered points, that are endpoints of a common deleted edge. By an easy case analysis it can be shown that this upper bound can be achieved only if one of the nearest branches consists of a single zig-zag.

In the rest of the paper, a *run* will be a maximal monochromatic interval of any color. In the following we will count the runs that are spanned by the paths P_i. The *weight* of a path P, $w(P)$, is the number of runs spanned by P. If P spans a whole run, it adds one unit to $w(P)$. If P partially spans a run, it adds half a unit to $w(P)$.

Observation 2. *The weight of a zig-zag or a rainbow is at most 1.5. A branch consists of at most two zig-zags, hence it weights at most three units.*

Lemma 9. *A path P_i that is not a leaf weights at most $3.5k + 3.5$ units where k is the number of branches in P_i.*

Proof. According to the above discussion, for each pair of uncovered points that are adjacent on P_i we can join one of them to the adjacent branch consisting of a single zig-zag. To each such branch we join at most two uncovered points, hence its weight increases by at most one unit to at most 2.5 units. The number of the remaining uncovered points is at most $k+1$. Therefore, $w(P_i) \leq 3k+3+0.5\cdot(k+1) = 3.5k+3.5$. $\square$

Lemma 10. *A leaf weights at most 3.5 units.*

Proof. Let L be a leaf spanning at least two points. Consider the interval spanned by L. Cut this interval out of C_1 and glue its endpoints together to form a circle. Take a line l that crosses the first and the last edge of L. Note that the line l doesn't separate any of the runs. Exactly one of the arcs determined by l contains the gluing point γ.

Each of the ending edges of L belongs to a rainbow, all of whose edges cross l. It follows that if L has only one rainbow, then this rainbow covers the whole leaf L and $w(L) \leq 1.5$. Otherwise L has exactly two rainbows, R_1 and R_2. We show that R_1 and R_2 cover all edges of L that cross the line l. Suppose there is an edge s in L that crosses l and does not belong to any of the rainbows R_1, R_2. Then one of these rainbows, say R_1, is separated from γ by s. Then the edge of L that is the second nearest to R_1 also has the same property as the edge s. This would imply that R_1 spans two whole runs, a contradiction. It follows that all the edges of L that are not covered by the rainbows are consecutive and connect adjacent points on the circle. There are at most three such edges; at most one connecting the points adjacent to γ, the rest of them being short on C_1. But this upper bound of three cannot be achieved since it would force both rainbows to span two whole runs. Therefore, there are at most two edges and hence at most one point in L uncovered by the rainbows. The lemma follows. $\square$

Lemma 11. $|\mathcal{L}| \geq \sum_{i, P_i \not\subseteq \mathcal{L}} (b(P_i) - 2) + 2$.

Proof. The number of runs in C_1 is at least 4. By Lemma 10, if all the paths P_i are leaves, then at least 2 of them are needed to cover C_1 and the lemma follows.

If not all the paths are leaves, we order the paths so that all the leaves come at the end of the ordering. The path P_1 spans $b(P_1)$ bases. Shrink these bases to points. These points divide the circle into $b(P_1)$ arcs each of which contains at least one leaf. If P_2 is not a leaf then continue. The path P_2 spans $b(P_2)$ intervals on one of the previous arcs. Shrink them to points. These points divide the arc into $b(P_2) + 1$ subarcs. At least $b(P_2) - 1$ of them contain leaves. This increased the number of leaves by at least $b(P_2) - 2$. The case of P_i, $i > 2$, is similar to P_2. The lemma follows by induction. $\square$

Corollary 12. $|\mathcal{B}| \leq |\mathcal{P}| - 2$.

Proof. Combining Lemma 11 and Observation 1 we get the following:

$$|\mathcal{B}| = \sum_{i, P_i \notin \mathcal{L}} (b(P_i) - 1) = \sum_{i, P_i \notin \mathcal{L}} (b(P_i) - 2) + |\mathcal{P}| - |\mathcal{L}| + 2 - 2 \leq |\mathcal{P}| - 2.$$

$\square$

Now we are in position to finish the proof of Theorem 8. If the whole C_1 is covered by the paths P_i, then $\sum_{i=1}^{t} w(P_i) \geq \frac{|C_1|}{4}$. Therefore,

$$|C_1| \leq 4 \cdot (3.5|\mathcal{B}| + 3.5(|\mathcal{P}| - |\mathcal{L}|) + 3.5|\mathcal{L}|) < 4 \cdot 7|\mathcal{P}| = 28|\mathcal{P}|.$$

$\square$

Acknowledgment

We thank Jakub Černý for his active participation at the earlier stages of our discussions.

References

1. Abellanas, M., García, J., Hernandez, G., Noy, M., Ramos, P.: Bipartite embeddings of trees in the plane. Discrete Appl. Math. 93, 141–148 (1999)
2. Abellanas, M., García, J., Hurtado, F., Tejel, J.: Caminos alternantes (in Spanish). In: Proc. X Encuentros de Geometría Computacional, Sevilla, pp. 7–12 (2003) (English version available on Ferran Hurtado's web page)
3. Brass, P., Moser, W., Pach, J.: Research Problems in Discrete Geometry. Springer, Heidelberg (2005)
4. García, A., Noy, M., Tejel, J.: Lower bounds on the number of crossing-free subgraphs of K_N. Comput. Geom. 16, 211–221 (2000)
5. Kaneko, A., Kano, M.: Discrete geometry on red and blue points in the plane - a survey. In: Aronov, B., et al. (eds.) Discrete and computational geometry, The Goodman-Pollack Festschrift. Algorithms Comb., vol. 25, pp. 551–570. Springer, Heidelberg (2003)
6. Kaneko, A., Kano, M., Suzuki, K.: Path coverings of two sets of points in the plane. Pach, J. (ed.), Towards a Theory of Geometric Graphs, Contemporary Mathematics 342, 99–111 (2004)
7. Kynčl, J., Pach, J., Tóth, G.: Long Alternating Paths in Bicolored Point Sets. In: Pach, J. (ed.) GD 2004. LNCS, vol. 3383, pp. 340–348. Springer, Heidelberg (2005); Also to appear in a special volume of Discrete Mathematics honouring the 60th birthday of M. Simonovits

Appendix: Proof in the Odd Case

In this appendix we prove Theorem 1 for the case when $|C_1| + |C_2|$ is odd. We set $\Delta = w_2 - b_2$ and proceed similarly as in the even case. On several places in the proof we will add one auxiliary point ω to get the even case (its color will be chosen to equalize the numbers of black and white points). We will be able to apply one of the

Lemmas 4–6 to obtain a good path. The point w will be at some end of the good path and by removing w we obtain a good path for (C_1, C_2).

Without loss of generality we may assume that $r_1 \geq r_2$. In the case $r_1 = r_2$, we add an auxiliary major point w, which is placed either as the left neighbor of the leftmost major point on C_1 or as the right neighbor of the rightmost major point on C_2. Then we get a good path by Lemma 5 and the removal of w gives us a good path for (C_1, C_2).

In the case $\Delta \geq r_1$, we add an auxiliary point w to the same place and we get a good path by Lemma 4. Again, the removal of w gives us a good path for (C_1, C_2).

Now, the only case left is $r_1 > r_2$, $r_1 > \Delta$. If there are any singletons on C_1, we make the contractions exactly the same way as in the proof of the even case. If Lemma 4 or 5 needs to be applied, we again add an auxiliary point w and proceed as above.

If there is no more singleton to contract on C_1 and still $r_1 > r_2$ and $r_1 > \Delta$, we have $|C_2| \geq \frac{|C_1|}{4} \geq r_1 - 1$ as in the proof of the even case and we can use Lemma 7 to get $r_1 - 1$ pairwise disjoint hedgehogs covering C_2. Now we need to consider two cases: (1) If $b_1 + b_2 > w_1 + w_2$, then we find a good path for (C_1, C_2) in the same way as in the proof of Lemma 6, except we do not add the auxiliary point σ. (2) If $b_1 + b_2 < w_1 + w_2$, we add an auxiliary point w as the right neighbor of the rightmost major point on C_1. The number r_1 didn't change so Lemma 6 gives us a good path. Again, the removal of w gives us a good path for (C_1, C_2).

There is a straightforward linear-time algorithm for finding a good path on (C_1, C_2) based on the above proof. $\square$

The Binary Stress Model for Graph Drawing

Yehuda Koren[1] and Ali Çivril[2]

[1] AT&T Labs — Research
[2] Rensselaer Polytechnic Institute

Abstract. We introduce a new force-directed model for computing graph layout. The model bridges the two more popular force directed approaches – the
stress and the electrical-spring models – through the *binary stress* cost function,
which is a carefully defined energy function with low descriptive complexity allowing fast computation via a Barnes-Hut scheme. This allows us to overcome
optimization pitfalls from which previous methods suffer. In addition, the binary
stress model often offers a unique viewpoint to the graph, which can occasionally add useful insight to its topology. The model uniformly spreads the nodes
within a circle. This helps in achieving an efficient utilization of the drawing area.
Moreover, the ability to uniformly spread nodes regardless of topology, becomes
particularly helpful for graphs with low connectivity, or even with multiple connected components, where there is not enough structure for defining a readable
layout.

1 Introduction

A popular approach to drawing graphs is based on measuring the quality of the layout
through a formal cost function. The layout of the graph is formed by an optimization
algorithm that finds a local minimum of the cost function. This family of algorithms is
known in the graph drawing literature as force-directed algorithms; see, e.g., [3,14].

Broadly speaking, force-directed cost functions (also known as *energies*) define a desired layout based on either the electric-spring metaphor or on a stress function. Electric
spring functions liken the graph to a physical system where nodes correspond to electrically charged particles, and edges correspond to springs with zero rest length. Repulsive
electric forces ensure that nodes are well separated, while attractive spring forces tend to
shorten edges and pack closely connected components. Two well known early versions
of this scheme are by Eades [4] and by Fruchterman and Reingold [6].

The stress function relates a nice drawing to good isometry. We have an ideal target
distance d_{ij} for every pair of nodes i and j. Given a 2-D layout, where node i is placed
at point p_i, the stress function is:

$$\sum_{i<j} w_{ij} \left(\|p_i - p_j\| - d_{ij} \right)^2 \tag{1}$$

We desire a layout that minimizes this function, thereby best realizing the target distances. Here, the distance d_{ij} is typically the graph-theoretical distance between nodes

I.G. Tollis and M. Patrignani (Eds.): GD 2008, LNCS 5417, pp. 193–205, 2009.
© Springer-Verlag Berlin Heidelberg 2009

i and j. The normalization constant w_{ij} equals $d_{ij}^{-\alpha}$. The function (1) appeared earlier as the stress function in multidimensional scaling [2], where it was applied to graph drawing [16]. It became a popular graph drawing tool by Kamada and Kawai [13].

Both electric-spring and stress approaches enjoy successful implementations and offer pleasing layouts to many graphs. In terms of layout appearance, there are distinct differences between the models, though they are hard to define. As for computational aspects, the two approaches induce different optimization processes, and each has a unique advantage. Electric-spring models have the advantage of a lower descriptive complexity compared to the stress model. This is because all repulsive forces are uniform, whereas attractive forces involve only the $|E|$ pairs of adjacent nodes. On the other hand, the stress function requires encoding a different target distance for each node pair. This fundamental difference bounds stress models to quadratic space complexity, while efficient implementations of electric-spring models scale to larger graphs.

On the other hand, the stress function has a mild landscape, which allows utilizing powerful optimization techniques such as majorization [7]. This way, good minima are usually achieved regardless of the initial positions. This is untrue for the electric-spring models, which induce an intricate landscape as repulsive forces make the energy go to infinity when nodes overlap. This causes serious convergence problems even for moderately sized graphs. Past works [9,11,19] used sophisticated initialization techniques through multilevel approximation to overcome these problems.

In this work we introduce the binary-stress model (*bStress*) for drawing graphs. Computationally, it is able to merge the advantages of both the electric-spring model and the stress model. Namely, it offers a low descriptive complexity, thus being scalable to very large graphs. At the same time, it is similar in its form to the known stress function, thus enabling the use of the majorization optimization scheme.

As for the quality of the layout, bStress frequently offers a unique perspective to the graph structure. More than other models, bStress emphasizes uniform spread of the nodes within a circular drawing area. This may lead to distinctive layouts, which can serve as useful addition to those produced by other algorithms. Moreover, the emphasis on uniform spread is advantageous for graphs with low connectivity, whose structure alone is not capable of defining a good layout. For example, bStress will naturally handle graphs with multiple connected components by packing all connected components together without requiring any post-processing or special treatment that alternative methods require. In addition, bStress is suitable for drawing large graphs, not only because of its improved scalability, but also because it achieves good area utilization that is important for placing a large number of nodes.

2   Basic Notions

We are seeking a layout for a graph $G(V = \{1, \ldots, n\}, E)$, where the position of node i is $p_i = (x_i, y_i)$. Sometimes, we will refer to the vectors $x, y \in \mathbb{R}^n$, which represent all x- or y-coordinates, respectively. Notice that while this work addresses the more common case of 2-D layouts, as usual with force-directed algorithms, extensions to 3-D are naturally possible.

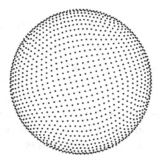

Fig. 1. A Layout of 1024 points that minimizes $G(p)$, by scattering the points within a circle

3   The Binary Stress Model

One of the earliest cost functions involved in defining a nice layout strives to shorten the squared edge lengths:

$$H(p) = \sum_{\langle i,j \rangle \in E} \|p_i - p_j\|^2 \qquad (2)$$

However, minimizing $H(p)$ on its own is not sufficient for defining a useful layout, as nothing prevents all nodes from collapsing at a single point. Thus, Tutte [18] and Hall [10] augmented $H(p)$ with simple constraints that prevented the formation of trivial layouts. Nonetheless, both solutions tend to generate layouts with very uneven sparsity, where many nodes are overcrowded together. Moreover, Tutte's and Hall's methods fail to produce adequate layouts for graphs of low connectivity such as tree-like graphs.

A hypothetical possible way to make $H(p)$ working for general graphs, is to lay out the graph over a grid and then minimize $H(p)$ while requiring that each node is positioned at a unique grid cell. This will ensure a uniform spread of the nodes and prevent nodes from getting too close to each other. However, practical implementation of such a strategy would be quite complicated. The primary issue is that constraining positions to grid cells transforms the problem into integer optimization, which would be much harder to solve and less scalable.

We avoid integer optimization by adopting a continuous relaxation of the grid layout strategy. The relaxation is based on the following cost function:

$$G(p) = \sum_{i \neq j \in V} (\|p_i - p_j\| - 1)^2 \qquad (3)$$

This function strives to place all nodes such that their pairwise distances are uniform. Notice that $G(p)$ is independent of the graph structure. The minimum of $G(p)$, as we have found experimentally, will position the nodes almost uniformly within a circle. For example, consider Fig. 1, where 1024 nodes are positioned so as to minimize $G(p)$.

The function $G(p)$ gives us the necessary tool to combat the over dense areas which are typical to minimization of $H(p)$. Thus, the binary stress function for computing a layout of a graph is defined as a linear combination of the two functions:

$$B(p) = \sum_{\langle i,j \rangle \in E} \|p_i - p_j\|^2 + \alpha \sum_{i \neq j \in V} (\|p_i - p_j\| - 1)^2 \qquad (4)$$

The first term relates the layout to the graph structure by ensuring that edges are short, whereas the second term makes the nodes spread uniformly within a circle. The constant α (discussed later) controls the balance between the two terms.

Our experience shows that bStress results in useful layouts for wide families of graphs. However, before we dwell into the quality of layouts generated by the bStress model, we would like to discuss computational aspects.

4 Minimizing the Binary Stress Function

The bStress function (4) is structured as a sum of two stress functions (Eq. (1)), one with target distances equal to 0, and the other with target distances equal to 1. This is the reason for choosing the "binary stress" name. Though, the particular value of 1 has no influence on the resulting layout and any other positive value could be used as well.

As sum of stress functions, the majorization optimization technique can be exploited to optimizing bStress. Derivation of the stress majorization was given by Gansner et al. [7]. The process used here is as follows:

Let us define two $n \times n$ matrices, L and M. The matrix L is the *Laplacian* of graph G, whose associated quadratic form is the sum of squared edge lengths $H(p)$. The other matrix, M, is associated with a quadratic form that bounds $G(p)$:

$$L_{i,j} = \begin{cases} -1 & \langle i,j \rangle \in E \\ \sum_{k \neq i} L_{ik} & i = j \\ 0 & \text{otherwise} \end{cases}, \qquad M_{i,j} = \begin{cases} -1 & i \neq j \\ n-1 & i = j \end{cases}$$

We also define two vectors, $b^x, b^y \in \mathbb{R}^n$, which sum all cosines and sines associated with each node:

$$b_i^x = \sum_{j \neq i} \frac{x_i - x_j}{\|(x_i, y_i) - (x_j, y_j)\|}, \qquad b_i^y = \sum_{j \neq i} \frac{y_i - y_j}{\|(x_i, y_i) - (x_j, y_j)\|} \qquad (5)$$

Given a current placement $p(t) = (x(t), y(t))$, an improved placement $p(t+1) = (x(t+1), y(t+1))$, which lowers $B(p)$, is computed by solving the system of equations:

$$(M + \alpha L)x(t+1) = b^{x(t)}, \qquad (M + \alpha L)y(t+1) = b^{y(t)} \qquad (6)$$

Now, let us consider computational complexity. The number of entries in matrix L is $n + |E|$. The other matrix – M – is, strictly speaking, dense. However its highly uniform structure makes it sparse for practical purposes. Typical to the stress majorization process is solving (6) by using the conjugate gradient method, which accesses $(M + \alpha L)$ as a linear operator. Thus, all we need to ensure is that the product $(M + \alpha L)x$, can be computed efficiently. This is indeed the case, as L is sparse, and $(Mx)_i = nx_i - \sum_j x_j$, which is computed in a constant time after precomputing $\sum_j x_j$. Thus, the product $(M + \alpha L)x$, is computed in time $O(n + |E|)$.

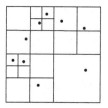

Fig. 2. A quad-tree hierarchical space decomposition

The more challenging operation is the computation of the b^x and b^y vectors of Eq. (5). This essentially involves computing the angles formed by all node pairs. Here we follow several recent graph drawing works [9,11,17] and use the Barnes-Hut scheme [1] for approximating the $O(n^2)$ interactions in practically $O(n \log n)$ time. Thus, we use a hierarchical geometric decomposition of the drawing area through a quad-tree data structure. The whole area is assigned to a square (or, a rectangle). Then, each square is subsequently partitioned into four identical squares, till each node is lying within a unique leaf square. See Fig. 2 for an illustration.

Computation of b_i^x and b_i^y is based on a top-bottom traversal of the quad-tree. Let v be a quad-tree vertex corresponding to square s with side length l. We compare l to d - the distance between node i and the center of square s. If $l/d > \theta$, then we continue the traversal recursively with the four children of v. Otherwise, we halt the traversal while taking the approximation that all graph nodes lying within square s are at the same location, and thus can be processed at once. Our default value for θ is 0.5.

In order to give a flavor of actual running times, we report our experience with graphs of varying sizes in Table 1. Times were measured on a Pentium 4 PC. We let the majorization process run for 200 iterations, while it was terminated earlier once $\|p(t+1) - p(t)\| / \|p(t)\| < 0.001$. Overall running time is divided among the two components of the algorithm: (1) solving Eq. (6) through the conjugate gradients iterative process. (2) Computing b^x and b^y (Eq. (5)) using a Barnes-Hut approximation. The table shows that the Barnes-Hut approximation is indeed closely following an $O(n \log n)$ running time. The conjugate gradient component takes $(n + |E|)$ time per internal iteration, but the number of those iterations is less consistent. Since the Barnes-Hut calculation is independent of the number edges, as graphs become denser the conjugate gradient component becomes more significant (see graphs 'plustk10' and 'gearbox'). Wall-clock measured running times are not directly comparable across different papers, due to differences in platforms and code optimization. However, we believe that the ability of bStress to lay out of 100,000 nodes in a few minutes, places it among the more efficient graph drawing techniques.

5   Results and Implementation Details

The binary stress model is based on unique principles, which in many cases lead to layouts quite different than those produced by other algorithms. Hence, a key to assessing the utility of the new model is a qualitative analysis of typical results. In the following subsections we discuss various aspects of bStress through concrete layout examples.

Table 1. Running time characteristics for graphs of varying sizes. We measure times for the two components of the algorithm: a conjugate gradient solver, and Barnes-Hut approximation of vectors b^x and b^y. The last two columns show the dependency of running time with graph size. Graphs are taken from [12].

name	nodes	edges	iterations	conjugate gradient time/it (sec.)	Barnes-Hut time/it (sec.)	$10^6 \times$ C.G. time $\frac{}{|E|+n}$	$10^6 \times$ B.H. time $\frac{}{n \cdot \log n}$
nopoly	10774	30034	133	0.019	0.182	0.477	4.181
skirt	12598	91961	109	0.082	0.272	0.784	5.264
tuma2	12992	20925	13	0.015	0.238	0.454	4.462
poli_large	15575	17468	200	0.106	0.305	3.199	4.666
powersim	15838	36430	200	0.045	0.357	0.869	5.366
ncvxqp9	16554	22493	200	0.023	0.405	0.598	5.797
lpl1	32460	147788	200	0.408	0.763	2.261	5.212
finance256	37376	130560	200	0.192	0.749	1.145	4.385
bcircuit	68902	153328	200	0.328	1.874	1.476	5.621
plustk10	80676	2114154	159	5.169	2.125	2.355	5.367
Ford2	100196	222246	33	0.582	2.230	1.806	4.450
gearbox	107624	3250488	200	5.874	3.317	1.749	6.124
lung2	109460	273646	137	0.272	3.477	0.710	6.304

5.1 Balancing the System

Recall that bStress is parametrized by α, which controls the balance between uniform spread and structure preservation. As α grows, the model will prefer shortening edges over uniformly spreading the nodes. This can significantly influence the appearance of the layout. For example, in Fig. 3 we show two layouts of the same graph, one computed with $\alpha = 1$ and the other with $\alpha = 1000$. When α is low (=1), the model emphasizes uniform spread, thus nodes are well separated and visible. On the other hand, when α is high (=1000), the model cares mostly about exposing the graph's structure through shortening edges. Thus, the different hubs that form the graph are clearly shown.

Notice that $G(p) = \sum_{i \neq j \in V} (\|p_i - p_j\| - 1)^2$ contains about $n^2/2$ terms, whereas the other part of bStress, $H(p) = \sum_{\langle i,j \rangle \in E} \|p_i - p_j\|^2$, contains only $|E|$ terms. Thus, $G(p)$ becomes more and more dominant as $n^2/|E|$ grows. This is undesirable, as it makes the determination of parameter α less stable across varying graphs. To offset some of this phenomenon, our experience shows that as $|E|/n$ grows, it is beneficial to overweight $H(p)$ over $G(p)$. In other words, for sparse graphs, there is no much structure in the graph and it is reasonable to pay much attention to uniform spread. However, for denser graphs, there is much structure to be captured from the connectivity information. Combining these considerations, we learned that a sensible choice to α is $c \cdot n$, for some positive constant c. Hence, the bStress model becomes:

$$B(p) = \sum_{\langle i,j \rangle \in E} \|p_i - p_j\|^2 + c \cdot n \sum_{i \neq j \in V} (\|p_i - p_j\| - 1)^2 \qquad (7)$$

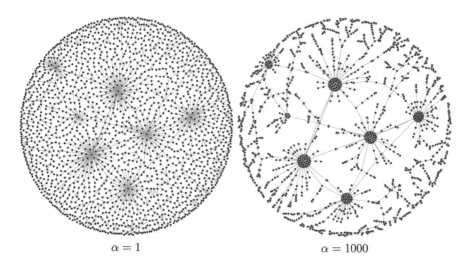

$$\alpha = 1 \qquad\qquad\qquad \alpha = 1000$$

Fig. 3. Two bStress layouts of a graph with 1933 nodes and 2043 edges. Setting $\alpha = 1$ achieves better separation of nodes and improved area utilization. However, some may prefer $\alpha = 1000$, for the better abstraction of the graph's structure.

Focusing on values of c is easier than focusing on values of α. In fact, our experiments show that $c = 1$ is a universally reasonable choice, being our default value. In some cases, better results are obtained with lower values of c.

There is another implication to the value of c, beyond layout appearance. We have found that the majorization optimization process may encounter bad local minima when c is too low. To avoid this, we first run the algorithm with higher values of c, and then use the resulting layout for seeding a process with a lower c value. That is, a typical run would start with $c=100$, and then restart with $c=1$. Usually, the number of majorization iterations after restarting the run is relatively low thanks to the improved initialization.

5.2 Drawing Trees

Prior adaptation of the $H(p)$ function to drawing graphs [10,18] could not handle trees and tree-like graphs adequately. The major issue was the inability to prevent many nodes from collapsing at the same location, thus resulting in a highly imbalanced layout with much unused area and a few overcrowded locations. Such an issue does not exist with bStress, as could be evident from the drawing of a tree-like graph given in Fig. 3. In fact, as graphs become sparser, results of bStress look increasingly different than those computed by alternative models such as the aforementioned stress and electric-spring models. This is because, the lack of sufficient connectivity information let the uniform spread component, $G(p)$, be more dominant in shaping the layout.

As an example, in Fig. 4–5 we present the drawings of two trees, which are derived from an Internet map and a BGP connectivity map. Results of bStress are compared to the results of the stress function. The known stress model seems to be better at exposing the decomposition of the tree, whereas bStress achieves more uniform node distribution.

bStress   stress

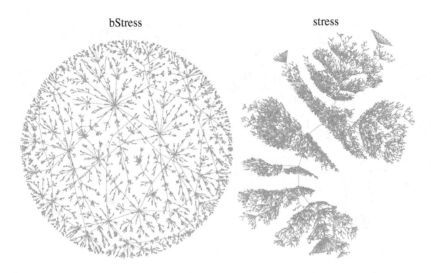

Fig. 4. Comparing stress to bStress in drawing an Internet map tree ($|V|$=9227, $|E|$= 9226)

bStress   stress

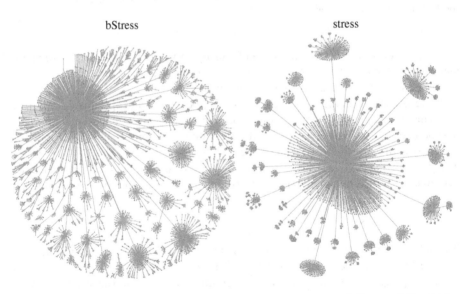

Fig. 5. Comparing stress to bStress in drawing a BGP connectivity tree ($|V|$=3487, $|E|$= 3486)

The uniform spread achieved by bStress becomes particularly useful when the number of nodes is large making area utilization a high priority.

5.3   Disconnected Graphs

Most force-directed methods cannot directly handle disconnected graphs. For example, the stress model requires defining the distance between each two nodes, which is

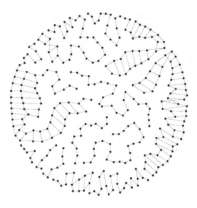

Fig. 6. A graph with 11 connected components ($|V|$=333, $|E|$=397)

not naturally defined for disconnected nodes. Likewise, the electric spring model assumes only repulsive forces among connected components, ultimately pushing them away from each other till infinity. Certainly, various modifications to those models can enable working with disconnected graphs. Most notably, each connected component can be drawn separately, and later a smart packing algorithm squeezes all components within the drawing area [5].

Interestingly, bStress handles disconnected graphs exactly the same way it handles connected graphs. Thus, unlike other methods, it does not require any modification or postprocessing when addressing disconnectivity. This is thanks to the uniform spread model ($G(p)$), which strives for a fairly uniform node distribution, regardless of connectivity. A small artificial example is brought in Fig. 6, where we draw a graph with 11 connected components. As can be seen, bStress could pack all components efficiently together within a circle, while no two components overlap, and each component is dawn reasonably. A larger, more realistic example is given in Fig. 7, where we show a graph consisting of many Internet traces. The graph contains 3743 connected components, which are all packed pretty well within the layout.

5.4 Filling a Circle

A notable feature of bStress is packing the graph within a circle. Admittedly, the circular shape of the layout is not a design goal but rather an outcome of the chosen cost function. However, filling the interior of the circle is indeed a design goal of the bStress model. In some cases this can lead to surprisingly looking layouts. For example, some layouts would be expected to lie on the periphery of a circle. However, bStress will "insist" on filling the circle with some of the nodes, due to the strict uniform spread requirement. This might look odd at first, but we argue that it has an advantage of enabling a better distinction between individual nodes.

We demonstrate this in Fig. 8. First simple example is a (topological) circle, which is twisted in order to spread nodes within the interior. Another example is the finan512 graph, which became a standard example in works aimed at drawing large graphs.

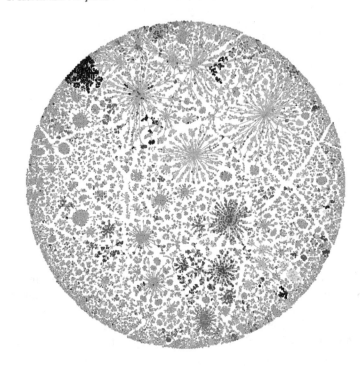

Fig. 7. An Internet map with 3743 connected components ($|V|$=33552, $|E|$=29809). Node colors indicate some known ISPs.

Previous works (e.g., [15,19]) placed all nodes on or close to the perimeter of a circle. On the other hand, bStress fills the interior of the circle. This enables a better view of the local details of this large graph, at the price of an inferior exhibition of symmetries. At this point, we would like to clarify that while frequently the outline of the layout is circular, this is not always the case; for example consider Fig. 9.

5.5   Distorting the Layout

The uniform spread component, $G(p)$, induces layouts where the periphery is denser than the central area. This effect can be seen in Fig. 1. Let us take a polar coordinates viewpoint, where the origin is the layout center. We observe that nodes are uniformly spread across different angular coordinates, but less so across different radial coordinates. Thus, we propose the following correction as an optional postprocessing phase.

We denote the layout density (or, sparsity) around node i by d_i. This way $d_i = 0$ for the densest possible area, while d_i is large when there is a lot of free area around i. One way to measure d_i is to set it to the average distance between i and its top k closest nodes in the layout. In our implementation, we compute a relative neighborhood graph (RNG), and define d_i as the average length of edges adjacent to i in the RNG.

We sort all nodes by their radial coordinates, which are distances from the center. Then, we smooth the computed densities, by averaging densities of nodes with similar

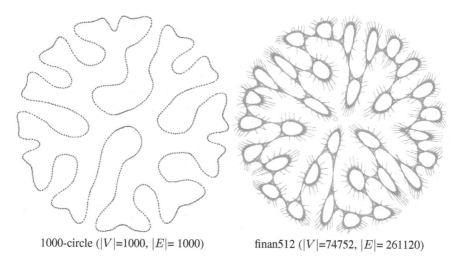

1000-circle ($|V|$=1000, $|E|$= 1000)   finan512 ($|V|$=74752, $|E|$= 261120)

Fig. 8. bStress tends to fill the interior of a circle

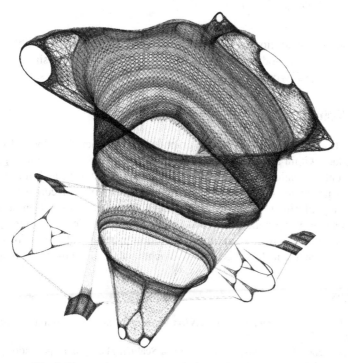

Fig. 9. The gearbox graph [12] ($|V|$=107624, $|E|$=3250488)

radial coordinates; see Sec. 6 of [8] for a similar procedure. Finally, for each node i, which comes immediately after node j in the sorted order, we modify the gap in radial coordinates between i and j by multiplying it by $1/d_i$. Thus, we shrink gaps in sparse areas, while widening gaps in dense areas.

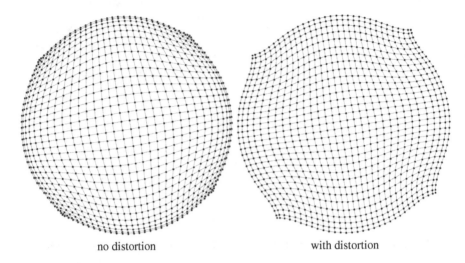

no distortion   with distortion

Fig. 10. The effect of post-processing the layout of a 32×32 grid with a radial distortion that makes node distribution more uniform

We include this distortion in our default settings, as it takes a negligible time, and occasionally leads to a modest improvement of layout appearance. A simple example is a square grid, whose layout improves when applying the distortion as shown in Fig. 10.

6   Conclusions

The binary stress model leads to unique graph layouts characterized by uniform distribution of nodes within a circular area. This is particularly beneficial for large graphs, where efficient utilization of the drawing area becomes vital. In addition, the model is capable of producing decent layouts even for graphs with low connectivity, where scant adjacency information cannot define a useful layout on its own. Computationally, it combines some of the benefits of both the stress and the electric-spring model, facilitating a simple, yet effective optimization procedure that scales well for very large graphs. We believe that it should coexist as a viable option along more familiar models.

References

1. Barnes, J.E., Hut, P.: A hierarchical $O(NlogN)$ force calculation algorithm. Nature 324(4), 446–449 (1986)
2. Borg, I., Groenen, P.: Modern Multidimensional Scaling: Theory and Applications. Springer, Heidelberg (1997)
3. Di Battista, G., Eades, P., Tamassia, R., Tollis, I.G.: Graph Drawing: Algorithms for the Visualization of Graphs. Prentice-Hall, Englewood Cliffs (1999)
4. Eades, P.: A heuristic for graph drawing. Cong. Numer. 42, 149–160 (1984)
5. Freivalds, K., Dogrusoz, U., Kikusts, P.: Disconnected graph layout and the polyomino packing approach. In: Mutzel, P., Jünger, M., Leipert, S. (eds.) GD 2001. LNCS, vol. 2265, pp. 378–391. Springer, Heidelberg (2002)

6. Fruchterman, T.M.G., Reingold, E.: Graph drawing by force-directed placement. Software-Practice Experience 21(11), 1129–1164 (1991)
7. Gansner, E., Koren, Y., North, S.: Graph drawing by stress majorization. In: Pach, J. (ed.) GD 2004. LNCS, vol. 3383, pp. 239–250. Springer, Heidelberg (2005)
8. Gansner, E., Koren, Y., North, S.: Topological fisheye views for visualizing large graphs. IEEE Trans. Vis. Comput. Graph. 11(4), 457–468 (2005)
9. Hachul, S., Junger, M.: Drawing large graphs with a potential-field-based multilevel algorithm. In: Pach, J. (ed.) GD 2004. LNCS, vol. 3383, pp. 285–295. Springer, Heidelberg (2005)
10. Hall, K.M.: An r-dimensional quadratic placement algorithm. Management Science 17(3), 219–229 (1970)
11. Hu, Y.F.: Efficient high quality force-directed graph drawing. The Mathematica Journal 10(1), 37–71 (2005)
12. Hu, Y.F.: A gallery of large graphs,
    http://www.research.att.com/~yifanhu/GALLERY/GRAPHS
13. Kamada, T., Kawai, S.: An algorithm for drawing general undirected graphs. Information Processing Letters 31(1), 7–15 (1989)
14. Kaufmann, M., Wagner, D. (eds.): Drawing Graphs. LNCS, vol. 2025. Springer, Heidelberg (2001)
15. Koren, Y.: Graph drawing by subspace optimization. In: Eurographics / IEEE TCVG Symposium on Visualization, pp. 65–74 (2004)
16. Kruskal, J., Seery, J.: Designing network diagrams. In: First General Conference on Social Graphics, pp. 22–50 (1980)
17. Quigley, A., Eades, P.: FADE: Graph drawing, clustering and visual abstraction. In: Marks, J. (ed.) GD 2000. LNCS, vol. 1984, pp. 197–210. Springer, Heidelberg (2001)
18. Tutte, W.T.: How to Draw a Graph. Proc. London Math. Soc. s3-13(1), 743–767 (1963)
19. Walshaw, C.: A multilevel algorithm for force-directed graph drawing. In: Marks, J. (ed.) GD 2000. LNCS, vol. 1984, pp. 171–182. Springer, Heidelberg (2001)

Efficient Node Overlap Removal Using a Proximity Stress Model

Emden R. Gansner and Yifan Hu

AT&T Labs, Shannon Laboratory, 180 Park Ave., Florham Park, NJ 07932
{erg,yifanhu}@research.att.com

Abstract. When drawing graphs whose nodes contain text or graphics, the non-trivial node sizes must be taken into account, either as part of the initial layout or as a post-processing step. The core problem is to avoid overlaps while retaining the structural information inherent in a layout using little additional area. This paper presents a new node overlap removal algorithm that does well by these measures.

1 Introduction

Most existing symmetric graph layout algorithms treat nodes as points. In practice, nodes usually contain labels or graphics that need to be displayed. Naively incorporating this can lead to nodes that overlap, causing information of one node to occlude that of others. If we assume that the original layout conveys significant aggregate information such as clusters, the goal of any layout that avoids overlaps should be to retain the "shape" of the layout based on point nodes.

The simplest and, in some sense, the best solution is to scale up the drawing [23] while preserving the node size until the nodes no longer overlap. This has the advantage of preserving the shape of the layout exactly, but can lead to inconveniently large drawings. In general, overlap removal is typically a trade-off between preserving the shape and limiting the area, with scaling at one extreme.

Many techniques to avoid overlapping nodes have been devised. One approach is to make the node size part of the model of the layout algorithm. It is assumed that whatever structure that would have been exposed using point nodes will still be evident in these more general layouts. Various authors [2, 13, 21, 26] have extended the spring-electrical model [4, 7] to take into account node sizes, usually as increased repulsive forces. Node overlap removal can also be built into the stress model [19] by specifying the ideal edge length to avoid overlap along the graph edges. Such heuristics, however, cannot guarantee all overlaps will be removed, so they rely on overly large repulsive forces, or the type of post-processing step considered next.

An alternative approach is to remove overlaps as a post-processing step after the graph is laid out. Here the trade-off between layout size and preserving the graph's shape is more explicit. A number of such algorithms have been proposed. For example, the Voronoi cluster busting algorithm [10, 22] works by iteratively forming a Voronoi diagram from the current layout and moving each node to the center of its Voronoi cell

I.G. Tollis and M. Patrignani (Eds.): GD 2008, LNCS 5417, pp. 206–217, 2009.

until no overlaps remain. Although roughly maintaining relative node positions, the overall affect is to lose much of the layout structure.

Another group of post-processing algorithms is based on maintaining the orthogonal ordering [25] of the initial layout as a way to preserve its shape. A force scan algorithm and variants were proposed [14, 17, 21, 25] based on these constraints. More recently, Marriott et al. [3, 23] have presented a quadratic programming algorithm which removes node overlaps while minimizing node displacement and keeping the orthogonal ordering. An orthogonal ordering invariant is fairly effective at preserving structure, but it still cannot ensure that relative proximity relations between nodes are preserved, while at other times, it is too restrictive. Also, some of these algorithm require, in practice, separate horizontal and vertical passes which often results in a layout with a distorted aspect ratio (e.g., Fig. 2, bottom right).

In this paper, we discuss (Sect. 2) metrics for the similarity between two layouts which we believe better quantifies the desired outcome of overlap removal than minimized displacement or such simpler measures as aspect ratio or edge ratio. We then present (Sect. 3) a node overlap removal algorithm based on a proximity graph of the nodes in the original layout. In Sect. 4, we evaluate our algorithm and others using the proposed similarity measures.

In the following, we use $G = (V, E)$ to denote an undirected graph, with V the set of nodes (vertices) and E edges. We use $|V|$ and $|E|$ for the number of vertices and edges, respectively. We let x_i represent the current coordinates of vertex i in Euclidean space.

2 Measuring Layout Similarity

The outcome of an overlap removal algorithm should be measured in two aspects. The first aspect is the overall bounding box area: we want to minimize the area taken by the drawing after overlap removal. The second aspect is the change in relative positions. Here we want the new drawing to be as "close" to the original as possible. It is this aspect that is hard to quantify.

One way to measure the similarity of two layouts is to measure the distance between all pairs of vertices in the original and the new layout. If the two layouts are similar, then these distances should match, subject to scaling. This is known as Frobenius metric in the sensor localization problem [5]. However, calculating all pairwise distances is expensive for large graphs, both in CPU time and in the amount of memory, so instead we form a *Delaunay triangulation* (DT) of the original graph, then measure the distance between vertices along the edges of the triangulation for the original and new layouts. If x^0 and x denote the original and the new layout, and E_P is the set of edges in the triangulation, we calculate the ratio of the edge length

$$r_{ij} = \frac{\|x_i - x_j\|}{\|x_i^0 - x_j^0\|}, \ \{i, j\} \in E_P,$$

then define a measure of the dissimilarity as the normalized standard deviation

$$\sigma_{\text{dist}}(x^0, x) = \frac{\sqrt{\frac{\sum_{\{i,j\} \in E_P} (r_{ij} - \bar{r})^2}{|E_P|}}}{\bar{r}},$$

where

$$\bar{r} = \frac{1}{|E_P|} \sum_{\{i,j\} \in E_P} r_{ij}$$

is the mean ratio. The reason we measure the edge length ratio along edges of the proximity graph, rather than along edges of the original graph, is that if the original graph is not rigid, then even if two layouts of the same graph have the same edge lengths, they could be completely different. For example, think of the graph of a square, and a new layout of the same graph in the shape of a non-square rhombus. These two layouts may have exactly the same edge lengths, but are clearly different. The rigidity of the triangulation avoids this problem.

Notice that $\sigma_{\text{dist}}(x^0, x)$ is not symmetric with regard to which layout comes first. Furthermore, in theory, this non-symmetric version could class a layout and a foldover of it (e.g., a square grid with one half folded over the other) as the same. We can symmetrize it by defining the dissimilarity between layout x and x^0 as $(\sigma_{\text{dist}}(x^0, x) + \sigma_{\text{dist}}(x, x^0))/2$. This also resolves the "foldover problem". The symmetric version may be more appropriate if we are comparing two unrelated layouts. Since, however, we are comparing a layout derived from an existing layout, we feel that the asymmetric version is adequate.

An alternative measure of similarity is to calculate the displacement of vertices of the new layout from the original layout [3]. Clearly a new layout derived from a shift, scaling and rotation should be considered identical. Therefore we modify the straight displacement calculation by discounting the aforementioned transformations. This is achieved by finding the optimal scaling, shift and rotation that minimize the displacement. The optimal displacement is then a measure of dissimilarity.

We define the displacement dissimilarity as

$$\sigma_{\text{disp}}(x^0, x) = \min_{p \in \mathbb{R}^2, \theta, r \in \mathbb{R}} \sum_{i \in V} \|rTx_i + p - x_i^0\|^2, \tag{1}$$

where r is the scaling, θ the rotation with $T = T(\theta)$ its rotation matrix, and $p \in \mathbb{R}^2$ is the translation. Solving this is a known problem in Procrustes analysis [1, 11] and the solution (the Procrustes statistic) is

$$\sigma_{\text{disp}}(x^0, x) = Tr(X^0 X^{0^T}) - (Tr((X^T X^0 X^{0^T} X)^{\frac{1}{2}})^2 Tr(X^T X), \tag{2}$$

where X is a matrix with columns $x_i - \bar{x}$, X^0 is a matrix with columns $x_i^0 - \bar{x}^0$, and $\bar{x}$ and $\bar{x}^0$ are the centers of gravity of the new and original layout. In the above we do not consider shearing, since we believe a layout derived from shearing of the original should not be considered identical to the latter.

3   A Proximity Stress Model for Node Overlap Removal

Our goal now is to remove overlaps while preserving the shape of the initial layout by maintaining the proximity relations. To do this, we first set up a rigid "scaffolding" structure so that while vertices can move around, their relative positions are maintained. This scaffolding is constructed using a proximity graph [18]. Here again, we work with the Delaunay triangulation.

Once we form a DT, we check every edge in it and see if there are any node overlaps along that edge. Let w_i and h_i denote the half width and height of the node i, and $x_i^0(1)$ and $x_i^0(2)$ the current X and Y coordinates of this node. If i and j form an edge in the DT, we calculate the *overlap factor* of these two nodes

$$t_{ij} = \max\left(\min\left(\frac{w_i + w_j}{|x_i^0(1) - x_j^0(1)|}, \frac{h_i + h_j}{|x_i^0(2) - x_j^0(2)|}\right), 1\right). \qquad (3)$$

For nodes that do not overlap, $t_{ij} = 1$. For nodes that do overlap, such overlaps can be removed if we expand the edge by this factor. Therefore we want to generate a layout such that an edge in the proximity graph has the ideal edge length close to $t_{ij}\|x_i^0 - x_j^0\|$. In other words, we want to minimize the following stress function

$$\sum_{(i,j)\in E_P} w_{ij}\left(\|x_i - x_j\| - d_{ij}\right)^2. \qquad (4)$$

Here $d_{ij} = s_{ij}\|x_i^0 - x_j^0\|$ is the ideal distance for the edge $\{i, j\}$, s_{ij} is a scaling factor related to the overlap factor t_{ij} (see (6)), $w_{ij} = 1/\|d_{ij}\|^2$ is a scaling factor, and E_P is the set of edges of the proximity graph. We call (4) the *proximity stress model* in obvious analogy with the standard stress model [19]

$$\sum_{i\neq j} w_{ij}\left(\|x_i - x_j\| - d_{ij}\right)^2, \qquad (5)$$

where d_{ij} is the graph theoretical distance between vertices i and j, and w_{ij} is a weight factor, typically $1/d_{ij}^2$.

Because DT is a planar graph, which has no more than $3|V| - 3$ edges, the above stress function has no more than $3|V| - 3$ terms. Furthermore, because DT is rigid, it provides a good scaffolding that constrains the relative position of the vertices and helps to preserve the global structure of the original layout.

It is important that we do not attempt to remove overlaps in one iteration by using the above model with $s_{ij} = t_{ij}$. Imagine the situation of a regular mesh graph, with one node i of particularly large size that overlaps badly with its nearby nodes, but the other nodes do not overlap with each other. Suppose nodes i and j form an edge in the proximity graph, and they overlap. If we try to make the length of the edge equal $t_{ij}\|x_i^0 - x_j^0\|$, we will find that t_{ij} is a number much larger than 1, and the optimum solution to the stress model is to keep all the other vertices at or close to their current positions, but move the large node i outside of the mesh, at a position that does not cause overlap. This is not desirable because it destroys the original layout. Therefore we damp the overlap factor by setting

$$s_{ij} = \min(t_{ij}, s_{\max}) \qquad (6)$$

and try to remove overlaps a little at a time. Here $s_{\max} > 1$ is a number limiting the amount of overlap we are allowed to remove in one iteration. We found that $s_{max} = 1.5$ works well.

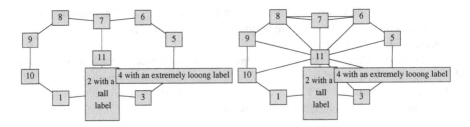

Fig. 1. (a): A graph layout where nodes 2 and 4 overlap. (b): the proximity graph (Delaunay triangulation) of the current layout. No two nodes linked by an edge of the proximity graph overlap.

After minimizing (4), we arrive at a layout that may still have node overlaps. We then regenerate the proximity graph using DT and calculate the overlap factor along the edges of this graph, and redo the minimization. This forms an iterative process that ends when there are no more overlaps along the edges of the proximity graph.

For many graphs, the above algorithm yields a drawing that is free of node overlaps. For some graphs, however, especially those with nodes having extreme aspect ratios, node overlaps may still occur. Such overlaps happen for pairs of nodes that are not near each other, and thus do not constitute edges of the proximity graph. Fig. 1(a) shows the drawing of a graph after minimizing (4) iteratively, so that no more node overlap is found along the edges of the Delaunay triangulation. Clearly, node 2 and node 4 still overlap. If we plot the Delaunay triangulation (Fig. 1(b)), it is seen that nodes 2 and 4 are not neighbors in the proximity graph, which explains the overlap. To overcome this situation, once the above iterative process has converged so that no more overlaps are detected over the DT edges, we apply a scan-line algorithm [3] to find all overlaps, and augment the proximity graph with additional edges, where each edge consists of a pair of nodes that overlap. We then re-solve (4). This process is repeated until the scan-line algorithm finds no more overlaps.

We call this algorithm PRISM (PRoxImity Stress Model). Concerning its complexity, Delaunay triangulation can be computed in $O(|V|log(|V|))$ time [6, 12, 20]. The scan-line algorithm can be implemented to find all the overlaps in $O(l|V|(log|V| + l))$ time [3], where l is the number of overlaps. Because we only apply the scan-line algorithm after no more node overlaps are found along edges of the proximity graph, l is usually a very small number, hence this step can be considered as taking time $O(|V|log|V|)$.

The proximity stress model (4), like the standard stress model (5), can be solved using the stress majorization technique [8] with a conjugate gradient algorithm. Because we use DT as our proximity graph and it has no more than $3|V| - 3$ edges, each iteration of the conjugate gradient algorithm takes a time of $O(|V|)$.

Overall, therefore, PRISM takes $O(t(mk|V| + |V|log|V|))$ time, where t is the total number of iterations in the two main loops, m is the average number of stress majorization iterations, and k the average number of iterations for the conjugate gradient algorithm.

4   Numerical Results

To evaluate the PRISM algorithm and other overlap removal algorithms, we apply them as a post-processing step to a selection of graphs from the Graphviz [9] test suite. This suite, part of the Graphviz source distribution, contains many graphs from users. As such, these are good examples of the kind of graphs actually being drawn.

Our baseline algorithm is Scalable Force Directed Placement (SFDP) [16], a multi-level, spring-electrical algorithm. Using the layout of SFDP, we then apply one of the overlap removal algorithms to get a new layout that has no node overlaps, and compare the new layout with the original in terms of dissimilarity and area.

In Table 1, we list the 14 test graphs, the number of vertices and edges, as well as CPU time[1] for PRISM and three other overlap removal algorithms. The graphs are selected randomly with the criteria that a graph chosen should be connected, and is of relatively large size. We compared PRISM with an implementation in Graphviz of the solve_VPSC algorithm [3][2], hereafter denoted as VPSC, as well as VORO, the Voronoi cluster busting algorithm [10,22]. The final algorithm is the ODNLS algorithm of Li et al. [21], which relies on varied edge lengths in a spring embedder.

The initial layout by SFDP is scaled so that the average edge length is 1 inch. From the table, it is seen that PRISM is usually faster, particularly for large graphs on which it scales much better. The others are slow for large graphs, with VORO the slowest.

Table 2 compares the dissimilarities and drawing area of the four overlap removal algorithms. The smaller the dissimilarities and area, the better. The ODNLS algorithm performs best in terms of smaller dissimilarity, followed by PRISM, VPSC and VORO. In terms of area, PRISM and VPSC are pretty close, and both are better than ODNLS and VORO, which can give extremely large drawings. Indeed, in terms of area, scaling outperformed ODNLS and VORO in 20%-30% of the examples.

Comparing PRISM with VPSC, Table 2 shows that PRISM gives smaller dissimilarities most of the time. The two dissimilarity measures, σ_{dist} and σ_{disp}, are generally correlated, except for ngk10_4 and root. Based on σ_{dist}, VPSC is better for these two graphs, while based on σ_{disp}, PRISM is better. The first row in Fig. 2 shows the original layout of ngk10_4, as well as the result after applying PRISM and VPSC. Through visual inspection, we can see that PRISM preserved the proximity relations of the original layout well. VPSC "packed" the labels more tightly, but it tends to line up vertices horizontally and vertically, and also produces a layout with aspect ratio quite different from the original graph. It seems that σ_{dist} is not as sensitive in detecting differences in aspect ratio. This is evident in drawings of the root graph (Fig. 2, second row). VPSC clearly produced a drawing that is overly stretched in the vertical direction, but its σ_{dist} is actually smaller than that of PRISM! Consequently, we conclude that σ_{disp} may be a better dissimilarity measure.

The fact that VPSC can produce very tall and thin, or very short and wide, layouts is not surprising, and has been observed often in practice. VPSC works in the vertical and

---

[1] All timings were derived on a 4 processor, 3.2 GHz Intel Xeon CPU, with 8.16 GB of memory, running Linux.

[2] A stand alone version of solve_VPSC by the authors of this algorithm has also been tried but was found to offer no advantage over VPSC. VPSC itself was also contributed originally by the same authors to Graphviz.

Table 1. Comparing the CPU time (in seconds) of several overlap removal algorithms. Initially the layout is scaled to an average edge length of 1 inch.

Graph	$|V|$	$|E|$	PRISM	VPSC	VORO	ODNLS
b100	1463	5806	1.44	14.85	350.7	258.9
b102	302	611	0.14	0.10	4.36	5.7
b124	79	281	0.03	0.01	0.02	0.5
b143	135	366	0.04	0.01	0.47	1.3
badvoro	1235	1616	0.54	71.15	351.51	73.6
mode	213	269	0.09	0.09	2.15	2.1
ngk10_4	50	100	0.01	0.00	0.02	0.14
NaN	76	121	0.01	0.01	0.11	0.27
dpd	36	108	0.01	0.01	0.02	0.1
root	1054	1083	0.89	7.81	398.49	46.9
rowe	43	68	0.00	0.00	0.04	0.1
size	47	55	0.01	0.00	0.06	0.09
unix	41	49	0.01	0.00	0.04	0.07
xx	302	611	0.13	0.10	8.19	5.67

Table 2. Comparing the dissimilarities and area of overlap removal algorithms. Results shown are σ_{dist}, σ_{disp} and area. Area is measured with a unit of 10^6 square points. Initially the layout is scaled to an average length of 1 inch.

Graph	PRISM			VPSC			VORO			ODNLS		
	σ_{dist}	σ_{disp}	area	σ_{dist}	σ_{disp}	area	σ_{dist}	σ_{disp}	area	σ_{dist}	σ_{disp}	area
b100	0.74	0.38	14.05	0.76	0.72	18.91	-	-	-	0.33	0.20	1.02E3
b102	0.44	0.25	2.45	0.58	0.8	2.71	0.8	0.3	31.79	0.30	0.16	53.13
b124	0.65	0.37	1.04	0.78	0.73	0.91	0.86	0.39	13.42	0.33	0.19	14.79
b143	0.59	0.35	1.5	0.78	0.83	2.16	0.99	0.45	22.91	0.49	0.34	23.79
badvoro	0.34	0.15	12.58	0.61	0.75	13.85	2.29	0.65	3.01E3	0.31	0.26	318.66
mode	0.59	0.37	0.79	1.02	0.77	1.29	0.97	0.54	10.84	0.38	0.27	49.45
ngk10_4	0.41	0.16	0.33	0.39	0.3	0.25	0.48	0.26	0.52	0.22	0.13	2.30
NaN	0.4	0.2	0.72	0.54	0.65	0.71	0.56	0.28	5.04	0.26	0.15	5.10
dpd	0.34	0.18	0.25	0.51	0.4	0.18	0.48	0.32	0.45	0.37	0.29	1.30
root	0.71	0.3	16.99	0.6	0.75	17.68	4.09	0.94	6.93E9	0.29	0.22	950.01
rowe	0.33	0.14	0.22	0.44	0.31	0.19	0.49	0.26	0.95	0.27	0.12	2.10
size	0.37	0.2	0.47	0.77	0.74	0.4	0.62	0.35	1.27	0.32	0.20	4.14
unix	0.39	0.23	0.39	0.51	0.67	0.36	0.6	0.35	0.85	0.26	0.13	2.35
xx	0.42	0.25	3.96	0.57	0.82	3.9	0.97	0.34	58.83	0.29	0.14	74.00

horizontal directions alternatively, each time trying to remove overlaps while minimizing displacement. As a result, when starting from a layout with severe node overlaps, it may move vertices significantly along one direction to resolve the overlaps, creating

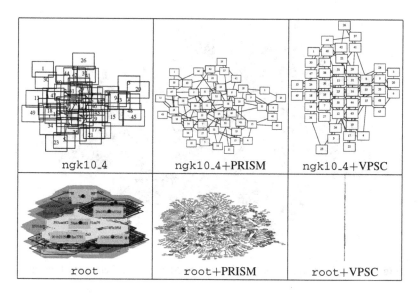

Fig. 2. Divergence of dissimilarity measures: for both graphs, σ_{dist} estimates that VPSC gives layout closer to the original, while σ_{disp} predicts the opposite

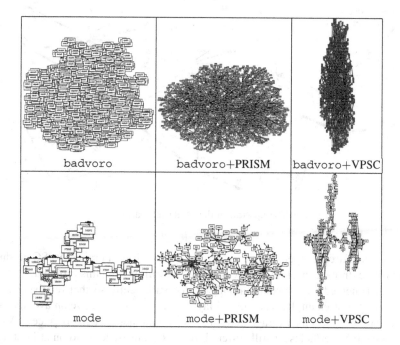

Fig. 3. Comparing PRISM and VPSC on two graphs. Original layouts are scaled to have an average edge length that equals 4 times the label size.

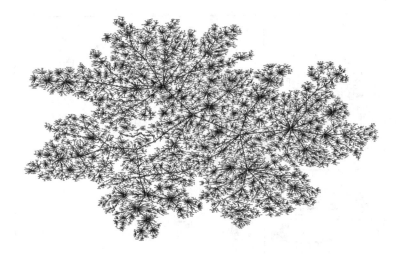

Fig. 4. The second largest component from the Mathematics Genealogy Project

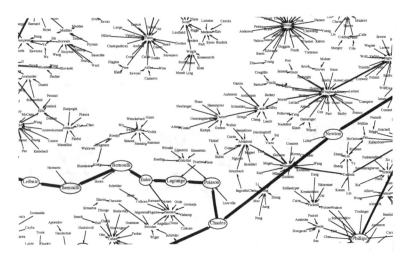

Fig. 5. Close-up view of the center-left part of Fig. 4

drawings with extreme aspect ratios. In fact, for 9 out of 14 test graphs, VPSC produces layouts with extreme aspect ratios. PRISM does not suffer from this problem.

We experimented with layouts initially scaled sufficiently so that relatively fewer nodes overlap. For example, when initial layouts were scaled to give an average edge length equal to 4 times the average node size, we found that the performance of VPSC was improved. Nevertheless it still suffered from extreme aspect ratio on at least 5 out of the 14 graphs. Figure 3 shows two of these graphs.

Overall, quantitative and visual comparison of the drawings of these 14 graphs, as well as drawings for graphs in the complete `Graphviz` test suite (a total of 204 graphs in March 2008), shows that PRISM performs very well, and is overall better and faster than VPSC and VORO. The ODNLS algorithm preserves similarity somewhat better than PRISM, but at much higher costs in term of speed and area.

As a demonstration of the scalability of PRISM, we consider its application to a large graph. This is a tree from the Mathematics Genealogy Project [24]. Each node is a mathematician, and an edge from node i to node j means that j is the first supervisor of i. The graph is disconnected and consists of thousands of components. Here we consider the second largest component with 11766 vertices. This graph took 31 seconds to layout using SFDP, and 15 seconds post-processing using PRISM for overlap removal. Important mathematicians (those with the most offspring) and important edges (those that lead to the largest subtrees) are highlighted with larger nodes and thicker edges. Figure 4 gives the overall layout, which shows that PRISM preserved the tree structure of the layout very well after node overlap removal. Figure 5 gives a close up view of the details of a small area in the center-left part of Fig. 4. Additional drawings of this and other components of the Mathematics Genealogy Project graph, including that of the largest component, are available [15].

5   Conclusions and Future Work

A number of algorithms have been proposed for removing node overlaps in undirected graph drawings. For graphs that are relatively large with nontrivial connectivities, these algorithms often fail to produce satisfactory results, either because the resulting drawing is too large (e.g., scaling, VORO, ODNLS), or the drawing becomes highly skewed (e.g., VPSC). In addition, many of them do not scale well with the size of the graph in terms of computational costs. The main contribution of this paper is a new algorithm for removing overlaps that is both highly effective and efficient. The algorithm is shown to produce layouts that preserve the proximity relations between vertices, and scales well with the size of the graph. It has been applied to graphs of tens of thousands of vertices, and is able to give aesthetic, overlap-free drawings with compact area in seconds, which is not feasible with any algorithm known to us.

It is possible that algorithms such as VPSC, which rely on separate passes in the X and Y directions, might be improved by randomizing which overlaps are removed in which pass or by gradually removing overlaps using many alternating X and Y passes. This would, however, further increase their computational cost, which is already much higher than the algorithm proposed in this paper.

For future work, we would like to extend the overlap removal algorithm to deal with edge node overlaps. We would also like to explore the possibility of using the proximity stress model for packing disconnected components.

Acknowledgments

We would like to thank Tim Dwyer and Wanchun Li for making their implementations of VPSC and ODNLS, respectively, available to us; Yehuda Koren and Stephen

North for helpful discussions; Stephen Kobourov for bringing our attention to the term "Frobenius metric"; and the referees for valuable suggestions and references.

References

1. Borg, I., Groenen, P.: Modern Multidimensional Scaling: Theory and Applications. Springer, Heidelberg (1997)
2. Chuang, J.H., Lin, C.C., Yen, H.C.: Drawing graphs with nonuniform nodes using potential fields. In: Stinson, D.R., Tavares, S. (eds.) SAC 2000. LNCS, vol. 2012, pp. 460–465. Springer, Heidelberg (2001)
3. Dwyer, T., Marriott, K., Stuckey, P.J.: Fast node overlap removal. In: Healy, P., Nikolov, N.S. (eds.) GD 2005. LNCS, vol. 3843, pp. 153–164. Springer, Heidelberg (2006)
4. Eades, P.: A heuristic for graph drawing. Congressus Numerantium 42, 149–160 (1984)
5. Erten, C., Efrat, A., Forrester, D., Iyer, A., Kobourov, S.G.: Force-directed approaches to sensor network localization. In: Raman, R., Sedgewick, R., Stallmann, M.F. (eds.) Proc. 8th Workshop Algorithm Engineering and Experiments (ALENEX), pp. 108–118. SIAM, Philadelphia (2006)
6. Fortune, S.: A sweepline algorithm for Voronoi diagrams. Algorithmica 2, 153–174 (1987)
7. Fruchterman, T.M.J., Reingold, E.M.: Graph drawing by force directed placement. Software - Practice and Experience 21, 1129–1164 (1991)
8. Gansner, E.R., Koren, Y., North, S.C.: Graph drawing by stress majorization. In: Pach, J. (ed.) GD 2004. LNCS, vol. 3383, pp. 239–250. Springer, Heidelberg (2005)
9. Gansner, E.R., North, S.: An open graph visualization system and its applications to software engineering. Software - Practice & Experience 30, 1203–1233 (2000)
10. Gansner, E.R., North, S.C.: Improved force-directed layouts. In: Whitesides, S.H. (ed.) GD 1998. LNCS, vol. 1547, pp. 364–373. Springer, Heidelberg (1999)
11. Gower, J.C., Dijksterhuis, G.B.: Procrustes Problems. Oxford University Press, Oxford (2004)
12. Guibas, L., Stolfi, J.: Primitives for the manipulation of general subdivisions and the computation of voronoi. ACM Trans. Graph. 4(2), 74–123 (1985)
13. Harel, D., Koren, Y.: A fast multi-scale method for drawing large graphs. J. Graph Algorithms and Applications 6, 179–202 (2002)
14. Hayashi, K., Inoue, M., Masuzawa, T., Fujiwara, H.: A layout adjustment problem for disjoint rectangles preserving orthogonal order. In: Whitesides, S.H. (ed.) GD 1998. LNCS, vol. 1547, pp. 183–197. Springer, Heidelberg (1999)
15. Hu, Y.F.: Drawings of the mathematics genealogy project graphs, http://www.research.att.com/~yifanhu/GALLERY/MATH_GENEALOGY
16. Hu, Y.F.: Efficient and high quality force-directed graph drawing. Mathematica Journal 10, 37–71 (2005)
17. Huang, X., Lai, W.: Force-transfer: A new approach to removing overlapping nodes in graph layout. In: Proc. 25th Australian Computer Science Conference, pp. 349–358 (2003), citeseer.ist.psu.edu/564050.html
18. Jaromczyk, J.W., Toussaint, G.T.: Relative neighborhood graphs and their relatives. Proc. IEEE 80, 1502–1517 (1992)
19. Kamada, T., Kawai, S.: An algorithm for drawing general undirected graphs. Information Processing Letters 31, 7–15 (1989)
20. Leach, G.: Improving worst-case optimal Delaunay triangulation algorithms. In: 4th Canadian Conference on Computational Geometry. pp. 340–346 (1992), citeseer.ist.psu.edu/leach92improving.html

21. Li, W., Eades, P., Nikolov, N.: Using spring algorithms to remove node overlapping. In: Proc. Asia-Pacific Symp. on Information Visualisation, pp. 131–140 (2005)
22. Lyons, K.A., Meijer, H., Rappaport, D.: Algorithms for cluster busting in anchored graph drawing. J. Graph Algorithms and Applications 2(1) (1998)
23. Marriott, K., Stuckey, P.J., Tam, V., He, W.: Removing node overlapping in graph layout using constrained optimization. Constraints 8(2), 143–171 (2003)
24. Department of mathematics at North Dekota State University: The mathematics genealogy project, http://genealogy.math.ndsu.nodak.edu/
25. Misue, K., Eades, P., Lai, W., Sugiyama, K.: Layout adjustment and the mental map. J. Vis. Lang. Comput. 6(2), 183–210 (1995)
26. Wang, X., Miyamoto, I.: Generating customized layouts. In: Brandenburg, F.J. (ed.) GD 1995. LNCS, vol. 1027, pp. 504–515. Springer, Heidelberg (1996)

An Experimental Study on Distance-Based Graph Drawing

(Extended Abstract)

Ulrik Brandes and Christian Pich

Department of Computer & Information Science, University of Konstanz
{Ulrik.Brandes,Christian.Pich}@uni-konstanz.de

Abstract. In numerous application areas, general undirected graphs need to be drawn, and force-directed layout appears to be the most frequent choice. We present an extensive experimental study showing that, if the goal is to represent the distances in a graph well, a combination of two simple algorithms based on variants of multidimensional scaling is to be preferred because of their efficiency, reliability, and even simplicity. We also hope that details in the design of our study help advance experimental methodology in algorithm engineering and graph drawing, independent of the case at hand.

1 Introduction

Graph drawing is concerned with the geometric representation of graphs. For general undirected graphs, force-directed and energy-based layout algorithms are commonly used, because they are often easy to implement and experience shows that they can result in undistorted and readable layouts which reveal structural features such as local clustering and symmetry [3].

Based on experimental evidence presented in this paper, we argue that approximate classical scaling with subsequent stress reduction should be used instead. The requirements leading to this argument are:

1. *quality*: pairwise distances between vertices are represented well,
2. *scalability*: the algorithm scales to very large graphs, and
3. *simplicity*: the algorithm is easy to understand and implement.

Note that the quality criterion is implicit on force-directed algorithms. Classical scaling and stress minimization are instances of the general concept of Multidimensional Scaling (MDS, see [1,8] for comprehensive references). MDS of graph-theoretic distances has been used early on for automatic layout of social networks [16], without explicit reference in the well-known algorithm of Kamada and Kawai [15], and in the wider context of data analysis (e.g.,[5,10]), but the use of advanced MDS algorithms well-known in other fields has gained momentum only after Gansner, Koren, and North applied majorization to stress minimization in graph drawing [12]. Stress minimization is generally assumed to be the

I.G. Tollis and M. Patrignani (Eds.): GD 2008, LNCS 5417, pp. 218–229, 2009.

method of choice for drawing general graphs, because of its intuitive and adaptable objective function and the visually pleasing layouts obtained. Yet, it is often found to be difficult to implement efficiently, and the presence of local minima is a serious concern.

Our study provides an assessment of layout quality and efficiency, and also yields a recommendation on how to implement the method to achieve reliability, efficiency, and simplicity at the same time. While a considerable number of experimental studies have been conducted to assess graph drawing criteria and algorithm performance, only two are closely related [2,13]. However, these compare implementations of suites of related algorithms which are treated as black boxes. The combination of our in-depth study with these more general comparisons provides additional support for our conclusion.

A methodological contribution of our study is the design of experiments along explicit hypotheses about the performance of algorithms. These guided our choice of experiments and structure argumentation.

The remainder of this paper is organized as follows: In Sect. 2, background on the relevant MDS variants and their application to graph drawing is given. The main hypotheses are stated in Sect. 3. The experimental setup is described in Sect. 4, and the actual experiments in Sect. 5. Section 6 discusses results with regard to our hypotheses. We conclude with a summary in Sect. 7.

2 Multidimensional Scaling

Let $V = \{1, \ldots, n\}$ be the set of n objects and let $D \in \mathbb{R}^{n \times n}$ be a square matrix of dissimilarities d_{ij} for each pair of objects $i, j \in V$. MDS yields a matrix $X = [x_1, \ldots, x_n]^T \in \mathbb{R}^{n \times d}$ of d-dimensional positions $x_1, \ldots, x_n \in \mathbb{R}^d$ such that

$$\|x_i - x_j\| \approx d_{ij} \quad \text{for all} \quad i, j \in V \tag{1}$$

is met as closely as possible; in our experiments, $d = 2$ throughout. We leave this somewhat informal for the moment and make it more precise in the following two subsections, where we describe the objective functions typically considered to assess compliance with (1). Straightforward implementations of these run in $\Theta(n^3)$ time, but we will discuss more efficient algorithms in Section 4.

Classical Scaling. The first approach to achieve (1) is based on linear algebra and is referred to as *classical* or *inner-product scaling*. Let $D \in \mathbb{R}^{n \times n}$ be defined as above, and let $D^{(2)}$ be matrix D with all entries squared. Classical scaling is based on a matrix $B \in \mathbb{R}^{n \times n}$ of *pseudo products* b_{ij} with

$$b_{ij} = -\frac{1}{2}\left(d_{ij}^2 - \frac{1}{n}\sum_{s=1}^{n} d_{is}^2 - \frac{1}{n}\sum_{r=1}^{n} d_{rj}^2 + \frac{1}{n^2}\sum_{r,s=1}^{n} d_{rs}^2\right) \tag{2}$$

or equivalently, written in matrix form, by double-centering $D^{(2)}$ with $B = -\frac{1}{2}J_n D^{(2)} J_n$, where $J_n = I_n - \frac{1}{n} \cdot \left(1_n 1_n^T\right) \in \mathbb{R}^{n \times n}$, I_n being the identity matrix and $1_n \in \mathbb{R}^n$ the all-ones vector of length n.

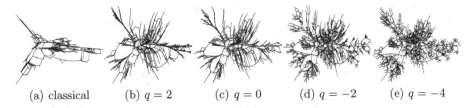

(a) classical   (b) $q = 2$ (c) $q = 0$ (d) $q = -2$ (e) $q = -4$

Fig. 1. Example drawings for the 1138bus graph. Drawing (a) is generated with classical scaling, drawings (b)–(e) with distance scaling and weights $w_{ij} = d_{ij}^q$.

Let $v_1 \dots, v_n \in \mathbb{R}^{n \times n}$ and $\lambda_1 \geq \cdots \geq \lambda_n \in \mathbb{R}$ be the sequence of eigenvectors and corresponding eigenvalues of B. Two-dimensional coordinates are then obtained by setting the configuration matrix $X \in \mathbb{R}^{n \times 2}$ to be

$$X = \left[\sqrt{\lambda_1} v_1, \sqrt{\lambda_2} v_2 \right], \tag{3}$$

which is optimal [1]the mismatch between the pseudo inner-products derived from the d_{ij}'s in (2) and the inner products $x_i^T x_j$, namely

$$\text{strain}(X) = \| B - X X^T \|_2 = \sum_{i,j} \left(b_{ij} - x_i^T x_j \right)^2 . \tag{4}$$

The advantage of this approach is that it gives analytic solutions which are essentially unique and optimal with respect to strain. A major drawback is the detour via inner products, sometimes leading to degenerate solutions.

Distance Scaling. Instead of achieving (1) by fitting inner products b_{ij} and $x_i^T x_j$, coordinates can be computed by directly fitting distances $\|x_i - x_j\|$ to dissimilarities d_{ij}. This leads to the objective function

$$\text{stress}(X) = \sum_{i,j} w_{ij} \left(d_{ij} - \|x_i - x_j\| \right)^2 , \tag{5}$$

where $w_{ij} \geq 0$ weights the contribution of pair i, j; frequently, $w_{ij} = d_{ij}^q$ for some $q \in \mathbb{R}$. Since there is no known method for directly computing a configuration X with minimal stress, the standard approach is iterative numerical optimization.

Graph Drawing and MDS. Most applications of MDS to graph drawing set the desired distances to be the shortest-path distances in the graph, which often spread nodes well over the drawing and display symmetries and clusterings.

While classical scaling was used for graph drawing [5]and made scalable to large graphs only recently [4,6], the distance scaling approach is pioneered much earlier [16]. Kamada and Kawai [15] used a layout energy equivalent to the objective function introduced independently by McGee [19] more than twenty years earlier (there termed *work*). In the framework of the more general weighted MDS, it corresponds to setting $w_{ij} = d_{ij}^{-2}$ in Eq. (5) Other weighting schemes and dissimilarities are discussed in [5,7]. Fig. 1 shows some example drawings.

3   Hypotheses

A combination of theoretical properties, previous experience, popular beliefs, and preliminary tests, led us to formulate and test the hypotheses below. These shall not be read as if they were results, but serve to focus attention and are formulated in such a way that they can be tested with algorithmic experiments. We therefore conducted a series of experiments described in the next section. See Section 6 for a discussion of the results.

The first hypothesis basically rules out force-directed methods.

Hypothesis 1. *For graph drawing representing graph-theoretic distances it is most appropriate to model this representation explicitly in the objective function.*

Given their objectives, both classical and distance scaling should represent graph-theoretic distances well in a geometric layout, and thus be useful for graph drawing. Because of the more direct influence on the objective function and a concave weighting of distance representation errors, it seems plausible that distance scaling would be the more suitable variant for graph drawing. While it is almost commonplace that classical scaling is better at representing global structure whereas distance scaling is better at representing fine details [5], we do not know of any systematic evaluation. We therefore provide experimental evidence for the following.

Hypothesis 2. *Distance scaling compares favorably with classical scaling in terms of layout quality, because local details are represented better.*

In our experience, based on many conversations with implementors and users of graph drawing systems, a main reservation against distance scaling is its assumed non-scalability, due to a multiude of local minima and high computational demand. The next two hypotheses focus on how to ensure that the layouts produced by implementations of distance scaling are actually those supporting H1.

Hypothesis 3. *Distance scaling is susceptible to poor local minima, because it is highly dependent on the initial layout.*

Hypothesis 4. *Classical scaling provides excellent initial layouts for distance scaling, because the better representation of large distances helps to avoid poor local minima.*

If H4 holds, we have complicated matters even more, because two demanding problems have to be solved rather than one. The final two hypotheses therefore regard the possibility of computing the initial and final layout efficiently.

Hypothesis 5. *Classical scaling layouts of very large graphs can be approximated efficiently using PivotMDS.*

Hypothesis 6. *Distance scaling is practical even on very large graphs.*

Table 1. Test set of graphs used in Experiments 1–3. n, m, D denote the number of nodes, the number of edges, and the diameter, respectively. The two rightmost columns contain plots for distance distributions and the 10 largest eigenvalues of B.

name	n	m	description	D	$\{d_{ij}\}$	$\lambda_{1,...,10}$
516	516	729	finite element mesh describing adjacencies between faces in a triangulation	61		
1138bus	1138	1458	network of high-voltage power distribution in the United States.	31		
qh882	882	2856	matrix derived from Quebec hydroelectric power system's small signal model	31		
plat1919	1919	15240	finite-difference model of shallow wave equations in Atlantic/Indian Ocean	43		
esslingen1	2075	4769	social network in the city of Esslingen in the 19th century	15		
sw0	500	1500	circle in which each node is adjacent to its 3 left and right neighbors	84		
sw002	500	1500	graph sw0, each edge rediretced randomly with probability 0.02	27		
sw01	500	1500	graph sw0, each edge rediretced randomly with probability 0.1	10		
btree	1023	1022	complete binary tree of height 10	18		
prot1	3025	3629	largest component of protein interaction network	27		

4   Experimental Design

Data. The experiments were run on a set of test graphs described in Table 1. The graphs were selected large enough to allow for extrapolation of the results to very large graphs, but also small enough to allow for, the exact computation of stress as given by (5) in a large number of experiments.

Note that the eigenvalues of the matrices B associated with each graph indicate the intrinsic dimensionality of the original distances d_{ij}. If, say, two dimensions suffice to reconstruct all the d_{ij}'s exactly, such that the strain criterion is zero, then $\lambda_1 \geq \lambda_2 > \lambda_3 = \cdots = \lambda_n = 0$, and inversely, few large and many (near-) zero eigenvalues indicate the existence of a good low-dimensional layout.

Environment. We implemented all MDS algorithms and speed-up techniques ourselves to avoid bias due to coding, system, or timing. The algorithms were implemented in Java using Sun's SDK 1.6.0 and the yFiles 2.5.0.1 graph library (www.yworks.com). All experiments were run on a standard 1.4 GHz Compaq NX 7000 notebook with 512 MB of RAM, using Windows XP Service Pack 2.

Implementation. A simple and convenient way of implementing classical scaling is by constructing matrix B in (2) and computing its two extremal eigenvalues λ_1, λ_2 and eigenvectors v_1, v_2 by power iteration.

The problem of drawing graphs with fixed edge lengths is $\mathcal{NP}$-hard in general [9], and for distance scaling no analytic solution is known, so layouts have to be computed iteratively. In Kruskal's original proposal [17], stress is evaluated for the current positions, and new positions are computed by gradient descent; this is also done in [15,19,20] with gradient terms specific to the weights w_{ij}. These approaches were superseded by majorization [18], which generates a sequence of layouts with decreasing stress and can handle arbitrary weights $w_{ij} \geq 0$. In our experiments we use a "local iteration" with node-by-node updates [12].

5   Experiments

The first experiments is to provides evidence for which method yields better layouts in principle (disregarding efficiency, ease of implementation, reliability, etc.), when graph-theoretic distances are to be represented by Euclidean distances. We use the following shorthand notation for the involved approaches:

- random: node coordinates drawn uniformly at random from $(0, 1)$,
- fm3: fast multipole multilevel method [13],
- grip: multilevel force-directed layout method [11],
- hde: high dimensional embedder [14] (50 pivots),
- cmds classical scaling.

Experiment 1 (Layout approach). *All test graphs are laid out with cmds, distance scaling with unweighted and weighted stress, fm3, hde, and grip.*

For convenience, most implementations of iterative layout algorithms start from a random initial configuration. It is, however, widely known that smart initialization is preferable. We here compare different initialization strategies for distance scaling and evaluate the resulting stress. Before the iteration all initial solutions X are scaled such that $\sum_{i,j} \|x_i - x_j\| = \sum_{i,j} d_{ij}$.

Experiment 2 (Distance scaling and initialization). *All test graphs are laid out using each of the following layout algorithms: random, fm3, hde, grip, cmds, and then minimizing weighted stress using local iteration.*

Classical scaling has running time at least quadratic in the number of nodes n for constructing distance matrix $D \in \mathbb{R}^{n \times n}$ and decomposing the derived matrix $B \in \mathbb{R}^{n \times n}$. Quick estimates for the eigenvectors v_1, v_2 corresponding to λ_1, λ_2 are obtained by using only parts of D by selecting a subset $W \subset V$ of $k \ll n$ *pivot* or *landmark* nodes and taking only $k \cdot n$ rather than n^2 distances into account. Once W is constructed, two approaches for this are considered:

- Pivot MDS [4] uses the singular value decomposition of a rectangular matrix: Let $D_k \in \mathbb{R}^{n \times k}$ be the matrix of k columns of distances from nodes in W, e.g. in k breadth-first searches. Then the right singular vectors u_1, u_2 of $C = -\frac{1}{2} J_n D_k^{(2)} J_k$ are estimates for the eigenvectors v_1, v_2 of $B = -\frac{1}{2} J_n D^{(2)} J_n$.
- Landmark MDS [22] places nodes in W by classical MDS. The each node in $V \setminus W$ is placed based on its k distances to nodes in W.

The k pivots should be well-scattered over the graph; intuitively, this is to represent as much of the full distance information D as possible. Assuming that W contains $k - 1$ selected nodes, our strategies to determine the k-th pivot are

- maxmin: $\text{argmax}_{i \in V \setminus W} \min_{j \in W} d_{ij}$, the node farthest from W;
- random: with uniform probability, from W;
- mixed: with maxmin, if k is even, with random otherwise;

combining them with the two estimation approaches above leads to six strategies.

Let $X, Y \in \mathbb{R}^{n \times 2}$ be the estimate and the actual solution, each centered at the origin. To find out how similar X is to Y we use the *Procrustes statistic*

$$R^2 = 1 - \left(\text{tr}(X^T Y Y^T X)^{1/2} \right)^2 / \left(\text{tr}\left(X^T X \right) \cdot \text{tr}\left(Y^T Y \right) \right) \qquad (6)$$

minimized by the *Procrustes rotation* $P \in \mathbb{R}^{2 \times 2}$ (see [21] for its formula) which, applied to each row in X, optimally dilates, scales, rotates, and reflects X to fit Y. It can be shown that $0 \leq R^2 \leq 1$; if $R^2 = 0$, X and Y can be perfectly matched, if $R^2 = 1$, they cannot be matched by any $P \in \mathbb{R}^{2 \times 2}$ at all.

Experiment 3 (Approximating classical scaling). *For each test graph, classical scaling is approximated using 6 strategies {maxmin, random, mixed} × {landmark, pivot}, and compared to the exact solutions using the Procrustes statistic.*

Experiments 2 and 3 were repeated 25 times, and to control for biases due to the internal representation of graphs and matrices, we used as many instances of each graph, each with randomly permuted vertices and edges.

Distance scaling by stress minimization is mostly used for improving the representation of local details; setting $w_{ij} = d_{ij}^{-2}$ assigns large weight to the representation of small distances and vice versa. Initializing distance scaling with cmds, we hope that large distances are fitted well; the subsequent fitting of smaller distances and local details is achieved by discarding the large distances from the stress term to be minimized, which we dub *sparse stress*

$$\text{stress}(X) = \sum_{\{i,j\} \in S}^{\bullet} w_{ij}(d_{ij} - \|x_i - x_j\|^2) \,, \qquad (7)$$

where $S \subseteq V \times V$ is a set of node pairs involved in the iteration, with $|S| \in \mathcal{O}(n)$. In our experiments we use *local neighborhoods* obtained by terminating the breadth-first searches after k neighbors have been found.

Experiment 4 (Sparse stress minimization). *For each of the test graphs the initial classical scaling configuration is subjected to sparse stress minimization using only local neighborhoods.*

We use another collection of larger graphs to examine the scalability of initialization and sparse stress minimization. Unlike the test graphs used earlier, their size prohibits methods using the full square matrices. The results are assessed visually with respect to the information known a priori.

Experiment 5 (Very large graphs). *Large graphs are laid out first using an approximation to classical scaling and then sparse stress minimization.*

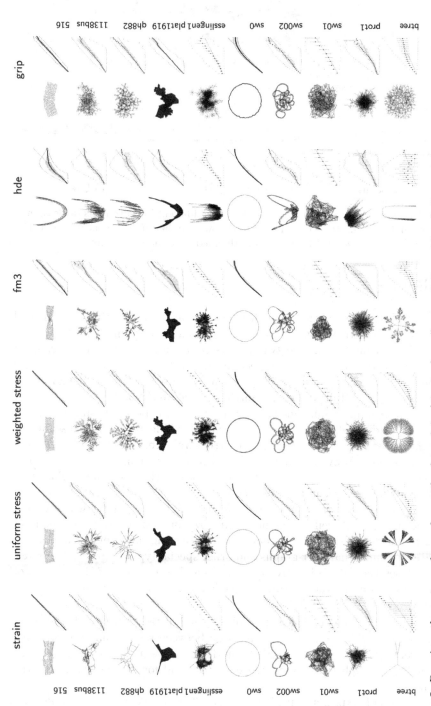

Fig. 2. Drawings the test graphs, and quartile plots of d_{ij} (abscissa) vs. $\|x_i - x_j\|$. Large dots indicate the median, small dots minimum and maximum, and black lines the range of the two middle quartiles (25–75 per cent). The thin blue line with slope 1 is a visual aid.

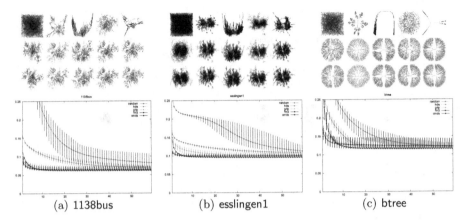

Fig. 3. Upper row: The majorization process with different initializations random, fm3, hde, grip, cmds after 0, 30, 60 iterations. Lower row: Number of iterations vs. stress. The bars indicate the range of values, the dots the median value, in 25 runs.

6   Results

Layout Quality. To assess layout quality both visually and quantitatively, aligned layouts and the distributions of layout distances are shown in Fig. 2 for each of the possible distance values between pairs of vertices, i.e. for values ranging from 1 to the diameter of the respective graph. The classical scaling layouts were generated with random initial positions and used as initial configurations for distance scaling. Initialization is further studied in Exp. 2.

The drawings for graphs qh882, 1138bus seem to confirm H1 and H2; using weights $w_{ij} = d_{ij}^{-2}$ helps to display local structures hidden by classical scaling or unweighted distance scaling. For regular structures 516, plat1919, sw0, distance scaling does not improve the quality of local representation. In a few cases classical scaling represents the overall structure better, such as the known clustering of esslingen1 into two densely connected parts.

In general, *H1 and H2 can be accepted* at least for graphs for which graph-theoretic distance is well representable in low dimensions. However, none of the MDS variants seems to be capable of representing both smaller and larger distances for small diameter graphs and other special types of graphs like btree. In such cases the MDS objective functions for distance representations is not always useful as an aesthetic criterion; see Section 7 for a discussion.

Initialization. For independence of graph size and distances we divide the stress by $\sum_{i,j} w_{ij} d_{ij}^2$, which allows for comparison between stress computations even for different graphs. We have carried out the iterative majorization process 25 times for each graph (with permuted edge list) and for each of the five initial placements.

The results of Exp. 2 are displayed in Fig. 3, which shows stress values over the majorization process for distance scaling, with weights $w_{ij} = d_{ij}^{-2}$ For almost

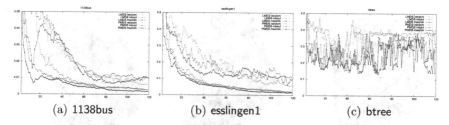

Fig. 4. Procrustes statistics measuring how well Pivot MDS (red) or Landmark MDS (blue) estimate the exact solution of classical scaling. Plotted are the median values of 25 runs with different node permutations, for $k \in \{3, \ldots, 120\}$ pivots.

all graphs we have tested, basically the same ranking resulted, with random being worst, followed by fm3, grip, hde. Initially, cmds solutions tend to have higher values, but overtakes the other initializations after some iterations.

All experiments indicate that *H3 is valid for all types of graphs*. Since large distances and thus global structures are represented well, classical scaling gives excellent initial configurations for distance scaling.

The bandwidth of stress values we observed for cmds-initialized layouts was almost always negligible, whereas stress values vary largely for all other methods in the 25 runs. Classical scaling gives reproducible initial configurations throughout, which are also robust against permutation of the input. All these observations *support H4*. Interestingly, btree is the only graph for which classical scaling resulted in some variation; we attribute this to the multiple occurrence of equal eigenvalues of matrix B (see Table 1).

Scalability. We computed estimates for the solution to classical scaling for all graphs, again in 25 runs with random node permutations. In each run, three sets of pivots were grown from $k = 3$ to 120 (following maxmin, random, and mixed) and used for Pivot MDS and Landmark MDS. The plots for the median values of three selected graphs are shown in Fig. 4.

For regular graphs like sw0, 516, the pivoting strategy is not crucial. In all other cases Pivot MDS is superior to Landmark MDS, regardless of the pivoting strategy. For Pivot MDS, the maxmin strategy performs better than random and slightly better than mixed. The corresponding plots seem to converge to zero faster and more smoothly than those for Landmark MDS. Once again, graph btree seems to be different from the others; estimating the full classical scaling solution appears to be unstable, no matter what pivoting strategy is used. Our observations indicate that *H5 is valid.*

We have conducted further experiments considering scalability, but omit them here due to space restrictions. One suite of experiments applies Pivot MDS to graphs with millions of nodes; we have observed that even those huge graphs, for which the full classical scaling is impractical, are laid out well with it, provided that two dimensions suffice, and, conversely, that increasing the number of pivots

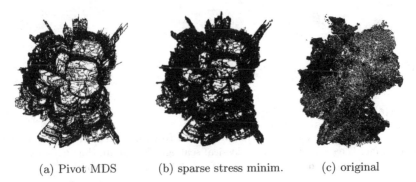

| (a) Pivot MDS | (b) sparse stress minim. | (c) original |

Fig. 5. Drawings for a large graph representing the street network in Germany (4 044 153 nodes, 9 564 235 edges, diameter 1 059)

does not improve layout quality if the graph is of higher intrinsic dimensionality; see also Sect. 4.

Another suite of experiments indicates that, technically, stress minimization scales even to very large graphs, but that *H6 is valid only with the limitation* that an appropriate sparsification scheme must be available.

7   Conclusion

We have studied different graph-layout approaches that aim at representing graph-theoretic distances by Euclidean distances. Our experiments suggest that minimization of weighted stress, an objective function that models the desired aesthetic properties explicitly, is to be preferred over force-directed placement. The recommended method for weighted stress minimization is to initialize with a fast approximation of classical scaling [4] and subsequent iterative improvement using localized stress reduction [12]. Both phases are easy to implement, but the second can be time-consuming. Approximation via sparse stress makes the algorithm scale to very large graphs, but further research on reliable sparsification schemes is needed.

The distance-based approach yields poor results on certain classes of graphs, which include small worlds and other graphs with many shortcuts or low diameter, and scale-free graphs with highly skewed degree distributions, large 1-shells, or other forms of structural imbalance. Some success has been obtained with stress weighting schemes based on graph invariants, but good characterizations of problematic graphs are missing and matching layout algorithms need to be developed further.

Using a hypotheses-based experimental design, we hope to foster clarity and reproducibility of our results, and to contribute to experimental evaluation of graph drawing algorithms in general.

References

1. Borg, I., Groenen, P.J.F.: Modern Multidimensional Scaling, 2nd edn. Springer, Heidelberg (2005)
2. Brandenburg, F.-J., Himsolt, M., Rohrer, C.: An experimental comparison of force-directed and randomized graph drawing algorithms. In: Brandenburg, F.J. (ed.) GD 1995. LNCS, vol. 1027, pp. 76–87. Springer, Heidelberg (1996)
3. Brandes, U.: Drawing on physical analogies. In: Kaufmann, M., Wagner, D. (eds.) Drawing Graphs. LNCS, vol. 2025, pp. 71–86. Springer, Heidelberg (2001)
4. Brandes, U., Pich, C.: Eigensolver methods for progressive multidimensional scaling of large data. In: Kaufmann, M., Wagner, D. (eds.) GD 2006. LNCS, vol. 4372, pp. 42–53. Springer, Heidelberg (2007)
5. Buja, A., Swayne, D.F.: Visualization methodology for multidimensional scaling. Journal of Classification 19, 7–43 (2002)
6. Civril, A., Magdon-Ismail, M., Bocek-Rivele, E.: SSDE: Fast graph drawing using sampled spectral distance embedding. In: Kaufmann, M., Wagner, D. (eds.) GD 2006. LNCS, vol. 4372, pp. 30–41. Springer, Heidelberg (2007)
7. Cohen, J.D.: Drawing graphs to convey proximity. ACM Transactions on Computer-Human Interaction 4(3), 197–229 (1997)
8. Cox, T.F., Cox, M.A.A.: Multidimensional Scaling, 2nd edn. CRC/Chapman and Hall, Boca Raton (2001)
9. Eades, P., Wormald, N.C.: Fixed edge-length graph drawing is NP-hard. Discrete Applied Mathematics 28(2), 111–134 (1990)
10. Freeman, L.C.: Graph layout techniques and multidimensional analysis. Journal of Social Structure 1 (2000)
11. Gajer, P., Kobourov, S.: GRIP – Graph drawing with intelligent placement. In: Marks, J. (ed.) GD 2000. LNCS, vol. 1984, pp. 222–228. Springer, Heidelberg (2001)
12. Gansner, E.R., Koren, Y., North, S.C.: Graph drawing by stress majorization. In: Pach, J. (ed.) GD 2004. LNCS, vol. 3383, pp. 239–250. Springer, Heidelberg (2005)
13. Hachul, S., Jünger, M.: An experimental comparison of fast algorithms for drawing general large graphs. In: Healy, P., Nikolov, N.S. (eds.) GD 2005. LNCS, vol. 3843, pp. 235–250. Springer, Heidelberg (2006)
14. Harel, D., Koren, Y.: Graph drawing by high-dimensional embedding. In: Goodrich, M.T., Kobourov, S.G. (eds.) GD 2002. LNCS, vol. 2528, pp. 207–219. Springer, Heidelberg (2002)
15. Kamada, T., Kawai, S.: An algorithm for drawing general undirected graphs. Information Processing Letters 31, 7–15 (1989)
16. Kruskal, J.B., Seery, J.B.: Designing network diagrams. In: Proc. First General Conference on Social Graphics, pp. 22–50 (1980)
17. Kruskal, J.B.: Multidimensional scaling by optimizing goodness of fit to a nonmetric hypothesis. Psychometrika 29(1), 1–27 (1964)
18. de Leeuw, J.: Applications of convex analysis to multidimensional scaling. In: Barra, J.R., Brodeau, F., Romier, G., van Cutsem, B. (eds.) Recent Developments in Statistics, pp. 133–145. North-Holland, Amsterdam (1977)
19. McGee, V.E.: The multidimensional scaling of "elastic" distances. Br. J. Math. Stat. Psychol. 19, 181–196 (1966)
20. Sammon, J.W.: A nonlinear mapping for data structure analysis. IEEE Transactions on Computers 18(5), 401–409 (1969)
21. Sibson, R.: Studies in the robustness of multidimensional scaling: Procrustes statistics. J. R. Stat. Sooc. 40(2), 234–238 (1978)
22. de Silva, V., Tenenbaum, J.B.: Sparse multidimensional scaling using landmark points. Tech. rep., Stanford University (2004)

Topology Preserving Constrained Graph Layout

Tim Dwyer, Kim Marriott, and Michael Wybrow

Clayton School of Information Technology,
Monash University, Clayton, Victoria 3800, Australia
{Tim.Dwyer,Kim.Marriott,Michael.Wybrow}@infotech.monash.edu.au

Abstract. Constrained graph layout is a recent generalisation of force-directed graph layout which allows constraints on node placement. We give a constrained graph layout algorithm that takes an initial feasible layout and improves it while preserving the topology of the initial layout. The algorithm supports poly-line connectors and clusters. During layout the connectors and cluster boundaries act like impervious rubber-bands which try to shrink in length. The intended application for our algorithm is dynamic graph layout, but it can also be used to improve layouts generated by other graph layout techniques.

1 Introduction

A core requirement of dynamic graph layout is stability of layout during changes to the graph so as to preserve the user's mental model of the graph. One natural requirement to achieve this is to preserve the topology of the current layout during layout changes. While topology preservation has been used for dynamic layout based on orthogonal graph layout, its use in force-directed approaches to dynamic layout is much less common.

Constrained graph layout [12,3,4] is a recent generalisation of the force-directed model for graph layout. Like force-directed methods, these techniques find a layout minimising a goal function such as the standard *stress* goal function which tries to place all pairs of nodes their ideal (graph-theoretic) distance apart. However, unlike force directed methods, constrained graph layout algorithms allow the goal to be minimised subject to placement constraints on the nodes. In this paper we detail a constrained graph layout algorithm that preserves the topology of the initial layout. The primary motivation for our development of this algorithm was to support dynamic layout but it can also be used to improve layouts generated by other graph layout techniques such as planarisation techniques [11].

Our algorithm supports network diagrams with poly-line connectors and arbitrary node clusters. It ensures that the nodes do not overlap and that additional constraints on the layout—such as alignment and downward pointing edges—remain satisfied. During layout optimisation the *paths*, i.e poly-line connectors and cluster boundaries, act like rubber-bands, trying to shrink in length and hence, in the case of connectors, straighten. Like physical rubber bands, the paths are impervious and do not allow nodes and other

I.G. Tollis and M. Patrignani (Eds.): GD 2008, LNCS 5417, pp. 230–241, 2009.

paths to pass through them. Thus, the initial layout topology is preserved. Figure 1 shows example layouts obtained with our algorithm.

Extending constrained graph layout to handle topology preservation is conceptually quite natural since topology preservation can be regarded as a kind of constraint. However, it was not possible to straightforwardly extend existing constrained graph layout algorithms to preserve topology. One issue is that previous algorithms were based on functional majorization whose use relied on particular properties of the stress goal function.

The main technical innovations in our new algorithm are fourfold. First, we utilise a new goal function, *P-stress*, that encodes the rubber band metaphor, measuring the stretch of paths as well as trying to place objects a minimum distance apart. Importantly, the *P-stress* is *bend-point invariant* in the sense that merging two consecutive collinear segments in a path does not change the value of the goal function. This aids convergence since it means that

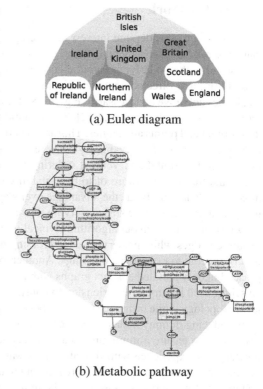

(a) Euler diagram

(b) Metabolic pathway

Fig. 1. Example layouts obtained with the topology preserving constrained graph layout algorithm. In the metabolic pathway, three vertical alignment constraints have been added to improve the layout.

the goal function behaves continuously as paths change during optimisation. Second, we utilise gradient projection rather than functional majorization. This approach is generic in the choice of goal function and so can be used to minimise *P-stress*. Third, we give a novel algorithm for updating paths in a layout given that nodes are moved in a single dimension. This maintains the relative order of nodes and paths in that dimension and so preserves the initial topology. The final innovation is our uniform treatment of connector routes and cluster boundaries as impervious paths. This allows our algorithm to handle arbitrary clusters.

The algorithm for topology preserving constrained graph layout given here underpins two dynamic graph layout applications we have developed. The first is a network diagram authoring tool, Dunnart, which uses the algorithm to provide continuous layout adjustment during user interaction [5]. The second is a network diagram browser which uses the algorithm to update the layout of a detailed view of part of the network as the user changes the focus node or collapses or expands node clusters [6]. The contribution of this paper is to detail the algorithm.

2   Related Work

There has been considerable interest in developing techniques for stable graph layout that preserve the user's mental model of the graph [14]. These techniques are quite specialised to the underlying layout algorithms. The standard approach for supporting stability in force-directed approaches is to simply add a "stay force" on each node so that it does not move unnecessarily, e.g. [9]. Stable dynamic layout has also been studied for orthogonal graph layout, e.g. [2]. There, stability is preserved by trying to preserve the current bend points and angles. This has the effect of preserving the layout topology. Finally, in the case of Sugiyama-style layered layout stability is achieved by preserving the current horizontal and vertical ordering between nodes, e.g. [15]. Our approach is the first that we are aware of to base stability on topology preservation in a force-directed style layout. It has the advantage over stay forces that the layout is better able to adjust to changes while still preserving the original structure.

Orthogonal graph layout algorithms typically feature a refinement step that attempts to shorten edges while preserving edge crossing topology [8]. However, the approach is very specific to orthogonal drawings. Another method, [1], used a force directed approach but only handled abstract graphs with point nodes and straight-line edges. Most closely related is our earlier extension to constrained stress majorization that preserves layout topology while trying to straighten bends in poly-line connectors [7]. This works by introducing dummy nodes in each connector at all possible bend points and adding constraints to ensure a minimum separation between objects and bend-points. Unfortunately, our experience with this algorithm was that straightening bends sometimes meant that connector length was increased and that the algorithm did not scale to moderately sized networks because of the large number of dummy nodes. Even worse it did not always converge because the goal function was not bend-point invariant. The algorithm given here is considerably simpler, convergent and faster.

3   Problem Definition

A graph $G = (V, E, C)$ consists of a set of nodes V, a set of edges $E \subseteq V \times V$, and a set of node clusters $C \subseteq \wp V$. We let $width(v)$ and $height(v)$ give the width and height of the bounding rectangle, r_v, of each node $v \in V$.

A 2-D drawing of a graph is specified by a tuple (x, y, P) where (x_v, y_v) gives the centre position for each node $v \in V$ and P is a set of *paths* specifying the edge routings and cluster boundaries. A path is a piecewise linear path through a sequence of points $p_1, \ldots, p_k$ where each point is either the center or one of the corners of a node's bounding rectangle and represented by a pair (v, i) where $v \in V$ and $i \in \{Centre, TL, TR, BL, BR\}$). In the case of a path giving the routing for an edge $e = (s, t) \in E$, p_1 is the centre of node s and p_k the centre of node t while the other points are node bounding rectangle corners. In the case that the path is for a cluster boundary, all points must correspond to node bounding rectangle corners and $p_1 = p_k$.

Separation constraints are inequality or equality constraints over pairs of position variables in either the horizontal or vertical axes of the drawing, e.g. for a pair of nodes $u, v \in V$ we might define a separation constraint over their $x-$positions: $x_u + g \leq x_v$ where g specifies a minimum spacing between them.

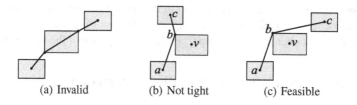

Fig. 2. Example of incorrect (a,b) and correct (c) paths

A *feasible* drawing of a graph (see Fig. 2) is one in which:

- all separation constraints are satisfied;
- no two node rectangles overlap;
- the nodes inside the region defined by the boundary of each cluster c are exactly the nodes in c;
- every path $p \in P$ is *valid* and *tight*.

A *valid* path is one in which no segment passes *through* a node rectangle, except the first and last segments in a path corresponding to an edge which must terminate at the centre of rectangles as specified above. A *tight* path is one where every bend (described by three consecutive points a, b, c in the path) is wrapped around the rectangle r_v associated with the bend point $b = (v, i)$. That is, the points a, b, c in order must constitute a turn in the same direction as the points a, b, v in order, and the points b, c, v must also constitute a turn in the same direction.

A common strategy for finding aesthetically pleasing drawings of graphs is to define a cost function over the positions of the nodes and then to minimise this cost function by adjusting these positions. In our case we are also interested in the lengths of paths. Therefore, we use a novel cost function *P-stress* which also takes the paths P of the layout into consideration:

$$\sum_{u<v\in V} w_{uv} \left((d_{uv} - ||(x_u, y_u), (x_v, y_v)||)^+ \right)^2 + \sum_{p\in P} w_p \left((||p|| - L_p)^+ \right)^2$$

where $(z)^+$ is z if $z \geq 0$ and 0 otherwise and $w_p = \frac{1}{L_p^2}$, $w_{uv} = \frac{1}{d_{uv}^2}$.

The first component of *P-stress* is a modification of the *stress* function used in the stress majorization [10] and Kamada and Kawai [13] layout methods. This considers the ideal distance d_{uv} between each pair of nodes which is proportional to the graph theoretic distance, i.e. shortest path, between the nodes. However, unlike the stress function, nodes that are more than their ideal distance apart are not penalised, thus eliminating long range attraction since this can cause issues in highly constrained problems.

The second component of *P-stress* tries to make the length of each path p in the network, no more than its ideal length L_p. The ideal length of the route for an edge e is simply a fixed constant while the desired length of the boundary for cluster c is $2\sqrt{\pi \sum_{v\in c} width(v)height(v)}$ (i.e. the ideal length is proportional to the perimeter of the circle of the same area as that of the constituent nodes). This second component is purely attractive, otherwise minimising *P-stress* could potentially *increase* bends.

Note that *P-stress* is *bend-point invariant* in the sense that merging two consecutive collinear segments in a path does not change the *P-stress* of layout since the overall path length does not change. This is important for convergence of the layout algorithm.

4   Minimising P-Stress Using Gradient Projection

Our layout problem is, therefore, given a feasible layout for a graph to find a new layout that is feasible, has the same topology as the original layout, and which locally minimises *P-stress*. In this section we give an algorithm to do this. An example of its operation is shown in Fig. 3.

Our algorithm works by alternately adjusting horizontal and vertical positions of all nodes to incrementally reduce *P-stress*. This makes the computation of the new positions considerably simpler than if both dimensions were considered together. Constrained stress majorization [4] also uses a similar approach to reduce stress. However, the useful Cauchy-Schwarz based expansion of the stress function into horizontal and vertical quadratic forms which strictly (upper-)bound the goal function, is no longer easily derived for *P-stress*. Instead, at each iteration we use a quadratic approximation based on the second order Taylor series expansion of *P-stress* around the current horizontal position x and compute a descent vector $-g$ and step size α from the first and second derivatives of this quadratic to compute a new position d for the horizontal position variables. We then use the function *project-x* to project d onto the horizontal constraints necessary to avoid overlap and to preserve topology and any other user specified separation constraints, C_x, on the horizontal variables. Next we perform an analogous operation to compute a new position for the vertical position variables y. The high-level algorithm is thus:

procedure *gradient-projection-x*(x, y, P, C)
   $g \leftarrow \nabla_x P\text{-}stress(x, y, P)$
   $H \leftarrow \nabla_x^2 P\text{-}stress(x, y, P)$
   $\alpha \leftarrow \frac{g^T g}{g^T H g}$
   $d \leftarrow x - \alpha g$
  return *project-x*(x, y, P, d, C)
procedure *improve*(x, y, P, C_x, C_y)
   $(x', y', P') \leftarrow (x, y, P)$
  repeat
     $(x, P'') \leftarrow$ *gradient-projection-x*(x, y, P, C_x)
     $(y, P) \leftarrow$ *gradient-projection-y*(x, y, P'', C_y)
  until $|P\text{-}stress(x', y', P') - P\text{-}stress(x, y, P)|$ sufficiently small
  return (x, y, P)

Before giving details of projection we must make precise what we mean by topology preservation. Considering just the horizontal case, since the vertical is symmetrical, we say that a horizontal adjustment of the nodes from feasible layout L to feasible L' is *topology preserving* if no node or line segment moves through another node or line segment. More exactly, let M and M' be the layouts obtained from L and L', respectively, by infinitesimally reducing the height of each node's bounding rectangle and

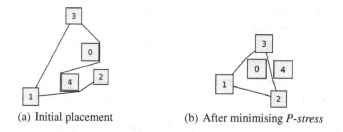

(a) Initial placement   (b) After minimising *P-stress*

Fig. 3. Example of how our layout algorithm improves the network layout by reducing *P-stress* (which shortens edge routes) while preserving the topology of the initial layout

appropriately modifying the paths. This means that rectangles whose top and bottom were aligned in the original layout now have a infinitesimal vertical separation between them. Then for any height h we must have that scanning left to right along the horizontal line $y = h$ encounters exactly the same sequence of edges, clusters and nodes in both M and M' where an edge is encountered whenever the line intersects a path segment for the edge, a cluster is encountered whenever the line intersects a path segment for its boundary and a node is encountered when the line intersects the node's bounding rectangle.

5   Topology Preserving Projection

The heart of the layout algorithm are the procedures *project-x* and *project-y* which perform a projection operation in the specified axis. We shall focus on *project-x*: procedure *project-y* is symmetric. The call *project-x*(x, y, P, d, C) returns a new x position and paths (x', P') s.t layout (x', y, P') is feasible and preserves the topology of (x, y, P) while ensuring x' is as close as possible to the desired position d. It has three main steps:

(1) Generate separation constraints C^{no} to ensure non-overlap of nodes and topology constraints TC to ensure topology preservation.
(2) Project d on to $SC = C \cup C^{no}$ giving $\bar{x}$. This is achieved by solving the quadratic program:

$$\min_{x} \sum_{v in V} (x_v - d_v)^2 \text{ subject to } SC$$

(3) Update the path routing P to give P' by moving the nodes smoothly from x to $\bar{x}$ appropriately adjusting the paths as the nodes move in order to satisfy the topology constraints TC.

In Step 2 of *project-x* we solve the quadratic program using the incremental active-set procedure *solveQPSC* given in [4]. Like most active-set methods it is difficult to prove that this has polynomial running time, but in practice it is very fast, as indicated by our experimental results. We now look at Steps 1 and 3 in more detail.

Non-overlap and topological constraints are generated for a horizontal move using a top-to-bottom scan of the drawing. At each step we keep the list of currently open node bounding rectangles and path line segments. To do so we process the vertical opening and closings of each rectangle OR, CR and line segment OS, CS of the given routing in order from top to bottom and, when two such events occur at the same vertical position, then with precedence:

- OS before CS so that horizontal segments are handled properly
- CR before OR to avoid unnecessary non-overlap constraints (assuming no zero height rectangles)
- CS before OR, CR before OS, OS before OR, and CR before CS to ensure all possible segment/rectangle interactions are considered.

For each rectangle opening (i.e. the top of each rectangle) we add to C^{no} a separation constraint between the rectangle and its immediate left and right neighbours in the list of open rectangles at that y-position (the scan position). Each separation constraint has the form $x_u + s \leq x_v$ over the x positions of nodes u and v and preserves the relative horizontal ordering of u and v and prevents the nodes from overlapping, where $s = (width(u) + width(v))/2$.

The scan also generates *topology* constraints between nodes and paths which ensure that the paths remain tight and valid. There are two types of topology constraints: *straight* constraints—between a node w and a path segment uv which ensures that the path remains valid, i.e. the node does not overlap the path segment and *bend* constraints associated with a bend point between two consecutive line segments uv and vw which ensures that the path remains tight around the bend point v.

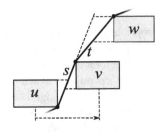

Fig. 4. Constraints generated during a vertical scan. There is one separation constraint $x_u + \frac{1}{2}(width(u) + width(v)) \leq x_v$ to prevent overlap, three bend constraints (the construction for the constraint ensuring the path remains tight around v is shown) and three straight constraints at the places where the segments s and t may potentially bend.

Both kinds of topology constraint give rise to a linear inequality over the three variables corresponding to u, v and w enforcing that the rectangle r_w associated with node w must be to the right or left of a line between the corners of two nodes u and v. We write this in the standard form $x_w + g \oplus x_u + p(x_v - x_u)$ where $\oplus$ is either $\leq$ or $\geq$. For straight constraints $0 < p \leq 1$ while for bend constraints $p > 1$. For instance, in the case of the bend constraint enforcing that the path remains tight around the bend point v in Fig. 4 we have that

$$x_{w^{TL}} \geq x_{u^{BR}} + \frac{y_{w^{TL}} - y_{u^{BR}}}{y_{v^{TL}} - y_{u^{BR}}}(x_{v^{TL}} - x_{u^{BR}})$$

where $x_{w^{TL}} = x_w - width(w)/2$ etc. This can be rewritten into the standard form. The procedures for creating each type of constraint is given in Fig. 5.

If $|P|$ denotes the number of path segments, the worst case complexity of Step 1 of *project-x* is $O(|V|(|P| + \log|V|))$ and up to $O(|V|)$ non-overlap constraints and $O(|P||V|)$ topological constraints can be generated.

We now consider Step 3 of *project-x*. This is performed by procedure *move* (Fig. 6). This updates the paths by moving the nodes horizontally from the initial feasible solution

procedure *createStraightConstraint*(s, w, y, TC)
  % for segment $s = uv$ and node w at scan pos y
  $p \leftarrow (y - y_u)/(y_v - y_u)$
  $x_p \leftarrow x_u + p(x_v - x_u)$
  $leftOf \leftarrow x_w < x_p$
  $corner \leftarrow$ **if** $y < y_w$ **then if** *leftOf* **then** BR **else** BL
                           else if *leftOf* **then** TR **else** TL
  $offset(w) \leftarrow width(w)/2$ (-ve if *leftOf*)
  $g \leftarrow offset(u) + p(offset(v) - offset(u)) - offset(w)$
  $TC \leftarrow TC \cup \{TopologyConstraint(straight, u, v, w, p, g, leftOf)\}$

procedure *createBendConstraint*(b, TC)
  % for bend point $b = (v, i)$, between segments ab and bc
  if i is the centre of v **then return**
  if existing bend constraint t on b **then** remove t
  $leftOf \leftarrow i \in \{TR, BR\}$
  if $|y_a - y_b| > |y_b - y_c|$ **then**
    $p \leftarrow (y_c - y_a)/(y_b - y_a)$
    $g \leftarrow offset(a) + p(offset(b) - offset(a)) - offset(c)$
    $t \leftarrow TopologyConstraint(bend, a, b, c, p, g, leftOf)$
  else
    $p \leftarrow (y_a - y_c)/(y_b - y_c)$
    $g \leftarrow offset(c) + p(offset(b) - offset(c)) - offset(a)$
    $t \leftarrow TopologyConstraint(bend, c, b, a, p, g, leftOf)$
  $TC \leftarrow TC \cup \{t\}$

Fig. 5. The procedures for creating straight constraints and bend constraints are used in both the initial scan to set up topology constraints and by the procedure *satisfy* (Fig. 6). The function *TopologyConstraint* creates a constraint of the form $x_w + g \leq x_u + p(x_v - x_u)$ if leftOf (or $\geq$ otherwise).

x for which the routing is correct towards $\bar{x}$ detecting violated topology constraints as they move. A violated *bend* constraint indicates that consecutive segments have become aligned and can be replaced with a single segment. A violated *straight* constraint indicates that a single segment needs be split into two new segments with a new bend point.

The maximum horizontal move γ that can be made along the line $x = a + \gamma(b - a)$ from a to b without violating topology constraint t is determined by solving the linear equation associated with the constraint. For example, if t is the constraint $x_w + g \leq x_u + p(x_v - x_u)$ then the maximum safe move is obtained by substituting $x_i = a_i + \gamma(b_i - a_i)$ for each node i and solving for γ:

$$\gamma = \frac{\alpha}{\beta} = \frac{a_w - g - a_u + p(a_u - a_v)}{b_u - a_u + p(a_u - b_u + b_v - a_v) + a_w - b_w}$$

The iterative process of finding the next such constraint and updating the paths P is accomplished in the *move* procedure, Fig. 6.

Note that the *satisfy* procedure shown in Fig. 6, which satisfies a topology constraint by either merging or splitting segments, must transfer or replace other bend and straight

procedure $satisfy(t, TC, P)$
  $TC \leftarrow TC \setminus \{t\}$
  if t is a *bend constraint* over points a, b, c **then**
    % $b = (v, i)$ is the bend point of t
    replace segments ab and bc in P with new segment ac
    $createStraightConstraint(ac, v, b_y)$
  else % t is a *straight constraint* over u, v, w
    replace segment uw in P with segments uv and vw
    transfer straight constraints on uw to either uv or vw
    $createBendConstraint(u)$
    $createBendConstraint(v)$

procedure $move(x, \bar{x}, TC, P)$
  repeat
    $\alpha \leftarrow \beta \leftarrow 1$
    $t* \leftarrow$ None
    for $t \in TC$
      % t is a Topology Constraint over u, v, w with constants p, g
      $a \leftarrow a_w - g - a_u + p(a_u - a_v)$
      $b \leftarrow b_u - a_u + p(a_u - b_u + b_v - a_v) + a_w - b_w$
      if $a\beta < \alpha b$ **then**
        $\alpha \leftarrow a, \beta \leftarrow b, t* \leftarrow t$
    $x \leftarrow x + \frac{\alpha}{\beta}(\bar{x} - x)$
    if $t* \neq$ None **then** $satisfy(t, TC, P)$
  until $\frac{\alpha}{\beta} = 1$

Fig. 6. The procedure $satisfy(t, TC, P)$ satisfies a topology constraint $t \in TC$ that is at equality, by modifying P with a *valid* and *tight* system of segments. Procedure $move(x, \bar{x}, TC, P)$ updates the path P by moving nodes in one dimension from position x to $\bar{x}$ to satisfy the topology constraints TC.

constraints associated with the affected segments. The detail is not shown, but an example of the difficult edge case of a horizontal path segment is shown in Fig. 7.

The *move* procedure used for updating the paths to preserve validity and topology can also be thought of as a kind of active-set process, and as such it is difficult to prove that it is polynomial. Again, however, please see our results section for actual running times which indicate that running times scale fairly well with the number of topology constraints generated. Note that the number of bend constraints is exactly the number of bend points in P, and the number of straight constraints—while the worst case is $O(|P||V|)$—is limited by only generating constraints for segments which are *visible* in the axis of movement from a given rectangle open/close.

Theorem 1. *Let (x, y, P) be a feasible layout with respect to the separation constraints C_x and C_y in the x and y dimensions, respectively. Then project-x(x, y, P, d, C_x) returns a new x position and paths (x', P') s.t layout (x', y, P') is feasible and preserves the topology of (x, y, P) while ensuring x' is as close as possible to the desired position d.*

Proof. (Sketch) Any feasible and topology preserving layout must satisfy $SC = C_x \cup C^{no}$. Step 2 ensures that x' is the projection of d onto SC, so it is the closest node

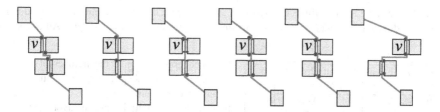

Fig. 7. The result of each iteration of *move* is shown for a path with a horizontal segment. The iterations progress from left to right. The node v is required to move to the right relative to the other nodes. The four central nodes are shown slightly separated for clarity, but we assume that the boundaries of these nodes are actually touching—hence creating, initially, a horizontal segment. The small circles represent bend points, while the '-'s represent straight topology constraints. Note that, to properly preserve topology as the segments are split to satisfy a straight constraint, the remaining straight constraints must be transferred to the correct sub-segments.

position that satisfies SC. Furthermore, one can prove by induction that the *satisfy* procedure returns updated paths P' that are topology preserving, tight and valid.

6 Finding a Feasible Topology

We can apply our topology preserving layout adjustment to a layout obtained by any graph drawing algorithm, assuming the generated layout is feasible as defined in §3. Although not the primary focus of this paper we have also developed an algorithm to find an initial feasible layout. This has two main steps:

(1) Perform standard stress majorization to find an initial position for the nodes. A position for the nodes satisfying the constraints is found by projecting this position on to the user specified separation constraints and then using a greedy heuristic to satisfy the non-overlap constraints and cluster containment constraints. We use the approach sketched in [4].

(2) Edge routing is performed using the incremental poly-line connector routing library libavoid [16] to compute poly-line routes for each edge, which minimise edge length and amount of bend. An initial cluster boundary is obtained by taking the convex hull of the nodes in the cluster.

We note that the edge routing library has been extended to handle clusters and finds routes for edges that do not unnecessarily pass through clusters. It also performs "nudging" on the final routes to separate paths with shared sub-routes.

7 Experimental Results

Table 1 gives some indicative run-times on various size graphs for finding an initial layout using the two-step algorithm given above, then using the topology-preserving constrained graph layout algorithm to find a locally optimal layout. The topology-preserving constrained graph layout algorithm[1] is quite fast with less than two seconds

---

[1] Implemented as part of the Adaptagrams project. http://adaptagrams.sf.net/

Table 1. Indicative running times for layout on an average (1GHz) PC for various size randomly generated directed networks with constraints imposing downward pointing edges. All times are in seconds.

$|V|$	$|E|$	Feasible layout		Optimise	Total
		Step 1	Step 2		
49	51	0.08	0.11	0.06	0.17
93	105	0.22	0.50	0.24	0.74
128	144	0.51	1.02	0.55	1.57
144	156	0.92	1.31	0.45	1.76
169	195	0.83	1.97	0.82	2.79
199	238	1.31	2.94	1.45	4.39
343	487	2.65	13.94	1.89	15.83

For each graph we give the number of nodes and edges. The number of separation constraints imposing downward edges is $|E|$. We give the time to find an initial feasible layout (Step 1 and Step 2) from a random starting configuration; and then to optimise the result using the topology preserving constrained graph layout algorithm. Optimisation algorithms were set to terminate when the change in P-stress or stress was $< 10^{-5}$.

required to layout networks of around 350 nodes. We have found that the main cost for each iteration is computation of the descent vector and step size. We also note that our experience with the algorithm in interactive applications is that it provides real-time updating of layout for graphs with up to 100 nodes.

Computing an initial layout is more expensive, and the dominating cost in finding the initial layout is finding the initial connector routing.

8   Conclusion

We have presented a constrained graph layout algorithm that preserves the topology of the initial layout. It supports network diagrams with poly-line connectors and arbitrary node clusters. It ensures that nodes do not overlap and that additional placement constraints on the layout remain satisfied. The algorithm is fast enough to support real-time layout of networks with up to 100 nodes in two dynamic graph layout applications we have developed: a network diagram authoring tool and a network diagram browser. While the primary motivation for our development of the algorithm was to support dynamic layout it can also be used to improve layouts generated by other graph layout techniques.

One of the strengths of the algorithm is that it can be straightforwardly modified to work with other goal function, so long as the second derivative is computable and the goal function is bend-point invariant. We plan to explore other goal functions. We also plan to explore generalising the algorithm to handle arbitrary linear constraints, not only separation constraints. As part of this we plan to modify the algorithm to perform minimization in both dimensions at once, rather than separately.

References

1. Bertault, F.: A force-directed algorithm that preserves edge crossing properties. In: Kratochvíl, J. (ed.) GD 1999. LNCS, vol. 1731, pp. 351–358. Springer, Heidelberg (1999)
2. Bridgeman, S.S., Fanto, J., Garg, A., Tamassia, R., Vismara, L.: InteractiveGiotto: An algorithm for interactive orthogonal graph drawing. In: DiBattista, G. (ed.) GD 1997. LNCS, vol. 1353, pp. 303–308. Springer, Heidelberg (1997)

3. Dwyer, T., Koren, Y., Marriott, K.: Drawing directed graphs using quadratic programming. IEEE Transactions on Visualization and Computer Graphics 12(4), 536–548 (2006)
4. Dwyer, T., Koren, Y., Marriott, K.: IPSep-CoLa: An incremental procedure for separation constraint layout of graphs. IEEE Transactions on Visualization and Computer Graphics 12(5), 821–828 (2006)
5. Dwyer, T., Marriott, K., Wybrow, M.: Dunnart: A constraint-based network diagram authoring tool. In: GD 2008. LNCS, vol. 5417. Springer, Heidelberg (to appear, 2009)
6. Dwyer, T., Marriott, K., Wybrow, M.: Exploration of networks using overview+detail with constraint-based cooperative layout. IEEE Transactions on Visualization and Computer Graphics (InfoVis 2008) (to appear 2008)
7. Dwyer, T., Marriott, K., Wybrow, M.: Integrating edge routing into force-directed layout. In: Kaufmann, M., Wagner, D. (eds.) GD 2006. LNCS, vol. 4372, pp. 8–19. Springer, Heidelberg (2007)
8. Eiglsperger, M., Fekete, S.P., Klau, G.W.: Drawing Graphs: Methods and Models, chap. Orthogonal graph drawing, pp. 121–171. Springer, London (2001)
9. Frishman, Y., Tal, A.: Online dynamic graph drawing. In: Eurographics/IEEE-VGTC Symp. on Visualization. Eurographics Association (2007)
10. Gansner, E., Koren, Y., North, S.: Graph drawing by stress majorization. In: Pach, J. (ed.) GD 2004. LNCS, vol. 3383, pp. 239–250. Springer, Heidelberg (2005)
11. Gutwenger, C., Mutzel, P., Weiskircher, R.: Inserting an edge into a planar graph. In: SODA 2001: Proc. of the 12th Annual ACM-SIAM Symp. on Discrete Algorithms, pp. 246–255. Society for Industrial and Applied Mathematics (2001)
12. He, W., Marriott, K.: Constrained graph layout. Constraints 3, 289–314 (1998)
13. Kamada, T., Kawai, S.: An algorithm for drawing general undirected graphs. Information Processing Letters 31, 7–15 (1989)
14. Misue, K., Eades, P., Lai, W., Sugiyama, K.: Layout adjustment and the mental map. Journal of Visual Languages and Computing 6(2), 183–210 (1995)
15. North, S.C., Woodhull, G.: Online hierarchical graph drawing. In: Mutzel, P., Jünger, M., Leipert, S. (eds.) GD 2001. LNCS, vol. 2265, pp. 232–246. Springer, Heidelberg (2002)
16. Wybrow, M., Marriott, K., Stuckey, P.J.: Incremental connector routing. In: Healy, P., Nikolov, N.S. (eds.) GD 2005. LNCS, vol. 3843, pp. 446–457. Springer, Heidelberg (2006)

Embeddability Problems for Upward Planar Digraphs[*]

Francesco Giordano[1], Giuseppe Liotta[1], and Sue H. Whitesides[2]

[1] Università degli Studi di Perugia, Italy
{giordano,liotta}@diei.unipg.it
[2] McGill University, Canada
sue@cs.mcgill.ca

Abstract. We study two embedding problems for upward planar digraphs. Both problems arise in the context of drawing sequences of upward planar digraphs having the same set of vertices, where the location of each vertex is to remain the same for all the drawings of the graphs. We develop a method, based on the notion of book embedding, that gives characterization results for embeddability as well as testing and drawing algorithms.

1   Introduction

In the *upward point-set embeddability problem with mapping* the input is an upward planar digraph G with n vertices, a set S of n distinct points in the plane, and a mapping Φ from the vertices of G to the points of S. The desired output is an upward planar drawing of G with the vertices located at the points of S assigned by the mapping. Not all instances of this problem admit a solution, as shown in Fig. 1(a) (there is no choice of upward direction with respect to which the location of vertex 1 is lowest).

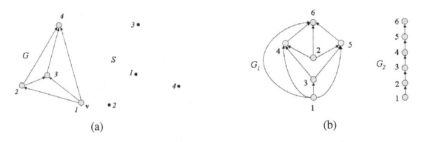

(a)   (b)

Fig. 1. (a) An upward planar digraph G, a set S of points, and a mapping of the vertices to the points of S such that an upward point-set embedding of G on S does not exist. (b) Two upward planar digraphs whose union is acyclic but that do not admit an upward consistent simultaneous embedding.

---

[*] This work is partially supported by the MIUR Project "MAINSTREAM: Algorithms for massive information structures and data streams", and by NSERC.

I.G. Tollis and M. Patrignani (Eds.): GD 2008, LNCS 5417, pp. 242–253, 2009.

In the *upward consistent simultaneous embeddability problem* the input is a sequence of upward planar digraphs that have the same vertex set. The desired output is a set S of points in the plane and a mapping from the vertices to the points such that all the digraphs have an upward point-set embedding on S with respect to a common, upward direction. Clearly, a solution exists only if the union of the digraphs is acyclic. However, this condition is not sufficient: the union of the two digraphs of Fig. 1(b) is acyclic, yet, as can be checked by straightforward case analysis, there is no simultaneous upward embedding with respect to a common direction.

Note that in the first problem, referred to as the *point-set embeddability* problem for short, the desired locations of the vertices are specified by Φ in the problem input, and only one graph is given. In the second problem, referred to as the *simultaneous embeddability* problem for short, the locations for the vertices are to be computed, and several graphs are given in the input. These problems arise in the context of computing drawings for a set or sequence of graphs under two different scenarios. In the first scenario, the graphs are specified one at a time, and the vertex locations for the drawing of the first graph determine the vertex locations for all the remaining drawings. Hence for each graph after the initial one, the locations for its drawings are specified. This gives rise to the point-set embeddability problem. An example of this scenario is provided by the visual analysis of self-modifiable code, based on computing a sequence of drawings whose edges are defined at run-time (see, e.g., [9]). In the second scenario, the graphs are all known from the outset. This gives rise to the simultaneous embeddability problem. This scenario occurs, for example, in the visual comparison of several phylogenetic trees proposed for the same organisms.

1.1   Summary of Main Results

Our first main result, of interest on its own, provides a tool for obtaining the others. Namely, in Sect. 3, we prove that a planar st-digraph together with any given topological numbering ρ admits an upward topological book embedding such that the ordering of the vertices along the spine is ρ. The number of spine crossings per edge is at most $2n - 4$, which is asymptotically worst-case optimal (n is the number of vertices).

For the point-set embeddability problem, we characterize in Sect. 4 those instances that admit a solution, providing an $O(n^2)$-time drawing algorithm that produces at most $2n - 3$ bends per edge, which is worst-case asymptotically optimal. Then in Sect. 5 we give an $O(n^3)$-time testing algorithm.

For the simultaneous embedding problem, in Sect. 6 we give a combinatorial characterization of instances that admit a solution.

1.2   Related Results

Both the embeddability problems we consider have mainly been studied for planar undirected graphs. In that case, Halton [9] proved that every instance of the point-set embeddability problem has a solution. Pach and Wenger [13] showed that solutions can require $\Omega(n)$ bends per edge and showed how to construct drawings with at most $O(n)$ bends per edge. See [1] for recent extensions and improvements. See Frati, Kaufmann, and Kobourov [5] for an extensive survey of simultaneous embeddability problems. Simultaneous embeddability for upward planar digraphs has been recently undertaken in [6], but for two digraphs, and without the requirement for the same choice of upward direction.

For book embeddings, see, e.g., [4,6] for the notion of an upward planar drawing where the vertices are aligned along a spine in a specified order, and edges are drawn as monotone curves that can cross the line. See [1] for results on book embeddings of undirected planar graphs.

For reasons of space, some proofs have been omitted and can be found in [7].

2 Preliminaries

We assume familiarity with basic graph drawing terminology [2,12]. A *digraph* is a directed graph. Let G be a digraph and let u, v be any two vertices of G; (u, v) denotes the directed edge from u to v. A *topological ordering* of a planar digraph G with n vertices is a mapping ρ of its vertices to distinct integers such that for every edge (u, v) we have $\rho(u) < \rho(v)$. A *topological numbering* is a topological ordering where the vertices are mapped to integers $1, \ldots, n$. Let u and v be two vertices of a digraph with a given topological numbering ρ; if $\rho(u) < \rho(v)$ we say that u *precedes* v. A topological numbering of a planar digraph with n vertices can be computed in $O(n)$ time using standard graph search techniques [3].

A *drawing* of a digraph G maps each vertex of G to a distinct point in the plane and each edge (u, v) of G to a simple Jordan curve oriented from the point representing u to the point representing v. A drawing of a digraph is *planar* if no two edges cross each other. A planar drawing Γ of a digraph G partitions the plane into topologically connected regions called the *faces*. The unbounded face is called the *external face*. A planar drawing of a digraph is *upward* if all of its edges are monotonically increasing in a common direction which is called the *upward direction* of the drawing. A digraph that admits an upward planar drawing is said to be *upward planar*. Let Γ be an upward planar drawing of an upward planar digraph G. Γ induces two linear lists of incoming and outgoing edges incident on each vertex v of G. An *upward planar embedding* of an upward planar digraph G is an equivalence class of upward planar drawings that induce the same two linear lists for each vertex of G and define the same external face. An upward planar digraph G with a given upward planar embedding is called an *upward planar embedded digraph*.

An *st-digraph* is a biconnected acyclic digraph with exactly one source s and exactly one sink t, and such that (s, t) is an edge of the digraph. A *planar st-digraph* is an *st*-digraph that is planar and embedded with vertices s and t on the boundary of the external face. A planar *st*-digraph is said to be *maximal* if all its faces are triangles, i.e. the boundary of each face has exactly three vertices and three edges. Given any planar *st*-digraph G with n vertices along with a topological ordering of its vertices, by using standard visit techniques, one can augment G by adding edges in $O(n)$ time such that the resulting digraph has the same vertex set as G, is a maximal planar st-digraph, and preserves the given topological ordering. Hence from now on, we assume without loss of generality that planar st-digraphs are maximal.

Lemma 1. [3] *Let G be a planar acyclic digraph. G is upward planar if and only if it is the spanning subgraph of a planar st-digraph.*

A planar st-digraph that includes G as a spanning subgraph is called an *including planar st-digraph* of G.

3 Upward Topological Book Embeddings

An *upward topological book embedding* of a planar st-digraph G is an upward planar drawing Γ of G such that: (i) The vertices of Γ lie on an oriented line called the *spine* of Γ; (ii) Each edge (u, v) of G is represented in Γ as a sequence of semi-circles $c_1, c_2, \ldots c_k$ such that consecutive semi-circles lie on different half-planes and share a point along the spine, called a *spine crossing* of the edge. An example of an upward topological book embedding with the spine oriented according to increasing y-coordinate is given in Fig. 2. In the figure, edge $(5, 7)$ consists of the concatenation of three semi-circles and has two spine crossings, while edge $(1, 3)$ does not have spine crossings.

In this section we show that for any given topological numbering ρ, a planar st-digraph always admits an upward topological book embedding such that a vertex v precedes a vertex w along the spine if and only if $\rho(v) < \rho(w)$. We call this type of drawing a *ρ-constrained* upward topological book embedding. This can be viewed as a constrained counterpart of [14].

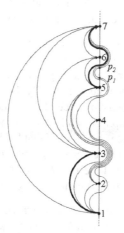

Fig. 2. An upward topological book embedding of the maximal planar st-digraph in Fig. 3(a). The vertices are ordered along the spine according to the indices of the vertices in Fig. 3(a). The drawing is computed by using the drawing algorithm of Theorem 1.

3.1   Dual Digraph and k-Facial Subgraph

Let G be a maximal planar st-digraph. For each edge $e = (u, v)$ of G, we denote by *left(e)* (resp. *right(e)*) the face to the left (resp. right) of e in G. Let s^* be the face $right((s, t))$, and let t^* be the face $left((s, t))$. In the rest of this section we assume that t^* is the external face of G. Faces s^* and t^* are highlighted in Fig. 3(a). Let G be a maximal planar st-digraph. The *dual* of G is the planar st-digraph denoted as G^* such that: (i) G^* has a vertex for each face of G; (ii) G^* has an edge $e^* = (left(e), right(e))$, for every edge $e \neq (s, t)$ of G; (iii) G^* has source s^*, sink t^*, and it has edge (s^*, t^*) on its external face. Figure 3(b) depicts with dashed edges the dual digraph of the digraph of Fig. 3(a).

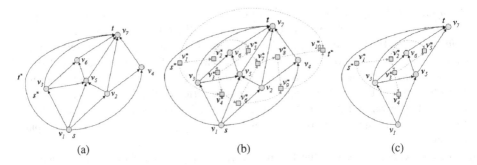

Fig. 3. (a) A planar st-digraph G with a topological numbering of its vertices. (b) Planar st-digraph G (solid) and its dual (dashed). The vertices of the dual are numbered according to a topological numbering. (c) The 5-facial subgraph of the maximal planar st-digraph in (a).

Property 1. Let G be a maximal planar st-digraph and let G^* be the dual digraph of G. Graph G^* is a planar st-digraph (without multiple edges) with source s^* and sink t^*.

Let G be a maximal planar st-digraph and let G^* be the dual of G. Let $\{v_1^* = s^*, v_2^*, \ldots, v_r^* = t^*\}$ be the set of vertices of G^* where the indices are given according to a topological numbering of G^*. See, for example, Fig. 3(b), where the vertices of the dual are numbered according to a topological numbering. By definition of dual st-digraph, a vertex v_i^* of G^* ($1 \leq i \leq r$) corresponds to a face of G; in the remainder of this section v_i^* both the vertex of the dual digraph G^* and its corresponding face in the primal digraph G. Let V_k be the subset of the vertices of G that belong to faces $v_1^*, v_2^*, \ldots, v_k^*$. The subgraph of G induced by the vertices in V_k is called the *k-facial subgraph* of G and is denoted as G_k. Face v_k^* is called the *k-th face* of G. Figure 3(c), for example, shows the 5-facial subgraph of the maximal planar st-digraph depicted in Figure 3(a).

Lemma 2. [6] *Let G be a maximal planar st-digraph with r faces, let G_{k-1} be the $(k-1)$-facial subgraph of G ($2 \leq k \leq r$) and let G_k be the k-facial subgraph of G. Let v_k^* be the k-th face of G consisting of edges (w, w'), (w', w''), and (w, w''). One of the following statements holds. (**S_1**): (w, w') and (w', w'') are edges of the external face of G_{k-1}; (w, w'') is an edge of the external face of G_k. (**S_2**): (w, w'') is an edge of the external face of G_{k-1}; (w, w') and (w', w'') are edges of the external face of G_k.*

Lemma 3. [6] *Let G be a maximal planar st-digraph with r faces and let G_k be the k-facial subgraph of G ($1 \leq k \leq r$). G_k is a planar st-digraph.*

3.2 ρ-Constrained Upward Topological Book Embeddings

Let Γ be an upward topological book embedding of a planar st-digraph G, let Λ be the spine of Γ, and let p_1 and p_2 be two vertices or two spine-crossings of Γ. Assume without loss that Λ is vertical. The notation $p_1 < p_2$ means that p_1 precedes p_2 along the spine of Γ. We denote with (p_1, p_2) a semi-circle in Γ (either an edge or a portion of an edge) with antipodal points p_1 and p_2 such that $p_1 < p_2$. We say that any point p of Λ is *covered* if there exists a semi-circle (p_1, p_2) in the half-plane on the right-hand side of Λ such that $p_1 < p < p_2$. Otherwise, we say that p is *visible*. For example, vertex 3 of Fig. 2 is covered while vertex 4 is visible.

Let p_1, p_2 be two points on the spine of Γ. We say that segment $\overline{p_1 p_2}$ is a *maximal covered segment* if every point p such that $p_1 < p < p_2$ is covered and there are no other segments $\overline{q_1 q_2}$ with $q_1 \leq p_1 < p_2 \leq q_2$ such that this same property holds. Similarly, $\overline{p_1 p_2}$ is a *maximal visible segment* if all of its points are visible and it is not a subset of another visible segment. For example, segment $\overline{p_1 p_2}$ in Fig. 2 is a maximal visible segment. Let v be a point of Λ that represents a vertex of G. We say that v has an *upper pocket* if there exists a maximal visible segment $\overline{p_1 p_2}$ such that: (i) $v \leq p_1 < p_2$, (ii) no semi-circle (either in the left or in the right half-plane defined by Λ) has an end-point in $\overline{p_1 p_2}$, and (iii) there is no vertex u such that $v < u \leq p_1 < p_2$. For example, segment $\overline{p_1 p_2}$ in Fig. 2 is the upper pocket of vertex 5 but it is not the upper pocket of vertex 4. Similarly the *lower pocket* of a vertex v in Γ is defined by considering maximal visible segments below v. Segment $\overline{p_1 p_2}$ in Fig. 2 is the lower pocket of vertex 6.

Theorem 1. *Let G be a maximal planar st-digraph and let ρ be a topological numbering of G. G admits a ρ-constrained upward topological book embedding with at most $2n - 4$ spine crossings per edge, which is asymptotically worst-case optimal. Also, such a ρ-constrained upward topological book embedding can be computed in $O(n^2)$ time.*

Proof. We compute a ρ-constrained upward topological book embedding Γ^ρ of G by maintaining the following invariant properties: (I_1): Every vertex has a lower pocket and an upper pocket. (I_2): For every maximal covered segment $\overline{p_1 p_2}$, there exists a vertex v of G such that $p_1 < v < p_2$.

We proceed by induction on the number of internal faces of G.

Base Case: Refer to Fig. 4(a). Suppose G has exactly one internal face v^* and let $\{s, t, w\}$ be the vertices of the boundary of face v^*. Let Λ be a vertical line in the plane. Draw s, t and w along Λ such that $s < w < t$. Draw edges (s, t), (s, w) and (w, t) as the semi-circles (s, t), (s, w) and (w, t) respectively, in the half-plane on the left-hand side of Λ. By construction, the resulting drawing is a ρ-constrained upward topological book embedding. Also, segment $\overline{sw}$ is a lower pocket for w and an upper pocket for s, while segment $\overline{wt}$ is an upper pocket for w and a lower pocket for t. The upper pocket for t and the lower pocket for s are the half lines above t and below s, respectively. Thus Property I_1 holds. Property I_2 holds since there are no covered segments in this case.

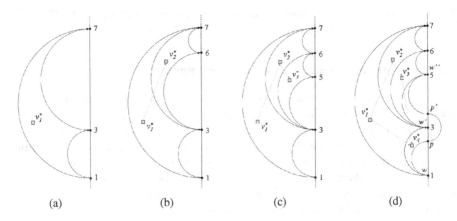

Fig. 4. Four steps of the algorithm in the proof of Theorem 1 applied to the digraph in Fig. 3(a)

Inductive case: Suppose by induction that a ρ-constrained upward topological book embedding of G satisfying Properties I_1 and I_2 can be computed when G has $k-1$ faces and assume that G has k faces ($k > 1$). Let G^* be the dual of G and let $\{v_1^* = s^*, v_2^*, \ldots, v_k^* = t^*\}$ be the vertex set of G^*, where the indices are given according to a topological numbering of G^*. Also, let G_{k-1} be the $(k-1)$-facial subgraph of G. By definition and by Lemma 3, G_{k-1} is a planar st-digraph with exactly $k-1$ internal faces. By the inductive hypothesis there exists a ρ-constrained upward topological book embedding Γ_{k-1}^ρ of G_{k-1} satisfying Properties I_1 and I_2. Since G has k internal faces, the k-facial subgraph of G is G itself. Let v_k^* be the k-th face of G consisting of edges (w, w'), (w', w''), and (w, w'') (see also Fig. 4). Let Λ be the spine of Γ_{k-1}^ρ. We show how to compute a ρ-constrained upward topological book embedding Γ^ρ of G satisfying Properties I_1 and I_2, by adding face v_k^* to Γ_{k-1}^ρ. We distinguish two cases depending on whether the k-th face of G satisfies Statement S_1 or Statement S_2 of Lemma 2.

Statement S_1 of Lemma 2 holds. Refer to Fig. 4(d) where v_k^* is face v_4^* and we need to insert edge $(1, 5)$. Let j be the number of vertices between w and w'' along Λ. Suppose $j = 1$. Since w, w' and w'' are on the external face of G_{k-1}, by Property I_1 there are no endpoints of semi-circles on segments $\overline{ww'}$ and $\overline{w'w''}$. Choose two arbitrary points p and p' such that $w < p < w' < p' < w''$. Draw edge (w, w'') as three semi-circles $(w, p), (p, p'), (p', w'')$, respectively in the half-planes on the left hand-side, right hand-side, and left hand-side of Λ. The resulting drawing Γ^ρ is a ρ-constrained upward topological book embedding of G. Also, Property I_1 holds, since $\overline{wp}$ is an upper pocket for w and a lower pocket for w' and $\overline{p'w''}$ is an upper pocket for w' and a lower pocket for w''. Furthermore, $\overline{pp'}$ is a maximal covered segment and indeed w is such that $p < w < p'$. Thus, Property I_2 holds.

If the number of vertices between w and w'' along Λ is $j = h - 1$ ($2 < h < n$) then edge (w, w'') can be added to Γ_{k-1}^ρ such that the resulting drawing is a ρ-constrained upward topological book embedding satisfying Properties I_1 and I_2. Assume $j = h$. Let p be the point of Λ above w such that $\overline{wp}$ is the upper pocket of w.

Two cases are possible: (i) p is the point representing vertex w'; (ii) p is an endpoint of a semi-circle (p, p') in the half-plane on the right-hand side of Λ. We can deal with cases (i) and (ii) at once, since case (i) can be seen as a special instance of case (ii), where p and p' coincide with vertex w. Thus, suppose case (ii) holds and let (q, q') be a semi-circle of Γ^ρ_{k-1} in the half plane on the right-hand side of Λ. Points q and q' cannot be such that $q < p < q'$ since every point of $\overline{wp}$ must be visible from the right-hand side, and they cannot be such that $p < q < p' < q'$ since in this case there would be a crossing between semi-circles (q, q') and (p, p'). Therefore, $\overline{pp'}$ is a maximal covered segment of Γ^ρ_{k-1}. Hence, by Property I_2, there must exist a vertex v of G_{k-1} such that $p < v < p'$ in Γ^ρ_{k-1}. Also, by Property I_1, there exists a point p'' of Λ such that $\overline{p'p''}$ is the upper pocket of v. Let $\tilde{p}$ and $\tilde{p}'$ be arbitrary points of Λ such that $w < \tilde{p} < p < p' < \tilde{p}' < p''$.

We draw edge (w, w'') by splitting it into two edges: An edge from w to $\tilde{p}'$ and an edge from $\tilde{p}'$ to w''. Since there are exactly $h - 1$ vertices between $\tilde{p}'$ and w'', edge $(\tilde{p}', w'')$ can be added to Γ^ρ_{k-1} so that the resulting drawing is a ρ-constrained upward topological book embedding satisfying Properties I_1 and I_2. Notice that, as a consequence of Property I_1, there must exist a point p'' with $\tilde{p}' < p''$ such that $\overline{p'p''}$ is an upper pocket for $\tilde{p}'$, which means that the first semi-circle of edge $(\tilde{p}', w'')$ is in the half-plane on the left hand-side of Λ. Now, draw edge $(w, \tilde{p}')$ as two semi-circles $(w, \tilde{p})$ and $(\tilde{p}, \tilde{p}')$ in the half planes on the left-hand side and on the right-hand side of Λ, respectively. Semi-circles $(w, \tilde{p})$ and $(\tilde{p}, \tilde{p}')$ do not cross any other semi-circle of Γ^ρ_{k-1}. Also, Properties I_1 and I_2 hold for w and v. Indeed, segment $\overline{wp}$ is an upper pocket for w and a lower pocket for v while segment $\overline{p'p''}$ is an upper pocket for v. Property I_2 holds for v as segment $\overline{pp'}$ is a maximal covered segment and v is such that $\tilde{p} < v < \tilde{p}'$. Therefore, the semi-circles we have drawn preserve planarity and respect Properties I_1 and I_2. It follows that edge (w, w'') has been drawn as a monotone curve from w to w'' formed by a sequence of semi-circles $c_1, c_2, \ldots c_{2h+1}$ such that consecutive semi-circles lie on different half-planes, share only a spine crossing along Λ, do not cross other semi-circles and Properties I_1 and I_2 hold. The resulting drawing Γ^ρ is thus a ρ-constrained upward topological book embedding of G satisfying Properties I_1 and I_2.

Statement S_2 of Lemma 2 holds. Refer also to Fig. 4(b) and Fig. 4(c). Let v be the vertex of G_{k-1} having the largest number in the topological numbering such that $\rho(v) < \rho(w')$. Let $\overline{p_1 p_2}$ be the upper pocket of v. Draw vertex w' such that $p_1 < w' < p_2$. Segment $\overline{p_1 w'}$ is both the new upper pocket of v and the lower pocket of w' while segment $\overline{w'p_2}$ is the upper pocket of w'. Thus, the drawing is a ρ-constrained upward topological book embedding satisfying Properties I_1 and I_2. Draw edge (w, w') by the same technique as in the previous case. The same reasoning proves that the resulting drawing is a ρ-constrained upward topological book embedding satisfying Properties I_1 and I_2. The same argument applies to edge (w', w''). The final drawing is thus a ρ-constrained upward topological book embedding of G satisfying Properties I_1 and I_2.

It remains to prove the time complexity of the algorithm and the number of spine crossings per edge of the drawing. The dual digraph of G and a topological numbering of its vertices can be computed in linear time. Indeed, by Property 1, the dual of G is a planar st-digraph without multiple edges. Also, each edge (u, v) is drawn by the algorithm so that for every vertex w with $u < w < v$ there are exactly two spine crossings p_1 and p_2 with $p_1 < w < p_2$. It follows that the number of spine crossings per edge is at most $2(n-2) = 2n - 4$; we also remark that $\Omega(n)$ bends are known to be necessary for constructing topological book embeddings with a fixed ordering of undirected planar graphs [1]. Finally, the time complexity of the described drawing algorithm is $O(m \cdot n)$ that is equal to $O(n^2)$ since G is a planar graph.

4   Upward Point-Set Embeddability with a Given Mapping Φ

We first study the special case that the points of S are collinear and that the input graph is a planar st-digraph, and then study general upward planar digraphs and points in general position.

4.1   Collinear Points and Planar st-Digraphs

Assume without loss that $p_1, \ldots, p_n$ are vertically aligned. We associate each point p_i of S with an integer in the set $\{1, \ldots, n\}$ such that point p_i is given integer k if p_i is the k-th point of S that we encounter moving along the increasing y-direction. We also consistently assign numbers to the vertices $v_1, \ldots, v_n$: If point $p_i = \Phi(v_j)$ has been given integer k, then also v_j is given integer k. See Fig. 5(a) for an example. We call such a numbering of the vertices of G the Φ-*numbering of G* and we call Φ-*number of v_j* the number assigned to vertex v_j. We say that mapping Φ *induces a topological numbering of G* if the Φ-numbering of G is also a topological numbering of G. For example, the Φ-numbering of Fig. 5(a) does not induce a topological numbering of G.

The characterization almost immediately follows from the result in Sect. 3 concerning ρ-constrained upward topological embeddability and from the observation that the y-coordinates of the vertices in an upward planar drawing induce a topological numbering of the graph.

Lemma 4. *Let G be an upward planar digraph with n vertices, $S = \{p_1, \ldots, p_n\}$ a set of vertically aligned points and Φ a mapping from G to S. Let Γ be an upward topological book embedding of G such that: (i) the maximum number of spine crossings per edge of Γ is k; (ii) for every pair of vertices u and v of G such that $u < v$ along the spine of Γ, the Φ-number of u is smaller than the Φ-number of v. Then G admits an upward point-set embedding on S consistent with Φ, with at most $k+1$ bends per edge.*

Theorem 2. *Let G be a planar st-digraph with n vertices; let S be a set of n distinct collinear points in the plane; let Φ be a mapping from G to S. G admits an upward point-set embedding on S consistent with Φ if only if Φ induces a topological numbering of G. Also, such an upward point-set embedding of G on S can be computed in $O(n^2)$ time with at most $2n - 3$ bends per edge, which is asymptotically worst-case optimal.*

4.2   Points in General Position and Upward Planar Digraphs

In this section we extend Theorem 2 by characterizing when mapping Φ guarantees the upward point-set embeddability of a (not necessarily st-) upward planar digraph G

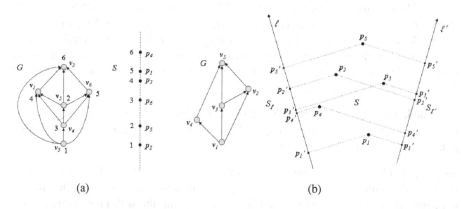

Fig. 5. (a) A planar st-digraph G, a set S of distinct collinear points in the plane and the Φ-numbering of G. (b) A digraph G and a set S of points in the plane. Mapping Φ induces a topological numbering of G on ℓ, whereas it does not induce a topological numbering of G on ℓ'.

on a set S of (not necessarily collinear) points. Let ℓ be a directed line. We denote as $S_\ell = \{p'_1, \ldots, p'_n\}$ the collinear set of points obtained by orthogonally projecting S onto ℓ; we assume that the direction of ℓ is such that when projecting S on ℓ no two projected points coincide. Also, let Φ_ℓ be the mapping from G to S_ℓ that associates each vertex v of G with the projection of $\Phi(v)$ on ℓ. We say that mapping Φ *induces a topological numbering of G on ℓ* if mapping Φ_ℓ induces a topological numbering of G. For example, Fig. 5(b) shows the digraph G, the set S and the mapping Φ from G to S defined by associating every vertex of G with the point of S having the same index; also, two directed lines ℓ and ℓ' are depicted such that Φ induces a topological numbering on ℓ while it does not induce a topological numbering on ℓ'.

Theorem 3. *Let G be an upward planar digraph with n vertices, S a set of n distinct points in the plane, and Φ a mapping from G to S. G admits an upward point-set embedding consistent with Φ if and only if there exists an including planar st-digraph G' of G and a directed line ℓ such that Φ induces a topological numbering of G' on ℓ. Also, such an upward point-set embedding of G on S can be computed in $O(n^2)$ time with at most $2n - 3$ bends per edge, which is asymptotically worst-case optimal.*

We remark that the number of bends per edge stated in Theorem 3 improves by a constant factor the best known upper bound of $3n + 2$ for the point-set embeddability with mapping of undirected planar graphs (Theorem 4 of [1]).

5   Testing Upward Point-Set Embeddability

Theorem 3 naturally raises the question about how to efficiently test whether an upward planar digraph G with n vertices admits an upward point-set embedding consistent with a given mapping Φ on a set S of n distinct points. By Theorem 3, it suffices to test whether there exist an including planar st-digraph G' of G and a directed line ℓ such that Φ induces a topological numbering of G' on ℓ. Therefore, we consider every directed line ℓ such that Φ induces a distinct Φ_ℓ-numbering. For each such Φ_ℓ-numbering, we check whether it is a topological numbering of G and, if so, we verify whether there exists an including planar st-digraph of G that preserves it. The following lemma strongly relies on results by [10,11] concerning level planarity testing and embedding; its proof is omitted from this abstract.

Lemma 5. *Let ρ be a Φ_ℓ-numbering. There exists an $O(n)$-time algorithm that tests whether ρ is a topological numbering of G and, if so, whether there exists an including planar st-digraph of G that preserves ρ.*

To compute all possible Φ_ℓ-numberings we must consider all possible directed lines such that the orthogonal projections of the points on these lines produce different permutations of the points. This is equivalent to computing the well-known circular sequence of permutations associated with point set S (see, e.g. [8]).

Lemma 6. [8] *Let S be a set of n distinct points in the plane. The circular sequence of permutations associated with S has cardinality $O(n^2)$ and can be computed in $O(n^2)$ time.*

Theorem 4. *Let G be an upward planar digraph, $S = \{p_1, \ldots, p_n\}$ a set of n distinct points in the plane, and Φ a mapping from G to S. There exists an $O(n^3)$-time algorithm that tests whether G admits an upward point-set embedding on S consistent with Φ.*

6   Upward Consistent Simultaneous Embeddability

The following theorem characterizes the upward simultaneous embeddability of a sequence of upward planar digraphs with respect to the same direction.

Theorem 5. *A sequence $G_1, G_2, \ldots, G_k$ of distinct upward planar digraphs sharing the same vertex set admits an upward consistent simultaneous embedding with respect to the same direction if and only if there exists a sequence $G'_1, G'_2, \ldots, G'_k$ such that: (i) G'_i is an including planar st-digraph of G_i, and (ii) $\bigcup_{i=1}^k G'_i$ is acyclic.*

Proof. If the sequence $G_1, G_2, \ldots, G_k$ admits the desired embedding, then each drawing is an upward planar drawing of G_i ($1 \leq i \leq k$) and has an including planar st-digraph by Lemma 1. Since all drawings have the same direction of upwardness, the union of these planar st-digraphs is acyclic.

   Let n be the number of vertices in each graph of the sequence and let G' be the union digraph, that is $G' = \bigcup_{i=1}^k G'_i$. Assume that G' is acyclic and let ρ be a topological numbering of G'. Note that ρ is a topological numbering of each G'_i. Compute a ρ-constrained upward topological book embedding for each G'_i by using Theorem 1. Define a set of n distinct points in the plane having consecutive y-coordinates from 1 to n. Define a mapping Φ that associates every vertex of G'_i having number h in the topological numbering with the point of S having y-coordinate equal to h. By Theorem 3 each G'_i has an upward point-set embedding consistent with Φ such that the edges are monotonically increasing with the y-direction.

7   Open Problems

We conclude with three open problems: (1) For upward point-set embedding with a given mapping, minimize the total number of bends; (2) Improve the time complexity in Theorem 4; and (3) Design a fast test for upward simultaneous embeddability with respect to the same direction (we have linear time results for the case of switch-regular digraphs).

References

1. Badent, M., Di Giacomo, E., Liotta, G.: Drawing colored graphs on colored points. In: Dehne, F., Sack, J.-R., Zeh, N. (eds.) WADS 2007. LNCS, vol. 4619, pp. 102–113. Springer, Heidelberg (2007)
2. Di Battista, G., Eades, P., Tamassia, R., Tollis, I.: Graph Drawing. Prentice-Hall, Englewood Cliffs (1999)
3. Di Battista, G., Tamassia, R.: Algorithms for plane representations of acyclic digraphs. Theoretical Computer Science 61(2-3), 175–198 (1988)
4. Enomoto, H., Miyauchi, M.: Embedding graphs into a three page book with $O(M \log N)$ crossings of edges over the spine. SIAM J. of Discrete Math. 12(3), 337–341 (1999)
5. Frati, F., Kaufmann, M., Kobourov, S.: Constrained simultaneous and near-simultaneous embeddings. In: Hong, S.-H., Nishizeki, T., Quan, W. (eds.) GD 2007. LNCS, vol. 4875, pp. 268–279. Springer, Heidelberg (2008)
6. Giordano, F., Liotta, G., Mchedlidze, T., Symvonis, A.: Computing upward topological book embeddings of upward planar digraphs. In: Tokuyama, T. (ed.) ISAAC 2007. LNCS, vol. 4835, pp. 172–183. Springer, Heidelberg (2007)
7. Giordano, F., Liotta, G., Whitesides, S.: Drawing a sequence of upward planar digraphs: Characterization results and testing algorithms. Tech. rep., Università degli Studi di Perugia, RT-008-01 (2008)
8. Goodman, J.E., Pollack, R.: On the combinatorial classification of nondegenerate configurations in the plane. J. of Combinatorial Theory, Ser. A 29(2), 220–235 (1980)
9. Halton, J.H.: On the thickness of graphs of given degree. Inform. Sciences 54, 219–238 (1991)
10. Jünger, M., Leipert, S.: Level planar embedding in linear time. J. of Graph Algorithms and Applications 6(1), 67–113 (2002)
11. Jünger, M., Leipert, S., Mutzel, P.: Level planarity testing in linear time. In: Whitesides, S.H. (ed.) GD 1998. LNCS, vol. 1547, pp. 224–237. Springer, Heidelberg (1999)
12. Kaufmann, M., Wagner, D. (eds.): Drawing Graphs. LNCS, vol. 2025. Springer, Heidelberg (2001)
13. Pach, J., Wenger, R.: Embedding planar graphs at fixed vertex locations. Graphs and Combin. 17(4), 717–728 (2001)

A Fully Dynamic Algorithm to Test the Upward Planarity of Single-Source Embedded Digraphs

Aimal Rextin and Patrick Healy

Computer Science Department, University of Limerick, Ireland
{aimal.tariq,patrick.healy}@ul.ie

Abstract. In this paper, we present a dynamic algorithm that checks if a single-source embedded digraph is upward planar in the presence of edge insertions and edge deletions. Let G_ϕ be an upward planar single-source embedded digraph and let $G'_{\phi'}$ be a single-source embedded digraph obtained by updating G_ϕ. We show that the upward planarity of $G'_{\phi'}$ can be checked in $O(\log n)$ amortized time when the external face is fixed.

1 Introduction

Assume we have a solution of a graph theoretic problem P on a graph G. A *dynamic graph algorithm* tries to solve P after G is updated in less time than recomputing P from scratch [5]. Dynamic graph algorithms are useful when a graph has discrete changes like the addition or deletion of vertices or edges. A practical example of a dynamic graph algorithm is the maintenance shortest paths in a communication network as links are added or deleted.

In this paper, we present a dynamic algorithm to check if a single-source embedded digraph remains upward planar after an edge is inserted or deleted. An *planar embedding* is an equivalence class of planar drawings for a graph G, such that each drawing of this class has the same circular order of edges around each vertex of G. A graph G with a given planar embedding is denoted by G_ϕ and we call it an *embedded* digraph. A digraph G is *upward planar* if it has a planar drawing with all edges pointing monotonically upward [6]. It is NP-hard to test if a digraph G is upward planar [9], hence upward planarity testing is either done for a fixed embedding [3,7], or for special classes of digraphs like single-source digraphs [10,4], series-parallel digraphs [8], and outer planar digraphs [11].

Let G_ϕ be an upward planar embedded digraph with the single-source s_G. We let $G'_{\phi'}$ be an embedded digraph with the single-source $s_{G'}$, such that $G'_{\phi'}$ is obtained from G_ϕ by performing one of the following update operations:

- **insert-edge**(e, u, v): Insert an edge $e = (u, v)$ between two existing vertices in G_ϕ.
- **attach-vertex**(e, u, v): Add a new vertex and insert an edge between an existing vertex and the new vertex.
- **delete-edge**(e): Delete the edge e from G_ϕ. We also delete a vertex if it results in no incident edge.

I.G. Tollis and M. Patrignani (Eds.): GD 2008, LNCS 5417, pp. 254–265, 2009.

An update operation is illegal if the resulting digraph is not single-source. In case of an edge insertion, this happens if an edge $e = (u,v)$ is inserted between existing vertices such that $v = s_G$, or when edge $e = (u,v)$ is inserted between a new vertex u and $v \neq s_G$. An edge deletion is illegal if an edge $e = (u,v)$ is deleted such that $G'_{\phi'}$ becomes disconnected.

It is generally believed that upward planar drawings of a digraph are more comprehensible to humans. Hence, it is reasonable to say that a non-upward planar digraph H is more readable if the largest possible subgraph of H is drawn in an upward planar fashion. In this paper, we present a dynamic algorithm to test the upward planarity of $G'_{\phi'}$. Our dynamic algorithm can be used to compute a maximal upward planar subgraph for a single-source digraph H by incrementally building an upward planar embedding of H and discarding a new edge if it results in a non-upward planar embedded digraph.

In the remainder of this section, we define some basic terminology and review some relevant results. In Sec. 2, we discuss how to obtain a bimodal and embedded $G'_{\phi'}$. In Sec. 3, we give a characterization of upward planarity of $G'_{\phi'}$ with respect to the update operations. In Sec. 4, we present our algorithm and its complexity analysis. We conclude by identifying some related open problems.

1.1 Preliminaries

We assume basic familiarity with graph theory. Let G be a graph. We denote the set of vertices of G by $V(G)$ and we denote the set of edges of G by $E(G)$. In a digraph, a *source* vertex has only outgoing edges, a *sink* vertex has only incoming edges, and an *internal* vertex has both incoming and outgoing edges. A planar drawing Γ divides the plane into non-overlapping regions called *faces*; the unique unbounded region is called the *external face* and each bounded region is called an *internal face*. The *facial boundary* of a face f is the path enclosing f in the clockwise direction, all drawings of an embedded graph have the same set of facial boundaries. An embedded digraph G_ϕ is bimodal when $\phi(v)$ can be partitioned into two sets of consecutive incoming and outgoing edges for every vertex $v \in G_\phi$.

In an embedded digraph G_ϕ, an *angle* is a triplet $\langle e_1, v, e_2 \rangle$ such that the edges are incident to the vertex v and edge e_1 is immediately before edge e_2 in $\phi(v)$. A vertex v is incident to the angle $\langle e, v, e \rangle$ when e is the only edge incident to v. A *switch* $\langle e_1, v, e_2 \rangle$ is an angle with both e_1 and e_2 pointing either toward or away from v: it is a *sink-switch* when e_1 and e_2 point toward v and it is a *source-switch* when e_1 and e_2 point away from v [7]. Switches were originally defined as nodes in an embedded biconnected digraph by Bertolazzi *et al.* [3], however Didimo generalized their concept to general embedded digraphs by defining them as angles [7].

We now show that both G_ϕ and $G'_{\phi'}$ have at most one sink-switch incident to a vertex v inside a particular face. This allows us to refer to a vertex v incident to a sink-switch $\langle e_1, v, e_2 \rangle$ in a face f as *sink-switch v incident to face f* for simplicity and clarity.

Lemma 1. *Let G_ϕ be an upward planar embedded digraph with a single source s_G, and let $G'_{\phi'}$ be the bimodal embedded digraph with a single source $s_{G'}$ obtained after adding an edge in G_ϕ. Both G_ϕ and $G'_{\phi'}$ have at most one sink-switch incident to a vertex v inside a particular face.*

The *face-sink graph* F of G_ϕ is an undirected graph such that the vertices of F are the faces of G_ϕ and all vertices of G_ϕ that are incident to a sink-switch; an edge (f, v) is in F if face f is incident to a sink-switch on a vertex v in G_ϕ. Bertolazzi *et al.* [4] presented an $O(n)$-time algorithm to test the upward planarity of a single-source embedded digraph G_ϕ. This algorithm is based on the following theorem:

Theorem 1 (Bertolazzi *et al.* [4]). *Let G_ϕ be a embedded digraph with a single-source s_G. G_ϕ is upward planar with face h as the external face if and only if the following conditions are satisfied.*

1. *The face-sink graph F of G_ϕ is a forest.*
2. *F has exactly one tree $\hat{T}$ with no internal vertices, while all other trees have exactly one internal vertex.*
3. *$\hat{T}$ contains the node corresponding to face h and s_G is incident to face h in G_ϕ.*

2 Maintaining Planarity and Bimodality

Theorem 1 requires a embedded single-source digraph, however bimodality is a necessary condition for upward planarity and hence $G'_{\phi'}$ will have more chances to be upward planar if it is already bimodal and planar. In this section, we see how a bimodal embedded digraph $G'_{\phi'}$ can be obtained after G_ϕ is updated. The embedded digraph G_ϕ will remain bimodal and planar after an edge is deleted, hence we only study the case when an edge is inserted.

When an edge is inserted, a planar and bimodal embedded digraph $G'_{\phi'}$ can be obtained, if it exists, by using the techniques of Bertolazzi *et al.* [2] and Tamassia [12]. Tamassia described a technique to incrementally build a planar embedding: it checks if an edge can be added to the current embedded graph without introducing a crossing in $O(\log n)$ time and it then adds the new edge to the current embedded graph in $O(\log n)$ amortized time [12]. A technique for constructing a bimodal embedding of a digraph $\mathcal{G}$ was discussed by Bertolazzi *et al.* [2]. It works by splitting all vertices of $\mathcal{G}$ with at least 2 incoming edges and at least 2 outgoing edges into a vertex v_a with all the incoming edges of v and a vertex v_b with all the outgoing edges of v, and adding the edge (v_a, v_b). We call the vertices that are split as *split-vertices* and we call the resulting digraph as the *split-digraph* $\tilde{\mathcal{G}}$. Bertolazzi *et al.* showed that $\mathcal{G}$ has a planar bimodal embedding if and only if $\tilde{\mathcal{G}}$ has a planar embedding. We get a planar and bimodal embedded $\mathcal{G}_\phi$ by merging back the split vertices in a planarly embedded $\tilde{\mathcal{G}}_{\tilde{\phi}}$ [2].

We obtain a bimodal and planarly embedded $G'_{\phi'}$, if it exists, by maintaining a corresponding planarly embedded split-digraph $\tilde{G}_{\tilde{\phi}}$. Figure 1 shows an embedded digraph G_ϕ and its corresponding split-digraph $\tilde{G}_{\tilde{\phi}}$. A vertex v in G has two corresponding vertices $\tilde{v}_a$ and $\tilde{v}_b$ in $\tilde{G}$ if it is a split-vertex, and it has one corresponding vertex $\tilde{v}$ otherwise. If v is a split-vertex, we let $\tilde{v}_b$ represent the vertex in $\tilde{G}$ with all corresponding outgoing edges of v, and we let $\tilde{v}_a$ represent the vertex in $\tilde{G}$ with all corresponding incoming edges of v. We define a function $o : V(G) \to V(\tilde{G})$, such that $o(v) = \tilde{v}_b$ when v is a split-vertex and $o(v) = \tilde{v}$ otherwise. Similarly, we define function $i : V(G) \to V(\tilde{G})$, such that $i(v) = \tilde{v}_a$ when v is a split-vertex and $i(v) = \tilde{v}$ otherwise. We also define a function

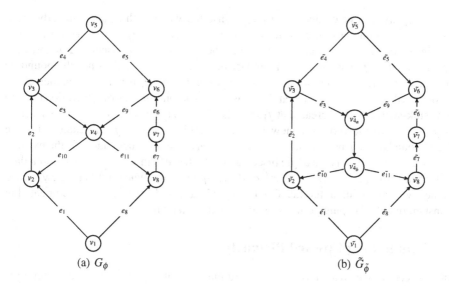

(a) G_ϕ (b) $\tilde{G}_{\tilde{\phi}}$

Fig. 1. An embedded digraph G_ϕ (a); and its embedded split digraph $\tilde{G}_{\tilde{\phi}}$ (b)

$e : E(G) \to E(\tilde{G})$, which maps the edges in G to their corresponding edges in $\tilde{G}$. When we want to add an edge $e = (u,v)$ in G_ϕ, we first try to add the edge $\tilde{e} = (o(u), i(v))$ in $\tilde{G}_{\tilde{\phi}}$. The embedded digraph $G'_{\phi'}$ is not planar or bimodal if $\tilde{e}$ cannot be added in $\tilde{G}_{\tilde{\phi}}$ using Tamassia's method. Lets assume that we get a planar $\tilde{G}'_{\tilde{\phi}'}$, with $\tilde{e}$ inserted between $\tilde{e}_1$ and $\tilde{e}_2$ at $o(u)$, and $\tilde{e}$ inserted between $\tilde{e}'_1$ and $\tilde{e}'_2$ at $i(u)$. In this case, we get a planar and bimodal $G'_{\phi'}$ by adding e between $e^{-1}(\tilde{e}_1)$ and $e^{-1}(\tilde{e}_2)$ at u, and adding e between $e^{-1}(\tilde{e}'_1)$ and $e^{-1}(\tilde{e}'_2)$ at v. Figure 1 shows that we can bimodally add the edge (v_5, v_4) in G_ϕ but not the edge (v_4, v_5).

The split-digraph $\tilde{G}_{\tilde{\phi}}$ takes $O(n)$ space. If the addition of edge e makes a vertex v a split-vertex then we will need to construct the corresponding $\tilde{v}_a$ and $\tilde{v}_b$ in $\tilde{G}_{\tilde{\phi}}$. This can be done in constant time because there will be either one incoming edge or one outgoing edge incident to v before the new edge is added. Hence we have the following lemma.

Lemma 2. *Let G_ϕ be an upward planar embedded digraph and let e be an edge that we want to insert in G_ϕ. We can perform the following two operations.*

1. *Check if an edge e can be added to G_ϕ such that the resulting graph has a bimodal and planar embedding in $O(\log n)$ time.*
2. *If the previous test is true then we can obtain a planar and bimodal embedded digraph $G'_{\phi'}$ in $O(\log n)$ amortized time.*

The insertion of an edge $e = (u,v)$ *bisects* an angle $\alpha_u = \langle e_1, u, e_2 \rangle$ at vertex u into two new angles $\langle e_1, u, e \rangle$ and $\langle e, u, e_2 \rangle$. The new edge e similarly bisects the angle α_v at vertex v into two new angles. The insert-face f is divided into two new faces

f_1 and f_2 when a new edge $e = (u,v)$ is inserted when both u and v already exist. Let the facial boundary of f be $w_0, e_0, \ldots, e_i, u, e_{i'}, \ldots, e_j, v, e_{j'}, \ldots, e_k, w_k = w_0$. After e is inserted, let $\langle e_i, u, e_{i'} \rangle$ and $\langle e_j, v, e_{j'} \rangle$ be the angles that are bisected at u and v respectively, then f_1 has the facial boundary $e, u, e_{i'}, \ldots, e_j, v, e$ and f_2 has the boundary $w_0, e_0, \ldots, u, e, v, \ldots, e_k, w_k = w_0$. If both u and v exist and α_v is not a switch then either f_1 or f_2 will have a sink-switch at v. We assume, without the loss of generality, that the new sink-switch will be created at f_1. Similarly, when a new edge $e = (u,v)$ is inserted in f and one of the vertices is new, then the facial boundary of f will change. Let the facial boundary of f be $w_0, e_0, \ldots, e_i, w', e_{i'}, \ldots, e_k, w_k = w_0$, such that w' is the existing vertex and $\langle e_i, w', e_{i'} \rangle$ is the angle bisected at w'. After e is inserted, the facial boundary will change to $w_0, e_0, \ldots, e_i, w', e, w'', e, w', \ldots, e_k, w_k = w_0$, where w'' is the new vertex. Hence, we can maintain the facial boundaries in a linked list which can be updated in constant time by keeping pointers to nodes in the linked list.

3   Maintaining Upward Planarity

In this section, we characterize the upward planarity of $G'_{\phi'}$ after an update operation. We will only study the case of inserting an edge because $G'_{\phi'}$ remains upward planar when an edge is deleted. We will however need to update our datastructures when an edge is deleted, this is discussed in the next section. We assume that $G'_{\phi'}$ is bimodal and planar because we construct it by using the method described in Sec. 2.

Let F be the face-sink graph corresponding to G_ϕ and let F' be the face-sink graph corresponding to $G'_{\phi'}$. Since G_ϕ is upward planar, F will satisfy Theorem 1. Further, $G'_{\phi'}$ will be upward planar if and only if F' satisfies Theorem 1. In this section, we show that we can check if F' satisfies Theorem 1 by considering a small subset of F'. This will lead to an efficient dynamic single-source upward planarity testing algorithm, which is presented in the next section.

We first present some definitions that will be used later in this section. An edge e is inserted in one particular face of G_ϕ, which we call the *insert-face* and denote it by f. Every face g in G_ϕ has a corresponding vertex $\bar{g}$ in F. Let $T_f = (V_{T_f}, E_{T_f})$ be the tree that contains $\bar{f}$, i.e. the vertex corresponding to f. Let T be a tree in F, we define $faces(T)$ to be the set of faces such that a face g is in $faces(T)$ if and only if $\bar{g} \in V(T)$. We also define a set of vertices, denoted by $nodes(T)$, that contains all vertices of G_ϕ that are in $V(T)$. Let $\hat{T}$ denote the tree of F with no internal vertices, and let H_{G_ϕ} denote the set of faces in G_ϕ that are incident to the single-source s_G. Then $H_{G_\phi} \cap faces(\hat{T})$ is the set of all possible external faces in an upward planar drawing of G_ϕ.

Our results in this section rely on observing how F changes into F'. We define a tree T_1 to be different from a tree T_2 if it has at least one different vertex or one different edge. We claim that either $F \setminus F' = \{T_f\}$ or $F \setminus F' = \emptyset$. This is because a tree T in F will transform to a new tree T' in F' only if a new sink-switch is added in a face $g \in faces(T)$ or a sink-switch is removed from g or when g is divided into two new faces. This can happen only for the insert-face f, hence at most T_f will be transformed by the edge insertion.

We now have a closer look at the structure of T_f. If we traverse f in the clockwise direction, we will encounter some vertices that are incident to a sink-switch in f. Let

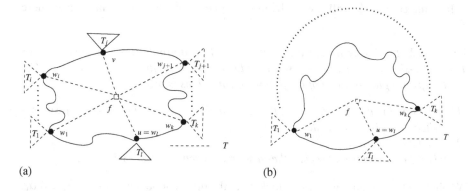

Fig. 2. The tree T_f with respect to face f such that both u and v already exist (a); and the tree T_f with respect to face f when only u is the existing vertex (b)

$W = \{w_1, w_2, \ldots, w_l\}$ be all such vertices. We define a subtree T_i as the part of T_f that is reachable from $\bar{f}$ through the vertex w_i, where $1 \leq i \leq l$. We call $w_i \in T_i$ the *access-vertex* of T_i with respect to $\bar{f}$. This is shown in Fig. 2 for both type of edge insertions discussed in Sec. 1. When both end vertices of $e = (u, v)$ already exist in G_ϕ then the partitioning of T_f is shown in Fig. 2(a): the access-vertex for $T_1, \ldots, T_i$ is between u and v in the clockwise direction, the access-vertex of $T_{j+1}, \ldots, T_k$ is between v and u in the clockwise direction, the access-vertex of T_j is v, and the access-vertex of T_l is u. Note that T_j and T_l will be empty subgraphs if u and v are not incident to a sink-switch in f. When $e = (u, v)$ has one existing vertex u then the partitioning is shown in Fig. 2(b): the access-vertex at u (if it exists) is T_l, and the access-vertices for $T_1, \ldots, T_k$ are encountered as we traverse f in the clockwise direction after u.

The next lemma is easily derived from the illegal operations described in Sec. 1 and the fact that $G'_{\phi'}$ is bimodal.

Lemma 3. *If edge $e = (u, v)$ is added in G_ϕ such that both u and v already exist, then α_v, the bisected angle at v, cannot be a source-switch.*

Proof. Assume that v is incident to a source-switch in f. We know from Sec. 1 that $v \neq s_G$ hence v has at least one incoming edge. This implies that $G'_{\phi'}$ is not bimodal, which is a contradiction.   □

We now come to the main results of this section, presented as a series of theorems. We divide the analysis into two main cases: $T_f \neq \hat{T}$ and $T_f = \hat{T}$. When $T_f \neq \hat{T}$, the tree $\hat{T}$ is in F' and all other trees in $F' \cap F$ have one internal vertex. In this case, $G'_{\phi'}$ will be upward planar if all trees of $F' \setminus F$ have one internal vertex. The single internal vertex of T_f is denoted by w_{T_f} when $T_f \neq \hat{T}$. On the other hand, all trees in $F' \cap F$ have one internal vertex when $T_f = \hat{T}$. In this case, $G'_{\phi'}$ will be upward planar if $F' \setminus F$ has one tree T with no internal vertex, all other trees trees in $F' \setminus F$ have exactly one internal vertex and $faces(T) \cap H_{G'_{\phi'}} \neq \emptyset$.

In some cases, $G'_{\phi'}$ will always be upward planar, the next theorem analyze these cases.

Theorem 2. *Let G_ϕ be an upward planar embedded digraph with a single source s_G and a face-sink graph F. If we insert an edge $e = (u,v)$ in the face $f \in G_\phi$, then $G'_{\phi'}$ will be upward planar if one of the following conditions is true.*

1. *Both u and v already exist, such that α_v is a sink-switch and α_u is either a source-switch or α_u is a non-switch angle;*
2. *u is the new vertex;*
3. *v is the new vertex and α_u is either a source-switch or a non-switch angle.*

We analyze the remaining cases by looking at the different possibilities for α_u and α_v. Both α_u and α_v can either be a sink-switch, a source-switch, or they can be a non-switch angle. If the new edge is added between two existing vertices, then we know from Lemma 3 that α_v cannot be a source-switch. The case when α_v is a sink-switch and α_u is either a source-switch or when α_v is a non-switch angle is already discussed in Theorem 2. Hence we need to analyze when α_v is a sink-switch or a non-switch angle, while α_u is any type of angle. These cases are discussed in Theorem 3 and Theorem 4. Theorem 3 discusses the case when α_v is not a switch while α_u can be any type of angle. The only case left for both end-vertices to be already existing is when both α_v and α_u are sink-switches, which is discussed in Theorem 4.

Theorem 3. *Let G_ϕ be an upward planar embedded digraph with a single source s_G and a face-sink graph F. If we insert an edge $e = (u,v)$ in the face $f \in G_\phi$ such that both u and v exists and α_v is not a switch then $G'_{\phi'}$ will be upward planar if and only if one of the following is true.*

1. *If $T_f \neq \hat{T}$ then $w_{T_f} \in nodes(T_{j+1} \cup \ldots \cup T_k)$.*
2. *If $T_f = \hat{T}$ then $s_{G'}$ is incident to at least one face in $faces(T_{j+1} \cup \ldots \cup T_k)$.*

While α_u can either be a sink-switch, or, a source-switch, or not a switch.

Proof. The tree T_f is transformed into two new trees $\mathscr{T}_1$ and $\mathscr{T}_2$, such that

$$V(\mathscr{T}_1) = V(T_1 \cup T_2 \ldots T_j)$$

$$E(\mathscr{T}_1) = E(T_1 \cup T_2, \ldots T_j) \cup \{(\bar{f}_1, w_1) \cup \ldots (\bar{f}_1, w_j)\})$$

$$V(\mathscr{T}_2) = V(T_{j+1} \cup T_{j+2} \ldots T_k)$$
$$E(\mathscr{T}_2) = E(T_{j+1} \cup T_{j+2} \ldots T_k) \cup \{(\bar{f}_2, w_{j+1}) \cup \ldots (\bar{f}_2, w_k)\})$$

where w_i is the access node for subtree T_i. When u is a sink-switch in f, there will also be a third tree

$$\mathscr{T}_3 = (V(T_l), E(T_l)).$$

We may observe that v is an internal vertex that is part of $\mathscr{T}_1$ and u is an internal vertex of $\mathscr{T}_3$ (when $\mathscr{T}_3$ exists). All possible cases are shown in Fig. 3.

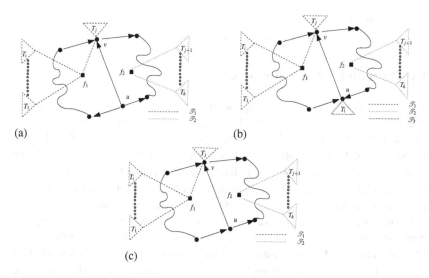

Fig. 3. Cases of Theorem 3: u and v already exist, α_u is source-switch and α_v is not a switch (a); u and v already exist, α_u is not a switch and α_v is sink-switch (b); u and v already exist, α_u is a sink-switch and α_v is not a switch (c)

If: If $T_f \neq \hat{T}$ and $w_{T_f} \in nodes(T_{j+1} \ldots \cup T_k)$ then each tree in $F' \setminus F$ has one internal vertex and $G'_{\phi'}$ is upward planar in this case. Similarly, if $T_f = \hat{T}$ then each of the generated new trees has one internal vertex except $\mathcal{T}_2$. Again, $G'_{\phi'}$ is upward planar because according to our assumption $faces(\mathcal{T}_2) \cap H_{G'_{\phi'}} \neq \emptyset$.

Only if: We show this by proving the contrapositive. If $T_f \neq \hat{T}$ and $w_{T_f} \in \{T_1, \ldots, T_j\}$ then F' has two trees $\mathcal{T}_2$ and $\hat{T}$ that have no internal vertices. Similarly, if $T_f = \hat{T}$ and $s_{G'}$ is not incident to a face in $faces(T_{j+1} \ldots \cup T_k)$ then $\mathcal{T}_2$ has no internal vertex but $faces(\mathcal{T}_2) \cap H_{G'_{\phi'}} = \emptyset$. Hence $G'_{\phi'}$ will not be upward planar. □

Theorem 4. *Let G_ϕ be an upward planar embedded digraph with a single source s_G and a face-sink graph F. We insert an edge $e = (u, v)$ in the face $f \in G_\phi$, such that both u and v already exists. If both α_u and α_v are sink-switches in f then $G'_{\phi'}$ will be upward planar if and only if one of the following is true.*

1. If $T_f \neq \hat{T}$ then $w_{T_f} \in nodes(T_1 \cup \ldots \cup T_k)$.

2. If $T_f = \hat{T}$ then $s_{G'}$ is incident to at least one face in $faces(T_1 \cup \ldots \cup T_k)$

When one of the end-vertices is a new vertices for the new edge $e = (u, v)$, then the case when u is the new vertex and when v is the new vertex and α_u is a sink-switch is already discussed in Theorem 2. The only remaining case is discussed in Theorem 5.

Theorem 5. *Let G_ϕ be an upward planar embedded digraph with a single source s_G and a face-sink graph F. We insert an edge $e = (u, v)$ in the face $f \in G_\phi$, such that v is a new vertex and α_u is a sink-switch, then $G'_{\phi'}$ will be upward planar if and only if one of the following is true.*

1. If $T_f \neq \hat{T}$ then $w_{T_f} \in nodes(T_1 \cup \ldots \cup T_k)$.
2. If $T_f = \hat{T}$ then $s_{G'}$ is incident to at least one face in $faces(T_1 \cup \ldots \cup T_k)$.

4   Algorithm and Time Complexity

We now present our algorithm for testing the upward planarity of a bimodal and pla-narly embedded $G'_{\phi'}$ with a fixed external face and discuss its complexity. The input to the algorithm is G_ϕ, the upward planar embedded digraph; e, the edge to be added or deleted; and $\tilde{G}_{\tilde{\phi}}$, the embedded split-digraph corresponding to G_ϕ. It first constructs a bimodal and planar $G'_{\phi'}$, if it exists. If we delete the edge e then the resulting $G'_{\phi'}$ will also be upward planar. The rest of the algorithm checks if $G'_{\phi'}$ satisfies the conditions of theorems from the previous section. The algorithm is shown in Algorithm 1.

We now show that dynamic upward planarity testing based on Theorem 1 requires $\Omega(n)$ time when we allow the external face to change and do not transform G_ϕ. We show it by assuming that $G'_{\phi'}$ is non-upward planar with h as its external face, where h was the external face of G_ϕ. The digraph $G'_{\phi'}$ will be upward planar if there is a face $g \neq h$, such that $g \in H_{G'_{\phi'}} \cap faces(\hat{T}')$, where $\hat{T}'$ is the tree in F' with no internal vertex. Recall that $H_{G'_{\phi'}}$ denotes the set of faces that are incident to the single-source in $G'_{\phi'}$. In order to find an alternative external face g in $o(n)$ time, we dynamically maintain $H_{G'_{\phi'}}$ by making appropriate additions or deletions in H_{G_ϕ} because recomputing $H_{G'_{\phi'}}$ from scratch will take $O(n)$ time. Now, if the new edge $e = (u, v)$ is between a new vertex u and an existing vertex $v = s_G$, then $H_{G'_{\phi'}} = \{f\}$. This results in a contradiction because removing the old faces will take $O(n)$ time. Hence it is not possible to design an efficient dynamic upward planarity testing algorithm for single-source embedded digraphs using Theorem 1.

We recall from Sec. 2 that finding a planar and bimodal $G'_{\phi'}$ requires $O(\log n)$ amor-tized time. We can check that an insertion satisfies Theorem 2 in constant time. Let μ represent the unique internal vertex w_{T_f} of T_f when $T_f \neq \hat{T}$ and represent the external face h of G_ϕ when $T_f = \hat{T}$. The overall time complexity of Algorithm 1 depends on how efficiently we can check if μ is in a particular subtree of T_f. The location of μ can be easily be determined in $O(n)$ time by traversing the nodes of T_f, but then the time complexity of Algorithm 1 will equal running the algorithm of Bertolazzi et al. from scratch. We propose instead an $O(1)$-time method. We maintain a directed version of F by rooting each tree $T \in F$ at its unique internal vertex or vertex corresponding to the external face, and then orienting all edges toward the root. Each vertex $v \neq \mu$ will have exactly one outgoing edge and if $v = \mu$ then it has no outgoing edge. Let $out(v)$ represent the outgoing edge for a vertex v and let $p(v)$ be the target node for $out(v)$. Note that, $p(\bar{f})$ is always an access-vertex w_i for a subtree T_i.

We can check if μ is in a subtree satisfying Theorems 3, 4 or 5 by finding the relative location of $p(\bar{f})$ in the facial boundary of f. This is done by maintaining a linked list L_f for every face $f \in G_\phi$, such that every vertex $v \in f$ has a corresponding real number $L_f[v]$. We construct $L_f = \{L_f[v_1], \ldots, L_f[v_k]\}$ such that: $L_f[v_i] < L_f[v_{i+1}]$, where $v_1, \ldots, v_k$ are consecutive vertices on the facial boundary of f in the clockwise direction.

Algorithm 1. Dynamic Upward Planarity Test $(G_\phi, e = (u,v), \tilde{G}_{\tilde{\phi}})$

1: Find a planar and bimodal embedding of G', $G'_{\phi'}$
2: **if** we cannot find a planar and bimodal $G'_{\phi'}$ **then**
3:     Return False
4: **end if**
5: **if** delete the edge e **then**
6:     Return True
7: **end if**
8: **if** Both u and v already exist and α_v is sink-switch; or u is a new vertex; or v is a new vertex
     and α_u is not sink-switch **then**
9:     Return True
10: **end if**
11: $\mu = w_{T_f}$ when $T_f \neq \hat{T}$ and $\mu = h$ when $T_f = \hat{T}$
12: **if** u and v exist and α_v is not a switch and $\mu \in \{T_{j+1}, \ldots, T_k\}$ **then**
13:     Return True
14: **else if** u and v exist and both α_u and α_v are sink-switches and $\mu \in \{T_1, \ldots, T_k\}$ **then**
15:     Return True
16: **else if** v is a new vertex and α_u is a sink-switch and $\mu \in \{T_1, \ldots, T_k\}$ **then**
17:     Return True
18: **else**
19:     Return False
20: **end if**

When the insertion of $e = (u,v)$ divides f into f_1 and f_2, we divide L_f to get L_{f_1} and L_{f_2} such that $L[u]$ and $L[v]$ are present in both of them. We also maintain pointers from each vertex incident to a face f to its entry in L_f. When we insert an edge e such that a new vertex v_j is added, we can choose a sufficiently small ε, letting $L_f[v_j] = L_f[v_{j-1}] + \varepsilon$. However, this can result in difficulties associated with high precision real numbers and hence increase the time complexity in comparing two elements of L_f. Instead, we suggest using the algorithm by Bender et $al.$ to assign $L_f[v_j]$ in $O(\log n)$ amortized time [1]. The algorithm by Bender et $al.$ maintains a dynamic list and allows a user to compare the order of any two elements in the list. This is done by assigning tags of $O(\log n)$ bits to each element in the list. Hence, any $L_f[v_j]$ and $L_f[v_i]$ can be efficiently compared. The following lemma shows that L_f can be used to efficiently check if μ is in the required subtree. This technique will also work when L_f is divided into two new lists L_{f_1} and L_{f_2} because the algorithm of Bender et $al.$ assigns a tag by locally relabeling a subset of a list.

Lemma 4. *We can check the conditions of Theorems 3, 4 5 in constant time.*

Proof. Theorem 4 and 5: We need to check if $p(\bar{f}) \in \{w_1, \ldots, w_k\}$, this will be true if $p(\bar{f}) \neq w_l$. Hence $p(\bar{f}) \in \{w_1, \ldots, w_k\}$ and $G'_{\phi'}$ will be upward planar if and only if $L_f[p(\bar{f})] \neq L_f[u]$.

Theorem 3: We can see from Fig. 3 that we need to check if $p(\bar{f}) \in \{w_{j+1}, \ldots, w_k\}$. We have the following 2 cases, based on the fact that $p(\bar{f})$ should be between v and u in the clockwise direction in order to satisfy the theorem.

1. When $L_f[v] < L_f[u]$ then $p(\bar{f}) \in \{w_{j+1}, \ldots, w_k\}$ will be true if $L_f[v] < L_f[p(\bar{f})] < L_f[u]$.
2. When $L_f[u] < L_f[v]$ then $p(\bar{f}) \in \{w_{j+1}, \ldots, w_k\}$ will be true if either $L_f[p(\bar{f})] < L_f[u] < L_f[v]$ or $L_f[u] < L_f[v] < L_f[p(\bar{f})]$.

□

We have yet to show that we maintain the correct orientation of the edges of F in the presence of updates. The following two lemmas shows that we can do this in constant time. We define the *splitting* of a vertex $\bar{f}$ with respect to the new edge $e = (u, v)$ as the creation of two new vertices $\bar{f}_1$ and $\bar{f}_2$, such that $\bar{f}_1$ has edges of $\bar{f}$ to and from $w_1, \ldots, w_i$ and $\bar{f}_2$ has all edges of $\bar{f}$ to and from $w_{j+1}, \ldots, w_k$. We also define the *merging* of a vertex $\bar{f}_1$ and a vertex $\bar{f}_2$ as the creation of a new vertex $\bar{f}$, such that $\bar{f}$ has all outgoing edges and incoming edges of both $\bar{f}_1$ and $\bar{f}_2$. We need to split $\bar{f}$ when as a result of edge insertion the face f splits into f_1 and f_2, and we need merging when two faces f_1 and f_2 combine to form the face f. Splitting and merging $\bar{f}$ can be done by splitting and merging the adjacency list of $\bar{f}$.

Lemma 5. *Let G_ϕ be an upward planar embedded digraph with a single-source s_G. If we add a new edge e to create an embedded digraph $G'_{\phi'}$ with a single source $s_{G'}$ such that $G'_{\phi'}$ is upward planar with the same external face G_ϕ then we can update F in constant time.*

The deletion of an edge e will either merge two faces f_1 and f_2 in G_ϕ to form a face f in $G'_{\phi'}$, or when one of the end vertices of e has a degree of 1 and is incident to a single face f then the facial boundary of f will change. Moreover, we let α_u and α_v represent the angle that is created at u and v respectively as a result of the edge deletion. We say that with the deletion of an edge e from G_ϕ, F will change to the face-sink graph F'. Let T'_f be the tree in F' that contains $\bar{f}$. $T'_{f'}$ is formed by merging trees in a set $\mathcal{M} \subset F, |\mathcal{M}| \geq 1$ and making some local changes in this merged tree. F' will always satisfy Theorem 1. When all trees in $\mathcal{M}$ have one internal vertex then the resulting tree T'_f will also have exactly one internal vertex. However, if $\mathcal{M}$ contains $\hat{T}$, the tree in F with no internal vertex, then T'_f will also have no internal vertex. We let μ' denote either the internal vertex in T'_f or the vertex $\bar{h}$ that corresponds to the external tree.

Lemma 6. *Let G_ϕ be an upward planar embedded digraph with a single-source s_G. If we delete an edge $e = (u, v)$ to create an embedded digraph $G'_{\phi'}$ with a single source $s_{G'}$ then we can update F in constant time.*

Hence we conclude that Algorithm 1 will take $O(\log n)$-time leading to the following theorem.

Theorem 6. *Let G_ϕ be an upward planar embedded digraph with a single-source s_G. If we add or delete an edge e to create an embedded digraph $G'_{\phi'}$ with a single source $s_{G'}$ then we can check the upward planarity of $G'_{\phi'}$ in $O(\log n)$ when the external face is fixed.*

5   Open Problems

As further work, we want to investigate if there is a dynamic upward planarity testing algorithm for embedded digraphs that allows for the external face to change. Moreover, it will be interesting to investigate the optimality of our algorithm. Our algorithm may also be relevant to finding a maximum upward planar subgraph of a single-source embedded digraph and we intend investigating this. A slightly more difficult open problem is to develop a dynamic upward planarity testing algorithm for a single-source digraph over all its embeddings.

References

1. Bender, M.A., Cole, R., Demaine, E.D., Farach-Colton, M., Zito, J.: Two simplified algorithms for maintaining order in a list. In: Möhring, R.H., Raman, R. (eds.) ESA 2002. LNCS, vol. 2461, pp. 152–164. Springer, Heidelberg (2002)
2. Bertolazzi, P., Battista, G.D., Didimo, W.: Quasi-upward planarity. Algorithmica 32(3), 474–506 (2002)
3. Bertolazzi, P., Battista, G.D., Liotta, G., Mannino, C.: Upward drawings of triconnected digraphs. Algorithmica 12(6), 476–497 (1994)
4. Bertolazzi, P., Battista, G.D., Mannino, C., Tamassia, R.: Optimal upward planarity testing of single-source digraphs. SIAM J. Comput. 27(1), 132–169 (1998)
5. Demetrescu, C., Finocchi, I., Italiano, G.: Handbook of Graph Theory. In: Yellen, J., Gross, J.L. (eds.) Dynamic Graph Algorithms. CRC Press Series, in Discrete Mathematics and Its Applications, vol. 10.2 (2003) ISBN 1-58488-090-2
6. Di Battista, G., Eades, P., Tamassia, R., Tollis, I.G.: Graph Drawing: Algorithms for the Visualization of Graphs. Prentice-Hall, Englewood Cliffs (1999)
7. Didimo, W.: Computing upward planar drawings using switch-regularity heuristics. In: SOFSEM, pp. 117–126 (2005)
8. Didimo, W., Giordano, F., Liotta, G.: Upward spirality and upward planarity testing. In: Healy, P., Nikolov, N.S. (eds.) GD 2005. LNCS, vol. 3843, pp. 117–128. Springer, Heidelberg (2006)
9. Garg, A., Tamassia, R.: On the computational complexity of upward and rectilinear planarity testing. SIAM J. Comput. 31(2), 601–625 (2001)
10. Hutton, M.D., Lubiw, A.: Upward planar drawing of single-source acyclic digraphs. SIAM J. Comput. 25(2), 291–311 (1996)
11. Papakostas, A.: Upward planarity testing of outerplanar DAGs. In: Proceedings Graph Drawing. pp. 298–306 (1994)
12. Tamassia, R.: On-line planar graph embedding. J. Algorithms 21(2), 201–239 (1996)

On the Hardness of
Orthogonal-Order Preserving Graph Drawing

Ulrik Brandes and Barbara Pampel*

Department of Computer & Information Science, University of Konstanz
`Barbara.Pampel@uni-konstanz.de`

Abstract. There are several scenarios in which a given drawing of a graph is to be modified subject to preservation constraints. Examples include shape simplification, sketch-based, and dynamic graph layout. While the orthogonal ordering of vertices is a natural and frequently called for preservation constraint, we show that, unfortunately, it results in severe algorithmic difficulties even for the simplest graphs. More precisely, we show that orthogonal-order preserving rectilinear and uniform edge length drawing is $\mathcal{NP}$-hard even for paths.

1 Introduction

In several scenarios, a graph drawing algorithm receives as input not only a graph, but also an initial (possibly partial) drawing. The task is to redraw the graph while maintaining selected features of the input drawing. Examples of this kind are embedding-constrained graph layout, shape simplification, sketch-based drawing, and dynamic graph layout.

A cartographic application of particular interest is the simplification of lines. Given a polygonal path, the task is to generate a simpler representation of the path, for instance by omitting vertices [8, 12] (level of detail) or by restricting the allowable types of segments [15, 14] (schematization).

Note that line simplification is also the base case in the design of schematic metro maps, where admissible slopes may be restricted and few bends are desired. Maintaining a user's mental map by preserving the orthogonal ordering [9] of stations and landmarks seems particularly appropriate in this scenario and has been tried, e.g., in [7]. For layout stability [3] and similarity [5] the relative position of vertices, strongly related to the orthogonal ordering, is considered, used [13] and tested [4] helpfull. Alternative constraints include preservation of the cyclic ordering of neighbors [16] and distance from original positions using various metrics [15, 14].

For two different drawing conventions we show that orthogonal ordering is $\mathcal{NP}$-hard to preserve, even for paths. This is in contrast to the direction-restricted models studied in [15] and [14], where paths or vertices must be within a given distance (according to the Fréchet or Euclidean metric) of the original and the number of bends can be minimized in polynomial time. For orthogonal-order preserving graph drawing, even the decision problems in the rectilinear and equal

---

* Corresponding author.

I.G. Tollis and M. Patrignani (Eds.): GD 2008, LNCS 5417, pp. 266–277, 2009.
© Springer-Verlag Berlin Heidelberg 2009

edge-length model are $\mathcal{NP}$-hard. The former implies, e.g., that bend-minimum orthogonal layout is hard under ordering constraints. The latter is also interesting, since drawing with given edge lengths is hard for general graphs [10], but easy for trees (see, e.g., [2]). With orthogonal ordering constraints, the problem is hard even for paths.

After some preliminaries, we treat the rectilinear case in Sect. 3 and the equal edge-length case in Sect. 4. For convenience, we give additional illustrations of gadgets in an appendix.

2 Preliminaries

We are interested in redrawing simple undirected paths $P = (v_1, \ldots, v_n)$ using straight line edges. An original geometric position (x_v, y_v) in the plane is given for each vertex $v \in P$. Let (x'_v, y'_v) be the position of a vertex v in the resulting layout. By preserving the orthogonal ordering of the vertices we mean that if for two vertices v_i, v_j it is $x_{v_i} \le x_{v_j}$ ($y_{v_i} \le y_{v_j}$) in the original layout, $x'_{v_i} \le x'_{v_j}$ ($y'_{v_i} \le y'_{v_j}$) holds also for the resulting layout.

For a (sub)-path P of $l \ge 1$ edges we call the area between the vertical line through P's rightmost vertex and the one through the leftmost vertex the x-range of P, and analogously the area between the horizontal line through P's highest vertex and the one through the lowest vertex P's y-range.

For the $\mathcal{NP}$-hardness proofs in this paper we use reductions from MONOTONE 3-SAT. In MONOTONE 3-SAT each clause contains exactly three literals either all negated or all non-negated. The problem is known to be $\mathcal{NP}$-hard [11]. Let I be an instance of the MONOTONE 3-SAT-problem with Boolean variables $X = \{x_1 \ldots x_n\}$ and clauses $C = \{C_1, C_2, \ldots, C_k\}$.

3 Rectilinear Drawings

The first problem we address is the following:

Orthogonal-order preserving rectilinear drawing problem: Given a graph in the plane, we want to decide whether we can draw each edge either horizontally or vertically, changing neither the horizontal nor the vertical order of endpoints, without introducing any intersection other than the common endpoint of two incident edges and keeping the edge-length positive for each edge.

Choosing the direction of an edge can force the direction of other edges. Figure 1 shows how an edge e_i can force the direction of another edge e_j. More formal: We say an edge e_i *pulls* another edge e_j horizontally, if e_j lies completely within e_i's y-range, hence, to keep the vertical order of endpoints, e_j has to be horizontal if e_i is horizontal. This of course also means that e_i cannot be drawn horizontally if e_j is vertical and we say e_j *pushes* e_i vertically. Analogously we say an edge e_i pulls another edge e_j vertically, if e_j lies within e_i's x-range

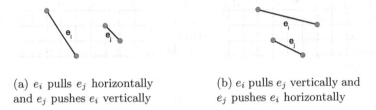

(a) e_i pulls e_j horizontally
and e_j pushes e_i vertically

(b) e_i pulls e_j vertically and
e_j pushes e_i horizontally

Fig. 1. Forcing to have the same direction

and therefore also e_j pushes e_i horizontally. We use this to construct the main elements of a gadget for the $\mathcal{NP}$-hardness proof.

Given a path $P = (e_1, e_2, e_3)$ of three edges as shown in Fig. 2. If there is a horizontal edge with one endpoint in the x-range of e_1 and one endpoint in the x-range of e_3, at least one of P's edges has to be drawn horizontally and we call P with the horizontal edge a *horizontal decision unit*. Analogously, if there is a vertical edge with one endpoint in the y-range of e_1 and one in the y-range of e_3 at least one of P's edges has to be drawn vertically and we call P with the vertical edge a *vertical decision unit*. We will later use these decision units to represent the 3-SAT clauses.

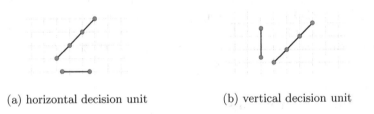

(a) horizontal decision unit

(b) vertical decision unit

Fig. 2. Decision units

In Fig. 3 two edges $e_i \neq e_j$ are linked by a third edge $l \neq e_i, e_j$. We call l the *horizontal link* for e_i and e_j, if e_j pulls l horizontally and l pushes e_i horizontally such that e_i, e_j and l are all horizontal if e_j is horizontal (see Fig. 3(a)). Of course this also means that if e_i is vertical, also l and e_j must be vertical. A *vertical link* is defined correspondingly and shown in Fig. 3(b). We will use these links for variables which occur in more the one clause.

3.1   Unions of Paths

We now use the described edge-dependency elements to create a gadget for a given instance of MONOTONE 3-SAT to prove the following:

Theorem 1. *The orthogonal-order preserving rectilinear drawing problem is $\mathcal{NP}$-hard for unions of paths.*

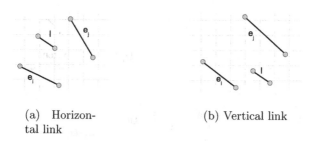

(a) Horizon-
tal link

(b) Vertical link

Fig. 3. Possible links l for e_i and e_j

For a given instance I of MONOTONE 3-SAT we create a union of paths as follows. Each variable will have several corresponding edges. For each positive clause C_i we place a horizontal decision unit $U(C_i)$ on the diagonal of the drawing and for each negative clause C_j we place a vertical one $U(C_j)$ as shown in Fig. 4. The horizontal and vertical edges of the decision units can be placed on a horizontal and a vertical line near the borders of the drawing. The diagonal edges in the decision units correspond to the literals in the decision unit's clause. We then place a *variable path* $(e_1, \ldots, e_n)$ (see Fig. 4) on the diagonal of the drawing with n edges corresponding to the n variables in X. For each diagonal edge in a horizontal decision unit we add a positive link between this edge and the edge in the variable path corresponding to the same variable and for each diagonal edge in a vertical decision unit we do the same with a negative link. Because of the links an edge in the variable path is horizontal in a valid orthogonal drawing if an edge corresponding to the same variable is drawn horizontally in a horizontal decision unit and vertical if drawn vertically in a vertical decision unit. We set a variable true if the corresponding edge is drawn horizontally in the variable path and false, if it is drawn vertically, such that for a valid drawing all clauses are satisfied and all other variables can be chosen arbitrary.

Analogously to this, setting the variables such that all clauses are satisfied will also induce a valid drawing, hence the edges in S can be drawn orthogonally without intersections keeping the horizontal and vertical order of their endpoints if and only if I is satisfiable. Thus the problem is proven to be $\mathcal{NP}$-hard.

The gadget is quite special but we can change it to a gadget with totally ordered vertices, i. e., no two vertices have the same x- or y-coordinates. In the horizontal decision units we used horizontal edges e_h that we can move away from their horizontal line, but force them to be later drawn horizontally again, by attaching at one endpoint a small edge e_f like shown in Fig. 5(a). e_f lies in e_h's x-range and e_h in e_f's y-range such that the only possibility of avoiding intersections is to draw e_h horizontally and e_f vertically. Since e_h must be drawn horizontally, still at least one of the diagonal edges in the decision unit must be drawn horizontally as well. We place e_f's endpoint that is not incident to e_h such that no vertex lies in e_f's x-range, so e_f cannot pull any other edge vertically. Because the horizontal edges of the decision units are placed near the borders of the drawing we can easily guarantee that e_f does not lie in any other edge's

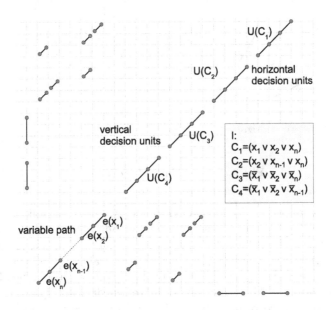

Fig. 4. Union of paths for a MONOTONE 3-SAT-instance I

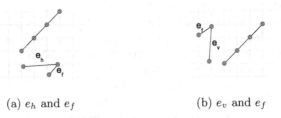

(a) e_h and e_f (b) e_v and e_f

Fig. 5. Decision units without horizontal or vertical edges

y-range, so it cannot push other edges vertically. The strategy for the vertical decision units is the same and shown in Fig. 5(b). After also making sure, that no two vertices of different horizontal (vertical) links for the same edge in the variable path lie on the same vertical (horizontal) line for example by making each horizontal link shorter than the one exactly above it (likewise for vertical links), we have a gadget with total ordering that is drawable if and only if the special gadget was drawable. Thus the problem is proven to be $\mathcal{NP}$-hard also for unions of paths with totally ordered vertices.

3.2   Single Path

Theorem 2. *Orthogonal-order preserving rectilinear drawing is $\mathcal{NP}$-hard for paths.*

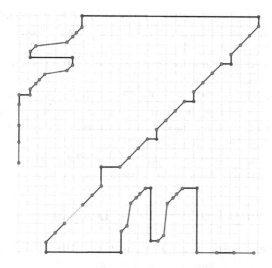

Fig. 6. Single path for the instance I

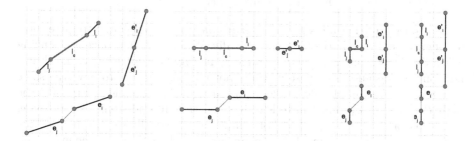

Fig. 7. l_c is connecting two positive links

To show $\mathcal{NP}$-hardness also for paths we have to connect the edges. We add horizontal and vertical edges connecting the decision units and the links for different decision units (see Fig. 6). They have no effect on the drawability because edges which are already horizontal or vertical cannot pull other edges. Additionally the horizontal and vertical edges added here do not lie in the range of any edge not yet horizontal or vertical and therefore cannot push other edges. Links belonging to the same decision unit are also connected. We take a closer look at these connecting edges. For two incident edges e'_i, e'_j in the same horizontal decision unit and the edges e_i, e_j in the variable path corresponding to the same variables, let $e_{ii'}$ and $e_{jj'}$ be the positive links. (See Fig. 7.) For e_i and e_j not incident let e_c be connecting the positive links. If e'_i and e'_j are both drawn horizontally they pull e_c (as well as $e_{ii'}$ and $e_{jj'}$ of course) horizontally. If e'_i, e'_j or even both edges are vertical, e_c can still be drawn horizontally. Furthermore e_c

can pull all edges $e_{i+1} \ldots e_{j-1}$ vertically so we draw it vertically if and only if all edges $e_i \ldots e_j$ must be vertical anyway and horizontally otherwise. The edges connecting the links have no effect on drawability. The edges in the union of paths are also kept in the connected path, hence the path is not drawable, if the union of paths had not been drawable.

With the connecting edges we again have vertices with the same x- or y-coordinates, but because they have no effect on drawability we can just slightly turn each horizonal or vertical line with two or more vertices. It can be turned back without conflicts when redrawing the path such that the new connected path is drawable if and only if the old one was. The problem is $\mathcal{NP}$-hard also for paths with double-totally ordered vertices.

4   Drawings with Uniform Edge Lengths

Orthogonal-order preserving equal edge lengths drawing problem:
    Given a graph in the plane, we want to decide whether we can draw each
    edge with length one changing neither the horizontal nor the vertical order
    of the edges' endpoints and without introducing any intersection other than
    the common endpoint of two connected edges.

The constraint that all edges have length one may be exploited tp force some of them to be drawn horizontally or vetically. We can use the concept of linking edges like in the previous section. Edge e_i in Fig. 8(a) forces e_j to be drawn horizontally, because otherwise it would not be short enough to have the same length as e_i. We can also define decision units. In the example in Fig. 8(b) the path $P = \{e_1, e_2, e_3\}$ must have length 3 and the only possibility of achieving this is to draw the framing edges horizontally and vertically to give the path the room of a 1×2-rectangle in which the longest possible path monotone in $x-$ and $y-$direction has length 3. It is easy to see that one edge of P has to be drawn horizontally and the other two edges vertically.

Let a *horizontal* decision unit be a 3-edge-path monotone in x- and y- direction contained into a 1x2-rectangle while a *vertical* decision unit is also a 3-edge-path monotone in x- and y- direction, but contained into a 2x1-rectangle

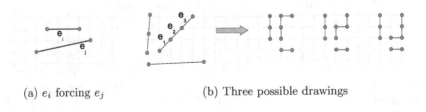

(a) e_i forcing e_j (b) Three possible drawings

Fig. 8.

4.1   Unions of Paths

We now use these edge-dependency elements to create a gadget for a given instance of MONOTONE 3-SAT to prove the following:

Theorem 3. *The orthogonal-order preserving equal edge lengths drawing problem is $\mathcal{NP}$-hard for unions of paths.*

For a given instance I of MONOTONE 3-SAT we create a union of paths as follows. Like in the proof for the rectilinear graph drawing we place decision units for the clauses with edges corresponding to the variables in the clause. For each positive clause we place a horizontal decision unit and for each negative clause we place a vertical one. Similar to the gadget in Sect. 3.1 we arrange the decision units on the diagonal such that all vertical and all horizontal decision units lie next to each other without being connected. The framing rectangle edges can be placed on an almost horizontal and an almost vertical line near the borders of the drawing. For each variable in a negative clause that also occurs in a positive clause we add a link between the edge in the vertical and the edge in the horizontal decision unit such that both edges are horizontal if drawn horizontally in the horizontal decision unit and vertical if drawn vertically in the vertical decision unit (see Fig. 9). In a valid drawing there is no edge drawn horizontally in a horizontal decision unit linked to an edge drawn vertically in a vertical decision unit. We choose the variable corresponding to an edge horizontal in a horizontal decision unit true and to an edge vertical in a vertical decision unit false. With this all clauses are satisfied and the other variables can be set arbitrarily.

For a given solution of I we can create a valid drawing as follows: For each positive clause we choose one of the variables set true and draw the corresponding edge in the corresponding decision unit horizontally, the other edges in this decision unit we draw vertically. From each negative clause we also choose one variable set false and draw the corresponding edge in the corresponding decision unit vertically and the other edges horizontally. Now all decision units have a

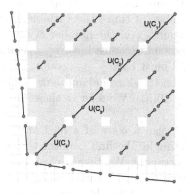

Fig. 9. The union of paths for the instance I from Sect. 3.1

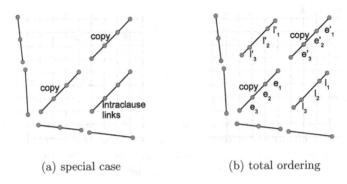

(a) special case   (b) total ordering

Fig. 10. Intraclause links

valid drawing. The links have a valid drawing, too, because the linked edges
are both either horizonal or vertical such that the link can have length one.
The union of paths can be drawn with equal edge lengths without intersections
keeping the horizontal and vertical order of their endpoints if and only if I is
satisfiable. Thus the problem is proven to be $\mathcal{NP}$-hard. Note that we did not use
horizontal or vertical edges and with our arrangement of decision units we can
guarantee total ordering by additionally avoiding links on the same horizontal
or vertical line like we did in Sect. 3.1.

4.2   Single Path

Theorem 4. *The orthogonal-order preserving equal edge lengths drawing is*
$\mathcal{NP}$*-hard for paths.*

To show $\mathcal{NP}$-hardness also for paths we have to add connecting edges and prove
that they have no effect on the drawability. We guarantee this by making sure
that the connected path stays drawable if the union of paths had been drawable.
To make it easier to connect the path segments, we copy the decision units such
that each copy has exactly one link to one copy of another decision unit. For
$\text{count}_{neg}(e)$ being the number of times the variable corresponding to e occurs
in a negative clause and $\text{count}_{pos}(e)$ the number of times it occurs in a positive
clause, each vertical decision unit U must have $\sum_{e \in U} \text{count}_{pos}(e)$ copies and each
horizontal decision unit U' must have $\sum_{e \in U'} \text{count}_{neg}(e)$ copies. The copies are
placed next to each other on the diagonal and linked such that each copy has to
be drawn equally (see Fig. 10(a)). We refer to these linking edges as *intraclause
links*.

An *interclause link* between a copy of a horizontal and a vertical decision
unit lies within a 2x2-rectangle. We first connect the link to two *inner anchor
vertices* outside of this rectangle. We take a close look at the case where the
third edge of a vertical decision unit is linked with the second edge of a vertical
decision unit (see Fig. 11). For the eight combinations of possible drawings of
the decision units, except for one combination the anchor vertices lie exactly

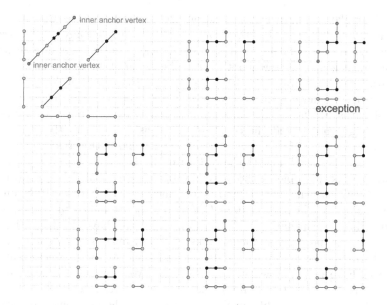

Fig. 11. Eight possible combinations

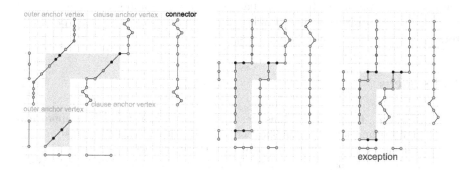

Fig. 12. Outer anchor vertices

at the same distance from the border of the horizontal decision units's y-range and always at the same distance from the border of the vertical decision units's x-range. Because of the one exception, we connect the inner anchor vertices to *outer anchor vertices* that can always be placed each at the same y-coordinates with a *clause anchor vertex* for the horizontal decision unit and at the same x-coordinates with a clause anchor point of the positive decision unit (see Fig. 12). This guarantees, that we can later connect these parts of the path, without having an effect on the drawability.

It is possible to move the horizontal decision unit and possible intraclause links by one edge length. Whenever we have to cross the x- or y-range between the outer anchor vertices when connecting other parts of the path, we can use a *connector* like shown in Fig. 12 and outside the ranges edges that are either horizontal or vertical with length already 1. Analogously for any interclause link such anchor points can be defined because only one of the decision units has to be movable by at least one edge length. Thus the union of paths can be connected to a path, that stays drawable, if the unions of paths was drawable.

We can also achieve total ordering for a single path. Before connecting the union of paths in the special case, we changed it by copying and linking the clauses. To guarantee that the copies of the decision units for the same clause are all drawn in the same way, we used intraclause links with vertices on the same horizontal and vertical line with the vertices of the decision units (Fig. 10(a)). We now have to link the copies without using vertices with equal x- or y-coordinates. We can do this by using shorter and longer links as shown in Fig. 10(b) on both sides of the units. The shorter links l_1, l'_1 and l_3, l'_3 are pulled by e_1 and e_3 but also push them, such that e_1, l_1, l'_1 and e'_1 must always have the same direction, just as e_3, l_3, l'_3 and e'_3. The longer links l_2 and l'_2 pull e_2 and e'_2 and are also pushed by them, hence also e_2, l_2, l'_2 and e'_2 all have the same direction and the two copies have to be drawn in the same way. We now have a new union of paths that is drawable if and only if the old one had been drawable as well. We connect the paths exactly like in Sect. 4.2 and turn the horizontal and vertical lines through more than one vertex like in Sect. 3. The connecting edges are always drawable and do not force any other edge, while the edges of the union of paths are still contained, hence the path is drawable if and only if the union of paths had been drawable. The problem is $\mathcal{NP}$-hard also for a single path with totally ordered vertices along each axis.

References

[1] Agarwal, P., Har-Peled, S., Mustafa, N., Wang, Y.: Near-linear time approximation algorithms for path simplification. Algorithmica 42, 203–219 (2000)

[2] Bachmaier, C., Brandes, U., Schlieper, B.: Drawing phylogenetic trees. In: Deng, X., Du, D.-Z. (eds.) ISAAC 2005. LNCS, vol. 3827, pp. 1110–1121. Springer, Heidelberg (2005)

[3] Boehringer, K.-F., Newbery Paulisch, F.: Using constraints to achieve stability in automatic graph algorithms. In: Proc. of the ACM SIGCHI Conference on Human Factors in Computer Systems, pp. 43–51. WA (1990)

[4] Bridgeman, S., Tamassia, R.: A user study in similarity measures for graph drawing. JGAA 6(3), 225–254 (2002)

[5] Bridgeman, S., Tamassia, R.: Difference metrics for interactive orthogonal graph drawing algorithms. In: Whitesides, S.H. (ed.) GD 1998. LNCS, vol. 1547, pp. 57–71. Springer, Heidelberg (1999)

[6] Carlson, J., Eppstein, D.: Trees with convex faces and optimal angles. In: Kaufmann, M., Wagner, D. (eds.) GD 2006. LNCS, vol. 4372, pp. 77–88. Springer, Heidelberg (2007)

[7] Dwyer, T., Koren, Y., Marriott, K.: Stress majorization with orthogonal order constraints. In: Healy, P., Nikolov, N.S. (eds.) GD 2005. LNCS, vol. 3843, pp. 141–152. Springer, Heidelberg (2006)

[8] Douglas, D., Peucker, T.: Algorithms for the reduction of the number of points required to represent a digitized line or its caricature. Canad. Cartog. 10(2), 112–122 (1973)

[9] Eades, R., Lai, W., Misue, K., Sugiyama, K.: Layout adjustment and the mental map. J. Visual Lang. Comput. 6, 183–210 (1995)

[10] Eades, P., Wormald, N.: Fixed edge-length graph drawing is $\mathcal{NP}$-hard. Discrete Appl. Math. 28, 111–134 (1990)

[11] Garey, M., Johnson, D.: Computers and Intractability. W. H. Freeman and Company, New York (1979)

[12] Imai, H., Iri, M.: An optimal algorithm for approximating a piecewise linear function. J. Inform. Process. 9(3), 159–162 (1986)

[13] Lee, Y.-Y., Lin, C.-C., Yen, H.-C.: Mental map preserving graph drawing using simulated annealing. In: Proc. of the 2006 Asia-Pacific Symposium on Inforamtion Visualisation, pp. 179–188. Australian Computer Science (2006)

[14] Merrick, D., Gudmundsson, J.: Path simplification for metro map layout. In: Kaufmann, M., Wagner, D. (eds.) GD 2006. LNCS, vol. 4372, pp. 258–269. Springer, Heidelberg (2007)

[15] Neyer, G.: Line simplification with restricted orientations. In: Dehne, F., Gupta, A., Sack, J.-R., Tamassia, R. (eds.) WADS 1999. LNCS, vol. 1663, pp. 13–24. Springer, Heidelberg (1999)

[16] Nöllenburg, M., Wolff, A.: A mixed-integer program for drawing high-quality metro maps. In: Healy, P., Nikolov, N.S. (eds.) GD 2005. LNCS, vol. 3843, pp. 321–333. Springer, Heidelberg (2006)

Generalizing the Shift Method for Rectangular Shaped Vertices with Visibility Constraints*

Seok-Hee Hong[1] and Martin Mader[2]

[1] School of IT, University of Sydney, NSW, Australia
[2] Department of Computer and Information Science, University of Konstanz, Germany

Abstract. In this paper we present a generalization of the *shift method* algorithm [4,6] to obtain a straight-line grid drawing of a triconnected graph, where vertex representations have a certain specified size. We propose vertex representations having a rectangular shape. Additionally, one may demand maintainance of the criterion of strong visibility, that is, any possible line segment connecting two adjacent vertices cannot cross another vertex' representation. We prove that the proposed method produces a straight-line grid drawing of a graph in linear time with an area bound, that is only extended by the size of the rectangles, compared to the bound of the original algorithm.

1 Introduction

The shift method [4] is a well-known method among several approaches to obtain a standard straight-line representation of planar graphs in the graph drawing literature [2,7,9]. Given a triangulated graph, the original algorithm calculates coordinates for each vertex on an 2D integer grid such that the final drawing has a quadratic area bound. A linear time variant is presented in [3], [6] provides a version for triconnected graphs, [5] for biconnected graphs.

The approach presented in the following sections is related to a version of the shift method given in [1], which allows square vertex representations. In this paper, the shift method for triconnected graphs [6] is generalized to have rectangular shaped vertex representations. Furthermore, we demand that the criterion of strong visibility between adjacent vertices is satisfied, that is, any possible line segment connecting two adjacent vertices does not cross another vertex' representation. To maintain the strong visibility criterion in the shift method, additional shifts have to be introduced. The main contribution is to prove that the proposed method produces a grid drawing with an area quadratic in the sum of number of vertices and the sizes of the vertex representations.

The generalized shift method can be used to draw clustered graphs having planar quotient graphs [8]. Other possible applications include drawing graphs that have arbitrary vertex representations by using the minimal bounding box, or drawing graphs with labeled vertices, where the positions of a vertex and its label are not known, but only the size of the region into which they are allowed to be drawn.

---

* This work was supported by DFG Research Training Group GK-1042 "Explorative Analysis and Visualization of Large Information Spaces", University of Konstanz.

I.G. Tollis and M. Patrignani (Eds.): GD 2008, LNCS 5417, pp. 278–283, 2009.

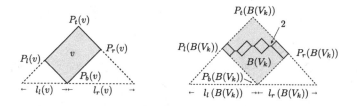

Fig. 1. Vertex representations. Left: singleton $V_k = \{v\}$. Right: $|V_k| > 1$.

2 Preliminaries

Let $G = (V, E)$ be a graph with $n = |V|$ and $m = |E|$. A graph is called *planar* if it has a crossing-free drawing in the plane. A *plane graph* is a planar graph with a fixed cyclic ordering of edges incident to each vertex and a fixed outer face. A plane graph divides the plane into which it is drawn into connected regions called *faces*. A *triconnected* graph is a graph where the removal of any pair of vertices does not disconnect the graph.

Let G be a triconnected plane graph. Let $\pi = (V_1, V_2, \ldots, V_K), K < n$, be a *lmc-ordering* of G as presented in [6]. It is shown that every triconnected plane graph has a *lmc*-ordering, and it can be computed in linear time. Let $G_k, k \leq K$, be the graph induced by $V_1 \cup \cdots \cup V_k$ according to π, particularly $G_K = G$. We denote by $C_0(G_k)$ the boundary of the outer face of G_k.

Vertices are represented as rectangles rotated by 45 degrees. For all $v \in V$, vertex lengths $l_l(v)$ and $l_r(v)$ are given according to the side lengths of a vertex representation, as illustrated in Fig.1. Let $l(v) = l_l(v) + l_r(v)$. Let $P_l(v), P_r(v), P_b(v)$ and $P_t(v)$ be the left, right, bottom and top corners of v's representation, with $P_l(v) = (x_l(v), y_l(v))$, etc. As illustrated in Fig.1, we represent a set $V_k = \{v_k^1, \ldots, v_k^j\}, j > 1$, as a chain of the single vertices, where $[P_r(v_k^i), P_l(v_k^{i+1})], 1 \leq i < j$, are horizontally aligned with distance two. Let $l(V_k) = \sum_{v \in V_k} l(v), l_l(V_k) = \sum_{v \in V_k} l_l(v)$, and $l_r(V_k)$ accordingly. Let $B(V_k)$ be the minimal bounding box of the representation of V_k. For a singleton $V_k = \{v_k\}$, the corner points of $B(V_k)$ are exactly the corner points of v_k. To obtain a grid drawing, we assume without loss of generality that $l_l(v), l_r(v) \in \mathbb{N}_0$ for all $v \in V$ and both are even.

For vertex representations having an area, as the representation given above, we can define the criterion of *strong visibility* for graph drawing algorithms:

Definition 1 (Strong visibility). *Let $v, w \in V$. Then v is* strongly visible *to w, if any line segment connecting a point within the representation of v to a point within the representation of w does not cross the representation of any other vertex $u \in V$ with $u \neq v, w$.*

Let P_1 and P_2 be two grid points on an integer grid and let $\mu(P_1, P_2)$ be the intersection point of the straight-line segment with slope $+1$ through P_1 and the straight-line segment with slope -1 through P_2. In the algorithm, vertices will be placed according to μ; hence the rotation of vertex representations by 45 degrees. Let $L(v)$ be a set of

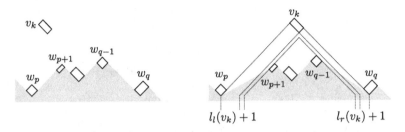

Fig. 2. Installing vertex v_k. Left: G_{k-1}. Right: G_k.

dependent vertices of v, that will later on contain the vertices which have to be rigidly moved with v when v itself is moved.

3 Algorithm

The algorithm starts by drawing G_2. We place $V_1 = \{v_1^1, v_1^2\}$ and V_2 with coordinates $P_r(v_1^1) \leftarrow (0,0)$, $P_l(v_1^2) \leftarrow (l(V_2)+\max\{l_l(v_1^1), l_r(v_1^2)\}+2 \cdot |V_2|, 0)$ and $P_t(B(V_2)) \leftarrow \mu\left(P_r\left(v_1^1\right), P_l\left(v_1^2\right)\right)$. The sets of dependent vertices are initialized with $L(v) \leftarrow \{v\}$ for $v \in G_2$. We proceed by placing the next set V_k in the *lmc*-ordering into G_{k-1}, one by one, starting with V_3. Let $C_0(G_{k-1}) = w_1, \ldots, w_t$, $w_1 = v_1$ and $w_t = v_2$. Assume that following conditions hold for G_{k-1}, $k \geq 3$:

(C1) $x_r(w_i) < x_l(w_{i+1})$, $1 \leq i \leq t - 1$.
(C2) each straight-line segment $(P_r(w_i), P_l(w_{i+1}))$, $1 \leq i \leq t - 1$, has either slope $+1$, 0 or -1.
(C3) every vertex in G_{k-1} is strongly visible to its adjacent vertices in G_{k-1}.

Obviously, these conditions hold for the initial Graph G_2. When inserting V_k, let $w_1, \ldots,$ $w_p, w_{p+1}, \ldots, w_q, \ldots, w_t$ be the vertices on $C_0(G_{k-1})$, where w_p is the leftmost and w_q the rightmost adjacent vertex of V_k in G_{k-1}. Similar to [3,6], install $V_k = \{v_k^1, \ldots, v_k^j\}$ by applying the following steps, see Fig.2.

Step 1. for all $v \in \bigcup_{i=p+1}^{q-1} L(w_i)$ do $x(v) \leftarrow x(v) + l_l(V_k) + |V_k|$
Step 2. for all $v \in \bigcup_{i=q}^{t} L(w_i)$ do $x(v) \leftarrow x(v) + l_l(V_k) + l_r(V_k) + 2 \cdot |V_k| + \Delta$
Step 3. $P_t(B(V_k)) \leftarrow \mu(P_r(w_p), P_l(w_q))$
Step 4. For one j', $1 \leq j' \leq j$ set $L(v_k^{j'}) \leftarrow \left\{v_k^{j'} \cup \left(\bigcup_{i=p+1}^{q-1} L(w_i)\right)\right\}$;
for all other $j'' \neq j'$, $1 \leq j'' \leq j$ set $L(v_k^{j''}) \leftarrow \{v_k^{j''}\}$

Actually, if V_k is not a singleton, the bottom corner of $B(V_k)$ is placed too low by $|V_k| - 1$. Nevertheless, this is sufficient since every vertex in V_k is separated by distance two, and therefore the lowest possible bottom corner of any $v \in V_k$ is at least $|V_k| - 1$ higher than $P_b(B(V_k))$. Assume for the moment that $\Delta = 0$ in step 2. Then all conditions are satisfied for G_k if $\{w_{p+1}, \ldots, w_{q-1}\} \neq \emptyset$, see [8]. However, if there are *no inner vertices* between w_p and w_q on the outer face of G_{k-1}, and $l_l(w_p), l_r(w_q) \neq 0$, condition (C3) is violated in G_k by placing V_k in steps 1 to 4, as w_q is not strongly visible to w_p anymore after insertion. Since step 1 will be omitted in this case, the problem can only be addressed by introducing an extra shift Δ in step 2, thus placing V_k high enough in step 3 such that the strong visibility between w_p and w_q is not violated in G_k. The following Lemma shows how much extra shift is needed, when installing V_k.

Lemma 1. Let $V_k = \{v_k\}$. Let $\{w_{p+1}, \ldots, w_{q-1}\} = \emptyset$ and $l_l(w_p), l_r(w_q) \neq 0$. Then w_p will be strongly visible to w_q in G_k, if an extra shift amount Δ is added in step 2 with

$$\Delta = \begin{cases} \left\lceil 2 \cdot \frac{l_l(w_p) \cdot l_r(w_q)}{l_l(w_p) + l_l(w_q) + l_r(w_q)} \right\rceil & \text{if } [P_r(w_p), P_l(w_q)] \text{ has slope } +1 \text{ in } G_{k-1} \\[2ex] \left\lceil 2 \cdot \frac{l_l(w_p) \cdot l_r(w_q) - 4}{l_l(w_p) + l_r(w_q) + 4} \right\rceil & \text{if } [P_r(w_p), P_l(w_q)] \text{ has slope } 0 \text{ in } G_{k-1} \\[2ex] \left\lceil 2 \cdot \frac{l_l(w_p) \cdot l_r(w_q)}{l_l(w_p) + l_r(w_p) + l_r(w_q)} \right\rceil & \text{if } [P_r(w_p), P_l(w_q)] \text{ has slope } -1 \text{ in } G_{k-1} \end{cases}$$

Proof. Let δ be the height, with which v_k must be lifted upwards to guarantee strong visibility. Assume $[P_r(w_p), P_l(w_q)]$ has slope $+1$ in G_{k-1}, as illustrated in Fig.3 (left). The gray rectangle indicates the position of v_k in G_k without introducing an extra shift. Let $\delta_{pq} = \sqrt{2} \cdot \overline{[P_r(w_p), P_l(w_q)]}$. Observe that δ is largest, if δ_{pq} has the smallest possible value, and that at the same time $\delta_{pq} \geq l_l(w_q)$. Thus, assume $\delta_{pq} = l_l(w_q)$. By the theorem on intersecting lines, we have

$$\frac{\delta}{l_l(w_p)} = \frac{l_r(w_q)}{l_l(w_p) + l_l(w_q) + l_r(w_q)} \quad \Leftrightarrow \quad \delta = \frac{l_l(w_p) \cdot |w_q|r}{l_l(w_p) + l_l(w_q) + l_r(w_q)}$$

It is easy to see that δ is analogous, if the line segment $[P_r(w_p), P_l(w_q)]$ has slope -1 in G_{k-1}. Assume $[P_r(w_p), P_l(w_q)]$ has slope 0 in G_{k-1}, as shown in Fig.3 (right). In this case, $P_r(w_p)$ and $P_l(w_q)$ are separated by a horizontal line segment with length two. Assume that $l_l(w_p) < l_r(w_q)$, then

$$\delta + 1 = \frac{l_l(w_p)}{2} + \frac{l_l(w_p)/2 + 1}{l_l(w_p)/2 + 2 + l_r(w_q)/2} \cdot \frac{l_r(w_q) - l_l(w_p)}{2}$$
$$\Leftrightarrow \quad \delta = \frac{l_l(w_p) \cdot l_r(w_q) - 4}{l_l(w_p) + l_r(w_q) + 4}$$

The same value is obtained, if $l_l(w_p) \geq l_r(w_q)$. Overall, if an extra shift $\Delta = \lceil 2\delta \rceil$ is introduced, v_k is lifted by at least δ, and hence w_p and w_q will be strongly visible to each other in G_k. $\square$

Observe that, if V_k is not a singleton, we have to add $2 \cdot (|V_k| - 1)$ to Δ, since $P_b(B(V_k))$ is $|V_k| - 1$ lower than the bottom corner of a singleton v_k, as indicated in Fig.3. Note also that, if Δ is an odd number, it has to be increased by one to maintain the grid drawing property.

4   Analysis

The following theorems state the bounds for the drawing area of the proposed method, and its time complexity.

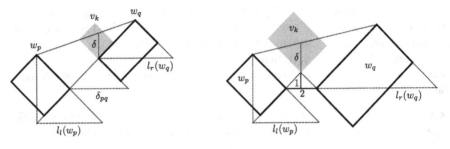

Fig. 3. Geometry for the case $\{w_{p+1}, \ldots, w_{q-1}\} = \emptyset$. Left: slope $+1$. Right: slope 0.

Theorem 1. *The total grid area of a drawing of a triconnected plane graph $G = (V, E)$ with given vertex lengths $l_l(v), l_r(v), v \in V$ produced by the proposed method is in* $O\left(|V| + \sum_{v \in V} l(v)\right)^2$.

Proof. The width of the initial layout of G_2 is clearly bounded by $2 \cdot |V_2| + \Delta_2 + \sum_{i=1}^{2} l(V_i)$, with $\Delta_2 = \max(l_l(v_1^1), l_r(v_1^2))$. Whenever a set V_k is added, the width increases by $2 \cdot |V_k| + \Delta_k + l(V_k)$, where Δ_k denotes the extra shift in step k. Thus, the total width is bounded by $2 \cdot |V| + \sum_{v \in V} l(v) + \sum_{i=2}^{K} \Delta_i$.

Assume that all $V_k, 2 < k \leq K$, are singleton, and that, instead of shifting exactly with $\Delta = \lceil 2\delta \rceil$ when installing V_k, we shift with either $\max(l_l(w_p), l_r(w_q))$ or $\min(l_l(w_p), l_r(w_q))$. If $[P_r(w_p), P_l(w_q)]$ has slope $+1$ in G_{k-1} and

$$
\begin{array}{lll}
1. \quad l_l(w_p) \geq \delta_{pq} + l_r(w_q) & \delta \leq l_l(w_p)/2, & \Delta = l_l(w_p) \\
2. \; l_r(w_q) \leq l_l(w_p) < \delta_{pq} + l_r(w_q) \Rightarrow & \delta \leq l_r(w_q)/2, \; \text{then} & \Delta = l_r(w_q) \\
3. \quad l_r(w_q) \geq l_l(w_p) + \delta_{pq} & \delta \leq l_r(w_q)/2, & \Delta = l_r(w_q) \\
4. \; l_l(w_p) < l_r(w_q) < l_l(w_p) + \delta_{pq} & \delta \leq l_l(w_p)/2, & \Delta = l_l(w_p)
\end{array}
$$

are sufficient to maintain strong visibility. If $[P_r(w_p), P_l(w_q)]$ has slope -1 in G_{k-1}, the bounds are analogous. If $[P_r(w_p), P_l(w_q)]$ has slope 0, δ is bounded by $\max(l_l(w_p), l_r(w_q))/2$, therefore we assume to shift with the maximum length in this case. To find an upper bound for $\sum \Delta$ we use amortized analysis.

Consider the part of $\sum \Delta$ which is contributed due to shifting with the maximum length of $l_l(w_p)$ and $l_r(w_q)$, i.e. cases 1 and 3, and the case where the slope of $[P_r(w_p), P_l(w_q)]$ is 0. It is easy to see that, after one of these cases occured on one side of a vertex v at step k, the length of v on the same side only contributes to another extra shift at step $k' > k$ as the minimum length of the two adjacent vertices of $V_{k'}$. Hence, this part of $\sum \Delta$ is bounded by $\sum_{v \in V} l(v)$.

For determining the part of $\sum \Delta$ which is contributed due to shifting with the minimum length, let each vertex v have two amounts $left(v)$ and $right(v)$, that it can spend to support one extra shift on its left side and one on its right side. Set $left(v) \leftarrow l_r(v)$ and $right(v) \leftarrow l_l(v)$. Let w_p and w_q be the neighbors of V_k on the outer face of G_{k-1} at step k with $\{w_{p+1}, \ldots, w_{q-1}\} = \emptyset$. Assume $[P_r(w_p), P_l(w_q)]$ has slope $+1$ in G_{k-1}. Since in this case w_q was inserted later than w_p, it cannot have spent $left(w_q)$, because otherwise there would be an inner vertex between w_p and w_q on the outer face. If $\min\{l_l(w_p), l_r(w_q)\} = l_r(w_q)$, then w_q pays for the extra shift with $left(w_q)$.

Suppose now that $\min\{l_l(w_p), l_r(w_q)\} = l_l(w_p)$. If w_p has not used $right(w_p)$ so far, then it just pays for the shift. If on the other hand $right(w_p)$ has already been spent (e.g. to insert w_q), then w_q uses $left(w_q) = l_r(w_q) \geq l_l(w_p)$ to pay the extra shift. The payment is analogous if $[P_r(w_p), P_l(w_q)]$ has slope -1 in G_{k-1}. Thus, the total amount of extra shift is sufficiently paid, and this part of $\sum \Delta$ is therefore also bounded by $\sum_{v \in V} l(v)$. The additional amount of extra shift which is contributed, if V_k are not singleton, is clearly bounded by $2 \cdot \sum_{2 \leq i \leq K}(|V_k| - 1) < 2 \cdot |V|$.

Since $G = G_K$ satisfies condition (c2), the height of the drawing is bounded by half of its width plus the part of vertices v_1^1 and v_1^2 beneath the x-axis.   □

If the strong visibility constraint has not to be maintained, the drawing area is exactly $\left(\frac{l(v_1^1)+l(v_1^2)}{2} + 2\omega\right) \times \left(\frac{\max(l_r(v_1^1), l_l(v_1^2))}{2} + \omega\right), \omega = |V|-2+\sum_{i=2}^{K} \frac{l(V_i)}{2}$, since no extra shift is needed in this case. It remains an open problem to give a worst-case scenario and sharp area bound if strong visibility has to be guaranteed.

The linear time implementation of the original shift method [3] can easily be extended to our problem. Since the determination of the extra shift amount takes only constant time, the overall asymptotic complexity is not changed.

Theorem 2. *Given a triconnected plane graph* $G = (V, E), n = |V|$, *the proposed method can be implemented with running time* $O(n)$.

References

1. Barequet, G., Goodrich, M.T., Riley, C.: Drawing planar graphs with large vertices and thick edges. J. Graph Algorithms Appl. 8, 3–20 (2004)
2. Battista, G.D., Eades, P., Tamassia, R., Tollis, I.G.: Graph Drawing: Algorithms for the Visualization of Graphs. Prentice-Hall, Englewood Cliffs (1999)
3. Chrobak, M., Payne, T.H.: A linear-time algorithm for drawing a planar graph on a grid. Information Processing Letters 54(4), 241–246 (1995)
4. de Fraysseix, H., Pach, J., Pollack, R.: How to draw a planar graph on a grid. Combinatorica 10(1), 41–51 (1990)
5. Harel, D., Sardas, M.: An algorithm for straight-line drawing of planar graphs. Algorithmica 20(2), 119–135 (1998)
6. Kant, G.: Drawing planar graphs using the canonical ordering. Algorithmica 16(1), 4–32 (1996)
7. Kaufmann, M., Wagner, D. (eds.): Drawing graphs: Methods and Models. Springer, London (2001)
8. Mader, M., Hong, S.: Drawing planar clustered graphs in 2.5 dimensions. Tech. rep., NICTA (2007), http://www.cs.usyd.edu.au/~visual/valacon/
9. Nishizeki, T., Rahman, S.: Planar Graph Drawing. Lecture Note Series on Computing, vol. 12. World Scientific, Singapore (2004)

Placing Text Boxes on Graphs*

A Fast Approximation Algorithm for Maximizing Overlap of a Square and a Simple Polygon

Sjoerd van Hagen and Marc van Kreveld

Department of Information and Computing Sciences
Utrecht University, The Netherlands

Abstract. In this paper we consider the problem of placing a unit square on a face of a drawn graph bounded by n vertices such that the area of overlap is maximized. Exact algorithms are known that solve this problem in $O(n^2)$ time. We present an approximation algorithm that—for any given $\epsilon > 0$—places a $(1+\epsilon)$-square on the face such that the area of overlap is at least the area of overlap of a unit square in an optimal placement. The algorithm runs in $O(\frac{1}{\epsilon} n \log^2 n)$ time. Extensions of the algorithm solve the problem for unit discs, using $O(\frac{\log(1/\epsilon)}{\epsilon\sqrt{\epsilon}} n \log^2 n)$ time, and for bounded aspect ratio rectangles of unit area, using $O(\frac{1}{\epsilon^2} n \log^2 n)$ time.

1 Introduction

The annotation of drawn graphs comes in different forms. Vertices can be labeled with their name or index, edges may be labeled with extra information, or faces of the embedded graph may receive a label. The analogy with cartographic label placement is clear: Here we have point feature labels, linear feature labels, and areal feature labels. Areal features are for instance lakes, national parks, provinces, and countries.

A related cartographic question is that of annotating regions of a map with extra information instead of names. These can be text boxes, pie charts, histograms, or other diagrams that show statistics about that region. If the annotation does not fit inside the region, it must obviously overlap parts of other regions. To achieve the best possible association of the correct region and the annotation, it is desirable to have the largest possible overlap in area of the annotation and that region. A possible positive side effect is that not too much of the region boundaries is covered by the annotation, and if more regions are annotated, that their annotations usually do not overlap.

One can abstract an annotation by a rectangle, square, or circle of some given size, which represents the bounding shape of the annotation. A region on a map is typically a simple polygon (although sometimes it has holes). The algorithmic

---

* This research is partially funded by the Netherlands Organisation for Scientific Research (NWO) under BRICKS/FOCUS grant number 642.065.503.

I.G. Tollis and M. Patrignani (Eds.): GD 2008, LNCS 5417, pp. 284–295, 2009.

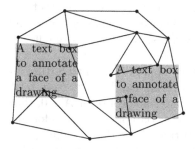
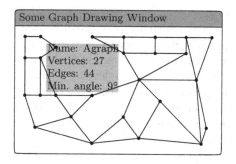

Fig. 1. Left, annotation of two faces by text boxes. Right, annotation in the outer face of a drawn graph.

question that arises is: How do we compute the placement of a simple shape with a simple polygon to maximize area of overlap efficiently? Van Kreveld *et al.* [16] studied this problem (along with some related problems) in the context of placing diagrams on maps. It was shown that the maximum overlap placement of a unit square on a simple polygon P with n vertices can be computed in $O(n^2)$ time. It was also shown that the placement problem with one degree of freedom—for example, the y-coordinate of the top of the unit square is fixed—can be solved in $O(n \log n)$ time.

The faces of a drawn graph are also simple polygons, and the annotation of a face is the same problem as the annotation of a region on a map. Hence, the problem we address in this paper is motivated by both automated cartography and graph drawing. Figure 1 gives two examples where faces are annotated, and maximizing overlap with a face appears reasonable for the best text box positioning. Annotation—or label placement—has been studied in the context of graph drawing various times, see for instance [6,7,14,19].

If one considers a quadratic time solution to the area of overlap maximization problem to be too slow, there are several approaches to deal with this. Firstly, one can argue that faces in typical graphs do not have large complexity, so an algorithm that takes time quadratic in the number of vertices of the face is no problem. In some cases this is obviously true, like drawings of triangulated graphs. In other cases it is not true, like drawings of trees with a few additional edges or other sparse planar graphs.

Secondly, one can make realistic input assumptions that allows one to show that for inputs satisfying those assumptions, a provably more efficient solution exists. This idea has led to a large body of research in computational geometry. For our problem, this idea does not seem to work. For standard definitions of realistic input polygons [8,20,15], the so-called *placement space* of a unit square remains combinatorially quadratic in size.

Thirdly, one can use approximation. For example, one could try to find the unit square placement that has area overlap with P of at least $c \cdot A$, for some fixed $c \leq 1$, where A is the area of overlap of the optimal placement. Then we have a c-approximation algorithm (which is the optimal algorithm if $c = 1$).

An approximation scheme is an algorithm that, for any $\epsilon > 0$, computes a placement with area overlap of at least $(1 - \epsilon) \cdot A$. Approximation algorithms and approximation schemes only make sense if they are significantly more efficient than the corresponding exact algorithm.

It appears hard to develop a subquadratic time approximation algorithm for our problem, due to the fact that the optimal overlap can be very close to 0. However, we can show that if the overlap is at least some constant $\hat{A} > 0$, then we can compute a placement that guarantees an overlap of $(1 - \epsilon) \cdot \hat{A}$, for any fixed $\epsilon > 0$. The algorithm runs in $O(\frac{1}{\epsilon} n \log^2 n)$ time. Note that assuming that the area of overlap is at least a constant is in a sense a realistic assumption: For instances where the optimal overlap is very small, the solution to the problem is not suitable for a good annotation anyway.

We solve our problem via a detour. We show that for any fixed $\delta > 0$, we can compute a placement of a square of size $1 + \delta$ whose area of overlap with the simple polygon P is at least A_{opt}, where A_{opt} is the maximum area of overlap that can be achieved for the placement of a unit square. When we shrink the $(1 + \delta)$-square to a unit square, we can lose an area of overlap of at most $2\delta + \delta^2$. Hence, given $\epsilon > 0$ and $\hat{A} > 0$, we choose $\delta = \hat{A} \cdot \epsilon / 3$ and compute a placement of a $(1 + \delta)$-square with the algorithm we present in this paper. Any unit square inside the $(1 + \delta)$-square we found will have area of overlap at least $(1 - \epsilon) \cdot A_{\text{opt}}$, so this implies a $(1 - \epsilon)$-approximation algorithm.

Our algorithm can be extended to compute the placement of a unit disc with maximum area of overlap approximately in $O(\frac{\log(1/\epsilon)}{\epsilon\sqrt{\epsilon}} n \log^2 n)$ time, again assuming that the area of overlap is at least some constant. We note that for this case, no exact algorithm exists at all, due to the algebraic complexity of maximizing the analytic form of the area-of-overlap function. The algorithm can also be extended to place a unit area rectangle with bounded aspect ratio in $O(\frac{1}{\epsilon^2} n \log^2 n)$ time (for rectangles with *fixed* aspect ratio one can use the algorithm for squares after scaling). This can be useful for *elastic labels*, an abstraction for text boxes of a fixed length text where the width of the text box is also free proposed by Iturriaga and Lubiw [12,13].

In computational geometry, there is a large body of research on optimal matching of two shapes [21]. One measure for similarity is the area of overlap, and hence, research has been done on maximizing this measure under various transformations. For translations only, Mount *et al.* [17] gave a $O((nm)^2)$ time algorithm for the maximum overlap of a simple n-gon and a simple m-gon. For two convex polygons, an $((n + m) \log(n + m))$ time algorithm exists [2].

There are also several papers that use approximation to find a unit square or disc that covers the maximum number of points of a given point set [9,10]. A main difference with our problem is that we optimize a (continuous) area measure instead of a (discrete) point count measure. Other related research is on finding a largest area rectangle inside a simple polygon, for which Daniels *et al.* [5] give an $O(n \log^2 n)$ time algorithm, and finding the largest similar copy of a convex polygon inside a simple polygon [1]. This would correspond to scaling such that the annotation just fits inside the face. For text boxes this implies changing the

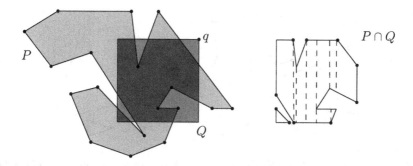

Fig. 2. Example of a simple polygon P intersecting a square Q; the area of intersection can be decomposed into trapezoids

font size, which may not be desirable. Finally, there are many papers on the topic of label placement in the algorithms and automated cartography research fields, but it is beyond the scope of this paper to review them.

We start with a brief description of the quadratic time exact algorithm from [17,16] in Sect. 2 since we will need ideas from it. In Sect. 3 we present the approximation algorithm. We first give a version whose running time is $O(\frac{1}{\epsilon} n \log^3 n)$. Then we show how to use Jordan sorting to improve the total running time to $O(\frac{1}{\epsilon} n \log^2 n)$. We present the extension for circles and bounded aspect ratio rectangles in Sect. 4. Concluding remarks are given in Sect. 5.

2   An Exact Quadratic-Time Algorithm

In this section we sketch the approach from [17,16] to compute an exact solution to the maximum overlap placement of a unit square on a simple polygon. It is based on the fact that there are quadratically many combinatorially distinct placements of a square Q on a simple polygon P.

The combinatorially distinct placements of Q on P are described by the different pairs of edges—one from Q and one from P—that can intersect. We use the top right corner of Q as a reference point q to characterize the possible positions of Q. When the pairs of intersecting edges are fixed, the reference point still has a little freedom to move, see Fig. 2. As long as the intersecting edges of P and Q remain the same, changing the position of q will change the area of $P \cap Q$, but in a prescribed manner. We can express the area of $P \cap Q$ as a quadratic function in the x- and y-coordinates of the reference point q. Specifically, it has the form:

$$ax^2 + bxy + cy^2 + dx + ey + f .$$

This is true because the overlap can be decomposed into a set of trapezoids whose vertices change linearly in x and y, see Fig. 2. Therefore the area changes as a quadratic function. If Q were a circle, the area-of-overlap function would have a non-constant description involving square roots, and maximizing the area of overlap would not be possible exactly.

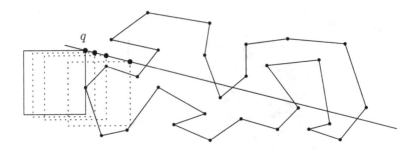

Fig. 3. The 1-dimensional problem of placing a unit square to maximize area of overlap. The first four positions where the area-of-overlap function changes are shown.

Suppose the reference point q and therefore the square Q translates in the plane. The quadratic function giving the area of overlap stops to be valid when the pairs of edges of Q and P that intersect change. Then a different quadratic function will describe the area of overlap, that is, the coefficients a, b, c, d, e, f are different. This happens when:

- an edge of Q passes over a vertex of P, or
- a vertex of Q passes over an edge of P.

Let Π be the subdivision obtained from all positions of q where an edge of Q coincides with a vertex of P, or vice versa. Π is called the placement space of Q with respect to P. In each face of Π, some fixed quadratic function describes the area of overlap of Q and P.

Theorem 1. (Adapted from [17,16]) *Given a simple polygon P with n edges and a square Q, the placement space of Q with respect to P can be constructed in $O(n \log n + N)$ time, where $N = O(n^2)$ is the number of combinatorially distinct placements.*

It can also be shown that the quadratic function that is valid in each cell of Π can be computed in quadratic time by a suitable traversal of the cells of Π. Given Π and the quadratic function for each cell, we can compute the placement of Q that maximizes the area of overlap in $O(N) = O(n^2)$ time.

In case we are only interested in square placements where the reference point is restricted to lie on a given line, the placement space is 1-dimensional and there are only $O(n)$ combinatorially distinct placements, see Fig. 3. The optimal placement can now be solved by a sweep of the square with its reference point on the line, and updating the quadratic function. Since we must sort the $O(n)$ events where the quadratic function changes, this takes $O(n \log n)$ time.

3   An Approximation Algorithm

In this section we compute a placement of a $(1 + \epsilon)$-square on a simple polygon P with n vertices so that the area of overlap is at least the maximum area of

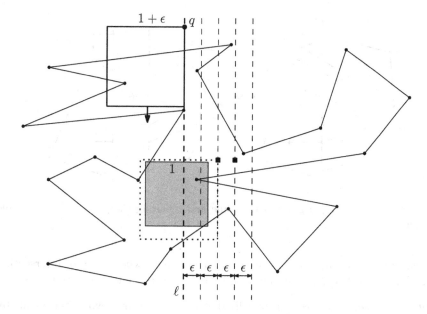

Fig. 4. If the optimal unit square intersects the split line ℓ, then one of the $1/\epsilon$ sweeps will consider $(1 + \epsilon)$-squares (like the one shown dotted) that contains it

overlap of a unit square and P. The solution is based on divide-and-conquer, where dividing gives rise to running the 1-dimensional exact algorithm for a $(1 + \epsilon)$-square a number of times. The algorithm will at some point consider a $(1 + \epsilon)$-square that contains the unit square in its optimal placement. This leads to the desired approximation guarantee.

Divide-and-conquer. The divide-and-conquer algorithm chooses vertical lines to partition the simple polygon into pieces. This will give vertical slabs in the plane. Every split will guarantee that the number of vertices in the interior of the slab is at least halved. So we determine the median of the x-coordinates of the vertices and choose a vertical line ℓ through this vertex.

We would like to create two subpolygons and recurse on them, but it may be that the optimal unit square intersects ℓ, and we must take this possibility into account. This is done by running the 1-dimensional algorithm $1+1/\epsilon$ times.[1] The 1-dimensional algorithm is run with the reference point q of the $(1 + \epsilon)$-square on ℓ, and on vertical lines at distances $\epsilon, 2\epsilon, 3\epsilon, \ldots, 1 + \epsilon$ to the right of ℓ, see Fig. 4. We observe:

Observation 1. *If the optimal unit square intersects ℓ, then at least one of the $1 + 1/\epsilon$ sweeps with a $(1 + \epsilon)$-square will give a position where a $(1 + \epsilon)$-square contains the optimal unit square.*

---

[1] With slight abuse of notation, we assume that $1 + 1/\epsilon$ is an integer, but technically we should use rounding. Asymptotically the running time is not affected.

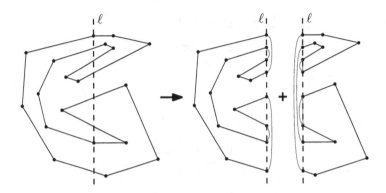

Fig. 5. Splitting a simple polygon by a line and repairing the parts into two simple polygons

Hence, in $O(\frac{1}{\epsilon} n \log n)$ time, we find a $(1 + \epsilon)$-square whose area of intersection with P is at least as large as the optimal unit square that intersects ℓ.

Splitting the polygon. Now we can divide the problem into two subproblems by splitting the simple polygon. We may split the polygon into more than two parts, but we can repair the situation while using only vertices on the splitting line, see Fig. 5. The resulting polygons may have edges that coincide on the split lines, but this degeneracy does not influence the algorithm. It is standard to perform such a split and repair in $O(n \log n)$ time. Observe that the number of vertices interior to each of the resulting slabs is at least halved. The divide-and-conquer algorithm will find a $(1 + \epsilon)$-square strictly left of ℓ recursively, a $(1 + \epsilon)$-square strictly right of ℓ recursively, and a $(1 + \epsilon)$-square that intersects ℓ. The one with largest area of overlap with P is returned.

Recall that a 1-dimensional sweep has two types of events: An edge of Q passes over a vertex of P, or a vertex of Q passes over an edge of P. In a slab, a subpolygon of P has interior vertices and boundary vertices. We call an edge of P that intersects a slab *short* if it has an endpoint that is an interior vertex, and we call it *long* if both endpoints are boundary vertices. Long edges cross the slab completely. The divide-and-conquer algorithm takes care of halving the number of interior vertices, and therefore the number of short edges is bounded as well. But the number of long edges can become large, and these also give rise to events in the 1-dimensional sweeps. Ultimately, the divide-and-conquer algorithm bottoms out when a slab has no more interior vertices, or when its width is at most unit. The final $O(n)$ slabs may all be crossed by a linear number of edges, leading to an algorithm that takes at least quadratic time in the worst case.

Free splits. To control the number of long edges in recursive subproblems, we use the concept of *free splits*, introduced by Patterson and Yao to prove bounds on the size of binary space partitions [18]. We will take measures to eliminate

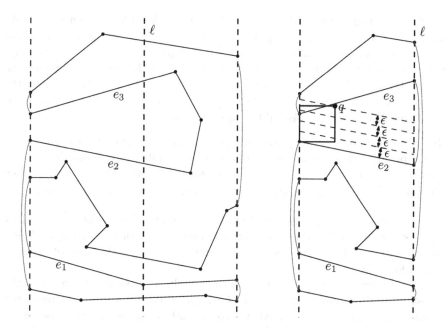

Fig. 6. After splitting along ℓ, the free split along e_2 is done in the left part, and then free splits along e_1 in the bottom left polygon and along e_3 in the top left polygon. Two free splits are performed in the right part as well. In the right figure, the sweeps are shown by dashed segments parallel to e_2.

long edges when they appear after a split with a line ℓ, so that there are no long edges when we choose a next split line.

Let S be a slab with no long edges. We determine the median x-coordinate of the interior vertices, which defines a vertical split line ℓ, and perform $1 + 1/\epsilon$ sweeps as described above. Then we split the simple polygon P into two polygons P_{left} and P_{right}, as described above as well. For each resulting polygon, say, P_{left}, we determine the long edges $e_1, \ldots, e_k$ from bottom to top, see Fig. 6. We use these edges to partition P_{left} further, also in a divide-and-conquer fashion. So we select $e_{\lceil k/2 \rceil}$, perform a number of sweeps parallel to this edge and then split P_{left} at $e_{\lceil k/2 \rceil}$ into two subpolygons that are handled recursively. Since the diameter of a unit square is $\sqrt{2}$, we must now perform up to $\sqrt{2}/\epsilon$ sweeps with q on lines that are a distance ϵ apart, see Fig. 6. Free splits are performed in the same way as splits along vertical lines; no additional cases occur.

Running time. To prove an upper bound on the running time of the algorithm, we first observe that the number of free splits is $O(n \log n)$ throughout the whole algorithm. This is standard; see for instance Chapter 10 in [3], or [4,18].

Lemma 1. *Assume that a slab contains m interior vertices and no long edges. The time needed to perform the vertical split and all necessary free splits is $O(\frac{1}{\epsilon} m \log^2 m)$, including the time for the sweeps.*

Proof. We already argued that the vertical split takes $O(\frac{1}{\epsilon} m \log m)$ time. We may have created $O(m)$ long edges while doing this. For each free split, each of the $O(m)$ interior and boundary vertices appears in only one of the two new polygons that is created. Each vertex creates an event in $O(\frac{1}{\epsilon} \log m)$ 1-dimensional sweeps, because the recursion depth of the free splits is $O(\log m)$. Hence, the $O(\frac{1}{\epsilon} m)$ sweeps due to free splits encounter $O(\frac{1}{\epsilon} m \log m)$ events together, leading to $O(\frac{1}{\epsilon} m \log^2 m)$ time for all free splits.   □

The lemma can be used to prove an $O(\frac{1}{\epsilon} n \log^3 n)$ time bound for the approximation problem: The recurrence that describes the efficiency of the algorithm is given by $T(n) = 2 \cdot T(n/2) + O(\frac{1}{\epsilon} n \log^2 n)$ for $n > 1$ (and $T(1) = O(1)$), which solves to $O(\frac{1}{\epsilon} n \log^3 n)$ time. However, we can remove a logarithmic factor.

Jordan sorting. We next improve the overall running time to $O(\frac{1}{\epsilon} n \log^2 n)$ by applying Jordan sorting to the 1-dimensional problems. Jordan sorting is a linear time algorithm which, given a simple polygon and a line, sorts their intersection points along the line [11].

The 1-dimensional sweep algorithm to find the $(1 + \epsilon)$-square that has the largest area of overlap with P takes $O(n \log n)$ time due to the sorting of the events. If the events were sorted, we could update the quadratic function in constant time because at most three trapezoids can appear or disappear during an event. We only have to perform some simple additions to the coefficients a, b, c, d, e, and f based on these changed trapezoids to get the new quadratic function that is valid. This fact was already used in [16] to generate the placement space with all quadratic functions.

To obtain a sorted list of events, recall that there are two types of events: an edge of Q crosses an vertex of P, and a vertex of Q passes a edge of P. The former type will be obtained in sorted order by pre-sorting and maintaining two sorted lists, the second type by Jordan sorting.

For the first type, we perform preprocessing for the algorithm by sorting all vertices of P by x-coordinate into a list L_x, and also by y-coordinate into a list L_y. Whenever we perform a split, by a vertical line or a free split, we traverse each list and generate two new sorted lists for the two subproblems that appear. This will take time linear in the length of the list, which is linear in the number vertices in the slab or trapezoid that is split.

For the second type, we compute the event just before performing the 1-dimensional sweep. We perform Jordan sorting four times, once for each path of a vertex of Q; this path is a line segment. We merge these sorted lists into one, and also merge them with the events of the first type. In total, we need six list merges to obtain all events in sorted order. Hence, a 1-dimensional sweep can be performed in linear time.

Summarizing the results of this section, we conclude:

Theorem 2. *Given a simple polygon P with n vertices and a constant $\epsilon > 0$, an $O(\frac{1}{\epsilon} n \log^2 n)$ time algorithm exists that computes a placement of an axis-aligned square with side length $1 + \epsilon$ of which the area of overlap with P is at least the area of overlap of any axis-aligned unit square with P.*

4   Extensions to Circles and Rectangles

Suppose we wish to place a circular annotation on a region of a map or a face drawn graph, like a pie chart. We can adapt the approximation algorithm given in previous section to this case. The idea is to choose a suitable integer k, and adapt the algorithm for squares to work for regular k-gons.

We will use a regular k-gon that is inside a diameter $(1 + \epsilon)$-disc but outside a $(1 + \frac{\epsilon}{2})$-disc. Instead of performing 1-dimensional sweeps with a distance ϵ in between, we must use a distance of $\epsilon/2$ in between. Finally, we need to maintain k sorted lists that give the order in which the edges of the regular k-gon will cross the vertices of subpolygons of P. The extensions are straightforward.

It is well known that the choice $k = \Theta(1/\sqrt{\epsilon})$ satisfies our requirement of approximating a disc well enough. Following the analysis for the square case, we notice that merging k sorted lists with $O(nk)$ events in total takes $O(nk \log k)$ time. Hence, we conclude:

Theorem 3. *Given a simple polygon P with n vertices and a constant $\epsilon > 0$, an $O(\frac{\log(1/\epsilon)}{\epsilon\sqrt{\epsilon}} n \log^2 n)$ time algorithm exists that computes a placement of a disc with diameter $1 + \epsilon$ of which the area of overlap with P is at least the area of overlap of any unit disc with P.*

Next we discuss the extension to placing an axis-aligned rectangle with aspect ratio $r : 1$ or less and unit area (assuming $r \geq 1$). If the aspect ratio were fixed, we could simply scale the input so that the problem reduces to placing an axis-aligned square. Our algorithm will test a fixed number of aspect ratios (depending on ϵ and r), scale the input appropriately, and run the square placement algorithm. We will asume that $r = O(1)$ since this will be true in any practical context.

Suppose that the optimal unit area rectangle is R_{opt}. Then we wish to find a rectangle with area at most $1 + \epsilon$, aspect ratio at most $r : 1$, and that has area of overlap with P that is at least as much as the area of overlap of R_{opt}. We must make sure to that our algorithm tries some rectangle during a sweep that contains the optimal rectangle R_{opt}.

We will try the following rectangle widths (or heights): $(1 + \frac{3\epsilon}{5r}), (1 + \frac{4\epsilon}{5r}), (1 + \frac{\epsilon}{r}), (1 + \frac{6\epsilon}{5r}), \ldots$ and the corresponding heights (resp. widths) to get an area of $1 + \epsilon$; these corresponding heights (resp. widths) increase by less than $\epsilon/(5r)$. We continue until the first value greater than $\sqrt{(1 + \epsilon)r}$.

One of the rectangles we try will be larger by $\epsilon/(5r)$ in height and width than R_{opt} but at most larger by $2\epsilon/(5r)$. If the optimal rectangle R_{opt} has size $h \times (1/h)$, then $1 \leq h \leq \sqrt{r}$, and a rectangle of size $(h + \frac{2\epsilon}{5r}) \times ((1/h) + \frac{2\epsilon}{5r})$ has area less than $1 + \epsilon$.

If we run the 1-dimensional sweeps with lines at distance $\epsilon/(5r)$ in between, then we will encounter a rectangle with the desired properties that contains R_{opt}. Since r is assumed to be constant, we run the algorithm for squares $O(1/\epsilon)$ times. We conclude:

Theorem 4. *Given a simple polygon P with n vertices, a constant $\epsilon > 0$, and a value $r \geq 1$, an $O(\frac{1}{\epsilon^2} n \log^2 n)$ time algorithm exists that computes a placement* *of an axis-aligned rectangle with area $1 + \epsilon$ and aspect ratio bounded by $r : 1$ of which the area of overlap with P is at least the area of overlap of any axis-aligned unit area rectangle and aspect ratio bounded by $r : 1$ (assuming $r = O(1)$).*

5 Concluding Remarks

We have studied the problem of annotating a region on a map or a face of a drawn graph by a square, circular, or rectangular shape while maximizing the area of overlap. This will give a good association between the face and the shape, may avoid unnecessary covering of edges by the annotation, and if more faces are annotated, may help to avoid overlap of different annotations. It was known that the problem can be solved in $O(n^2)$ time if the face has n boundary vertices. We showed that a placement of a shape that is larger by a factor $1 + \epsilon$ can be found in $O(n \log^2 n)$ time that has at least the area of overlap of the optimal placement of the original shape (ignoring factors depending on the constant $\epsilon > 0$).

With the same approach, we can compute a placement whose length of overlap with the boundary of the face is minimized. For this problem to make sense, we must restrict the space of all placements somehow, otherwise the optimum can lie fully outside the face. We can for instance require that the center of the shape lies inside the proper face. For each combinatorially distinct placement, the length of overlap changes linearly in the coordinates of the reference point, but otherwise, the solution approach is the same, and we get the same running time bounds.

The main open problems are improvements in the running time. Firstly, we suspect that it must be possible to remove one log-factor from the running time, but it is even conceivable that both log-factors can be removed. For the disc and rectangle versions, we may be able to improve the dependency on ϵ, or generalize to rectangles of unbounded aspect ratio.

Acknowledgements

The authors thank Sariel Har-Peled for sharing his ideas that lead to the solution.

References

1. Baker, B., Fortune, S., Mahaney, S.: Polygon containment under translation. J. Algorithms 7(4), 532–548 (1986)
2. de Berg, M., Cheong, O., Devillers, O., van Kreveld, M., Teillaud, M.: Computing the maximum overlap of two convex polygons under translations. Theory Comput. Syst. 31(5), 613–628 (1998)
3. de Berg, M., Cheong, O., van Kreveld, M., Overmars, M.: Computational Geometry – Algorithms and Applications, 3rd edn. Springer, Berlin (2008)
4. Chazelle, B., Edelsbrunner, H., Guibas, L., Sharir, M.: Algorithms for bichromatic line-segment problems and polyhedral terrains. Algorithmica 11(2), 116–132 (1994)

5. Daniels, K., Milenkovic, V., Roth, D.: Finding the largest area axis-parallel rectangle in a polygon. Comput. Geom. Theory Appl. 7, 125–148 (1997)
6. Dogrusöz, U., Feng, Q.W., Madden, B., Doorley, M., Frick, A.: Graph visualization toolkits. IEEE Computer Graphics and Appl. 22(1), 30–37 (2002)
7. Dogrusöz, U., Kakoulis, K., Madden, B., Tollis, I.: On labeling in graph visualization. Inf. Sci. 177(12), 2459–2472 (2007)
8. Efrat, A., Sharir, M.: On the complexity of the union of fat convex objects in the plane. Discr. & Comput. Geometry 23(2), 171–189 (2000)
9. Gudmundsson, J., van Kreveld, M., Speckmann, B.: Efficient detection of patterns in 2D trajectories of moving points. GeoInformatica 11, 195–215 (2007)
10. Har-Peled, S., Mazumdar, S.: Fast algorithms for computing the smallest k-enclosing circle. Algorithmica 41(3), 147–157 (2005)
11. Hoffman, K., Mehlhorn, K., Rosenstiehl, P., Tarjan, R.: Sorting jordan sequences in linear time using level-linked search trees. Information and Control 68(1–3), 170–184 (1986)
12. Iturriaga, C., Lubiw, A.: Elastic labels: the two-axis case. In: DiBattista, G. (ed.) GD 1997. LNCS, vol. 1353, pp. 181–192. Springer, Heidelberg (1997)
13. Iturriaga, C., Lubiw, A.: Elastic labels around the perimeter of a map. J. Algorithms 47(1), 14–39 (2003)
14. Kakoulis, K., Tollis, I.: A unified approach to automatic label placement. Int. J. Comput. Geometry Appl. 13(1), 23–60 (2003)
15. van Kreveld, M.: On fat partitioning, fat covering, and the union size of polygons. Comput. Geom. Theory Appl. 9, 197–210 (1998)
16. van Kreveld, M., Schramm, E., Wolff, A.: Algorithms for the placement of diagrams on maps. In: GIS 2004: Proc. 12th annu. ACM Int. Symp. on Advances in Geographic Information Systems, pp. 222–231. ACM Press, New York (2004)
17. Mount, D., Silverman, R., Wu, A.: On the area of overlap of translated polygons. Computer Vision and Image Understanding 64(1), 53–61 (1996)
18. Paterson, M., Yao, F.: Binary partitions with applications to hidden surface removal and solid modelling. In: Proc. 5th annual ACM Symp. on Computational Geometry, pp. 23–32 (1989)
19. Ryall, K., Marks, J., Shieber, S.: An interactive system for drawing graphs. In: North, S.C. (ed.) GD 1996. LNCS, vol. 1190, pp. 387–394. Springer, Heidelberg (1997)
20. van der Stappen, A.: Motion Planning amidst Fat Obstacles. Ph.D. thesis, Department of Computer Science, Utrecht University (1994)
21. Veltkamp, R., Hagedoorn, M.: State of the art in shape matching. In: Lew, M. (ed.) Principles of Visual Information Retrieval, pp. 87–119. Springer, Heidelberg (2000)

Removing Node Overlaps Using Multi-sphere Scheme[*]

Takashi Imamichi[1], Yohei Arahori[1], Jaeseong Gim[1], Seok-Hee Hong[2], and Hiroshi Nagamochi[1]

[1] Department of Applied Mathematics and Physics, Kyoto University
{ima,arahori,jaeseong,nag}@amp.i.kyoto-u.ac.jp
[2] School of Information Technologies, University of Sydney
shhong@it.usyd.edu.au

Abstract. In this paper, we consider the problem of removing overlaps of labels in a given layout by changing locations of some of the overlapping labels, and present a new method for the problem based on a packing approach, called *multi-sphere scheme*. More specifically, we study two *new* variations of the label overlap problem, inspired by real world applications, and provide a solution to each problem. Our new approach is very *flexible* to support various operations such as translation, translation with direction *constraints*, and *rotation*. Further, our method can support labels with *arbitrary shapes* in both 2D and *3D layout* settings. Our extensive experimental results show that our new approach is very effective for removing label overlaps.

1 Introduction

Graph Drawing has been extensively studied over the last twenty years due to its popular application for visualization in VLSI layout, computer networks, software engineering, social networks and bioinformatics. As a result, many algorithms and method are available. Note that most algorithms in Graph Drawing deal with abstract graph layout, where each node is represented as a point. However, in many real world applications, nodes may have labels with different size and shape. For example, some nodes have very long text labels or large images, and they can be represented as boxes or circles as in UML diagrams. Consequently, direct use of layout algorithm for abstract graph often leads to overlapping of nodes (i.e. labels) in the resulting visualization.

In order to visualize graphs with different node sizes, the following three steps approach is used in general: (1) a reasonably good initial layout is created using a graph layout algorithm without considering node size; (2) labels of nodes are added in the layout; (3) the post processing step to remove node overlapping is performed. The problem of removing node overlaps has been well studied

---

[*] This research was partially supported by Research Fellowships of the Japan Society for the Promotion of Science for Young Scientists and a Scientific Grant in Aid from the Ministry of Education, Science, Sports and Culture of Japan.

I.G. Tollis and M. Patrignani (Eds.): GD 2008, LNCS 5417, pp. 296–301, 2009.

by the Graph Drawing community. These can be classified into three different approaches: force-directed method [5,6,9,11], Voronoi Diagram method [5,11], and constrained optimization method [4,12]. Further, they differ in their optimization criteria considered. The variations of Force Scan algorithm based on the force-directed method [6,9] preserves *orthogonal ordering*, the top-down and left-right relationship between nodes. Note that the problem of transforming a given layout of a graph with overlapping rectangular nodes into a *minimum area* layout without node overlapping which preserves the orthogonal order is proved as NP-complete [6]. The constrained optimization techniques using a quadratic programming approach minimizes the *total change* of node positions while satisfying non-overlap constraints [4,12]. Note that most of the methods solve the problem of overlap removal of *rectangular* labels with *translation* only.

We present a new method for removing overlap of labels based on *multi-sphere scheme* [8], a general algorithmic framework for solving the problem of *packing* objects both in two and three dimensions. Based on this scheme, each label in a given layout is approximated by a set of circles or spheres and a penalty function of the overlap between two labels is introduced. By minimizing the penalty function using a quasi-Newton method [10], we compute a layout of the set of circles or spheres as an approximate solution to the original problem. Our new approach is very *flexible*, and has the following three advantages over previous work:

1. Our approach can handle labels with *arbitrary shapes*. Note that previous methods can deal with only rectangular labels. However, in our approach, we can treat any non-rectangular-shaped labels by approximating each of them as a set of circles. We can also place given labels inside a specified area with a non-rectangular boundary.

2. Our algorithm can use three types of operation: *translation, translation with direction constraints* (i.e. move along the specified line), and *rotation*. Note that the previous methods deal with only translation.

3. Our method can be used for *both 2D and 3D layouts*. Note that previous study can only deal with 2D layout.

In order to demonstrate our three advantages, we consider two new variations of the label overlap problem, each inspired by real world applications, and design an algorithm for each problem setting. More specifically, we consider following two types of label overlap removal problems.

Problem 1: Rectangular Label with Direction-Constrained Translation
Input: A set of overlapping rectangles, where each rectangle is located on its initial position with a specified direction constraints (i.e. a line segment) in the plane.
Output: A set of new positions of rectangles such that no two rectangles overlap and the change of new positions from the initial positions is small, where the new position of each rectangle is obtained by restricted translation along the specified direction only.

Problem 2: 3D Multi-attribute Label with Translation and Rotation
Input: A set of overlapping spiked sphere (i.e. a sphere with several small cones on its surface), where each spiked sphere is located on its initial position in the 3D space.
Output: A set of new positions of spiked spheres in 3D such that no two spiked spheres overlap and the change of new positions from the initial positions is small, where the new position of each spiked sphere is obtained by both translation and rotation in 3D.

Problem 1 appears in applications such as placing labels of street names in a road map layout [2]. Problem 2 appears in applications such as visualization of network data with multiple attributes in three dimensions. For example, a spiked sphere was used to represent an author of the Information Visualization community, where each sphere represents an author, the size of sphere represents the number of research papers published by the author in the conference proceedings, and the length of each spike attached to the sphere represents special attributes such as the number of papers in specific research area [3]. We implemented our algorithm and evaluated with two different types of data sets. Our extensive experimental results show that our new approach is very fast and effective for removing label overlaps. For the full version of this paper, see [7].

2   Algorithm Based on Multi-sphere Scheme

In the multi-sphere scheme, we first approximate each label by a set of spheres, and then search for positions of all the spheres that minimize an appropriate penalty function. Approximating objects by spheres makes it easy to check collisions of objects and handle rotations of objects by arbitrary angles.

To find a layout of sets of spheres, we formulate *penalized rigid sphere set packing problem* of as an unconstrained optimization program and apply an algorithm RIGIDQN, which moves the labels simultaneously and modifies the entire layout gradually. Given an initial layout of labels, where the labels are approximated by sphere sets, RIGIDQN returns a locally optimal layout computed by applying the quasi-Newton method to the penalized rigid sphere set packing problem. Although we do not use an explicit criteria to minimize the total change between the initial and final layouts, RIGIDQN obtains the final positions of labels are close to the initial positions in most cases because RIGIDQN moves sphere sets gradually.

We formulate the penalized rigid sphere set packing problem for $\mathbb{R}^d$, which asks to move a collection $\mathcal{O} = \{O_1, \ldots, O_m\}$ of m objects so that no two objects overlap each other. Each object O_i consists of n_i spheres $\{S_{i1}, \ldots, S_{in_i}\}$. Let c_{ij} be the vector that represents the center of spheres S_{ij}, r_{ij} be the radius of S_{ij} and $N = \sum_{i=1}^{m} n_i$. We let $r_i = \sum_{j=1}^{n_i} c_{ij}/n_i$, which represents the center of O_i. For a set S of points, let ∂S be the boundary of S, and $\text{int}(S) = S \setminus \partial S$ be the interior of S. After translating object O by a translation vector $v \in \mathbb{R}^d$, the resulting object is described as $O \oplus v = \{x + v \mid x \in O\}$. The *penetration depth* [1] of two shapes S and T is defined by $\delta(S,T) = \min\{\|\boldsymbol{x}\| \mid \text{int}(S) \cap (T \oplus \boldsymbol{x}) = \emptyset, \boldsymbol{x} \in \mathbb{R}^d\}$, where $\|\cdot\|$ denotes the Euclidean norm. Let $\Lambda_i(\boldsymbol{x}, \boldsymbol{v}) : \mathbb{R}^{d \times \lambda_i} \to \mathbb{R}^d$ $(i = 1, \ldots, m)$ be a motion function that moves a point $\boldsymbol{x} \in \mathbb{R}^d$ by λ_i variables $\boldsymbol{v} \in \mathbb{R}^{\lambda_i}$. For a set of points $S \subseteq \mathbb{R}^d$, let $\Lambda_i(S, \boldsymbol{v}) = \{\Lambda_i(\boldsymbol{x}, \boldsymbol{v}) \mid \boldsymbol{x} \in S\}$. For simplicity, we let $\boldsymbol{c}_{ij}(\boldsymbol{v}) = \Lambda_i(\boldsymbol{c}_{ij}, \boldsymbol{v})$ and $S_{ij}(\boldsymbol{v}) = \Lambda_i(S_{ij}, \boldsymbol{v})$. The penalized rigid sphere set packing problem is formally defined by

$$
\begin{aligned}
\text{minimize} \quad & F_{\text{pen}}(\boldsymbol{v}) = \sum_{1 \le i < k \le m} \sum_{j=1}^{n_i} \sum_{l=1}^{n_k} f_{ijkl}^{\text{pen}}(\boldsymbol{v}), \\
\text{subject to} \quad & \boldsymbol{v} = (\boldsymbol{v}_1, \ldots, \boldsymbol{v}_m) \in \mathbb{R}^{\sum_{i=1}^{m} \lambda_i}, \\
& \boldsymbol{v}_i \in \mathbb{R}^{\lambda_i}, \quad i = 1, \ldots, m,
\end{aligned}
\tag{1}
$$

where $f_{ijkl}^{\text{pen}}(\boldsymbol{v}) = [\delta(S_{ij}(\boldsymbol{v}_i), S_{kl}(\boldsymbol{v}_k))]^2$ denotes the penetration penalty of two spheres S_{ij} and S_{kl}. This problem is an unconstrained nonlinear program. If the motion functions $\boldsymbol{c}_{ij}(\boldsymbol{v})$ of the centers of objects are chosen to be differentiable, F_{pen} is also differentiable and we can apply the quasi-Newton method to (1). In this paper, we consider following two types of motions.

Translations with a Fixed Direction in 2D for Problem 1. We first consider the case where object O_i is allowed to translate only in a prescribed direction in $\mathbb{R}^2$, but not allowed to rotate. Assume that the reference point $\boldsymbol{r}_i$ of object O_i lies on a line $\boldsymbol{d}_i + t_i \boldsymbol{e}_i$, where $\boldsymbol{d}_i, \boldsymbol{e}_i \in \mathbb{R}^2$ are given and t_i is a variable. Then

$$
\Lambda_i(\boldsymbol{x}, t_i) = \boldsymbol{x} - \boldsymbol{r}_i + \boldsymbol{d}_i + t_i \boldsymbol{e}_i, \qquad \frac{\partial \boldsymbol{c}_{ij}(t_i)}{\partial t_i} = \frac{\partial \Lambda_i(\boldsymbol{c}_{ij}, t_i)}{\partial t_i} = \boldsymbol{e}_i.
$$

Translations and Rotations in 3D for Problem 2. We next consider the case where each object O_i in $\mathbb{R}^3$ is allowed to translate and rotate around its reference point $\boldsymbol{r}_i$. Let $(x_i, y_i, z_i)^{\mathsf{T}}$ be the translation vector, $(\phi_i, \theta_i, \psi_i)$ be the z-x-z Euler angles, and $R_3(\phi_i, \theta_i, \psi_i)$ be the rotation matrix. Given variables $\boldsymbol{v}_i = (x_i, y_i, z_i, \phi_i, \theta_i, \psi_i)^{\mathsf{T}}$, we define the resulting position of a point $\boldsymbol{x} \in \mathbb{R}^3$ after the motion by $\Lambda_i(\boldsymbol{x}, \boldsymbol{v}_i) = R_3(\phi_i, \theta_i, \psi_i)(\boldsymbol{x} - \boldsymbol{r}_i) + (x_i, y_i, z_i)^{\mathsf{T}} + \boldsymbol{r}_i$. Then,

$$
\frac{\partial \boldsymbol{c}_{ij}(\boldsymbol{v}_i)}{\partial x_i} = (1, 0, 0)^{\mathsf{T}}, \qquad \frac{\partial \boldsymbol{c}_{ij}(\boldsymbol{v}_i)}{\partial \phi_i} = \frac{\partial R_3(\phi_i, \theta_i, \psi_i)}{\partial \phi_i}(\boldsymbol{c}_{ij} - \boldsymbol{r}_i).
$$

The other derivatives of $\boldsymbol{c}_{ij}(\boldsymbol{v}_i)$ with respective to y_i, z_i, θ_i, and ψ_i can be calculated analogously.

3 Experimental Results

We conducted computational experiments of RigidQN by generating instances of both Problems 1 and 2 randomly. We implemented RigidQN in C++, compiled it by GCC 4.1 and conducted experiments on a PC with an AMD Sempron 3000+ 1.8 GHz processor and 450 MB memory. We adopted a quasi-Newton method package L-BFGS [10].

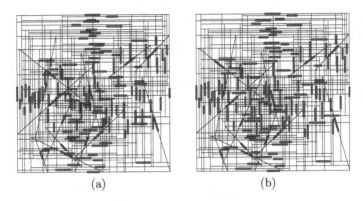

(a)   (b)

Fig. 1. An example of a road map layout with 147 labels ($\ell_{\text{label}} = 100, \ell_{\text{grid}} = 150$)

The data set of Problem 1 was generated as follows. We first start with a square with size $\ell_{\text{map}} \times \ell_{\text{map}}$ which consists of four lines as a drawing area, where we set $\ell_{\text{map}} = 10000$, and place a square grid on the square, where the minimum distance between two grid lines is ℓ_{grid}. Next we draw horizontal and vertical line segments one after another on the grid lines. Then, we draw some slanted line segments by choosing two arbitrary points in the drawing. Finally, we place a rectangle in the middle of each line segment in the drawing, where the height of a rectangle is ℓ_{label} and the length of a rectangle varies over a range $[5\ell_{\text{label}}, 10\ell_{\text{label}}]$. For example, Fig. 1(a) shows an initial layout with labels for $\ell_{\text{label}} = 100$, where a line segment represents an edge (i.e. street) and a rectangle represents a label (i.e. street name). Figure 1(a) is generated for $\ell_{\text{grid}} = 150$, which has 147 labels and 4818 circles. See Fig. 1(b) for the resulting layout. It took 0.43 seconds for Fig. 1(b).

To observe the influence of the density of the road map layout and the number of labels on the efficiency of our algorithm, we varied two parameters ℓ_{label} and ℓ_{grid} from 100 to 1000 with a step size 50 and from 50 to 1000 with a step size 50, respectively, and conducted experiments. For each setting, we generated 100 instances and applied RIGIDQN to them. We observed that our algorithm removed almost all overlaps in less than one second for the instances for $\ell_{\text{grid}} \geq 2\ell_{\text{label}}$. For details on the experimental results, see [7].

For the data set of Problem 2, we created instances which resemble the spiked spheres used in [3]. We generated an instance as follows. A spiked sphere has a sphere of radius 10 together with attached 10 spikes. Each spike consists of 20 spheres and the length varies on a range $[10, 70]$. To create an instance, we place the spiked spheres randomly in a cube with edge length 300, where the number of spiked spheres is a parameter. See Fig. 2 for magnified pictures of the initial layout and the resulting layout of an instance with 100 spiked spheres. We can see a spiked sphere in Fig. 2(a) penetrating another spiked sphere, and the removal of overlap in Fig. 2(b). RIGIDQN run in 1.7 seconds for Fig. 2(b).

To observe the influence of the number of spiked spheres on the efficiency of our algorithm, we varied the number of spiked spheres from 50 to 250 with a step

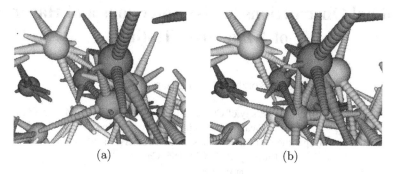

(a)   (b)

Fig. 2. Magnified figures of an instance with 100 labels of Problem 2

size 50, generated 10 instances for each setting, and measured the computation time. We observed that our algorithm removed almost all overlaps in less than 10 seconds for the instances with the number of spikes spheres less than or equal to 200. For details on the experimental results, see [7].

References

1. Agarwal, P.K., Guibas, L.J., Har-Peled, S., Rabinovitch, A., Sharir, M.: Penetration depth of two convex polytopes in 3D. Nordic Journal of Computing 7(3), 227–240 (2000)
2. Agrawala, M.: Visualizing Route Maps. Ph.D. thesis, Stanford University (2002)
3. Ahmed, A., Dwyer, T., Hong, S.H., Murray, C., Song, L., Wu, Y.X.: Visualisation and analysis of large and complex scale-free networks. In: EUROVIS 2005: Eurographics / IEEE VGTC Symposium on Visualization, pp. 239–246 (2005)
4. Dwyer, T., Marriott, K., Stuckey, P.J.: Fast node overlap removal. In: Healy, P., Nikolov, N.S. (eds.) GD 2005. LNCS, vol. 3843, pp. 153–164. Springer, Heidelberg (2006)
5. Gansner, E.R., North, S.C.: Improved force-directed layouts. In: Whitesides, S.H. (ed.) GD 1998. LNCS, vol. 1547, pp. 364–373. Springer, Heidelberg (1999)
6. Hayashi, K., Inoue, M., Masuzawa, T., Fujiwara, H.: A layout adjustment problem for disjoint rectangles preserving orthogonal order. Systems and Computers in Japan 33(2), 31–42 (2002)
7. Imamichi, T., Arahori, Y., Gim, J., Hong, S.H., Nagamochi, H.: Removing overlaps in label layouts using multi-sphere scheme. Tech. Rep. 2008-006, Dept. of Applied Mathematics and Physics, Kyoto University (2008)
8. Imamichi, T., Nagamochi, H.: A multi-sphere scheme for 2D and 3D packing problems. In: Stützle, T., Birattari, M., Hoos, H.H. (eds.) SLS 2007. LNCS, vol. 4638, pp. 207–211. Springer, Heidelberg (2007)
9. Li, W., Eades, P., Nikolov, N.: Using spring algorithms to remove node overlapping. In: APVis 2005. CRPIT, vol. 45, pp. 131–140 (2005)
10. Liu, D.C., Nocedal, J.: On the limited memory BFGS method for large scale optimization. Mathematical Programming 45(3), 503–528 (1989)
11. Lyons, K.A., Meijer, H., Rappaport, D.: Algorithms for cluster busting in anchored graph drawing. Journal of Graph Algorithms and Applications 2(1), 1–24 (1998)
12. Marriott, K., Stuckey, P., Tam, V., He, W.: Removing node overlapping in graph layout using constrained optimization. Constraints 8(2), 143–171 (2003)

Minimal Obstructions for 1-Immersions and Hardness of 1-Planarity Testing

Vladimir P. Korzhik[1,*] and Bojan Mohar[2,**,***]

[1] National University of Chernivtsi and Institute of APMM
National Academy of Science, Lviv, Ukraine
[2] Department of Mathematics, Simon Fraser University
Burnaby, B.C. V5A 1S6, Canada
mohar@sfu.ca

Abstract. A graph is 1-*planar* if it can be drawn on the plane so that each edge is crossed by no more than one other edge. A non-1-planar graph G is *minimal* if the graph $G - e$ is 1-planar for every edge e of G. We construct two infinite families of minimal non-1-planar graphs and show that for every integer $n \geq 63$, there are at least $2^{\frac{n}{4} - \frac{54}{4}}$ non-isomorphic minimal non-1-planar graphs of order n. It is also proved that testing 1-planarity is NP-complete. As an interesting consequence we obtain a new, geometric proof of NP-completeness of the crossing number problem, even when restricted to cubic graphs. This resolves a question of Hliněný.

1 Introduction

A graph is 1-*immersed* in the plane if it can be drawn in the plane so that each edge is crossed by no more than one other edge. A graph is 1-*planar* if it can be 1-immersed into the plane. It is easy to see that if a graph has 1-immersion in which two edges e, f with a common endvertex cross, then the drawing of e and f can be changed so that these two edges no longer cross. Consequently, we may assume that adjacent edges are never crossing each other and that no edge is crossing itself. We take this assumption as a part of the definition of 1-immersions since this limits the number of possible cases when discussing 1-immersions.

The notion of 1-immersion of a graph was introduced by Ringel [11] when trying to color the vertices and faces of a plane graph so that adjacent or incident elements receive distinct colors.

Little is known about 1-planar graphs. Borodin [1,2] proved that every 1-planar graph is 6-colorable. Some properties of maximal 1-planar graphs are considered in [12]. It was shown in [3] that every 1-planar graph is acyclically 20-colorable. The existence of subgraphs of bounded vertex degrees in 1-planar graphs is investigated in [7]. It was shown in [4,5] that a 1-planar graph with n vertices has at most $4n - 8$ edges and that this upper bound is tight. In the paper [6] it was observed that the class of 1-planar graphs is not closed under the operation of edge contraction.

---

* This work was done while the first author visited Simon Fraser University.
** Supported in part by ARRS (Slovenia), P1-0297, and by NSERC (Canada).
*** On leave from Dept. Math., University of Ljubljana, Ljubljana, Slovenia.

I.G. Tollis and M. Patrignani (Eds.): GD 2008, LNCS 5417, pp. 302–312, 2009.

Much less is known about non-1-planar graphs. The basic question is how to recognize 1-planar graphs. This problem is clearly in NP, but it is not clear at all if there is a polynomial time recognition algorithm. We shall answer this question by proving that 1-planarity testing problem is NP-complete.

The recognition problem is closely related to the study of minimal obstructions for 1-planarity. A graph G is said to be a *minimal* non-1-planar graph (MN-*graph*, for short) if G is not 1-planar, but $G - e$ is 1-planar for every edge e of G. An obvious question is:

How many MN-graphs are there? Is their number finite? If not, can they be characterized?

The answer to the first question is not hard: there are infinitely many. This was first proved in [10]. Here we present two additional simple arguments implying the same conclusion.

Example 1. Let G be a graph such that and $t = \lceil \mathrm{cr}(G)/|E(G)| \rceil - 1 \geq 1$, where $\mathrm{cr}(G)$ denotes the crossing number of G. Let G_t be the graph obtained from G by replacing each edge of G by a path of length t. Then $|E(G_t)| = t|E(G)| < \mathrm{cr}(G) = \mathrm{cr}(G_t)$. This implies that G_t is not 1-planar. However, G_t contains an MN-subgraph H. Clearly, H contains at least one subdivided edge of G in its entirety, so $|V(H)| > t$. Since t can be arbitrarily large, this shows that there are infinitely many MN-graphs.

Example 2. Let $K \in \{K_5, K_{3,3}\}$ be one of Kuratowski graphs. For each edge $xy \in E(K)$, let L_{xy} be a 5-connected triangulation of the plane and u, v be adjacent vertices of L_{xy} whose degree is at least 6. Let $L'_{xy} = L_{xy} - uv$. Now replace each edge xy of K with L'_{xy} by identifying x with u and y with v. It is not hard to see that the resulting graph G is not 1-planar (since two of graphs L'_{xy} must "cross each other", but that is not possible since they come from 5-connected triangulations). Again, one can argue that they contain large MN-graphs.

The paper [10] and the above examples prove the existence of infinitely many MN-graphs but do not give any concrete examples. In [10], two specific MN-graphs of order 7 and 8, respectively, are given. One of them, the graph $K_7 - E(K_3)$, is the unique 7-vertex MN-graph and since all 6-vertex graphs are 1-planar, the graph $K_7 - E(K_3)$ is the MN-graph with the minimum number of vertices. Surprisingly enough, the two MN-graphs in [10] are the only explicit MN-graphs known in the literature.

The main problem when trying to construct 1-planar graphs is that we have no characterization of 1-planar graphs. The set of 1-planar graphs is not closed under taking minors, so 1-planarity can not be characterized by forbidding some minors.

In the present paper we construct two explicit infinite families of MN-graphs and, correspondingly, we give two different approaches how to prove that a graph has no plane 1-immersion.

In Sect. 2 we construct MN-graphs based on the Kuratowski graph $K_{3,3}$. To obtain the MN-graphs, we replace six edges of $K_{3,3}$ by some special subgraphs. The non-1-planarity of the obtained MN-graphs follows from the nonplanarity of $K_{3,3}$. Using these MN-graphs, we show that for every integer $n \geq 63$, there are at least $2^{\frac{n}{4} - \frac{54}{4}}$ non-isomorphic minimal non-1-planar graphs of order n. In Sect. 3 we describe a class of 3-connected planar graphs that have no plane 1-immersions with at least one crossing point (PN-*graphs*, for short). Every 3-connected PN-graph has a unique plane

1-immersion, namely, the unique plane embedding of the graph. Hence, if a 1-planar graph G contains as a subgraph a PN-graph H, then in every plane 1-immersion of G the subgraph H is 1-immersed in the plane in the same way. Having constructions of PN-graphs, we can construct 1-planar and non-1-planar graphs with some desired properties: 1-planar graphs that have exactly $k > 0$ different plane 1-immersions; MN-graphs, etc.

In Sect. 4 we construct MN-graphs based on PN-graphs. Each of these MN-graphs G has as a subgraph a PN-graph H and the unique plane 1-immersion of H prevents to draw the remaining part of G on the plane when trying to obtain a plane 1-immersion of G.

Despite the fact that minimal obstructions for 1-planarity (i.e., the MN-graphs) have diverse structure, and despite the fact that discovering 1-immersions of specific graphs can be very tricky, it turned out to be a hard problem to establish hardness of 1-planarity testing. A solution is outlined in Sect. 5, where we show that 1-planarity testing is NP-complete, see Theorem 4. The proof is geometric in the sense that the reduction is from 3-colorability of planar graphs (or similarly, from planar 3-satisfiability). As an interesting consequence we obtain a new, geometric proof of NP-completeness of the crossing number problem, even when restricted to cubic graphs. Hardness of the crossing number problem for cubic graphs was established recently by Hliněný [9], who asked if one can prove this result by a reduction from an NP-complete geometric problem instead of the Optimal Linear Arrangement problem used in his proof.

2   MN-Graphs Based on the Graph $K_{3,3}$

Two cycles of a graph are *adjacent* if they share a common edge. If a graph G is drawn in the plane, then we say that a vertex x lies *inside* (resp. *outside*) a non-self-intersecting embedded cycle C, if x lies in the interior (resp. exterior) of C, and does not lie on C. Having two embedded adjacent cycles C and C', we say that C lies inside (resp. outside) C' if every point of C either lies inside (resp. outside) C' or lies on C'. We assume that in 1-immersions, adjacent edges do not cross each other and no edge crosses itself. Thus, every 3-cycle of a 1-immersed graph is embedded in the plane. Hence, given a 3-cycle of a 1-immersed graph, we can speak about its interior and exterior.

In what follows, throughout the paper, given a 1-immersion of a graph, when we speak about vertices, paths and cycles of the graph, we usually mean (the exact meaning will be always clear from the context) immersed vertices, paths and cycles of the 1-immersed graph.

Now we begin describing a family of MN-graphs based on the graph $K_{3,3}$.

By a *link* $L(x, y)$ connecting two vertices x and y we mean any of the graphs shown in Fig. 1 where $\{z, \bar{z}\} = \{x, y\}$.

By an A-*chain* of length $n \geq 2$ we mean the graph shown in Fig. 2(a). By a B-*chain* of length $n \geq 2$ we mean the graph shown in Fig. 2(c) and every graph obtained from this graph in the following way: for some integers $h_1, h_2, \ldots, h_t$, where $t \geq 1$ and $1 \leq h_1 < h_2 < \cdots < h_t \leq n - 2$, for every $i = 1, 2, \ldots, t$, we replace the subgraph at the left of Fig. 2(e) by the subgraph shown at the right of the figure. Note that, by definition, A- and B-chains have length at least 2. We say that the chains in

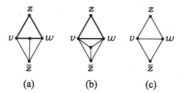

Fig. 1.

Figs. 2(a) and (c) connect the vertices $v(0)$ and $v(n)$ which are called the *end vertices* of the chain. Two chains are *adjacent* if they share a common end vertex. The A-chain in Fig. 2(a) and the B-chain in Fig. 2(c) will be designated in later figures by a single directed (broken) edge, as shown in Figs. 2(b) and (d), respectively, where the arrow points to the end vertex incident with the base link. The vertices $v(0), v(1), \ldots, v(n)$ are the *core vertices* of the chains.

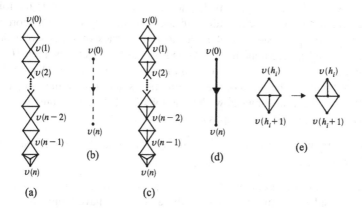

Fig. 2.

By a *chain graph* we mean the graph obtained from $K_{3,3}$ as shown in Fig. 3(a), where the three A-chains and three B-chains can have arbitrary lenghts ≥ 2. The vertices $\Omega(1)$, $\Omega(2)$, and $\Omega(3)$ are the *base vertices* of the chain graph. The edges joining the vertex Ω to the base vertices are called the Ω-edges.

We will show that every chain graph is an MN-graph.

Lemma 1. *If G is a chain graph and $e \in E(G)$, then the graph $G - e$ is 1-planar.*

Proof. It is easy to see that $G - e$ is 1-planar for every Ω-edge e. Let us now consider a plane embedding f of $G - \Omega$ of Fig. 3(a) after we delete the vertex Ω. If e is not an Ω-edge, then, because of the symmetry, it suffices to prove that $G - e$ is 1-planar for every edge e belonging to the A- or B-chain incident to $\Omega(2)$. Figs. 3(b) and (c) show how f can be modified to obtain a 1-immersion of $G - e$ for every edge e belonging to the chains incident to $\Omega(2)$ (the edge e is represented by the dotted line). ∎

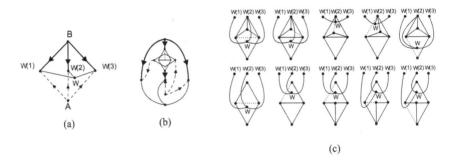

Fig. 3.

We are not aware of a simple argument showing that a chain graph G is not 1-planar. We prove it by *reductio ad absurdum* – assuming that G has a 1-immersion φ, we show that φ has the following properties that eventually yield a contradiction. If Π and Π' are nonadjacent A- and B-chain, respectively, then for every 3-cycle C of Π the following holds: The core vertices of Π' either all lie inside or all lie outside C. If all core vertices of Π' lie inside (resp. outside) C, then at most one vertex of Π' lies outside (resp. inside) C. If Π and Π' are nonadjacent A- and B-chain, respectively, then Π does not cross Π' in φ. The Ω-edges do not cross all three edges of a link incident to the same core vertex of the link. The proof of these properties is deferred for the full paper.

The following theorem shows how chain graphs can be used to construct exponentially many nonisomorphic MN-graphs of order n.

Theorem 1. *For every integer $n \geq 63$, there are at least $2^{\frac{n}{4} - \frac{54}{4}}$ non-isomorphic MN-graphs of order n.*

Proof. The A-chain of length t has $3t + 2$ vertices and a B-chain of length t has $4t + 1$ vertices. Consider a chain graph whose three A-chains have length 2, 2, and $\ell \geq 2$, respectively, and whose B-chains have length 2, 3, and $t \geq 4$, respectively. The graph has $35 + 3\ell + 4t$ vertices. One can apply the modification shown in Fig. 2(e) to an arbitrary subset of the links of the B-chains of the graph, and thus obtain 2^{t-1} nonisomorphic chain graphs of order $35 + 3\ell + 4t$, where $\ell \geq 2$ and $t \geq 4$. We claim that for every integer $n \geq 63$, there are integers $2 \leq \ell \leq 5$ and $t \geq 4$ such that $n = 35 + 3\ell + 4t$. Indeed, if $m \equiv 0, 1, 2, 3 \pmod 4$, put $\ell = 3, 2, 5, 4$, respectively. If $n = 35 + 3\ell + 4t$, where $2 \leq \ell \leq 5$, then $t \geq n/4 - 50/4$. Hence, there are at least $2^{\frac{n}{4} - \frac{54}{4}}$ non-isomorphic chain graphs of order $n \geq 63$. Since every chain graph is an MN-graph, the theorem follows.

3   PH-Graphs

By a *proper* 1-immersion of a graph we mean a 1-immersion with at least one crossing point. Let us recall that a PN-*graph* is a planar graph that does not have proper 1-immersions. In this section we describe a class of PN-graphs and construct some graphs of the class. They will be used in Sect. 4 to construct MN-graphs.

Two disjoint edges vw and $v'w'$ of a graph are *paired* if the four vertices v, w, v', w' are all four vertices of two adjacent 3-cycles. For every cycle C of a graph denote by $N(C)$ the set of all vertices of the graph not belonging to C but adjacent to vertices of C.

Consider a 3-connected plane graph. By a *basic k-cycle* of the graph we mean the boundary cycle of a k-gonal face of the embedding. By a *nontriangular* basic cycle we mean every basic k-cycle, $k \geq 4$.

Theorem 2. *Suppose that a 3-connected plane graph G satisfies the following conditions:*

(C1) *Every vertex has degree at least 4 and at most 6.*

(C2) *Every edge belongs to at least one 3-cycle.*

(C3) *Every 3-cycle is basic.*

(C4) *Every 3-cycle is adjacent to at most one other 3-cycle.*

(C5) *No vertex belongs to three mutually edge-disjoint 3-cycles.*

(C6) *Every 4-cycle is either basic or is the boundary of two adjacent triangular faces.*

(C7) *For every 3-cycle C, any two vertices of $V(G) \setminus (V(C) \cup N(C))$ are connected by 4 edge-disjoint paths not passing through the vertices of C.*

(C8) *If an edge vw of a nontriangular basic cycle C is paired with an edge $v'w'$ of a nontriangular basic cycle C', then C and C' have no vertices in common and any two vertices a and a' of C and C', respectively, such that $\{a, a'\} \not\subseteq \{v, w, v', w'\}$ are non-adjacent and are not connected by a path a, b, a' of length 2, where b does not belong to C and C'.*

(C9) *G does not contain the subgraphs shown in Fig. 4 (in this figure, 4-valent (resp. 5-valent) vertices of G are encircled (resp. encircled twice)).*

Then G has no proper 1-immersion.

The proof of Theorem 2 is long and will be given in the full paper.

Fig. 4.

Denote by $\mathcal{A}$ the class of all 3-connected plane graphs G satisfying the conditions (C1)–(C9) of Theorem 2. In the full paper we show how to construct graphs of the class $\mathcal{A}$. Figure 5 shows two graphs of $\mathcal{A}$, one of which (in Fig. 5(a)) will be used in Sect. 4 to construct MN-graphs. To simplify checking conditions (C1)–(C9) we construct the graphs to be symmetrical so that, for example, to check the condition (C7) we need to consider only two 3-cycles of a graph.

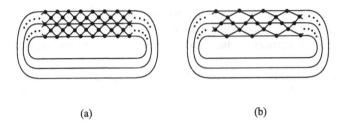

(a)   (b)

Fig. 5.

4 MN-Graphs Based on PN-Graphs

In this section we construct MN-graphs based on the PN-graphs G_n described in Sect. 3.

Denote by S_m, $m \geq 2$, the graph shown in Fig. 6. The graph has $m + 1$ cycles of length $12m - 2$ labelled by $B_0, B_1, \ldots, B_m$ as shown in the figure. The vertices of B_0 are called the *central vertices* of S_m and are labelled by $1, 2, \ldots, 12m - 2$ (see Fig. 6). For every central vertex $x \in \{1, 2, \ldots, 12m - 2\}$, denote by x^* the vertex $6m - 1 + x$ if $x \in \{1, 2, \ldots, 6m - 1\}$ and the vertex $x - (6m - 1)$ if $x \in \{6m, 6m + 1, \ldots, 12m - 2\}$. In S_m any pair $\{x, x^*\}$ of central vertices is connected by a *central path* $P(x, x^*)$ of length $6m - 3$ with $6m - 4$ two-valent vertices.

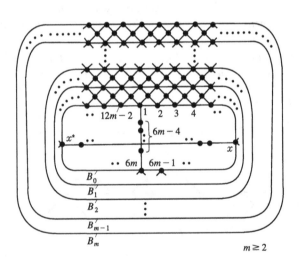

Fig. 6.

For any integers $m \geq 4$ and $n \geq 0$, denote by $\Phi_m(n)$ the set of all $(12m - 2)$-tuples $n_1, n_2, \ldots, n_{12m-2}$ of nonnegative integers such that $n_1 + n_2 + \cdots + n_{12m-2} = n$. For every $\lambda \in \Phi_m(n)$, denote by $S_m(\lambda)$ the graph obtained from S_m if for every central vertex $x \in \{1, 2, \ldots, 12m - 2\}$, we replace the 8 edges marked by transverse stroke in Fig. 7(a) by $8(1 + n_x)$ new edges marked by transverse stroke in Fig. 7(b) (here $x + 1 = 1$

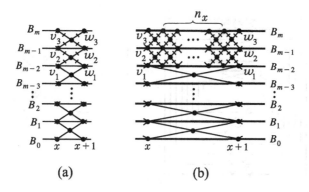

Fig. 7.

for $x = 12m - 2$). The graph $S_m(\lambda)$ has $m - 2$ $(12m - 2)$-cycles $B_0, B_1, \ldots, B_{m-3}$ and three $(12m - 2 + n)$-cycles B_{m-2}, B_{m-1}, B_m; all the $m + 1$ cycles are depicted in Fig. 7(b) in thick line.

We want to show that for every $m \geq 4$ and for every $\lambda \in \Phi_m(n)$, $n \geq 0$, the graph $S_m(\lambda)$ is an MN-graph.

Lemma 2. *The graph $S_m(\lambda) - e$, where $m \geq 4$, $\lambda \in \Phi_m(n)$, is 1-planar for every edge e.*

Proof. If we delete an edge of a central path, then the remaining $6m - 2$ central paths, each with $6m - 3$ edges, can be 1-immersed inside B_0 in Fig. 6. If we delete one of the edges depicted in Fig. 8(a) in thick line, then the central path $P(x, x^*)$ can be drawn outside B_0 with $6m - 3$ crossing points as shown in the figure (where the path is depicted in thin line) and then the remaining $6m - 2$ central paths can be 1-immersed inside B_0. If we delete one of the two edges depicted in Fig. 8(a) in dotted line, then Fig. 8(b) shows how to place the central vertex x so that the path $P(x, x^*)$ can be drawn outside B_0 with $6m - 3$ crossing points and then the remaining $6m - 2$ central paths can be 1-immersed inside B_0.   ∎

Lemma 3. *The graph obtained from the graph $S_m(\lambda)$, where $m \geq 4$ and $\lambda \in \Phi_m(n)$, by deleting the two-valent vertices of all central paths is a PN-graph.*

Theorem 3. *The graph $S_m(\lambda)$, where $m \geq 4$ and $\lambda \in \Phi_m(n)$, is not 1-planar.*

The proofs of Lemma 3 and Theorem 3 are deferred for the full paper.

We have shown that every graph $S_m(\lambda)$, where $m \geq 4$ and $\lambda \in \Phi_m(n)$, is an MN-graph (the graph has order $(5m - 1)(12m - 2) + 5n$). Clearly, graphs $S_{m_1}(\lambda_1)$ and $S_{m_2}(\lambda_2)$, where $\lambda_1 \in \Phi_{m_1}(n_1)$ and $\lambda_2 \in \Phi_{m_2}(n_2)$, are nonisomorphic for $m_1 \neq m_2$ and for $m_1 = m_2$ and $n_1 \neq n_2$.

Corollary 1. *For any integers $m \geq 4$ and $n \geq 0$, there are at least $\frac{1}{2(12m-2)}\binom{n+12m-3}{12m-3}$ non-isomorphic MN-graphs $S_m(\lambda)$, where $\lambda \in \Phi_m(n)$.*

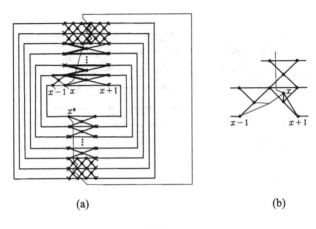

(a)   (b)

Fig. 8.

Proof. It is well known that $|\Phi_m(n)| = \binom{n+12m-3}{12m-3}$. The automorphism group of the graph S_m is the automorphism group of a regular $(12m-2)$-gonal, that is, the dihedral group D_{12m-2} of order $2(12m-2)$. Now the claim follows. ∎

5 Testing 1-Immersibility Is Hard

In this section we prove that it is NP-complete to decide if a given input graph is 1-immersible. This shows that it is extremely unlikely that there exists a nice classification of MN-graphs.

The reduction showing completeness for the class NP is from 3-colorability of planar graphs. It is worth mentioning that our method also yields a similar reduction of planar 3-colorability to the problem of computing the crossing number of cubic graphs. NP-completeness of the crossing number problem on cubic graphs was proved recently by Hliněný [9]. The author has observed in [9] that his proof is non-geometric and asked for an accessible proof based on geometric reduction. Our construction, correspondingly adapted, in particular answers the question of Hliněný.

Theorem 4. *It is NP-complete to decide if a given graph is 1-immersible in the plane.*

Proof (sketch). Since 1-immersions can be represented combinatorially, it is clear that 1-immersability is in NP. To prove its completeness, we shall make a reduction from a known NP-complete problem, that of 3-colorability of planar graphs of maximum degree at most four [8].

Let G_0 be a given planar graph of maximum degree 4 whose 3-colorability is to be tested. We shall show how to construct, in polynomial time, a related graph $\hat{G}$ such that $\hat{G}$ is 1-immersible if and only if G_0 is 3-colorable. We may assume that G_0 has no vertices of degree less than three.

The construction of $\hat{G}$ involves replacement of each vertex v of G_0 by a *vertex-block* L_v, and replacement of each edge $uv \in E(G_0)$ by an *edge-block* F_{uv} which is

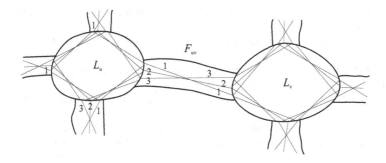

Fig. 9.

henceforth attached to L_u and L_v. Each building block has constant size, so the whole construction can be carried over in linear time. The building blocks L_v and F_{uv} are 1-planar but there is very little flexibility among their 1-immersions. They are pasted together so that their 1-immersions influence each other in such a way that globally consistent choices exist if and only if G_0 has a 3-coloring.

The vertex block essentially consists of a PN-graph L together with several subdivided edges, called *legs*. The legs can "pass through" L in a unique way since the number of degree-two vertices on the legs (the *lengths* of the legs) allow crossing it through a part of L that is not too dense. The legs are connecting vertices of L in a way as shown in Fig. 9, where they are represented by thin lines. Where the legs attach to the "boundary", there is an additional crossing edge, which can be turned outside to cross the leg in the edge-block instead. Each edge-block contains three legs that correspond to three colors 1,2,3, and we say that the leg i is active if it is crossed by the additional edge at the boundary part of the edge-block. A leg i that is active at the connection of L_u and F_{uv} corresponds to the choice of color i for the vertex u of G_0. The construction is made in such a way that an active leg i cannot be active at the other end of the edge-block (so we have proper coloring), that around u at least one leg is active, and that being active in the edge-block F_uv, the ith leg is also active in other edge-blocks F_{uw}, for other edges uw of G_0 incident with u. The details are cumbersome and are left for the full version of the paper.   ■

References

1. Borodin, O.V.: Solution of Ringel's problem about vertex bound colouring of planar graphs and colouring of 1-planar graphs [in Russian]. Metody Discret. Analiz. 41, 12–26 (1984)
2. Borodin, O.V.: A new proof of the 6-color theorem. J. Graph Theory 19, 507–521 (1995)
3. Borodin, O.V., Kostochka, A.V., Raspaud, A., Sopena, E.: Acyclic colouring of 1-planar graphs. Discrete Analysis and Operations Researcher 6, 20–35 (1999)
4. Chen, Z.-Z.: Approximation algorithms for independent sets in map graphs. Journal of Algorithms 41, 20–40 (2001)
5. Chen, Z.-Z.: New bounds on the number of edges in a k-map graph. In: Chwa, K.-Y., Munro, J.I.J. (eds.) COCOON 2004. LNCS, vol. 3106, pp. 319–328. Springer, Heidelberg (2004)

6. Chen, Z.-Z., Kouno M.: A linear-time algorithm for 7-coloring 1-plane graphs. Algorithmica 43, 147–177 (2005)
7. Fabrici, I., Madaras, T.: The structure of 1-planar graphs. Discrete Math. 307, 854–865 (2007)
8. Garey, M.R., Johnson, D.S., Stockmeyer, L.: Some simplified NP-complete graph problems. Theor. Comp. Sci. 1, 237–267 (1976)
9. Hliněný, P.: Crossing number is hard for cubic graphs. J. Combin. Theory, Ser. B 96, 455–471 (2006)
10. Korzhik, V.P.: Minimal non-1-planar graphs. Discrete Math. 308, 1319–1327 (2008)
11. Ringel, G.: Ein Sechsfarbenproblem auf der Kugel. Abh. Sem. Univ. Hamburg 29, 107–117 (1965)
12. Schumacher, H.: Zur Struktur 1-planarer Graphen. Math. Nachr. 125, 291–300 (1986)

Connected Rectilinear Graphs on Point Sets

Maarten Löffler[1,*] and Elena Mumford[2]

[1] Dept. Information and Computing Sciences, Utrecht University, The Netherlands
loffler@cs.uu.nl
[2] Dept. of Mathematics and Computer Science, TU Eindhoven, The Netherlands
e.mumford@tue.nl

Abstract. Given n points in d-dimensional space, we would like to connect the points with straight line segments to form a connected graph whose edges use d pairwise perpendicular directions. We prove that there exists at most one such set of directions. For $d = 2$ we present an algorithm for computing these directions (if they exist) in $O(n^2)$ time.

1   Introduction

Given a set V of n points in d-dimensional space, we would like to connect the points of V with straight line segments to form a connected rectilinear graph G. A *rectilinear graph* is an embedded straight-line graph such that any two edges in the graph are either parallel or perpendicular. We define the *orientation* of such a graph as the set of d pairwise perpendicular directions used by its edges. Two orientations are said to be *different* if there is a direction e in one of them and a direction e' in the other such that e and e' are neither the same nor perpendicular. We say that an orientation O *allows* for a connected rectilinear graph on V if there exists such a graph G that uses V as its vertices and has orientation O.

At the Canadian Conference on Computational Geometry in 2007, Therese Biedl asked whether a given set of points in the plane can be the vertex set of two rectilinear polygons that have different orientations. In this paper we show that the answer is no, and more generally, for a set V of points in $\mathbb{R}^d$ there exists at most one orientation that allows for a connected rectilinear graph on V. Figure 1 shows an example of two rectilinear graphs on the same point set, but note that G' is not connected.

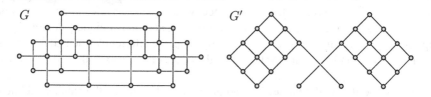

Fig. 1. Two rectilinear graphs with the same vertex set, but different orientations

* Supported by the Netherlands Organisation for Scientific Research (NWO) through the project GOGO.

I.G. Tollis and M. Patrignani (Eds.): GD 2008, LNCS 5417, pp. 313–318, 2009.

A special case of this problem has been considered by Fekete and Woeginger [4]. They show that for a set of points that have rational coordinates in the plane at most one orientation is possible. Problems on rectilinear polygons have been studied extensively. O'Rourke [10] proves that there is at most one way to connect a given point set into a rectilinear polygon that makes a 90° turn at each vertex, and gives a simple algorithm to compute it. On the other hand, if turns of 180° are allowed, Rappaport [12] shows that the problem is NP-hard. Durocher and Kirkpatrick [1] study the problem of finding a collection of rectilinear tours that use the given points as vertices, where the tours are allowed to have different orientations. They prove that this is NP-hard as well.

A number of papers address the problem of drawing a graph on a fixed point set. For example, Pach and Wenger [11] show that to make such a graph planar, a linear number of bends per edge may be necessary. Efrat *et al.* [3] study the possibility of drawing a crossing-free graph with circular arcs as edges. Rectilinear graphs also received a lot of attention from a graph drawing perspective. Vijayan and Wigderson [13] show how to embed an abstract graph with an additional "direction" associated to each edge as a rectilinear graph in $O(n^2)$ time; Hoffman and Kriegel [7] improve this to $O(n)$ time. Garg and Tamassia [6] show that without such associated directions, it is NP-hard to decide if a graph has a rectilinear embedding.

The remainder of the paper is organised as follows. In Sec. 2 we show that any point set allows for a connected rectilinear graph in at most one orientation. Then, in Sec. 3, we discuss the related algorithmic question of finding such an orientation for a point set in the plane. We conclude in Sec. 4.

2   Existence of Orientations

In this section, we prove that a point set cannot be the vertex set of two differently oriented rectilinear graphs. We first study the situation in the plane, then we extend the result to any dimension. We use several algebraic concepts, which we try to define briefly when we use them, but we refer to the full version [9] for a more complete and formal discussion.

2.1   Points in the Plane

Let V be a set of points in the plane, and let X and Y be the sets of all x-coordinates and y-coordinates of the vertices in V respectively. Assume w.l.o.g. that $\min(X) = \min(Y) = 0$. In this section for convenience we are going to refer to an orientation using the slope on one of its directions since the orientation is uniquely defined by it. Suppose for a contradiction that there are two connected differently oriented rectilinear graphs G and G' on V. Assume w.l.o.g. that the edges of G are axis-aligned and G' has edges of slopes s and $-\frac{1}{s}$.

Let $\mathbb{Q}(s)$ be the field generated by adjoining s to $\mathbb{Q}$; that is, the smallest subfield of $\mathbb{R}$ that contains both $\mathbb{Q}$ and s. Consider the vector space $\mathbb{Q}(s)\langle X \cup Y \rangle$; that is, the set of all sums of products of an element from $\mathbb{Q}(s)$ and an element from X or Y. Let $E = (e_1, \ldots, e_k)$ be a basis for this vector space. We can now denote this vector space by $\mathbb{Q}(s)\langle E \rangle = \mathbb{Q}(s)\langle e_1, \ldots, e_k \rangle$. We now have $X, Y \subset \mathbb{Q}(s)\langle E \rangle$ so we can write $x_i = \sum_j x_{ij} e_j$ for all $x_i \in X$, and $y_i = \sum_j y_{ij} e_j$ for all $y_i \in Y$, where $x_{ij}, y_{ij} \in \mathbb{Q}(s)$. We use $[\mathbb{Q}(s) : \mathbb{Q}]$ to denote the *degree of the extension* of field $\mathbb{Q}(s)$ over $\mathbb{Q}$, which is defined as the dimension of $\mathbb{Q}(s)$ as a vector space over $\mathbb{Q}$. We consider the following cases:

$[\mathbb{Q}(s) : \mathbb{Q}] = 1$: s is rational.
$[\mathbb{Q}(s) : \mathbb{Q}] < \infty$: s is algebraic over $\mathbb{Q}$.
$[\mathbb{Q}(s) : \mathbb{Q}] = \infty$: s is transcendental over $\mathbb{Q}$.

In fact, the rational case follows directly from the algebraic case since rational numbers are also algebraic, but we have separated them to allow the reader to follow the main argument without needing too much algebraic machinery yet.

Rational Slopes. When s is rational, $\mathbb{Q}(s) = \mathbb{Q}$, therefore $x_{ij}, y_{ij} \in \mathbb{Q}$. We can assume w.l.o.g. that $x_{ij}, y_{ij} \in \mathbb{Z}$ and that their greatest common divisor *(GCD)* is 1 (if not, scale the input by the appropriate factor). Now x_i, y_i are elements of the $\mathbb{Z}$-module $\mathbb{Z}\langle E \rangle$; that is, the set of all sums of integer multiples of elements from E. Consider any pair of vertices $v, v' \in V$, and the horizontal and vertical distances Δx and Δy between them. These vertices are connected by a path $v_1, v_2, \ldots, v_m$ in G, where $v_1 = v$ and $v_m = v'$. Denote by $(\Delta x_i, \Delta y_i)$ the horizontal and vertical distance between v_i and v_{i+1}. We know that there exists a path in G' from v_i to v_{i+1}, see Fig. 2. This path uses edges with slope s or $-\frac{1}{s}$, so when following this path we move over distances (a, sa) or $(sb, -b)$. Since all vertex coordinates are in $\mathbb{Z}\langle E \rangle$, we know that $a, b \in \mathbb{Z}\langle E \rangle$. In total we move from v_i to v_{i+1} over a distance $(a_i + sb_i, sa_i - b_i)$ where $a_i, b_i \in \mathbb{Z}\langle E \rangle$. Since G is axis-parallel, every edge between two points v_i and v_{i+1} is either horizontal or vertical. If it is horizontal, $\Delta y_i = sa_i - b_i = 0$, thus $\Delta x_i = a_i + sb_i = a_i + s^2 a_i = (1 + s^2)a_i$. If it is vertical, then $\Delta x_i = a_i + sb_i = 0$, thus $\Delta y_i = sa_i - b_i = -s^2 b_i - b_i = -(1 + s^2)b_i$.

Now write $\Delta x_i = \sum_j \Delta x_{ij} e_j$ and $\Delta y_i = \sum_j \Delta y_{ij} e_j$, and also write $a_i = \sum_j a_{ij} e_j$ and $b_i = \sum_j b_{ij} e_j$. Clearly $\Delta x_{ij}, \Delta y_{ij}, a_{ij}, b_{ij} \in \mathbb{Z}$. Since the elements of E are linearly independent over $\mathbb{Q}$, it follows that $\Delta x_{ij} = (1 + s^2)a_{ij}$ for horizontal segments and $\Delta y_{ij} = -(1 + s^2)b_{ij}$ for vertical segments for all i, j.

Now $s^2 \in \mathbb{Q}$, so we can write $s^2 = p/q$ with p and q co-prime. This means that $\Delta x_{ij} = (1 + p/q)a_{ij} = (p + q)a_{ij}/q$ or $\Delta y_{ij} = -(p + q)b_{ij}/q$. Since q does not divide $p + q$ (unless it is 1), $p + q$ is in $\mathbb{Z}$ and divides Δx_{ij} and Δy_{ij}. Since $\Delta x = \sum_{i,j} \Delta x_{ij} e_j$, it follows that $p + q$ divides Δx, and similarly Δy. So, any two vertices v and v' are a $\mathbb{Z}\langle E \rangle$-multiple of $p + q$ away from each other in both horizontal and vertical direction, which contradicts the fact that all their coordinates had GCD 1.

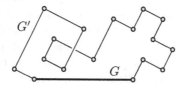

Fig. 2. For any edge of G, there is also a path in G' connecting its vertices

Algebraic Slopes. When s is algebraic over $\mathbb{Q}$, the argument described in the previous section still goes through when we replace all instances of $\mathbb{Q}$ by $\mathbb{Q}(s)$ and $\mathbb{Z}$ by $O_{\mathbb{Q}(s)}$. Here $O_{\mathbb{Q}(s)}$ is the ring of integers of $\mathbb{Q}(s)$, which consists of all algebraic integers that are in $\mathbb{Q}(s)$. Intuitively, the ring of integers $O_{\mathbb{Q}(s)}$ behaves towards the field $\mathbb{Q}(s)$ as $\mathbb{Z}$ behaves towards $\mathbb{Q}$. Every element of $\mathbb{Q}(s)$ can be written as p/q, where $p, q \in O_{\mathbb{Q}(s)}$. Every element of $O_{\mathbb{Q}(s)}$ can be written as the product of a finite sequence of irreducible elements of $O_{\mathbb{Q}(s)}$, though this factorisation is not necessarily unique. This means we can divide out common divisors in $O_{\mathbb{Q}(s)}$, and that we can have irreducible fractions p/q in $\mathbb{Q}(s)$.

Transcendental Slopes. When s is transcendental, every element $w \in \mathbb{Q}(s)$ can be written in the form

$$w = \frac{\sum_{0 \leq l \leq h} w_l s^l}{\sum_{0 \leq l' \leq h'} w'_{l'} s^{l'}}$$

for some $h, h' \in \mathbb{N}$, and $w_l, w'_{l'} \in \mathbb{Q}$. Assume w.l.o.g. that we can write

$$x_{ij} = \sum_{0 \leq l \leq h} x_{ijl} s^l \quad \text{and} \quad y_{ij} = \sum_{0 \leq l \leq h} y_{ijl} s^l$$

where $h \in \mathbb{N}$ and $x_{ijl}, y_{ijl} \in \mathbb{Z}$ (otherwise scale the input).

Now $x_{ij}, y_{ij} \in \mathbb{Z}\langle s, \ldots, s^h \rangle$. Assume $h \geq 2$ (if it is smaller, just add some 0's to the descriptions of the coordinates). We now also know that $(1 + s^2) \in \mathbb{Z}\langle s, \ldots, s^h \rangle$. Assume w.l.o.g. that not all of x_{ij}, y_{ij} can be written as $(1 + s^2)w$ for some $w \in \mathbb{Z}\langle s, \ldots, s^h \rangle$ (otherwise divide everything by $(1 + s^2)$).

However, in the same way as before, we argue that $\Delta x_{ij} = (1 + s^2)a_{ij}$ for horizontal segments and $\Delta y_{ij} = -(1 + s^2)b_{ij}$ for vertical segments for all i, j, where now $a_{ij}, b_{ij} \in \mathbb{Z}\langle s, \ldots, s^h \rangle$. This clearly contradicts our assumption. Thus we arrive at the following theorem:

Theorem 1. *Given a set of points in the plane, there can be at most one orientation that allows for a connected rectilinear graph that has these points as its vertices.*

2.2   Points in Higher Dimensions

Let V be a set of n points in $\mathbb{R}^d$. We will show that there is at most one orientation that allows for a connected rectilinear graph that uses V as its vertex set. Suppose for a contradiction there are two connected rectilinear graphs G and G' on V. And let E be the orientation of G and E' be the orientation of G'. Let $e \in E$ and $e' \in E'$ be two distinct directions that are not perpendicular.

Let α be a plane spanned by e and e'. Let V_α be the projection of V on α, and let G_α and G'_α be the projections of G and G' on α. We ignore any duplicate points in V_α and edges that were reduced to single points in G_α and G'_α. Note that G_α and G'_α are still connected graphs. Moreover, since α contains e, all edges of G_α map either to an edge in α parallel to e, or to one perpendicular to e, so G_α is a rectilinear graph. Similarly, G'_α is a rectilinear graph. However, these are two graphs on the same vertex set in 2-dimensional space, which is not possible by Theorem 1. We have proven the following theorem:

Theorem 2. *Given a set of points in $\mathbb{R}^d$, there exists at most one orientation that allows for a connected rectilinear graph that has these points as its vertices.*

3 Finding the Right Orientation

We now discuss the algorithmic side of the problem: given a set V of points in the plane, can we *find* a slope s such that the graph on V with edges of slope s and $-\frac{1}{s}$ is connected? A trivial approach takes $O(n^2 \log n)$ time. Consider all pairs of points and the line segment connecting them, and sort those segments by slope. For each slope that has at least $n - 1$ segments (together with its perpendicular slope), we can test whether they form a connected graph in linear time. Note that the most expensive step here is sorting the directions: a long-standing open problem is whether this can be done any faster than in $O(n^2 \log n)$ time [5].

However, it is not necessary to sort all directions, since many of them are uninteresting. Namely, since our graph has to be connected, an arbitrary point p has to share an edge with at least one other point of V. Thus we only need to consider $n - 1$ (possibly non-distinct) directions obtained by connecting p to all other points in V. Now consider the problem in dual space. Our set of points becomes a set of lines, our slope an x-coordinate, and two points are connected by a line segment of slope s if the two corresponding lines intersect at x-coordinate s. We sweep two vertical lines (at $x = -\frac{1}{s}$ and $x = s$) simultaneously over the dual plane, and keep track of the intersection points on those lines. The arrangement of the lines can be computed in $O(n^2)$ time [2]. We can inspect the potentially interesting slopes, and process the events in between in $O(n^2)$ time in total. The details are not hard, and can be found in the full version [9].

Deciding whether there is an orientation that allows for a *planar* rectilinear graph (a simple polygon, for example) on a given set of points is NP-hard: Since there is at most one possible orientation, we can use the algorithm sketched above to find it. Then we can take the maximal rectilinear graph in that orientation. However, now we need to decide whether this graph has a non-crossing subgraph, which is NP-complete [8].

4 Conclusion

We have proven that given a point set in $\mathbb{R}^d$, there exists at most one orientation such that the maximal rectilinear graph on the points in that orientation is connected. However, finding this orientation remains an interesting challenge. We have shown that this can be done in $O(n^2)$ time for a 2-dimensional point set, but we see no reason for this bound to be tight. Furthermore, finding such an orientation in higher dimensions is still open. We have also shown that deciding whether the points can be connected into a planar rectilinear graph is NP-hard.

Acknowledgments

The authors would like to thank Therese Biedl for introducing them to the problem, and Oswin Aichholzer, Chris Gray and Rodrigo Silveira for fruitful discussions on the subject.

References

1. Durocher, S., Kirkpatrick, D.: On the hardness of turn-angle-restricted rectilinear cycle cover problems. In: CCCG 2002, pp. 13–16 (2002)
2. Edelsbrunner, H., Guibas, L.: Topologically sweeping an arrangement. J. Comput. Syst. Sci. 38, 165–194 (1989)
3. Efrat, A., Erten, C., Kobourov, S.: Fixed-location circular-arc drawing of planar graphs. In: Liotta, G. (ed.) GD 2003. LNCS, vol. 2912, pp. 147–158. Springer, Heidelberg (2004)
4. Fekete, S., Woeginger, G.: Angle-restricted tours in the plane. Comput. Geom. Theory Appl. 8(4), 195–218 (1997)
5. Fredman, M.: How good is the information theory bound in sorting? Theoret. Comput. Sci. 1, 355–361 (1976)
6. Garg, A., Tamassia, R.: On the computational complexity of upward and rectilinear planarity testing. SIAM J. on Computing 31(2), 601–625 (2002)
7. Hoffman, F., Kriegel, K.: Embedding rectilinear graphs in linear time. Inf. Process. Lett. 29(2), 75–79 (1988)
8. Jansen, K., Woeginger, G.: The complexity of detecting crossingfree configurations in the plane. BIT 33(4), 580–595 (1993)
9. Löffler, M., Mumford, E.: Connected rectilinear polygons on point sets (2008), http://www.cs.uu.nl/research/techreps/UU-CS-2008-028.html
10. O'Rourke, J.: Uniqueness of orthogonal connect-the-dots. In: Toussaint, G. (ed.) Computational Morphology, pp. 97–104 (1988)
11. Pach, J., Wenger, R.: Embedding planar graphs at fixed vertex locations. In: Whitesides, S.H. (ed.) GD 1998. LNCS, vol. 1547, pp. 263–274. Springer, Heidelberg (1999)
12. Rappaport, D.: On the complexity of computing orthogonal polygons from a set of points. Technical Report TR-SOCS-86.9, McGill Univ., Montreal, PQ (1986)
13. Vijayan, G., Wigderson, A.: Rectilinear graphs and their embeddings. SIAM J. on Computing 14(2), 355–372 (1985)

3-Regular Non 3-Edge-Colorable Graphs with Polyhedral Embeddings in Orientable Surfaces*

Martin Kochol

MÚ SAV, Štefánikova 49, 814 73 Bratislava 1, Slovakia and FPV ŽU
kochol@savba.sk

Abstract. The Four Color Theorem is equivalent with its dual form stating that each 2-edge-connected 3-regular planar graph is 3-edge-colorable. In 1968, Grünbaum conjectured that similar property holds true for any orientable surface, namely that each 3-regular graph with a polyhedral embedding in an orientable surface has a 3-edge-coloring. Note that an embedding of a graph in a surface is called polyhedral if its geometric dual has no multiple edges and loops. We present a negative solution of this conjecture, showing that for each orientable surface of genus at least 5, there exists a 3-regular non 3-edge-colorable graph with a polyhedral embedding in the surface.

1 Introduction

Edge-coloring of cubic (3-regular) graphs is an important topic in graph theory and theoretical computer science. By Tait [11], a cubic planar graph is 3-edge-colorable if an only if its geometric dual is 4-colorable. Since geometric dual of a 2-edge-connected planar cubic graph is a planar triangulation, the Four Color Theorem (see [2]) is equivalent to the statement that every 2-edge-connected planar cubic graph has a 3-edge-coloring.

Nonplanar cubic graphs do not need to be 3-edge-colorable. The best know example is the Petersen graph (see Fig. 1). In fact, by Holyer [8], the problem to decide whether a cubic graph is 3-edge-colorable is NP-complete.

An embedding of a graph in a surface is called *polyhedral* if its dual has no multiple edges and loops. In 1968, Grünbaum [7] presented a conjecture that each 3-regular graph with a polyhedral embedding in an orientable surface has a 3-edge-coloring. If this is true, it would generalize the dual form of the Four Color Theorem for any orientable surface.

In this paper we disprove the Grünbaum's conjecture and for every orientable surface of genus at least 5, we construct non 3-edge-colorable cubic graphs with a polyhedral embedding in the surface.

Note that Petersen graph has a polyhedral embedding in projective plane. Thus Grünbaum's conjecture has a sense only for orientable surfaces. More details about this conjecture and related results can be found in [1,3,4,12]. Basic facts about embeddings of graphs into surfaces can be found in [5,6].

---

* Supported by grant VEGA 2/7037/7 and by A. v. Humboldt Fellowship.

I.G. Tollis and M. Patrignani (Eds.): GD 2008, LNCS 5417, pp. 319–323, 2009.

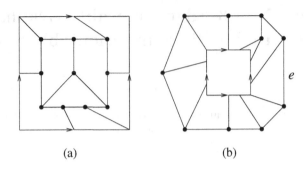

(a)   (b)

Fig. 1.

2   Snarks and Superposition

By a *snark* we mean a cubic graph without a 3-edge-coloring. It is well known
(see, e.g., [10]) that any cubic graph with a bridge (1-edge-cut) is a snark. Such
snarks are considered to be trivial. A nontrivial snark is the Petersen graph (see
Fig. 1).

Suppose v is a vertex of a graph G. Let G' arise from G in the following
process. Replace v by a graph H_v so that each edge e of H having one end v has
one end from H_v. If e is a loop having both ends v, then both ends of e become
vertices of H_v. Then G' is called *v-superposition* or a *vertex superposition* of G.

Suppose e is an edge of G with ends u and v. Let G' arises from G in the
following process. Replace e by a graph H_e having at least two vertices, i.e., we
delete e, pick up two distinct vertices u', v' of H_e and identify u' with u and
v' with v. Then G' is called an *e-superposition* or an *edge superposition* of G.
Furthermore, if H_e is a snark, then G' is called a *strong e-superposition* or a
strong edge superposition of G.

We say that a graph G' is a (strong) superposition of G if G' arises from
G after finitely many vertex and (strong) edge superpositions. The following
statement was proved in [10, Lemma 4.4] (see [9,10] for more details).

Lemma 1. *Let G be a snark and G' be a strong superposition of G. Furthermore,
suppose that G' is cubic. Then G' is a snark.*

3   Constructions

Clearly, a graph has an embedding in an orientable surface of genus n if and
only if it has an embedding in the plane with n handles. In parts (a) and (b) of
Fig. 1 are embeddings of the Petersen graph in the torus and in the plane with
one handle, respectively. (If we identify the opposite segments of the square in
part (b) of Fig. 1, we get a handle on the plane.)

Replacing edge e by another copy of Petersen graph we get graph G_{18} from
Fig. 2. Replacing in G_{18} the vertices of degree 5 by paths of length 2, we get a

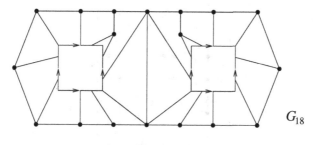

Fig. 2.

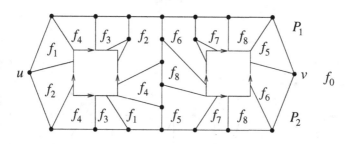

Fig. 3.

cubic graph G indicated in Fig. 3. By Lemma 1, G is a snark. The boundary of the infinite face f_0 is a circuit, composed from paths P_1 and P_2 with ends u and v. The following holds true.

(1) any two faces f_i, f_j, $i, j \in \{1, \ldots, 8\}$, share at most one edge,
(2) the infinite face f_0 share exactly two edges with each f_i, $i \in \{1, \ldots, 8\}$ so that P_1 and P_2 contain exactly one of them.

Properties (1) and (2) are important. We can take a nonpolyhedral embedding of a snark in an orientable surface, and replacing some of its edges by copies of G and some vertices by suitable graphs, we can get snarks with polyhedral embeddings in orientable surface. By replacing an edge of a copy of G, we identify the ends of e by u and v, respectively. For example, in Fig. 4 is a snark constructed in [4, Fig. 8]. Let us note that this is not a polyhedral embedding in the torus, because the pairs of faces a_1, a_2 and b_1, b_2 have two edges in common (i.e., its geometric dual has two pairs of parallel edges). In order to remove this obstacle, we replace edges e_1 and e_2 by two copies of G and we get graph G_{66} indicated in Fig. 5. Replacing in G_{66} the vertices of degree 5 by paths of length 2 we get the graph indicated in Fig. 6. By Lemma 1, this is a snark. Furthermore, by (1) and (2), any two faces of this graph have at most one edge in common (i.e., the pairs of faces a_1, a_2 and b_1, b_2 are "separated" by the copies of graph G). Thus the geometric dual has no parallel edges and loops, i.e., we have a polyhedral embedding of a snark in orientable surface of genus 5.

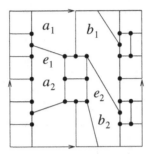

Fig. 4.

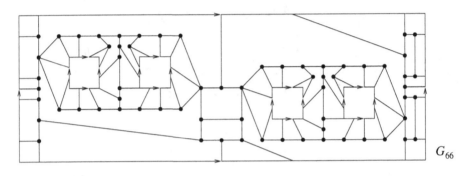

Fig. 5.

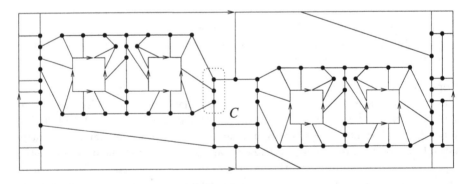

Fig. 6.

In order to get snarks with embeddings in orientable surfaces of genus > 5, it suffices to replace a vertex of degree 5 from G_{66} by suitable graphs with embeddings in a plane with handles. For example, consider the three vertices of the graph from Fig. 6 contained inside of the disc C indicated by dotted line. Replacing them by the graph indicated in Fig. 7, we get a snark with polyhedral

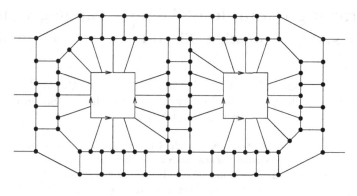

Fig. 7.

embedding in surface of genus 7. This snarks is a strong superposition of the snark from Fig. 4. In this way we can prove the following statement.

Theorem 1. *For any orientable surface of genus ≥ 5, there exists a 3-regular non-3-edge-colorable graph with a polyhedral embedding in this surface.*

References

1. Albertson, M.O., Alpert, H., Belcastro, S.-M., Haas, R.: Grünbaum colorings of toroidal triangulations (manuscript) (April 2008)
2. Appel, K., Haken, W.: Every Planar Map Is Four Colorable. Contemp. Math., vol. 98. Amer. Math. Soc., Providence, RI (1989)
3. Archdeacon, D.: Problems in topological graph theory: Three-edge-coloring planar triangulations, http://www.emba.uvm.edu/~archdeac/problems/grunbaum.htm
4. Belcastro, S.-M., Kaminski, J.: Families of dot-product snarks on orientable surfaces of low genus. Graphs Combin. 23, 229–240 (2007)
5. Diestel, R.: Graph Theory, 3rd edn. Springer, Heidelberg (2005)
6. Gross, J.L., Tuker, T.W.: Topological Graph Theory. Wiley, New York (1987)
7. Grünbaum, B.: Conjecture 6. In: Tutte, W.T. (ed.) Recent Progress in Combinatorics, Proceedings of the Third Waterloo Conference on Combinatorics, May 1968, p. 343. Academic Press, New York (1969)
8. Holyer, I.: The NP-completeness of edge-coloring. SIAM J. Comput. 10, 718–720 (1981)
9. Kochol, M.: Snarks without small cycles. J. Combin. Theory Ser. B 67, 34–47 (1996)
10. Kochol, M.: Superposition and constructions of graphs without nowhere-zero k-flows. European J. Combin. 23, 281–306 (2002)
11. Tait, P.G.: Remarks on the colouring of maps. Proc. Roy. Soc. Edinburgh 10, 729 (1880)
12. Vodopivec, A.: On embedding of snarks in the torus. Discrete Math. 308, 1847–1849 (2008)

Drawing (Complete) Binary Tanglegrams
Hardness, Approximation, Fixed-Parameter Tractability[*]

Kevin Buchin[1,**], Maike Buchin[1,**], Jaroslaw Byrka[2,3],
Martin Nöllenburg[4,***], Yoshio Okamoto[5,†], Rodrigo I. Silveira[1,**],
and Alexander Wolff[2]

[1] Dept. Computer Science, Utrecht University, The Netherlands
{buchin, maike, rodrigo}@cs.uu.nl
[2] Faculteit Wiskunde en Informatica, TU Eindhoven, The Netherlands
http://www.win.tue.nl/algo
[3] Centrum voor Wiskunde en Informatica (CWI), Amsterdam, The Netherlands
j.byrka@cwi.nl
[4] Fakultät für Informatik, Universität Karlsruhe, Germany
noellenburg@iti.uka.de
[5] Grad. School of Infor. Sci. and Engineering, Tokyo Inst. of Technology, Japan
okamoto@is.titech.ac.jp

Abstract. A *binary tanglegram* is a pair $\langle S, T \rangle$ of binary trees whose leaf sets are in one-to-one correspondence; matching leaves are connected by inter-tree edges. For applications, for example in phylogenetics, it is essential that both trees are drawn without edge crossings and that the inter-tree edges have as few crossings as possible. It is known that finding a drawing with the minimum number of crossings is NP-hard and that the problem is fixed-parameter tractable with respect to that number.

We prove that under the Unique Games Conjecture there is no constant-factor approximation for general binary trees. We show that the problem is hard even if both trees are complete binary trees. For this case we give an $O(n^3)$-time 2-approximation and a new and simple fixed-parameter algorithm. We show that the maximization version of the dual problem for general binary trees can be reduced to a version of MaxCut for which the algorithm of Goemans and Williamson yields a 0.878-approximation.

1   Introduction

In this paper we are interested in drawing so-called *tanglegrams* [16], that is, pairs of trees whose leaf sets are in one-to-one correspondence. The need to

---

[*] This work was started at the 10th Korean Workshop on Computational Geometry, organized by H. Haverkort and Ch. Knauer in Schloss Dagstuhl, Germany, July 2007.

[**] Supported by the Netherlands' Organisation for Scientific Research (NWO) under BRICKS/FOCUS project no. 642.065.503 and under the project GOGO.

[***] Supported by grant WO 758/4-3 of the German Research Foundation (DFG).

[†] Partially supported by Grant-in-Aid for Scientific Research and Global COE Program from Ministry of Education, Science and Culture, Japan, and Japan Society for the Promotion of Science.

I.G. Tollis and M. Patrignani (Eds.): GD 2008, LNCS 5417, pp. 324–335, 2009.

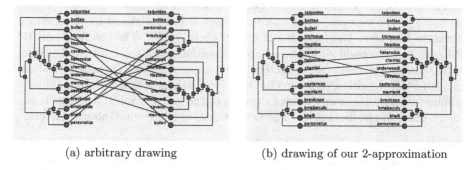

(a) arbitrary drawing   (b) drawing of our 2-approximation

Fig. 1. A binary tanglegram showing two evolutionary trees for pocket gophers [9]

visually compare pairs of trees arises in applications such as the analysis of software projects, phylogenetics, or clustering. In the first application, trees may represent package-class-method hierarchies or the decomposition of a project into layers, units, and modules. The aim is to analyze changes in hierarchy over time or to compare human-made decompositions with automatically generated ones. Whereas trees in software analysis can have nodes of arbitrary degree, trees from our second application, that is, (rooted) phylogenetic trees, are binary trees. This makes binary tanglegrams an interesting special case, see Fig. 1. Hierarchical clusterings, our third application, are usually visualized by a binary tree-like structure called *dendrogram*, where elements are represented by the leaves and each internal node of the tree represents the cluster containing the leaves in its subtree. Pairs of dendrograms stemming from different clustering processes of the same data can be compared visually using tanglegrams.

In this paper we consider binary tanglegrams if not stated otherwise. From the application point of view it makes sense to insist that (a) the trees under consideration are drawn plane (namely, without edge crossings), (b) each leaf of one tree is connected by an additional edge to the corresponding leaf in the other tree, and (c) the number of crossings among the additional edges is minimized. As in the bioinformatics literature (e.g., [13, 16]), we call this the *tanglegram layout* (TL) problem; Fernau et al. [7] refer to it as *two-tree crossing minimization*. Note that we are interested in the minimum number of crossings for visualization purposes. The number is not intended to be a tree-distance measure. Examples for such measures are nearest-neighbor interchange and subtree transfer [3].

Related problems. In graph drawing the so-called *two-sided crossing minimization problem* (2SCM) is an important problem that occurs when computing layered graph layouts. Such layouts have been introduced by Sugiyama et al. [17] and are widely used for drawing hierarchical graphs. In 2SCM, vertices of a bipartite graph are to be placed on two parallel lines (*layers*) such that vertices on one line are incident only to vertices on the other line. As in TL the objective is to minimize the number of edge crossings provided that edges are drawn as straight-line segments. In one-sided crossing minimization (1SCM) the order of the vertices on one of the layers is fixed. Even 1SCM is NP-hard [6]. In contrast to TL, a vertex in 1SCM or 2SCM can have several incident edges and the linear order of the vertices in the non-fixed layer is not restricted by the internal structure of a tree. The following is known about 1SCM. The median heuristic of Eades and Wormald [6] yields a 3-approximation and a randomized algorithm of Nagamochi [14] yields an expected 1.4664-approximation. Dujmović et al. [4] gave an FPT algorithm that runs in $O^\star(1.4664^k)$ time, where k is the minimum number of crossings in any 2-layer drawing of the given graph that respects the vertex order of the fixed layer. The $O^\star(\cdot)$-notation ignores polynomial factors.

Previous work. Dwyer and Schreiber [5] studied drawing a series of tanglegrams in 2.5 dimensions, i.e., the trees are drawn on a set of stacked two-dimensional planes. They considered a one-sided version of TL by fixing the layout of the first tree in the stack, and then, layer-by-layer, computing the leaf order of the next tree in $O(n^2 \log n)$ time each. Fernau et al. [7] showed that TL is NP-hard and gave a fixed-parameter algorithm that runs in $O^\star(c^k)$ time, where c is a constant estimated to be 1024 and k is the minimum number of crossings in any drawing of the given tanglegram. They showed that the problem can be solved in $O(n \log^2 n)$ time if the leaf order of one tree is fixed. This improves the result of Dwyer and Schreiber [5]. They also made the simple observation that the edges of the tanglegram can be directed from one root to the other. Thus the existence of a planar drawing can be verified using a linear-time upward-planarity test for single-source directed acyclic graphs [1]. Later, apparently not knowing these previous results, Lozano et al. [13] gave a quadratic-time algorithm for the same special case, to which they refer as *planar tanglegram layout*. Holten and van Wijk [10] presented a visualization tool for general tanglegrams that heuristically reduces crossings (using the barycenter method for 1SCM on a per-level base) and draws inter-tree edges in bundles (using Bézier curves).

Our results. Let us call the restriction of TL to (complete) binary trees the *(complete) binary TL problem*. We first analyze the complexity of binary TL, see Sect. 2. We show that binary TL is essentially as hard as the MINUNCUT problem. If the (widely accepted) Unique Games Conjecture holds, it is NP-hard to approximate MINUNCUT—and thus TL—within any constant factor [12]. This motivates us to consider complete binary TL. It turns out that this special case has a rich structure. We start our investigation by giving a new reduction from MAX2SAT that establishes the NP-hardness of complete binary TL.

The main result of this paper is a simple recursive factor-2 approximation algorithm for complete binary TL, see Sect. 3. It runs in $O(n^3)$ time and extends to d-ary trees. Our algorithm can also process general binary tanglegrams—without guaranteeing any approximation ratio. It works very well in practice and is quite fast when combined with a branch-and-bound procedure [15].

Next we consider a dual problem: maximize the number of edge pairs that do *not* cross. We show that this problem (for *general* binary trees) can be reduced to a version of MAXCUT for which the algorithm of Goemans and Williamson yields a 0.878-approximation.

Finally, we investigate the parameterized complexity of complete binary TL. Our parameter is the number k of crossings in an optimal drawing. We give a new FPT algorithm for complete binary TL that is much simpler and faster than the FPT algorithm for *general* binary TL by Fernau et al. [7]. The running time of our algorithm is $O(4^k n^2)$, see Sect. 4. An interesting feature of the algorithm is that the parameter does *not* drop in each level of the recursion.

Formalization. We denote the set of leaves of a tree T by $L(T)$. We are given two rooted trees S and T with n leaves each. We require that S and T are *uniquely leaf-labeled*, that is, there are bijective labeling functions $\lambda_S : L(S) \to \Lambda$ and $\lambda_T : L(T) \to \Lambda$, where Λ is a set of labels, for example, $\Lambda = \{1, \ldots, n\}$. These labelings define a set of new edges $\{uv \mid u \in L(S), v \in L(T), \lambda_S(u) = \lambda_T(v)\}$, the *inter-tree edges*. The TL problem consists of finding plane drawings of S and T that minimize the number of induced crossings of the inter-tree edges, assuming that edges are drawn as straight-line segments. We insist that the leaves in $L(S)$ are placed on the line $x = 0$ and those in $L(T)$ on the line $x = 1$. The trees S and T themselves are drawn to the left of $x = 0$ and to the right of $x = 1$, respectively. For an example see Fig. 1. Given uniquely leaf labeled trees S and T, we denote the resulting instance of TL by $\langle S, T \rangle$.

The TL problem is purely combinatorial: Given a tree T, we say that a linear order of $L(T)$ is *compatible* with T if for each node v of T the nodes in the subtree of v form an interval in the order. Given a permutation π of $\{1, \ldots, n\}$, we call (i, j) an *inversion* in π if $i < j$ and $\pi(i) > \pi(j)$. For fixed orders σ of $L(S)$ and τ of $L(T)$ we define the permutation $\pi_{\tau,\sigma}$, which for a given position in τ returns the position in σ of the leaf having the same label. Now the TL problem consists of finding an order σ of $L(S)$ compatible with S and an order τ of $L(T)$ compatible with T such that the number of inversions in $\pi_{\tau,\sigma}$ is minimum.

2 Complexity

In this section we consider the complexity of binary TL, which Fernau et al. [7] have shown to be NP-complete for general binary tanglegrams. We strengthen their findings in two ways. First, we show that it is unlikely that an efficient constant-factor approximation for general binary TL exists. Second, we show that TL remains hard even when restricted to *complete* binary tanglegrams.

We start by showing that binary TL is essentially as hard as the MINUNCUT problem. This relates the existence of a constant-factor approximation for TL to the Unique Games Conjecture (UGC) by Khot [11]. The UGC became famous when it was discovered that it implies optimal hardness-of-approximation results for problems such as MAXCUT and VERTEXCOVER, and forbids constant factor-approximation algorithms for problems such as MINUNCUT and SPARSESTCUT. We reduce the MINUNCUT problem to the TL problem, which, by the result of Khot and Vishnoi [12], makes it unlikely that an efficient constant-factor approximation for TL exists.

The MINUNCUT problem is defined as follows. Given an undirected graph $G = (V, E)$, find a partition (V_1, V_2) of the vertex set V that minimizes the number of edges that are not cut by the partition, that is, $\min_{(V_1,V_2)} |\{uv \in E : u,v \in V_1 \text{ or } u,v \in V_2\}|$. Note that computing an optimal solution to MinUncut is equivalent to computing an optimal solution to MaxCut. Nevertheless, the MinUncut problem is more difficult to approximate.

Theorem 1. *Under the Unique Games Conjecture it is NP-hard to approximate the TL problem for general binary trees within any constant factor.*

Proof. As mentioned above we reduce from the MinUncut problem. Our reduction is similar to the one in the NP-hardness proof by Fernau et al. [7].

Consider an instance $G = (V, E)$ of the MinUncut problem. We construct a TL instance $\langle S, T \rangle$ as follows. The two trees S and T are identical and there are three groups of edges connecting leaves of S to leaves of T. For simplicity we define multiple edges between a pair of leaves. In the actual trees we can replace each such leaf by a binary tree with the appropriate number of leaves.

Suppose $V = \{v_1, v_2, \ldots, v_n\}$, then both S and T are constructed as follows. There is a *backbone* path $(v_1^1, v_1^2, v_2^1, v_2^2, \ldots, v_n^1, v_n^2, a)$ from the root node v_1^1 to a leaf a. Additionally, there are leaves $l_S(v_i^j)$ and $l_T(v_i^j)$ attached to each node v_i^j for $i \in \{1, \ldots, n\}$ and $j \in \{1, 2\}$ in S and T, respectively. The edges form the following three groups.

Group A contains n^{11} edges connecting $l_S(a)$ with $l_T(a)$.
Group B contains for every $v_i \in V$ n^7 edges connecting $l_S(v_i^1)$ with $l_T(v_i^2)$, and n^7 edges connecting $l_S(v_i^2)$ with $l_T(v_i^1)$.
Group C contains for every $v_i v_j \in E$ a single edge from $l_S(v_i^1)$ to $l_T(v_j^1)$.

Next we show how to transform an optimal solution of the MinUncut instance into a solution of the corresponding TL instance. Suppose that in the optimal partition (V_1^*, V_2^*) of G there are k edges that are not cut. Then we claim that there exists a drawing of $\langle S, T \rangle$ such that $k \cdot n^{11} + O(n^{10})$ pairs of edges cross. It suffices to draw, for each vertex $v_i \in V_1^*$ ($v_i \in V_2^*$), the leaves $l_S(v_i^1)$ and $l_T(v_i^2)$ above (below) the backbones, and the nodes $l_S(v_i^2)$ and $l_T(v_i^1)$ below (above) the backbones. It remains to count: there are $k \cdot n^{11}$ A–C crossings, no A–B crossings, $O(n^{10})$ B–C crossings, and $O(n^4)$ C–C crossings.

Now suppose there exists an α-approximation algorithm for the TL problem with some constant α. Applying this algorithm to the instance $\langle S, T \rangle$ defined above yields a drawing $D(S, T)$ with at most $\alpha \cdot k \cdot n^{11} + O(n^{10})$ crossings. Let us assume that n is much larger than α. We show that from such a drawing $D(S, T)$ we would be able to reconstruct a cut (V_1, V_2) in G with at most $\alpha \cdot k$ edges uncut. First, observe that if a node $l_S(v_i^1)$ is drawn above (below) the backbone in $D(S, T)$, then $l_T(v_i^2)$ must be drawn on the same side of the backbone, otherwise it would result in n^{18} A–B crossings. Similarly $l_S(v_i^2)$ must be on the same side as $l_T(v_i^1)$. Then observe that if a node $l_S(v_i^1)$ is drawn above (below) the backbone in $D(S, T)$, then $l_S(v_i^2)$ must be drawn below (above) the backbone, otherwise there would be $O(n^{14})$ B–B crossings. Finally, observe that if we interpret the set of vertices v_i for which $l_S(v_i^1)$ is drawn above the backbone as a set V_1 of a partition of G, then this partition leaves at most $\alpha \cdot k$ edges from E uncut.

Hence, an α-approximation for the TL problem provides an α-approximation for the MINUNCUT problem, which contradicts the UGC.   □

The above negative result for (general) binary TL is our motivation to investigate the complexity of complete binary TL. It turns out that even this special case is hard. Unlike Fernau et al. [7] who show hardness of binary TL by a reduction from MAXCUT using extremely unbalanced trees, we use a quite different reduction from a variant of MAX2SAT (see full version for the proof [2]).

Theorem 2. *The TL problem is NP-hard even for complete binary tanglegrams.*

3 Approximation

We now present our main result, a 2-approximation algorithm for complete binary TL that runs in $O(n^3)$ time. The idea is to split a given tanglegram recursively at the roots of the two trees into two subinstances, each again consisting of a pair of complete binary trees. Let $\langle S, T \rangle$ be a subinstance of $\langle S_0, T_0 \rangle$ with subtrees $S \subseteq S_0$ and $T \subseteq T_0$ rooted at nodes $v_S \in S_0$ and $v_T \in T_0$, respectively (see Fig. 2). When treating $\langle S, T \rangle$, we use the following pieces of information.

Firstly, associated with v_S and v_T we have labels ℓ_S and ℓ_T that indicate what choices in the recursion so far led to the current subinstances. A label is a bit string that represent the choices (swap/do not swap children) made at each node, from the first recursive step to the current one (see Fig. 3).

We also assign labels to some other subtrees of $\langle S_0, T_0 \rangle$ apart from S and T. Given a leaf $v \in T_0 \setminus T$, we define the *largest T-avoiding tree* of v to be the largest complete binary subtree of T_0 that contains v, but not T. Largest S-avoiding trees are defined analogously for leaves in S_0. Each largest S- or T-avoiding tree receives a label in the same way as S and T. Note that the labels of the avoiding trees are relative to the labels of v_S and v_T, that is, a different subinstance leads to different labels. If we refer (in the context of a subinstance $\langle S, T \rangle$) to the label of a leaf $v \in T_0$, we mean the label of the largest T-avoiding tree of v.

Secondly, since S and T are part of a larger tree, some leaves of S may not have the matching leaf in T (and vice versa). This means that at some previous step such leaves were matched to leaves in some other subtrees, above or below $\langle S, T \rangle$. We do not know exactly to which leaves they are matched, but we do know, for each leaf, the label of the subtree that contains the matching leaf.

At each level of the recursion we have to choose between one out of four configurations. Let the current subinstance be given by $\langle S, T \rangle = \langle (S_1, S_2), (T_1, T_2) \rangle$. At each node v_S on the left side, we must choose between having S_1 above S_2 or the other way around. On the right side for v_T, there are also two different ways of placing T_1 and T_2. For each of the four configurations we invoke the algorithm twice recursively: for the top half and for the bottom half. We return the configuration with the smallest number of crossings.

When counting the crossings that a configuration creates, we distinguish two types: *current-level* and *lower-level* crossings.

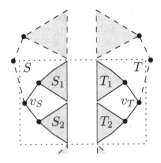
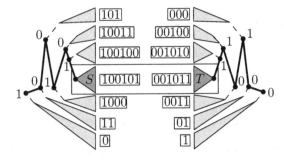

Fig. 2. Context of subinstance $\langle S,T \rangle = \langle (S_1, S_2), (T_1, T_2) \rangle$

Fig. 3. Labels for a particular subinstance $\langle S,T \rangle$. The numbers at the nodes show the choices taken (swap/do not swap children) that led to S and T.

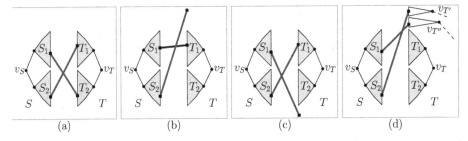

Fig. 4. Different types of current-level crossings. Type (d) is considered current-level only if the right leaves of the crossing edges have different labels, that is, if $\ell_{T'} \neq \ell_{T''}$.

Current-level crossings are crossings that can be avoided at this level by choosing one of the four configurations for the subtrees, independently of the choices to be done elsewhere in the recursion. Figure 4 illustrates the four different types. For type (d), we remark that crossings are considered to be *current-level* only if the largest S- and T-avoiding trees that contain the endpoints of the edges outside S and T are different. Crossings of type (d) where that is not the case cannot be counted at this point. We call them *indeterminate crossings*.

Lower-level crossings are crossings that appear based on choices taken by solving the subinstances of S and T recursively. We cannot do anything about them at this level, but we know their exact number after solving the subinstances.

Here is a sketch of the algorithm.

1. For all four choices of arranging $\{S_1, S_2\}$ and $\{T_1, T_2\}$, compute the total number of lower-level crossings recursively. Before each recursive call $\langle S_i, T_j \rangle$, we assign proper labels to some of the leaves of S and T, as follows. All leaves in S_i that connect to T_{3-j} (that is, T_1 if $j = 2$, T_2 otherwise) get the label ℓ_T with a 0 or 1 appended depending on whether T_j is above or below T_{3-j}. Then we do the analogue for all leaves of T_j connected to S_{3-i}.

2. For each choice $\langle S_i, T_j \rangle$ compute the number of current-level crossings (details below).
3. Return the choice that has the smallest sum of lower-level and current-level crossings.

The labels are needed to propagate as much information as possible to the smaller subinstances. For example, even though at this stage of the recursion it is clear that the leaves of, say T_{3-j}, are above the leaves of the subtrees below T, once we recurse into the top subinstance, this information will be lost, implying that what was a current-level crossing at this stage, will become an indeterminate crossing later. The labeling allows to prevent this loss of information.

The number of current-level crossings can be computed in linear time as follows. We go through all inter-tree edges incident to leaves of S and put each edge into one of at most $O(\log n)$ different classes, depending on the labels of the endpoints outside S. Then we repeat the same for T. This takes linear time. Depending on where the largest S- or T-avoiding trees go (above or below), all edge pairs belonging to a specific pair of labels do or do not intersect. Hence we can count the total number of current-level crossings by multiplying the cardinalities of the $O(\log^2 n)$ pairs of classes whose edges all intersect each other.

The running time of the algorithm satisfies the recurrence $T(n) \leq 8T(n/2) + O(n)$, which solves to $T(n) = O(n^3)$. We now prove that the algorithms yields a 2-approximation. In the full version [2] we show that our analysis is tight.

Theorem 3. *Given a complete binary tanglegram $\langle S_0, T_0 \rangle$ with n inter-tree edges, the recursive algorithm computes in $O(n^3)$ time a drawing of $\langle S_0, T_0 \rangle$ that has at most twice as many crossings as an optimal drawing.*

Proof sketch. Fix an optimal drawing δ of $\langle S_0, T_0 \rangle$. The algorithm tries, for a given subinstance $\langle S, T \rangle$ of $\langle S_0, T_0 \rangle$, all four possible layouts of $S = (S_1, S_2)$ and $T = (T_1, T_2)$. Assume that in δ, $\langle S, T \rangle$ is drawn as $\langle (S_1, S_2), (T_1, T_2) \rangle$. We distinguish between four different areas for the endpoints of the edges: above $\langle S, T \rangle$, in $\langle S_1, T_1 \rangle$, in $\langle S_2, T_2 \rangle$, and below $\langle S, T \rangle$. We number these regions from 0 to 3 (see Fig. 5(a)). This allows us to classify the edges into 16 groups (two of which, 0–0 and 3–3, are not relevant). We denote the number of i–j edges, that is, edges from area i to area j, by n_{ij} (for $i, j \in \{0, 1, 2, 3\}$). Figures 5(b) and 5(c) show the 14 relevant groups of edges.

The only edge crossings that our recursive algorithm cannot take into account are the indeterminate crossings, which occur when the two edges connect to leaves above or below $\langle S, T \rangle$ that are in the same largest S- or T-avoiding tree. This is the case if both leaves have the same label. Such crossings cannot be predicted from the current subinstance because they depend on the relative position of the other two endpoints of the edges. We can, however, bound the number of these crossings.

We observe that any crossing of that type at the current subinstance was, in some previous step of the recursion, a crossing between two 1–2 edges or two 2–1 edges. We can upper-bound the number of these crossings by $\binom{n_{12}}{2} + \binom{n_{21}}{2}$. Let c_{alg} be the number of crossings in the solution produced by the algorithm, and let c_{opt} be the number of crossings of δ. Then

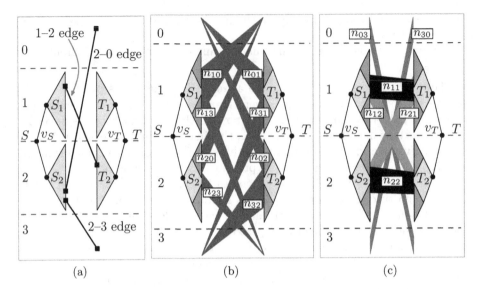

Fig. 5. For an instance $\langle (S_1, S_2), (T_1, T_2) \rangle$ the locations of the edge endpoints are divided into four areas (numbered 0–3); each edge is classified accordingly (a). This defines 14 groups of relevant edges, where n_{ij} denotes the number of i–j edges (b & c).

$$c_{\mathrm{alg}} \le c_{\mathrm{opt}} + \binom{n_{12}}{2} + \binom{n_{21}}{2} \le c_{\mathrm{opt}} + (n_{12}^2 + n_{21}^2)/2. \tag{1}$$

Since our (sub)trees are complete, we have $n_{10} + n_{12} + n_{13} = n_{01} + n_{21} + n_{31}$ and $n_{01} + n_{02} + n_{03} = n_{10} + n_{20} + n_{30}$. These two equalities yield $n_{12} \le n_{01} - n_{10} + n_{21} + n_{31}$ and $n_{01} - n_{10} \le n_{20} + n_{30}$, respectively, and thus we obtain $n_{12} \le n_{20} + n_{30} + n_{21} + n_{31}$ or, equivalently, $n_{12}^2 \le n_{12} \cdot (n_{20} + n_{30} + n_{21} + n_{31})$.

It is easy to verify that all the terms on the right-hand side of the last inequality count crossings that cannot be avoided and must be present in the optimal solution as well. Hence $n_{12}^2 \le c_{\mathrm{opt}}$, and symmetrically $n_{21}^2 \le c_{\mathrm{opt}}$. Plugging this into (1) yields $c_{\mathrm{alg}} \le 2 \cdot c_{\mathrm{opt}}$. □

General binary trees. Our recursive algorithm can also be applied to general, non-complete tanglegrams. Then, however, the approximation factor does not hold any more. Nöllenburg et al. [15] have evaluated several heuristics for TL; our recursive algorithm turned out to be a successful method for both complete and general binary tanglegrams.

Generalization to d-ary trees. The algorithm can also be generalized to complete d-ary trees. The recurrence relation of the running time changes to $T(n) \le d \cdot (d!)^2 \cdot T(n/d) + O(n)$ since we need to consider all $d!$ subtree orderings of both trees, each triggering d subinstances of size n/d. This resolves to

$T(n) = O(n^{1+2\log_d(d!)})$. At the same time the approximation factor increases to $1 + \binom{d}{2}$.

Maximization version. Instead of the original TL problem, we now consider the dual problem TL* of maximizing the number of pairs of edges that do not cross. The tasks of finding optimal solutions for these problems are equivalent, but from the perspective of approximation it makes quite a difference which of the two problems we consider. Now we do not assume that we draw *binary* trees. Instead, if an internal node has more than two children, we assume that we may only choose between a given permutation of the children and the reverse permutation obtained by flipping the whole block of children.

In contrast to the TL problem, which is hard to approximate as we have shown in Theorem 1, the TL* problem has a constant-factor approximation algorithm. We show this (see full version [2]) by reducing TL* to a constrained version of the MAXCUT problem, which can be approximately solved with the semidefinite programming rounding algorithm of Goemans and Williamson [8].

Theorem 4. *There exists a 0.878-approximation for the TL* problem.*

4   Fixed-Parameter Tractability

We consider the following parameterized problem. Given a complete binary TL instance $\langle S, T \rangle$ and a non-negative integer k, decide whether there exists a TL of S and T with at most k induced crossings. Our algorithm for this problem uses a labeling strategy, just as our algorithm in Sect. 3. However, here we do not select the subinstance that gives the minimum number of lower-level crossings, but we consider all subinstances and recurse on them. Thus, our algorithm traverses a search tree of branching factor 4. For the search tree to have bounded height, we need to ensure that whenever we go to a subinstance, the parameter value decreases at least by one. For efficient bookkeeping we consider current-level crossings only. At first sight this seems problematic: if a subinstance does not incur any current-level crossings, the parameter will not drop. The following key lemma—which does not hold for general binary trees—shows that there is a way out. It says that if there is a subinstance without current-level crossings, then we can ignore the other three subinstances and do not have to branch.

Lemma 1. *Let $\langle S, T \rangle$ be a complete binary TL instance, and let v_S be a node of S and v_T a node of T such that v_S and v_T have the same distance to their respective root. Further, let (S_1, S_2) be the subtrees incident to v_S and let (T_1, T_2) be the subtrees incident to v_T. If the subinstance $\langle (S_1, S_2), (T_1, T_2) \rangle$ does not incur any current-level crossings, then each of the subinstances $\langle (S_1, S_2), (T_2, T_1) \rangle$, $\langle (S_2, S_1), (T_1, T_2) \rangle$, and $\langle (S_2, S_1), (T_2, T_1) \rangle$ has at least as many crossings as $\langle (S_1, S_2), (T_1, T_2) \rangle$, for any fixed ordering of the leaves of S_1, S_2, T_1 and T_2.*

Proof. If the subinstance $\langle (S_1, S_2), (T_1, T_2) \rangle$ does not incur any current-level crossings, there are no edges between S_1 and T_2 or between S_2 and T_1. We

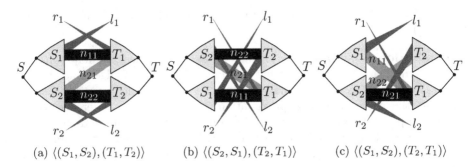

Fig. 6. Edge types and crossings of the instance $\langle S, T \rangle$

only consider the first case; the second is symmetric. We categorize the inter-tree edges originating from the four subtrees according to their destination—see Fig. 6(a)—and denote the numbers of edges of the various types by n_{11}, n_{21}, n_{22}, l_1, l_2, r_1, and r_2. Since we consider complete binary trees, we obtain $l_1 = r_1 + n_{21}$, $r_2 = l_2 + n_{21}$, and $r_1 + n_{11} = l_2 + n_{22}$.

We fix an ordering σ of the leaves of the subtrees S_1, S_2, T_1, T_2. We first compare the number of crossings in $\langle (S_1, S_2), (T_1, T_2) \rangle$ with the number of crossings in $\langle (S_2, S_1), (T_2, T_1) \rangle$, see Fig. 6(b). The subinstance $\langle (S_1, S_2), (T_1, T_2) \rangle$ can have at most $n_{21}(n_{11} + n_{22})$ crossings that do not occur in $\langle (S_2, S_1), (T_2, T_1) \rangle$. However, $\langle (S_2, S_1), (T_2, T_1) \rangle$ has at least $l_1(l_2 + n_{21} + n_{22}) + l_2 n_{11} + r_2(r_1 + n_{21} + n_{11}) + r_1 n_{22}$ crossings that do not appear in $\langle (S_1, S_2), (T_1, T_2) \rangle$. Plugging in the above equalities for l_1 and r_2, we get $(r_1 + n_{21})(l_2 + n_{21} + n_{22}) + l_2 n_{11} + (l_2 + n_{21})(r_1 + n_{21} + n_{11}) + r_1 n_{22} \geq n_{21}(n_{11} + n_{22})$. Thus, the subinstance $\langle (S_2, S_1), (T_2, T_1) \rangle$ has at least as many crossings with respect to σ as $\langle (S_1, S_2), (T_1, T_2) \rangle$ has.

Next, we compare the number of crossings in $\langle (S_1, S_2), (T_1, T_2) \rangle$ with the number of crossings in $\langle (S_1, S_2), (T_2, T_1) \rangle$, see Fig. 6(c). Now the number of additional crossings of $\langle (S_1, S_2), (T_1, T_2) \rangle$ is at most $n_{21} n_{22}$, and the subinstance $\langle (S_1, S_2), (T_2, T_1) \rangle$ has at least $(r_1 + n_{11})(r_2 + n_{22}) + r_2 n_{21}$ crossings more. With the equality $r_1 + n_{11} = l_2 + n_{22}$ and the inequality $r_2 + n_{22} \geq n_{21}$ we get $(r_1 + n_{11})(r_2 + n_{22}) + r_2 n_{21} \geq n_{22} n_{21}$. Thus, the subinstance $\langle (S_1, S_2), (T_2, T_1) \rangle$ has at least as many crossings with respect to σ as $\langle (S_1, S_2), (T_1, T_2) \rangle$ has.

By symmetry, the same holds for $\langle (S_2, S_1), (T_1, T_2) \rangle$.   □

Thus, to decompose the instance into four subinstances we spend $O(n^2)$ time. Therefore we spend $O(4^k n^2)$ time to produce all leaves of our bounded-height search tree (omitting details). At each leaf of the search tree, we obtain a certain layout of $\langle S, T \rangle$, and the accumulated number of current-level crossings is at most k. This, however, does not mean that the total number of crossings is at most k since we did not keep track of the indeterminate crossings. Therefore, at each leaf we still need to check how many crossings the corresponding layout has. This can be done in $O(n \log n)$ time. If one of the leaves yields at most k crossings, the algorithm outputs "Yes" and the layout; otherwise it outputs "No".

Theorem 5. *The algorithm sketched above solves the parameterized version of complete binary TL in $O(4^k n^2)$ time.*

References

1. Bertolazzi, P., Di Battista, G., Mannino, C., Tamassia, R.: Optimal upward planarity testing of single-source digraphs. SIAM J. Comput. 27(1), 132–169 (1998)
2. Buchin, K., Buchin, M., Byrka, J., Nöllenburg, M., Okamoto, Y., Silveira, R.I., Wolff, A.: Drawing (complete) binary tanglegrams: Hardness, approximation, fixed-parameter tractability (2008), http://arxiv.org/abs/0806.0920 Arxiv report
3. DasGupta, B., He, X., Jiang, T., Li, M., Tromp, J., Zhang, L.: On distances between phylogenetic trees. In: Proc. 18th Annu. ACM-SIAM Sympos. Discrete Algorithms (SODA 1997), pp. 427–436 (1997)
4. Dujmović, V., Fernau, H., Kaufmann, M.: Fixed Parameter Algorithms for ONE-SIDED CROSSING MINIMIZATION Revisited. In: Liotta, G. (ed.) GD 2003. LNCS, vol. 2912, pp. 332–344. Springer, Heidelberg (2004)
5. Dwyer, T., Schreiber, F.: Optimal leaf ordering for two and a half dimensional phylogenetic tree visualization. In: Proc. Australasian Sympos. Inform. Visual (In-Vis.au 2004). CRPIT, vol. 35, pp. 109–115. Australian Comput. Soc. (2004)
6. Eades, P., Wormald, N.: Edge crossings in drawings of bipartite graphs. Algorithmica 10, 379–403 (1994)
7. Fernau, H., Kaufmann, M., Poths, M.: Comparing trees via crossing minimization. In: Ramanujam, R., Sen, S. (eds.) FSTTCS 2005. LNCS, vol. 3821, pp. 457–469. Springer, Heidelberg (2005)
8. Goemans, M.X., Williamson, D.P.: Improved approximation algorithms for maximum cut and satisfiability problems using semidefinite programming. J. ACM 42(6), 1115–1145 (1995)
9. Hafner, M.S., Sudman, P.D., Villablanca, F.X., Spradling, T.A., Demastes, J.W., Nadler, S.A.: Disparate rates of molecular evolution in cospeciating hosts and parasites. Science 265, 1087–1090 (1994)
10. Holten, D., van Wijk, J.J.: Visual comparison of hierarchically organized data. In: Proc. 10th Eurographics/IEEE-VGTC Sympos. Visualization (EuroVis 2008), pp. 759–766 (2008)
11. Khot, S.: On the power of unique 2-prover 1-round games. In: Proc. 34th Annu. ACM Sympos. Theory Comput (STOC 2002), pp. 767–775 (2002)
12. Khot, S., Vishnoi, N.K.: The unique games conjecture, integrality gap for cut problems and embeddability of negative type metrics into l_1. In: Proc. 46th Annu. IEEE Sympos. Foundat. Comput. Sci. (FOCS 2005), pp. 53–62 (2005)
13. Lozano, A., Pinter, R.Y., Rokhlenko, O., Valiente, G., Ziv-Ukelson, M.: Seeded tree alignment and planar tanglegram layout. In: Giancarlo, R., Hannenhalli, S. (eds.) WABI 2007. LNCS (LNBI), vol. 4645, pp. 98–110. Springer, Heidelberg (2007)
14. Nagamochi, H.: An improved bound on the one-sided minimum crossing number in two-layered drawings. Discrete Comput. Geom. 33(4), 565–591 (2005)
15. Nöllenburg, M., Holten, D., Völker, M., Wolff, A.: Drawing binary tanglegrams: An experimental evaluation. Arxiv report (2008), http://arxiv.org/abs/0806.0928
16. Page, R.D.M. (ed.): Tangled Trees: Phylogeny, Cospeciation, and Coevolution. University of Chicago Press (2002)
17. Sugiyama, K., Tagawa, S., Toda, M.: Methods for visual understanding of hierarchical system structures. IEEE Transactions on Systems, Man, and Cybernetics 11(2), 109–125 (1981)

Two Polynomial Time Algorithms for the Metro-line Crossing Minimization Problem[*]

Evmorfia Argyriou[1], Michael A. Bekos[1], Michael Kaufmann[2],
and Antonios Symvonis[1]

[1] School of Applied Mathematical & Physical Sciences,
National Technical University of Athens, Greece
{fargyriou,mikebekos,symvonis}@math.ntua.gr
[2] University of Tübingen, Institute for Informatics, Germany
mk@informatik.uni-tuebingen.de

Abstract. The *metro-line crossing minimization (MLCM)* problem was recently introduced as a response to the problem of drawing metro maps or public transportation networks, in general. According to this problem, we are given a planar, embedded graph $G = (V, E)$ and a set L of simple paths on G, called *lines*. The main task is to place the lines on G, so that the number of crossings among pairs of lines is minimized.

Our main contribution is two polynomial time algorithms. The first solves the general case of the MLCM problem, where the lines that traverse a particular vertex of G are allowed to use any side of it to either "enter" or "exit", assuming that the endpoints of the lines are located at vertices of degree one. The second one solves more efficiently the restricted case, where only the left and the right side of each vertex can be used.

To the best of our knowledge, this is the first time where the general case of the MLCM problem is solved. Previous work was devoted to the restricted case of the MLCM problem under the additional assumption that the endpoints of the lines are either the topmost or the bottommost in their corresponding vertices, i.e., they are either on top or below the lines that pass through the vertex. Even for this case, we improve a known result of Asquith et al. from $O(|E|^{5/2}|L|^3)$ to $O(|V|(|E| + |L|))$.

1 Introduction

A metro map can be modeled as a tuple (G, L), which consists of a connected graph $G = (V, E)$, referred to as the *underlying network*, and a set L of simple paths on G. The nodes of G correspond to train stations, an edge connecting two nodes implies that there exists a railway track connecting them, whereas the paths illustrate the lines connecting terminal stations. Then, the process of constructing a metro map consists of a sequence of steps. Initially, one has to draw the underlying network nicely. Then, the lines have to be properly added

---

[*] This work has been funded by the project PENED-2003. PENED-2003 is co-funded by the European Social Fund (75%) and Greek National Resources (25%).

I.G. Tollis and M. Patrignani (Eds.): GD 2008, LNCS 5417, pp. 336–347, 2009.

into the visualization and, finally, a labeling of the map has to be performed over the most important features.

In the graph drawing and computational geometry literature, the focus so far has been nearly exclusively on the first and the third step. Closely related to the first step are the works of Hong et al. [5], Merrick and Gudmundsson [6], Nöllenburg and Wolff [7] and Stott and Rodgers [8]. The map labeling problem has also attracted the interest of several researchers. An extensive bibliography on map labeling is maintained on-line by Strijk and Wolff [9]. Interestingly enough, the intermediate problem of adding the line set into the underlying network was recently introduced by Benkert et al. [3], followed by [2]. Since crossings within a visualization are often considered as the main source of confusion, the main goal is to draw the lines, so that they cross each other as few times as possible. This problem is referred to as the *metro-line crossing minimization problem (MLCM)*.

1.1 Problem Definition

The input of the metro-line crossing minimization problem consists of a connected, embedded, planar graph $G = (V, E)$ and a set $L = \{l_1, l_2 \ldots l_k\}$ of simple paths on G, called *lines*. We will refer to G as the underlying network and to the nodes of G as *stations*. We also refer to the endpoints of each line as its *terminals*. In this paper, we study the case where all line terminals are located at stations of degree one, which are referred to as *terminal stations*. Stations of degree greater than one are referred to as *internal stations*. The stations are represented as particular shapes (usually as rectangles but in general as polygons). The sides of each station that each line may use to either "enter" or "exit" the station are also specified as part of the input. Motivated by the fact that a line cannot make a $180°$ turn within a station, we do not permit a line to use the same side of a station to both "enter" and "exit".

The output of the MLCM problem should specify an ordering of the lines at each side of each station, so that the number of crossings is minimized.

Each line l_i consists of a sequence of edges $e_1 = (v_0, v_1), \ldots, e_d = (v_{d-1}, v_d)$. Stations v_0 and v_d are the terminals of line l_i. Equivalently, we say that l_i *terminates* or *has terminals* at v_0 and v_d. By $|l_i|$ we denote the length of line l_i.

Each line that traverses a station u has to touch two of the sides of u at some points (one when it "enters" u and one when it "exits" u). These points are referred to as *tracks* (see the dark-gray colored bullets on the boundary of each station in Fig. 1b). In general, we may permit tracks to all sides of each station, (see Fig. 1a). In the case where the stations are represented as rectangles, this model is referred to as the *4-side* model. In the general case where the stations are represented as polygons of at most k sides, this model is referred to as the *k-side* model. A more restricted model, referred to as the *2-side* model, is the one where i) the stations are represented as rectangles and ii) all lines that traverse a station may use only its left and right side (see Fig. 1b).

A particularly interesting case that arises under the 2-side model is the one where the lines that terminate at a station occupy its topmost and bottommost tracks, in the following referred to as *top* and *bottom station ends*, respectively

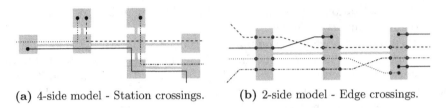

(a) 4-side model - Station crossings.   **(b)** 2-side model - Edge crossings.

Fig. 1. The underlying network is the gray colored graph

(see Fig. 1b). This is to emphasize that the line terminates at that station. The variant of the MLCM problem that fulfills this restriction is referred to as the *metro-line crossing minimization problem with station ends (MLCM-SE)*. If additionally, the information whether a line terminates at a top or at a bottom station end in its terminal station is specified as part of the input, the corresponding problem is referred to as *metro-line crossing minimization problem with fixed station ends (MLCM-FixedSE)*.

A further refinement of the MLCM problem concerns the location of the crossings among pairs of lines. If the relative order of two lines changes between two consecutive stations, then the two lines must intersect between these stations (see Fig. 1b). We call this an *edge crossing*. As opposed to an edge crossing, a *station crossing* occurs inside a station. For aesthetic reasons, we want to avoid station crossings whenever this is possible (e.g. in the case of 4-side model this is not always feasible; see Fig. 1a).

1.2   Previous Work and Our Results

The first results on the MLCM problem were presented by Benkert et al. in [3], who devised a dynamic-programming algorithm that runs in $O(|L|^2)$ time for the restricted case where the crossings are minimized along a single edge of G. Bekos et al. [2] proved that the MLCM-SE problem is NP-complete even in the case where the underlying network is a path. They also proved that the MLCM-FixedSE problem can be solved in $O(|V| + \log d \sum_{i=1}^{|L|} |l_i|)$, in the case where the underlying network is a tree of degree d. Extending the work of Bekos et al., Asquith et al. [1] proved that the MLCM-FixedSE problem is also solvable in polynomial time in the case where the underlying network is an arbitrary planar graph. The time complexity of their algorithm was $O(|E|^{5/2}|L|^3)$. They also proposed an integer linear program which solves the MLCM-SE problem.

This paper is structured as follows: In Section 2, we present a polynomial time algorithm, which runs in $O((|E| + |L|^2)|E|)$ time for the MLCM problem under the k-side model, assuming that the line terminals are located at stations of degree one. To the best of our knowledge no results are currently known regarding this general model. In Section 3, we present a faster algorithm for the special case of 2-side restriction. The time complexity of the proposed algorithm is $O(|V||E| + \sum_{i=1}^{|L|} |l_i|)$. It can also be employed to solve the MLCM-FixedSE problem, which drastically improves the running time of the algorithm of Asquith et al. [1] from $O(|E|^{5/2}|L|^3)$ to $O(|V||E| + |V||L|)$. We conclude in Section 4 with open problems and future work.

2   The MLCM Problem under the k-Side Model

To simplify the description of our algorithm and to make the accompanying figures simpler, we restrict our presentation to the MLCM problem under the 4-side model, i.e., we assume that each station is represented as a rectangle and we permit tracks to all four sides of each station. Our algorithm for the case of k-side model is identical, since it is based on recursion over the edges of the underlying network. Recall that all line terminals are located at stations of degree one, the lines can terminate at any track of their terminal stations, and, finally, the sides of each station that each line may use to either "enter" or "exit" are specified as part of the input. We further assume that an internal station always exists within the underlying network, otherwise the problem can be solved trivially.

The basic idea of our algorithm is to decompose the underlying network by removing an arbitrary edge out of the edges that connect two internal stations (and, consequently, appropriately partitioning the set of lines that traverse this edge), then recursively solve the subproblem and, finally, derive a solution of the initial problem by i) re-inserting the removed edge and ii) connecting the partitioned lines along the re-inserted edge.

2.1   Base of Recursion

The base of the recursion corresponds to the case of a graph G_B consisting of a "central station" u containing no terminals and a particular number of terminal stations, say $v_1, v_2 \ldots v_d$, incident to u (see Fig. 2c). To cope with this case, we first group all lines that have exactly the same terminals into a single line, which is referred to as *bundle*. The notion of bundles corresponds to the fact that lines with same terminals are drawn in a uniform fashion, i.e., occupying consecutive tracks at their common stations. So, in an optimal solution once a bundle is drawn, it can be safely replaced by its corresponding lines without affecting the optimality of the solution. In Fig. 2c, lines belonging to the same bundle have been drawn with the same type of non-solid line. Note that single lines are also treated as bundles in order to maintain a uniform terminology (refer to the solid lines of Fig. 2c). Then, the number of bundles of each terminal station is bounded by the degree of the "central station" u.

In order to route the bundles along the edges of G_B, we will make use of the *Euler tour numbering* that was proposed by Bekos et al. [2]. Let v be a terminal station of G_B. The Euler tour numbering of the terminal stations $v_1, v_2, \ldots, v_d$ of G_B with respect to v is a function $\text{ETN}_v : \{v_1, v_2, \ldots, v_d\} \to \{0, 1, \ldots, d-1\}$. More precisely, given a terminal station v of G_B, we number all terminal stations of G_B according to the order of first appearance when moving clockwise along the external face of G_B starting from station v, which is assigned the zero value. Note that such a numbering is unique with respect to v and we refer to it as

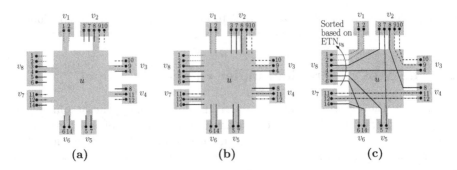

Fig. 2. Illustration of the base of the recursion. The numbering of the lines is arbitrary.

the Euler tour numbering starting from station v or simply as ETN_v. Also, note that the computation of only one numbering is enough in order to compute the corresponding Euler tour numberings from any other terminal station of G_B, since $\text{ETN}_{v'}(w) = (\text{ETN}_v(w) - \text{ETN}_v(v')) \mod d$.

Our approach is outlined as follows: We first sort in ascending order the bundles at each terminal station v based on the Euler tour numbering ETN_v of their destinations (see Fig. 2a). This implies the desired ordering of the bundles along the side of each terminal station that is incident to the "central station" u. We will denote by $\text{BND}(v)$ the ordered set of bundles of each terminal station v. Then, we pass these bundles from each terminal station to the "central station" u along their common edge without introducing any crossings (see Fig. 2b). This will also imply an ordering of the bundles at each side of the "central station" u. To complete the routing procedure, it remains to connect equal bundles in the interior of the "central station" u, which may imply crossings (see Fig. 2c). Note that only necessary station crossings are created, since the underlying network is planar and from the Euler tour numbering it follows that no edge crossings will eventually occur. So, the optimality of the solution follows trivially.

2.2   Description of the Recursive Algorithm

Having specified the base of the recursion, we now proceed to describe our recursive algorithm in detail. Let $e = (v, w)$ be an edge which connects two internal stations v and w of the underlying network. If no such edge exists, then the problem can be solved by employing the algorithm of the base of the recursion.

Let L_e be the set of lines that traverse e. Any line $l_{e,i} \in L_e$ originates from a terminal station, passes through a sequence of edges, then enters station v, traverses edge e, exits station w and, finally, passes through a second sequence of edges until it terminates at another terminal station. Let $p : E \times L \to \mathbb{N}$ be a function, such that $p(e, l)$ denotes the position of edge e along line l. Formally, $L_e = \{l_{e,1}, l_{e,2}, \ldots, l_{e,|L_e|}\}$, where $l_{e,i}$ denotes the i-th line of L_e. Since each line of L_e consists of a sequence of edges, set L_e can be written in the form $\{l_{e,i} = e_{e,i}^1 \, e_{e,i}^2 \, \cdots \, e_{e,i}^{k-1} \, e \, e_{e,i}^{k+1} \, \cdots \, e_{e,i}^{|l_{e,i}|}; \ k = p(e, l_{e,i}), \ i = 1, 2, \ldots, |L_e|\}$. We proceed by removing edge e from the underlying network and by inserting two

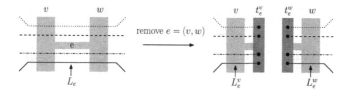

Fig. 3. Illustration of the removal of an edge that connects two internal stations

new terminal stations t_e^v and t_e^w incident to the stations v and w, respectively (see the dark-gray colored stations of the right drawing of Fig. 3). Let $G^* = (V \cup \{t_e^v, t_e^w\}, (E - \{e\}) \cup \{(v, t_e^v), (t_e^w, w)\})$ be the new underlying network obtained in this manner.

Since the edge e has been removed from the underlying network, the lines of L_e cannot traverse it any more. So, we force them to terminate at t_e^v and t_e^w, as it is depicted in the right drawing of Fig. 3. This is done by splitting the set L_e into two new sets L_e^v and L_e^w (see Fig. 3), which are formally defined as follows:

- $L_e^v = \{e_{e,i}^1 \, e_{e,i}^2 \, \cdots \, e_{e,i}^{k-1} \, (v, t_e^v); \; k = p(e, l_{e,i}), \; i = 1, 2, \ldots, |L_e|\}$
- $L_e^w = \{(t_e^w, w) \, e_{e,i}^{k+1} \, \cdots \, e_{e,i}^{|l_{e,i}|}; \; k = p(e, l_{e,i}), \; i = 1, 2, \ldots, |L_e|\}\}$

The new set of lines that it is obtained after the removal of the edge e is $L^* = (L - L_e) \cup (L_e^v \cup L_e^w)$. Observe that the removal of edge e from the underlying network may disconnect it. In the case where G^* is connected, we recursively solve the MLCM problem on (G^*, L^*). Otherwise, since G^* was obtained from G by the removal of a single edge, it has exactly two connected components, say G_1^* and G_2^*. Let $L(G_i^*)$ denotes the lines of L^* induced by G_i^*. In this case, we recursively solve the MLCM problem on $(G_1^*, L(G_1^*))$ and $(G_2^*, L(G_2^*))$.

The recursion will lead to a solution of (G^*, L^*). Part of the solution consists of two ordered sets of bundles $\text{BND}(t_e^v)$ and $\text{BND}(t_e^w)$ at each of the terminal stations t_e^v and t_e^w, respectively. Recall that, in the base of the recursion, all lines in a bundle have exactly the same terminals. In the recursive step, a bundle corresponds to a set of lines whose relative positions cannot be determined. In order to obtain a solution of (G, L), we restore the removed edge e and remove the terminal stations t_e^v and t_e^w. The bundles $\text{BND}(t_e^v)$ and $\text{BND}(t_e^w)$ of t_e^v and t_e^w have also to be connected appropriately along the edge e. Note that the order of the bundles of t_e^v and t_e^w is equal to those of v and w, due to the base of the recursion. Thus, the removal of t_e^v and t_e^w will not produce unnecessary crossings.

We now proceed to describe the procedure of connecting the ordered bundle sets $\text{BND}(t_e^v)$ and $\text{BND}(t_e^w)$ along edge e. We say that a *bundle is of size k* iff it contains exactly k lines. We also say that two bundles are *equal* iff they contain the same set of lines, i.e., the parts of the lines that each bundle contains correspond to the same set of lines. First, we connect all equal bundles. Let $b \in \text{BND}(t_e^v)$ and $b' \in \text{BND}(t_e^v)$ be two equal bundles. The connection of b and b' will result into a new bundle which i) contains the lines of b (or equivalently of b') and ii) its terminals are the terminals of b and b' that do not participate in

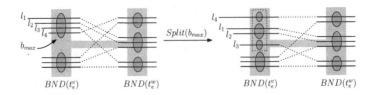

Fig. 4. Splitting the largest bundle. Note that no equal bundles exist.

the connection. Note that a bundle is specified as a set of lines and a pair of stations, that correspond to its terminals. When the connection of b and b' is completed, we remove both b and b' from $BND(t_e^v)$ and $BND(t_e^w)$.

If both $BND(t_e^v)$ and $BND(t_e^w)$ are empty, all bundles are connected. In the case where they still contain bundles, we determine the largest in size bundle, say b_{max}, of $BND(t_e^v) \cup BND(t_e^w)$. W.l.o.g. we assume that $b_{max} \in BND(t_e^v)$ (see the left drawing of Fig. 4). Since b_{max} is the largest bundle among the bundles of $BND(t_e^v) \cup BND(t_e^w)$ and all equal bundles have been removed from both $BND(t_e^v)$ and $BND(t_e^w)$, b_{max} contains at least two lines that belong to different bundles of $BND(t_e^w)$. So, it can be split into smaller bundles, each of which contains a set of lines belonging to the same bundle in $BND(t_e^w)$ (see the right drawing of Fig. 4). Also, the order of the new bundles in $BND(t_e^v)$ should follow the order of their corresponding bundles in $BND(t_e^w)$ in order to avoid unnecessary crossings (refer to the order of the bundles within the dotted rectangle of Fig. 4). In particular, the information that a bundle was split should be propagated to all stations that this bundle traverses, i.e., splitting a bundle is not a local procedure that takes place along a single edge but it requires greater effort. Note that the crossings between lines of b_{max} and bundles in $BND(t_e^w)$ cannot be avoided. In addition, no crossings among lines of b_{max} occur.

We repeat these two steps (i.e. connection of equal bundles and splitting the largest bundle) until both $BND(t_e^v)$ and $BND(t_e^w)$ are empty. Since we always split the largest bundle into smaller ones, this guarantees that our algorithm regarding the connection of the bundles along the edge e will eventually terminate.

Theorem 1. *Given a graph $G = (V, E)$ and a set of lines L on G that terminate at stations of degree 1, the metro-line crossing minimization problem under the 4-side model can be solved in $O((|E| + |L|^2)|E|)$ time.*

Proof. The base of the recursion trivially takes $O(|V| + \sum_{i=1}^{|L|} |l_i|)$, or simply, $O(|V||L|)$ total time. The complexity of our algorithm is actually determined by the connection of the bundles along a particular edge, which is performed at most $O(|E|)$ times, since we always remove an edge that connects two internal stations. The previous steps of our algorithm (i.e., the construction of graph G^* and the necessary recursive calls) need a total of $O((|V| + |E|)|E| + |V||L|)$ time.

In order to connect equal bundles, we initially sort the lines of $BND(t_e^w)$ using counting sort [4] in $O(|L| + |L_e|)$ time, assuming that the lines are numbered from 1 to $|L|$, and we store them in an array, say B, such that the i-th numbered line occupies the i-th position of B. Then, all equal bundles can be connected by performing a single pass over the lines of each bundle of $\text{BND}(t_e^v)$. Note that, given a line l that belongs to a particular bundle of $\text{BND}(t_e^v)$, say b, we can determine in constant time to which bundle of $\text{BND}(t_e^w)$ it belongs by employing array B. So, in a total of $O(|b|)$ time, we decide whether b is equal to one of the bundles of $\text{BND}(t_e^w)$, which yields into an $O(|L_e|)$ total time for all bundles of $\text{BND}(t_e^v)$. Thus, the connection of equal bundles can be accomplished in $O(|L| + |L_e|)$ time.

Having connected all equal bundles, the largest bundle is then determined in $O(|m_e|)$ time, where $m_e = \text{BND}(t_e^v) \cup \text{BND}(t_e^w)$. Using counting sort, we can split the largest bundle in $O(|L| + |L_e|)$ time. The propagation of the splitting of the largest bundle needs $O(|V||L_e|)$ time. The connection of the equal bundles and the splitting of the largest bundle will take place at most $O(|L_e|)$ times. Since $|m_e| \leq 2|L_e|$ and $|L_e| \leq |L|$, the total time needed for our algorithm is $O((|E| + |V||L|^2)|E| + |V||L|)$.

Note that the above straight-forward analysis can be improved by a factor of $|V|$. This is accomplished by propagating the splitting of each bundle only to its endpoints (i.e., not to all stations that each individual bundle traverses). This immediately implies that some stations of G may still contain bundles after the termination of the algorithm. So, we now need an extra post-processing step to fix this problem. We use the fact that the terminals of G do not contain bundles, since they are always at the endpoints of each bundle, when it is split. This suggests that we can split –up to lines– all bundles at stations incident to the terminal stations. We continue in the same manner until all bundles are eventually split. Note that this extra step needs a total of $O(|E||L|)$ time and consequently does not affect the total complexity, which is now reduced to $O((|V|+|E|+|L|^2)|E|+|V||L|)$. Since G is connected, $|E| \geq |V|-1$ and therefore our algorithm needs $O((|E| + |L|^2)|E|)$ time, as desired.   □

Corollary 1. *Given a graph $G = (V, E)$ and a set of lines L on G that terminate at stations of degree 1, the metro-line crossing minimization problem under the k-side model can be solved in $O((|E| + |L|^2)|E|)$ time.*

3   The MLCM Problem under the 2-Side Model

In this section, we adopt the scenario of Section 2 under the 2-side model, i.e., we study the MLCM problem assuming that each station is represented as a rectangle and we permit tracks to the left and the right side of each station, i.e., one of the rectangle's sides is devoted to "incoming" edges/lines while the other is devoted to "outgoing" edges/lines (see Fig. 5a). This assumption, combined with the fact that we do not permit a line to use the same side of a station to both "enter" and "exit", implies that all lines should be x-monotone.

Since the lines are x-monotone, we refer to the leftmost (rightmost) terminal of each line as its *origin* (*destination*). We also say that a line uses the left side of a station to *enter* it and the right side to *exit* it. Furthermore, we refer to the edges incident to the left (right) side of each station u in the embedding of G as

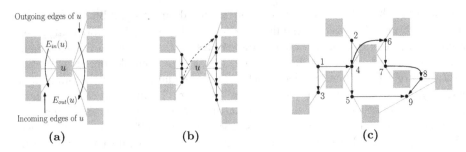

Fig. 5. (a) Incoming/outgoing edges of u. (b) Construction of graph G' when G consists of a single internal station u. (c) An edge numbering of G.

incoming (*outgoing*) edges of station u (see Fig. 5a). For each station u of G, the embedding of G also specifies an order of both the incoming and outgoing edges of u. We denote these orders by $E_{in}(u)$ and $E_{out}(u)$, respectively (see Fig. 5a).

A key component of our algorithm is a numbering of the edges of G, i.e., a function EN : $E \rightarrow \{1, 2, \ldots, |E|\}$. In order to obtain this numbering, we first construct a directed graph $G' = (V', E')$, as follows: For each edge $e \in E$ of G, we introduce a new vertex v_e in G' (refer to the black-colored bullets of Figures 5b and 5c). Therefore, $|V'| = |E|$. Also, for each pair of edges e_i and e_{i+1} of G that are consecutive in that order in $E_{in}(u)$ or $E_{out}(u)$, where $u \in V$ is an internal station of G, we introduce an edge $(v_{e_i}, v_{e_{i+1}})$ in G' (refer to the black-colored solid edges of Fig. 5b). Finally, we introduce an edge connecting the vertex of G' associated with the last edge of $E_{in}(u)$ to the vertex of G' associated with the first edge of $E_{out}(u)$ (refer to the black-colored dashed edge of Fig. 5b). Then, $|E'| = O(|E|)$. An illustration of the proposed construction is depicted in Fig. 5c. Note that all edges of G' are either directed "downward" or "left-to-right" w.r.t. an internal station. Thus there exist no cycles within the constructed graph (no "right-to-left" edges exist to form cycles). The desired numbering of the edges of G is then implied by performing a topological sorting on G' (see Fig. 5c).

Since each line is a sequence of edges, it can be expressed as a sequence of numbers based on the edge numbering EN : $E \rightarrow \{1, 2, \ldots, |E|\}$. We refer to the sequence of numbers assigned to each line as its *numerical representation*. Note that the numerical representation of each line is sorted in ascending order.

Let l and l' be two lines that share a common path of the underlying network G. Let also $a_1 \ldots a_k c_1 \ldots c_m b_1 \ldots b_n$ and $g_1 \ldots g_q c_1 \ldots c_m h_1 \ldots h_r$ be their numerical representations, respectively, where the subsequence $c_1 c_2 \ldots c_m$ corresponds to their common path. Then, l and l' inevitably cross iff $(a_k - g_q) \times (b_1 - h_1) < 0$ (see Fig. 6a). Note that their crossing can be placed along any edge of their common path. This is because we aim to avoid unnecessary station crossings.

Consider now two lines l and l' that share only a single internal station u of G. We assume that u is incident to –at least– four edges, say e_v^1, e_v^2, e_v^3 and e_v^4, where e_v^1 and e_v^2 are incoming edges of u, whereas e_v^3 and e_v^4 outgoing. We further assume that l enters u using e_v^1 and exits u using e_v^4. Similarly, l' enters u using e_v^2 and exits u using e_v^3 (see Fig. 6b). Then, l and l' form a station crossing which

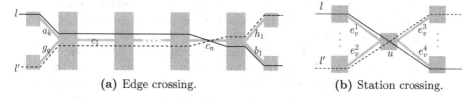

(a) Edge crossing.   **(b)** Station crossing.

Fig. 6. Crossings that cannot be avoided. Note that in Fig. 6a, $a_k < g_q < h_1 < b_1$, whereas in Fig. 6b, $\text{EN}(e_v^1) < \text{EN}(e_v^2) < \text{EN}(e_v^3) < \text{EN}(e_v^4)$.

cannot be avoided iff $(\text{EN}(e_v^1) - \text{EN}(e_v^2)) \times (\text{EN}(e_v^4) - \text{EN}(e_v^3)) < 0$. In this case, the crossing of l and l' can only be placed in the interior of station u.

Our intention is to construct a solution where only crossings that cannot be avoided are present. We will draw the lines of G incrementally by appropriately iterating over the stations of G and by extending the lines from previously iterated stations to the next station. Assuming that the edges of G are directed from left to right in the embedding of G, we first perform a topological sorting of the stations of G. Note that since all edges are directed from left to right, the graph does not contain cycles (no right to left edges exist to form cycles) and therefore a topological order exists. We consider the stations of G in their topological order. This ensures that whenever we consider the next station, its incoming lines have already been routed up to its left neighbors. Let u be the next station in the order. We distinguish the following cases:

Case (a) : $indegree(u) = 0$ *(i.e. terminal station)*.
A station u with $indegree(u) = 0$ corresponds to a station which only contains the origins of some lines. In this case, we simply sort in ascending order these lines lexicographically with respect to their numerical representations. This implies the desired ordering of the lines along the right side of station u. It also ensures that these lines do not cross along their first common path.

Case (b) : $indegree(u) > 0$.
Let $e_u^1, e_u^2, \ldots, e_u^k$ be the incoming edges of station u, where $k = indegree(u)$ and $e_u^i = (u_i, u)$, $i = 1, \ldots, k$. W.l.o.g. we assume that $\text{EN}(e_u^i) < \text{EN}(e_u^j)$, $\forall i < j$. The lines that enter u from e_u^1 will occupy the topmost tracks of the left side of station u. Then, the lines that enter u from e_u^2 will occupy the next available tracks and so on. This ensures that the lines that enter u from different edges will not cross with each other, when entering u.

Let L_u^i be the lines that enter u from edge e_u^i, $i = 1, 2, \ldots, k$, ordered according to the order of the lines along the right side of station u_i. In order to specify the order of all lines along the left side of station u, it remains to describe how the lines of L_u^i are ordered when entering u, for each $i = 1, 2, \ldots, k$. We stably sort in ascending order the lines of L_u^i based on the numbering of the edges that they use when exit station u. Note that in order to perform this sorting we only consider the number following $\text{EN}(e_u^i)$ in the numerical representation of each line. Also, the stable sorting ensures that only unavoidable edge crossings will occur along e_u^i. To see this consider two lines $l, l' \in L_u^i$ which use the same edge to exit station u. Since the sorting is stable their relative position will not change when they enter u, which implies that they will not cross along the edge e_u^i.

Up to this point, we have specified the order of the lines along the left side of station u, say L_{in}^u. In order to complete the description of this case it remains to specify the order, say L_{out}^u, of these lines along the right side of u. Again, the desired order L_{out}^u is implied by stably sorting the lines of L_{in}^u based on the numbering of the edges that they use when they exit station u. Note that also in this case the sorting of the lines is performed by considering only the EN-number of the edges used by the lines when exit station u. Again, the stable sorting ensures that only unavoidable station crossings will occur in the interior of station u.

Note that the stable sortings that are performed at each terminal station ensure that only unavoidable station and edge crossings eventually occur. Also, an unavoidable edge crossing between two lines is always placed along the last edge of their common path.

Theorem 2. *Given a graph $G = (V, E)$ and a set of lines L on G that terminate at stations of degree 1, the metro-line crossing minimization problem under the 2-side model can be solved in $O(|V||E| + \sum_{i=1}^{|L|} |l_i|)$ time.*

Proof. The topological sorting on G needs $O(|V| + |E|)$ time. The construction of graph G' and the computation of a topological sorting on it need $O(|E|)$ time, since both the number of nodes and the number of edges of G' are bounded by $|E|$. Having computed the EN-number of each individual edge of the underlying network, the numerical representations of all lines can be computed in $O(\sum_{i=1}^{|L|} |l_i|)$ time. Using radix sort [4], we can lexicographically sort all lines at each terminal station v of indegree zero in $O((|E| + |L_v|)|l_{max}^v|)$ total time, where L_v is the set of lines that originate at v and l_{max}^v is the longest line of L_v. Therefore, the sorting of all lines at stations of indegree zero needs a total of $O((|E| + |L|)|V|)$ time, since the length of the longest line of L is at most $|V|$. We can –stably– sort the lines of each set L_u^i, $i = 1, 2, \ldots, k$ based on the numbering of the edges that they use when exit u, using counting sort. This can be done in $O(|E| + |L_u|)$ total time, where L_u denotes the set of lines that traverse station u. Recall that counting sort is stable. Similarly, can –stably– sort the lines of each set L_{in}^u based on the numbering of the edges they use when exit station u in $O(|E| + |L_u|)$ time. Summing over all internal stations, our algorithm needs $O(|V||E| + \sum_{i=1}^{|L|} |l_i|)$. $\qquad\square$

As already stated, our algorithm can be employed to solve the MLCM-FixedSE problem. Our approach is as follows: For each station u of G, we introduce four new stations, say u_l^t, u_l^b, u_r^t and u_r^b, adjacent to u. Station u_l^t (u_l^b) is placed on top (below) and to the left of u in the embedding of G and contains all lines that originate at u's top (bottom) station end. Similarly, station u_r^t (u_r^b) is placed on top (below) and to the right of u in the embedding of G and contains all lines that are destined for u's top (bottom) station end. In the case where some of the newly introduced stations contain no lines, we simply ignore their existence. So, instead of restricting each line to terminate at a top or at a bottom station end in its terminal stations, we equivalently assume that it terminates to one of the newly introduced stations. The following theorem summarizes this result.

Theorem 3. *Given a graph $G = (V, E)$ and a set of lines L on G, the metro-line crossing minimization problem with fixed station ends under the 2-side model can be solved in $O(|V||E| + \sum_{i=1}^{|L|} |l_i|)$ time.*

4   Conclusions

In this paper, we studied the MLCM problem under the k-side model for which we presented an $O((|E| + |L|^2)|E|)$ algorithm, and a more efficient algorithm for the special case of 2-side model. Possible extensions would be to study the problem where the lines are not simple, and/or the underlying network is not planar. Our first approach seems to work even for these cases, although the time complexity is harder to analyze and cannot be estimated so easily. The focus of our work was on the case where all line terminals are located at specific stations of the underlying network. Allowing the line terminals anywhere within the underlying network would hinder the use of the proposed algorithms in both models. Therefore, it would be of particular interest to study the computational complexity of this problem.

References

1. Asquith, M., Gudmundsson, J., Merrick, D.: An ILP for the metro-line crossing problem. In: 14th Computing: The Australian Theory Symposium, pp. 49–56 (2008)
2. Bekos, M.A., Kaufmann, M., Potika, K., Symvonis, A.: Line crossing minimization on metro maps. In: Hong, S.-H., Nishizeki, T., Quan, W. (eds.) GD 2007. LNCS, vol. 4875, pp. 231–242. Springer, Heidelberg (2008)
3. Benkert, M., Nöllenburg, M., Uno, T., Wolff, A.: Minimizing intra-edge crossings in wiring diagrams and public transport maps. In: Kaufmann, M., Wagner, D. (eds.) GD 2006. LNCS, vol. 4372, pp. 270–281. Springer, Heidelberg (2007)
4. Cormen, T.H., Leiserson, C.E., Rivest, R.L., Stein, C.: Introduction to Algorithms, 2nd edn. MIT Press, Cambridge (2001)
5. Hong, S.H., Merrick, D., Nascimento, H.: The metro map layout problem. In: Churcher, N., Churcher, C. (eds.) Invis.au 2004. CRPIT, vol. 35, pp. 91–100 (2004)
6. Merrick, D., Gudmundsson, J.: Path simplification for metro map layout. In: Kaufmann, M., Wagner, D. (eds.) GD 2006. LNCS, vol. 4372, pp. 258–269. Springer, Heidelberg (2007)
7. Nöllenburg, M., Wolff, A.: A mixed-integer program for drawing high-quality metro maps. In: Healy, P., Nikolov, N.S. (eds.) GD 2005. LNCS, vol. 3843, pp. 321–333. Springer, Heidelberg (2006)
8. Stott, J.M., Rodgers, P.: Metro Map Layout Using Multicriteria Optimization. In: 8th International Conference on Information Visualisation (IV 2004), pp. 355–362. IEEE, Los Alamitos (2004)
9. Wolff, A., Strijk, T.: The Map-Labeling Bibliography, maintained since (1996), http://i11www.ira.uka.de/map-labeling/bibliography

Cyclic Leveling of Directed Graphs

Christian Bachmaier, Franz J. Brandenburg,
Wolfgang Brunner, and Gergö Lovász

University of Passau, Germany
{bachmaier,brandenb,brunner,lovasz}@fim.uni-passau.de

Abstract. The Sugiyama framework is the most commonly used concept for visualizing directed graphs. It draws them in a hierarchical way and operates in four phases: cycle removal, leveling, crossing reduction, and coordinate assignment.

However, there are situations where cycles must be displayed as such, e. g., distinguished cycles in the biosciences and processes that repeat in a daily or weekly turn. This forbids the removal of cycles. In their seminal paper Sugiyama et al. also introduced recurrent hierarchies as a concept to draw graphs with cycles. However, this concept has not received much attention since then.

In this paper we investigate the leveling problem for cyclic graphs. We show that minimizing the sum of the length of all edges is $\mathcal{NP}$-hard for a given number of levels and present three different heuristics for the leveling problem. This sharply contrasts the situation in the hierarchical style of drawing directed graphs, where this problem is solvable in polynomial time.

1 Introduction

The Sugiyama framework [8] is among the most intensively investigated algorithms in graph drawing. It is the standard technique to draw directed graphs, and displays them in an hierarchical manner. This is well-suited particularly for directed acyclic graphs, which are drawn top-down (or left to right) and level by level. These drawings reflect the underlying graph as a partial order. Typical applications are schedules, UML diagrams and flow charts.

In the general case, the Sugiyama framework first destroys cycles. In the decycling phase it removes or redirects some edges until the resulting graph is acyclic. However, there are many situations, where this procedure is inacceptable. For example, there are well-known cycles in the biosciences, and it is a common standard there to display these cycles as such. These cycles often serve as a landmark [7]. Another application for cycles are repeating processes, such as daily, weekly or monthly schedules with almost the same tasks. Here again it is important that these cycles are clearly visible in a "nice" drawing.

In their original paper from 1981 [8], Sugiyama et al. have proposed a solution for both the hierarchical and the cyclic style. The latter is called *recurrent hierarchy*. A recurrent hierarchy is a level graph with additional edges from the last

I.G. Tollis and M. Patrignani (Eds.): GD 2008, LNCS 5417, pp. 348–359, 2009.
© Springer-Verlag Berlin Heidelberg 2009

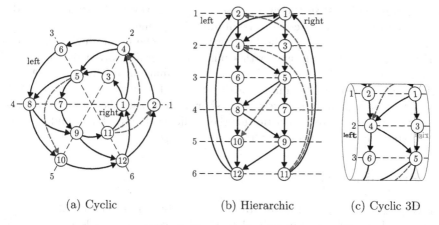

(a) Cyclic   (b) Hierarchic   (c) Cyclic 3D

Fig. 1. Example drawings

to the first level. Here, two drawings are natural: The first is a 2D drawing, where the levels are rays from a common center, and are sorted counterclockwise by their number, see Fig. 1(a). All nodes of one level are placed at different positions on their ray and an edge $e = (u, v)$ is drawn as a monotone counterclockwise curve from u to v wrapping around the center at most once. The second is a 3D drawing on a cylinder, see Fig. 1(c). A combination of the two drawing methods would be the best of both worlds: An interactive 2D view which shows horizontal levels. This view can be scrolled upwards and downwards infinitely and always shows a different part of the cylinder, e. g., the front view of Fig. 1(c).

Recurrent hierarchies are known to most graph drawers – but unnoticed. A planar recurrent hierarchy is shown on the cover of the book by Kaufmann and Wagner [6]. There it is stated that recurrent hierarchies are "unfortunately [. . .] still not well studied". The reason is that they are much harder. Intuitively, there is no start and no end, there are no top and bottom levels. Formally, we pinpoint a problem which is tractable in the hierarchical style and is intractable in the cyclic style.

In cyclic drawings edges are irreversible and cycles are represented in a direct way. Thus, the cycle removal phase disappears from the common Sugiyama framework. This saves much effort, since the underlying problem is the $\mathcal{NP}$-hard feedback arc set problem [5]. Another advantage are short edges. The sum of the edge length can be smaller than in the hierarchical case: Consider a cycle consisting of three nodes. The only way to draw this graph in the Sugiyama framework is to reverse one edge which will then span two levels. Therefore, the sum of the edge length will be four. In the cyclic case this graph can be drawn on three levels s. t. each edge has span one. Moreover, the cyclic style reduces the number of crossings in general. See Fig. 1(a) and (b) as an example. At the threshold with no crossings [2], there are cyclic level planar graphs which are not level planar. Here, consider Fig. 1 with the solid edges only.

Note that any Sugiyama drawing is a cyclic Sugiyama drawing which discards the option to draw edges between the last and first level. Therefore, all benefits of such drawings exist in the cyclic case as well. However, the sum of the edge length and the number of crossings will often be smaller.

In this paper we consider the leveling phase for the cyclic Sugiyama framework. In the hierarchical version this phase is generally solved by topological sorting, or more advanced, by the Coffman-Graham algorithm [6, 3]. As our main result we show that minimizing the sum of the length of all edges is $\mathcal{NP}$-hard for a given number of levels. This sharply contrasts the hierarchical case. Then we introduce three heuristics for the cyclic leveling problem, and evaluate them experimentally within the Gravisto system [1].

2   Preliminaries

Let $G = (V, E)$ be a directed graph. For a given $k \in \mathbb{N}$ we call $\phi: V \to \{1, 2, \ldots, k\}$ a level assignment and $G = (V, E, \phi)$ a cyclic k-level graph. We denote with $\deg(v)$ the degree of a node $v \in V$. For two nodes $u, v \in V$ let $\mathrm{span}(u, v) = \phi(v) - \phi(u)$ if $\phi(u) < \phi(v)$, and $\mathrm{span}(u, v) = \phi(v) - \phi(u) + k$ otherwise. For an edge $e = (u, v) \in E$ we define $\mathrm{span}(e) = \mathrm{span}(u, v)$ and $\mathrm{span}(G) = \sum_{e \in E} \mathrm{span}(e)$. For a set of edges $E' \subseteq E$ we define $\mathrm{span}(E') = \sum_{e \in E'} \mathrm{span}(e)$. $\mathrm{next}(l) = (l \bmod k) + 1$ denotes the level after l. For a node $v \in V$ and a subset $V' \subset V$ we set $E(v, V') = \{ (u, v) \in E \mid u \in V' \} \cup \{ (v, w) \in E \mid w \in V' \}$.

3   Complexity of Cyclic Leveling

In this section we consider different leveling problems and compare their complexity in the hierarchical and the cyclic style. The graphs $G = (V, E)$ are directed in both cases and are acyclic in the hierarchical case.

Definition 1 (Height and Width). *Let G be a directed graph which is drawn s. t. the edges connect vertices on different levels and are uni-directed from the start level to a successive level. Let the* height *be the number of levels and let the* width *be the maximal number of nodes on a level.*

We can now state our leveling problems, both for the common hierarchical style and for the cyclic style of recurrent hierarchies.

Problem 1. Let $k \in \mathbb{N}$. Does there exist a leveling of G with height at most k?

Problem 2. Let $\omega \in \mathbb{N}$. Does there exist a leveling of G with width at most ω?

Problem 3. Let $k, \omega \in \mathbb{N}$. Does there exist a leveling of G with height at most k and width at most ω?

In the hierarchical case problems 1 and 2 are easy: The former can be solved in linear time by the longest path search algorithm [6], whereas, the latter is trivial as each graph has a leveling with width 1 by placing each node on its own level according to a topological sorting of G. Problem 3 is $\mathcal{NP}$-hard as this corresponds to precedence constrained scheduling [5].

In the cyclic case all these problems are easy. Note that an edge $e = (u, v)$ does not impose any constraint on the leveling of the nodes u and v. u can have a smaller level, a larger level, and even the same level as v. Therefore, the answer to Problems 1 and 2 is yes (if $k, \omega > 0$). For Problem 3 there is a cyclic leveling if $|V| \leq k \cdot \omega$ by arbitrarily placing vertices in a $k \times \omega$ grid.

Problem 4. Let $l \in \mathbb{N}$. Does there exist a leveling of G with span$(G) \leq l$?

In the hierarchical case minimizing the span can be formulated as an ILP:

$$\min \sum_{(u,v)\in E} (\phi(v) - \phi(u)) \tag{1}$$

$$\forall v \in V : \phi(v) \in \mathbb{N} \tag{2}$$
$$\forall e = (u, v) \in E : \phi(v) - \phi(u) \geq 1 \tag{3}$$

This ILP can be solved in polynomial time, since the constraint matrix is totally unimodular [6]. Therefore, Problem 4 has a polynomial time complexity in the hierarchical case as well. In the cyclic case the span can no longer be formulated by a system of linear equations, as a case differentiation or the modulo operation is needed.

As a degenerated case we may place all nodes on a single level. Then all edges have span 1 which is obviously minimal. Therefore, we sharpen Problem 4:

Problem 5. Let $l, k \in \mathbb{N}$. Does there exist a leveling of G with exactly k levels with span$(G) \leq l$?

Problem 5 is simple for $k = 1$, as such a leveling exists if $l \geq |E|$. For $k > 1$ we now show that the problem is $\mathcal{NP}$-hard. We use two different reductions for $k = 2$ and $k > 2$. For $k = 2$ we use the $\mathcal{NP}$-hard bipartite subgraph problem [5]:

Problem 6 (Bipartite subgraph). Let $G = (V, E)$ be an undirected graph and $k \in \mathbb{N}$. Does there exist a bipartite subgraph G' of G with at least k edges?

Lemma 1. *Let $G = (V, E)$ be an undirected graph and $l \in \mathbb{N}$. Let $G^* = (V, E^*)$ be a directed version of G with an arbitrary direction for each edge. G contains a bipartite subgraph G' with at least l edges if and only if there exists a leveling of G^* on two levels with span$(G^*) \leq 2|E| - l$.*

Proof. "$\Rightarrow$": Let $G' = (V', E')$ be a bipartite subgraph of G with at least l edges. Let $V_1 \dot\cup V_2 = V'$ be the partition of the node set with all edges of E' between V_1 and V_2. We construct the following leveling for G^*: For each node $v \in V_1$ we set $\phi(v) = 1$, for each node $v \in V_2$ we set $\phi(v) = 2$, and for all nodes $v \in V \setminus (V_1 \cup V_2)$ we set $\phi(v) = 1$. Then each edge in E' has span 1 and all other edges have span 1 or 2. Thus, span$(G^*) \leq |E'| + 2(|E| - |E'|) = 2|E| - E' \leq 2|E| - l$.

"$\Leftarrow$": Let ϕ be a leveling of G^* with span$(G^*) \leq 2|E| - l$. Let V_1 and V_2 be the nodes of V on level 1 and 2, respectively. Let $E' \subseteq E$ be the set of edges e s.t. one end node is in V_1 and the other is in V_2. Then, $G' = (V, E')$ is bipartite.

All edges in E' have span 1 in the leveling ϕ and all other edges have span 2. As $\text{span}(G^*) \leq 2|E| - l = 1 \cdot l + 2(|E| - l)$, there are at least l edges with span 1. As these edges are in E', G' is a bipartite subgraph of G with at least l edges. □

For $k > 2$ we use graph k-colorability, which is $\mathcal{NP}$-hard for a fixed $k > 2$ [5]:

Problem 7 (Graph k-colorability). Let $G = (V, E)$ be an undirected graph and let $k \in \mathbb{N}$. Does there exist a coloring $c : V \rightarrow \{1, \ldots, k\}$, s.t. $c(u) \neq c(v)$ for every edge $e = \{u, v\} \in E$?

Lemma 2. *Let $G = (V, E)$ be an undirected graph and let $k \in \mathbb{N}$. Let $G' = (V, E')$ with E' containing the edges (u, v) and (v, u) for each edge $\{u, v\} \in E$. G is k-colorable if and only if G' has a leveling on k levels with $\text{span}(G') \leq k \cdot |E|$.*

Proof. Let $e = \{u, v\} \in E$. Note that for each leveling ϕ of G' and each edge $e = (u, v) \in E'$ the sum of the spans of (u, v) and (v, u) is either k (if $\phi(u) \neq \phi(v)$) or $2k$ (if $\phi(u) = \phi(v)$). Thus, $\text{span}(G') \geq k \cdot \frac{|E'|}{2} = k \cdot |E|$.

"$\Rightarrow$": Let c be a coloring of G. Set $\phi = c$. Then, for each edge with end nodes u and v in G (and G') $\phi(u) \neq \phi(v)$ holds. Thus, each pair of edges (u, v) and (v, u) in sum has span k and $\text{span}(G') = k \cdot |E|$ holds.

"$\Leftarrow$": Let ϕ be a leveling of G' with $\text{span}(G') \leq k \cdot |E|$. Then, $\text{span}(G') = k \cdot |E|$ and for each edge $(u, v) \in E'$ $\phi(u) \neq \phi(v)$ holds. Thus, $c = \phi$ is a correct coloring.
□

Theorem 1. *Let $G = (V, E)$ be a directed graph and $l, k \in \mathbb{N}$ ($k \geq 2$). The problem whether there exists a leveling of G on k levels with $\text{span}(G) \leq l$ is $\mathcal{NP}$-complete.*

Proof. Lemma 1 and Lemma 2 show that the problem is $\mathcal{NP}$-hard for $k = 2$ and $k > 2$, respectively. The problem is obviously in $\mathcal{NP}$.
□

4   Heuristics

As minimizing the span of a graph in a cyclic leveling with k levels is $\mathcal{NP}$-complete, we have to use heuristics. Known approaches from the hierarchical case as the longest path method [6] or the Coffman-Graham algorithm [3] cannot be easily adapted to the cyclic case. They heavily rely on the fact that the graph is acyclic and start the leveling process at nodes with no incoming edges. As it is not guaranteed that such nodes exist in the cyclic case at all, we introduce three new heuristics. They are evaluated experimentally in Sect. 5.

The input to the algorithms are the number of levels k and the maximum number of nodes on a level ω. The output is the leveling $\phi : V \rightarrow \{1, \ldots, k\}$. The parameter k is either given by the user or it is pre-computed, e.g., as the average length of simple cycles detected by a depth first search of the graph.

Table 1. Complexity of leveling (k as height and ω as width)

	hierarchical	cyclic
Minimizing k	$\mathcal{O}(|V| + |E|)$, by longest path	Set $\phi : V \to \{1\}$
Minimizing ω	$\mathcal{O}(|V| + |E|)$, by ϕ = topsort	Choose injective ϕ
Leveling with k and ω given	$\mathcal{NP}$-hard, precedence constrained scheduling	Test $k \cdot \omega \geq |V|$
Minimizing k with ω given	$\mathcal{NP}$-hard for arbitrary $\omega > 2$	Set $k = \lceil \frac{|V|}{\omega} \rceil$
Minimizing ω with k given	$\mathcal{NP}$-hard for $k > 2$	Set $\omega = \lceil \frac{|V|}{k} \rceil$
Minimizing span(G) with k given	$\mathcal{P}$, by LP	$\mathcal{NP}$-hard for $k > 1$

4.1 Breadth First Search

The breadth first search (BFS) heuristic (Algorithm 1) is rather simple: We choose an arbitrary start node v, set $\phi(v) = 1$ and perform a directed BFS from v. When we reach a node w for the first time using an edge (u, w), we set $\phi(w) = \text{next}(\phi(u))$ if this level does not contain ω nodes already. Otherwise, we move w to the first non-full level.

Using this heuristic the tree edges will have a rather short span. But the back edges are not taken into account for the leveling at all. Thus, these edges can be arbitrarily long.

Lemma 3. *The BFS leveling heuristic needs* $\mathcal{O}(|V| + |E| + k^2)$ *time.*

Proof. BFS runs in $\mathcal{O}(|V| + |E|)$ time. In addition we must keep and update an array N of size k. $N[i]$ denotes the first non-full level from level i. At most all k levels can get full which costs $\mathcal{O}(k)$ time for each. □

4.2 Minimum Spanning Tree

This heuristic has similarities to the algorithm of Prim [4], which computes the minimum spanning tree (MST) of a graph. We sequentially level the nodes by a greedy algorithm. Let $V' \subset V$ be the set of already leveled nodes. When we level a node v, all edges in $E(v, V')$ get a fixed span. Therefore, we set $\phi(v)$ s. t. span($E(v, V')$) is minimized. Note that there are possibly more edges incident to v which are also incident to not yet leveled nodes. These edges will be considered when the second end node is leveled.

We decide in which order to add the nodes by using a distance function $\delta(v)$. We discuss four options:

Algorithm 1. breadthFirstSearchLeveling

Input: G: a directed graph, k: the number of levels,
        ω: the maximum number of nodes on each level

Output: ϕ: a cyclic leveling of G

1 Queue $Q \leftarrow \emptyset$
2 Leveling $\phi \leftarrow \emptyset$
3 **foreach** $u \in V$ **do** $u.marked \leftarrow false$
4 **foreach** $l \in \{1, \ldots, k\}$ **do** $N[l] \leftarrow l$
5 **foreach** $u \in V$ **do**
6   **if** $\neg u.marked$ **then**
7        $Q.append(u)$
8        $u.marked \leftarrow true$
9        $\phi(u) \leftarrow N[1]$
10      $updateN(N[1])$
11   **while** $\neg Q.isEmpty()$ **do**
12         $v \leftarrow Q.removeFirst()$
13   **foreach** $neighbor\ w\ of\ v$ **do**
14   **if** $\neg w.marked$ **then**
15             $w.marked \leftarrow true$
16             $\phi(w) \leftarrow N[next(\phi(v))]$
17             $updateN(\phi(w))$
18             $Q.append(w)$

19 **return** ϕ

Minimum Increase in Span (MST_MIN). We choose the node which will create the minimum increase in span in the already leveled graph:

$$\delta_{\text{MIN}}(v) = \min_{\phi(v) \in \{1, \ldots, k\}} \text{span}(E(v, V')) \tag{4}$$

Minimum Average Increase in Span (MST_MIN_AVG). Using the distance function δ_{MIN} will place nodes with a low degree first, as nodes with a higher degree will almost always cause a higher increase in span. Therefore, considering the increase in span per edge is reasonable:

$$\delta_{\text{MIN\_AVG}}(v) = \min_{\phi(v) \in \{1, \ldots, k\}} \frac{\text{span}(E(v, V'))}{|E(v, V')|} \tag{5}$$

We distribute isolated nodes evenly on the non-full levels in the end.

Maximum (Average) Increase in Span (MST_MAX(_AVG)). Choose the node which causes the maximum (average) increase in span per edge:

$$\delta_{\text{MAX}}(v) = \frac{1}{\delta_{\text{MIN}}(v)}, \qquad \delta_{\text{MAX\_AVG}}(v) = \frac{1}{\delta_{\text{MIN\_AVG}}(v)} \tag{6}$$

The idea behind this is the following: A node which causes a high increase in span will cause this increase when leveled later as well. But if we level this node now, we can possibly level other adjacent, not yet leveled nodes in a better way.

Note that we only use the distance function $\delta(v)$ to determine which node to level next. When we level a node v, we set $\phi(v)$ s. t. the increase in span will be minimized. In some cases several levels for v will create the same increase in span. We will then choose a level for v which minimizes $\sum_{e \in E(v,V')} \text{span}(e)^2$ as well. Thus, we assign v a level which is more centered between its leveled adjacent nodes. In each case we can only use a level which has not yet ω nodes on it. Nodes with already leveled neighbors block a place on their optimal level s. t. they can later be placed on the level. Algorithm 2 shows the complete heuristic.

Algorithm 2. minimumSpanningTreeLeveling

Input: G: a directed graph, k: the number of levels,
       ω: the maximum number of nodes on each level
Output: ϕ: a cyclic leveling of G

```
1  Heap H ← ∅
2  Leveling φ ← ∅
3  foreach u ∈ V do
4  │   u.status ← white
5  │   δ(u) ← ∞
6  foreach u ∈ V do
7  │   if u.status = white then
8  │   │   δ(u) ← 0
9  │   │   H.insert(u)
10 │   │   while ¬H.isempty() do
11 │   │   │   v ← H.removeMin()
12 │   │   │   v.status ← black
13 │   │   │   φ(v) ← getOptimalLevel(v)
14 │   │   │   foreach neighbor w of v with w.status ≠ black do
15 │   │   │   │   δ(w) ← computeDistance(w)
16 │   │   │   │   φ(w) ← getOptimalLevel(w)
17 │   │   │   │   if w.status = gray then
18 │   │   │   │   │   H.update(w)
19 │   │   │   │   else
20 │   │   │   │   │   w.status ← gray
21 │   │   │   │   │   H.insert(w)

22 return φ
```

Lemma 4. *The MST heuristic needs* $\mathcal{O}(|V| \log |V| + k \cdot \deg(G) \cdot |E|)$ *time.*

Proof. The time complexity is dominated by the while loop. Here, removing each node from the heap costs $\mathcal{O}(|V| \log |V|)$. Each edge $e = (w, z) \in E$ may change its span whenever a neighbor v of w (or z) is fixed on a level. In this case each of the k levels is tested for w (or z). Thus, we get $\mathcal{O}(k \cdot (\deg(w) + \deg(z)))$ for e and $\mathcal{O}(k \cdot \deg(G) \cdot |E|)$ for all edges. Finally, updating all neighbors in the heap costs $\mathcal{O}(|E| \log |V|)$ (or $\mathcal{O}(|E|)$ using a Fibonacci heap).   □

4.3   Force Based

Spring embedders use a physical model to simulate the edges as springs [6]. Forces between nodes are computed and the nodes are moved accordingly. Transferring this idea to the cyclic leveling problem, we could use a force function similar to conventional energy based placement algorithms as follows:

$$\text{force}(v) = \sum_{(v,w)\in E} (\text{span}(v,w) - 1)^2 - \sum_{(u,v)\in E} (\text{span}(u,v) - 1)^2 \qquad (7)$$

However, moving a node to its energy minimum using this force will not minimize the span of the graph, i. e., (7) minimizes the deviation between the edge lengths, e. g., see Fig. 2. Furthermore, the span may increase when moving a node towards its energy minimum, as some edges can flip from span 1 to span k. We solve this problem by using directly the span as the (undirected) force which is minimized:

$$\text{force}(v) = \text{span}(E(v, V)) \qquad (8)$$

We move the node with the maximum impacting force. And we directly move the node to its energy minimum, which is the level s.t. the span is minimized. For this, we test all possible (non-full) levels. Note that moving all nodes at once would not decrease time complexity here. Algorithm 3 shows the pseudo code.

As an initial leveling we either use a random leveling (SE_RND) or the result of the best minimum spanning tree heuristic MST_MIN_AVG (SE_MST).

Lemma 5. *In the force based heuristic $\mathcal{O}(|V| \log |V| + k \cdot |E|)$ time is needed for each iteration.*

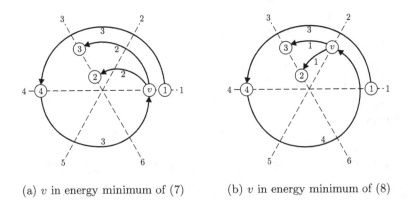

(a) v in energy minimum of (7)   (b) v in energy minimum of (8)

Fig. 2. Force based placement of node v

Algorithm 3. forceBasedLeveling

Input: G: a directed graph, k: the number of levels,
        ω: the maximum number of nodes on each level
Output: ϕ: a cyclic leveling of G

```
1  Heap H ← ∅
2  φ ← computeInitialLeveling()
3  foreach v ∈ V do
4  |   computeForce(v)

5  while improvement ∧ iterations < limit do
6  |   foreach v ∈ V do
7  |   |   H.insert(v)

8  |   while ¬H.isEmpty() do
9  |   |   v ← H.removeMax()
10 |   |   φ(v) ← energyMinimalLevel(v)
11 |   |   foreach neighbor w of v do
12 |   |   |   updateForce(w, v)

13 return φ
```

Proof. Inserting all nodes in the heap can be implemented in $\mathcal{O}(|V|)$ time. Removing each node from the heap has time complexity $\mathcal{O}(|V| \log |V|)$. Computing the energy minimal level for v costs $\mathcal{O}(k \cdot \deg(v))$, which is $\mathcal{O}(k \cdot |E|)$ for all nodes. Computing the new force is possible in time $\mathcal{O}(1)$ for each neighbor of v, in $\mathcal{O}(\deg(v))$ for all neighbors and $\mathcal{O}(|E|)$ in total. The $\mathcal{O}(|E|)$ updates in the heap cost $\mathcal{O}(|E| \log |V|)$ (or $\mathcal{O}(|E|)$ using a Fibonacci heap).   □

5   Empirical Results

In this section we evaluate and compare the heuristics with each other and with an optimal leveling. The optimal leveling is computed by a branch and bound algorithm which can be used for graphs up to 18 nodes.

In Fig. 3 the running times of the algorithms are shown. Figure 4 compares the calculated spans of the heuristics with the optimal span. For a better pairwise comparison of the heuristics, Fig. 5 only shows their results.

For Fig. 3 and 5 the number of nodes $|V|$ was increased by steps of 50 each time. For each size 10 graphs with $|E| = 5|V|$ were created randomly. For Fig. 4 10 graphs for each size $|V|$ and $|E| = 2|V|$ were used. In all three diagrams k and ω were set to $\sqrt{2|V|}$, s.t. there were $2|V|$ possible node positions. Each heuristic was applied to each graph $\min(|V|, 30)$ times using different start nodes resp. initial levelings and choosing the average.

The benchmarks show the practical performance of the algorithms. All tests were run on a 2.8 GHz Celeron PC under the Java 6.0 platform from Sun Microsystems, Inc. within the Gravisto framework [1].

As expected the force based heuristics with MST initialization computes the best leveling and the results are close to the optimum. All MST variants do

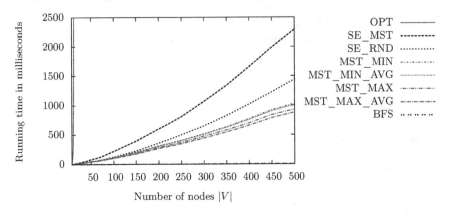

Fig. 3. Running times of the algorithms

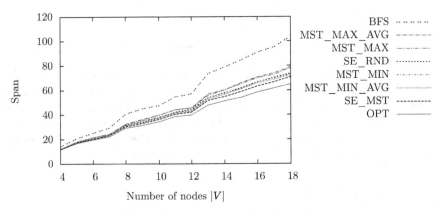

Fig. 4. Average spans of small graphs

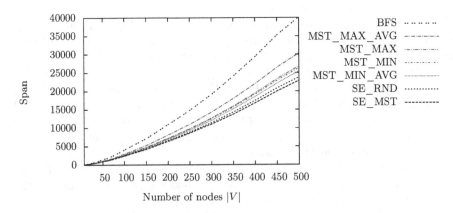

Fig. 5. Average spans of large graphs

not differ very much, but MST_MIN_AVG seems to be the best. The results can be improved by applying the heuristics i times to the same graph with i different start nodes or different initial levelings, respectively, and choosing the best result. However, the price is an i times higher running time.

6 Summary and Open Problems

The leveling problem has turned out to be essentially different in the hierarchical and cyclic style. We have shown different optimization goals for the cyclic leveling compared to the goals of the hierarchic leveling. For the reasonable minimization of the sum of the edge lengths we have shown the $\mathcal{NP}$-hardness and presented three practical heuristics for the problem.

Open problems are the approximation ratios of our heuristics, other quality measures for cyclic drawings, the best number of levels, and the completion of the cyclic style to a cyclic Sugiyama framework.

References

1. Bachmaier, C., Brandenburg, F.J., Forster, M., Holleis, P., Raitner, M.: Gravisto: Graph visualization toolkit. In: Pach, J. (ed.) GD 2004. LNCS, vol. 3383, pp. 502–503. Springer, Heidelberg (2005)
2. Bachmaier, C., Brunner, W.: Linear time planarity testing and embedding of strongly connected cyclic level graphs. In: Halperin, D., Mehlhorn, K. (eds.) ESA 2008. LNCS, vol. 5193, pp. 136–147. Springer, Heidelberg (2008)
3. Coffman, E.G., Graham, R.L.: Optimal scheduling for two processor systems. Acta Informatica 1(3), 200–213 (1972)
4. Cormen, T.H., Leiserson, C.E., Rivest, R.L., Stein, C.: Introduction to Algorithms, 2nd edn. MIT Press, Cambridge (2001)
5. Garey, M.R., Johnson, D.S.: A Guide to the Theory of NP-Completeness. W. H. Freemann, New York (1979)
6. Kaufmann, M., Wagner, D. (eds.): Drawing Graphs. LNCS, vol. 2025. Springer, Heidelberg (2001)
7. Michal, G. (ed.): Biochemical Pathways: An Atlas of Biochemistry and Molecular Biology. Wiley, New York (1999)
8. Sugiyama, K., Tagawa, S., Toda, M.: Methods for visual understanding of hierarchical system structures. IEEE Transactions on Systems, Man, and Cybernetics 11(2), 109–125 (1981)

Constrained Point-Set Embeddability of Planar Graphs[*]

Emilio Di Giacomo[1], Walter Didimo[1], Giuseppe Liotta[1],
Henk Meijer[2], and Stephen Wismath[3]

[1] Dip. di Ingegneria Elettronica e dell'Informazione, Università degli Studi di Perugia
{digiacomo,didimo,liotta}@diei.unipg.it
[2] Roosevelt Academy, The Netherlands
h.meijer@roac.nl
[3] Department of Mathematics and Computer Science, University of Lethbridge
wismath@cs.uleth.ca

Abstract. This paper starts the investigation of a constrained version of the point-set embeddability problem. Let $G = (V, E)$ be a planar graph with n vertices, $G' = (V', E')$ a subgraph of G, and S a set of n distinct points in the plane. We study the problem of computing a point-set embedding of G on S subject to the constraint that G' is drawn with straight-line edges. Different drawing algorithms are presented that guarantee small curve complexity of the resulting drawing, i.e. a small number of bends per edge. It is proved that: (i) If G' is an outerplanar graph and S is any set of points in convex position, a point-set embedding of G on S can be computed such that the edges of $E \setminus E'$ have at most 4 bends each. (ii) If S is any set of points in general position and G' is a face of G or if it is a simple path, the curve complexity of the edges of $E \setminus E'$ is at most 8. (iii) If S is in general position and G' is a set of k disjoint paths, the curve complexity of the edges of $E \setminus E'$ is $O(2^k)$.

1 Introduction

The problem of computing a planar drawing of a graph on a given set of points in the plane is a classical subject of investigation both in the graph drawing and in the computational geometry literature. The input is a planar graph G with n vertices and a set of n distinct points in the plane. The output is a drawing of G such that each vertex is mapped to a distinct point of S and no two edges cross each other. Besides the intrinsic theoretical interest of studying the interplay between the topology of the graph and the geometry of the given set of points, the question is in part justified by the variety of graph drawing applications where some or all of the vertices are constrained at fixed locations (see, e.g., [17]).

Different versions of the problem have been investigated. In the *point-set embeddability problem with given mapping* the function that associates each vertex of G with a distinct point of S is given as part of the input. Halton [10] shows that a planar graph always admits a point-set embedding with given mapping on any set of n points but he

[*] Research partially supported by the MIUR Project "MAINSTREAM: Algorithms for massive information structures and data streams" and by NSERC.

I.G. Tollis and M. Patrignani (Eds.): GD 2008, LNCS 5417, pp. 360–371, 2009.

does not give any result about the number of bends per edge. Pach and Wenger [16] describe an algorithm to compute a point-set embedding with given mapping of a planar graph with at most $120n$ bends per edge; they also prove that $\frac{m}{40^3}$ bends per edge may be necessary in some cases (m is the number of edges of the graph). Both the upper and the lower bounds of Pach and Wenger have been recently reduced by Badent et al. [1] to $3n + 2$ and $\frac{n}{8}$, respectively.

The *point-set embeddability problem without mapping* allows the drawing algorithm to choose for each vertex v of G the point of S that represents v. Algorithms to compute straight-line point-set embedding without mapping of trees [3,11,15] and outerplanar graphs [2,9] have been presented. Since outerplanar graphs are the largest class of graphs admitting a straight-line point-set embedding without mapping on any set of points [9], Kaufmann and Wiese [14] study the point-set embeddability problem with bends. They show that every planar graph admits a point-set embedding without mapping with at most two bends per edge and that this bound is worst-case optimal.

The above two versions of the problem have also been studied in a unifying framework. In a *k-colored point set embedding* each vertex of G and each point of S is given one of k colors and the drawing algorithm can map a vertex of color i to any point of S having the same color. If $k = 1$ the problem coincides with the point-set embeddability problem without mapping; if $k = n$ we have the point-set embeddability problem with mapping; if the input specifies a mapping of n_1 vertices to n_1 points while there is no mapping for the remaining $n - n_1$, we have a k-colored point-set embeddability problem with $k = n_1 + 1$. A limited list of papers about the k-colored point set embeddability includes [1,4,5,8,11,12,13].

This paper studies a natural extension of the point-set embeddability problem without mapping. It is assumed that a subgraph of G is given whose edges are required to be drawn as straight-line segments. Our input is a graph $G = (V, E)$ with n vertices, a set S of n distinct points in the plane, and a subgraph $G' = (V', E')$ of G. We want to compute a point-set embedding of G with small *curve complexity* (i.e. a small number of bends per edge) and such that G' is drawn with straight-line edges. It may be worth recalling that a recent paper [6] has proved that if G is a tree, G' is a tree with k edges, and a mapping from the vertices of G' to a subset of the points of S is given, then a point-set embedding where the edges of G' are straight-line can have $O(k)$ curve complexity. A fundamental difference between the setting of [6] and this paper is that here we do not assume that the mapping from the subgraph to the point set is given as part of the input. A high level description of the results in the paper is as follows.

- We prove that if S is a set of points in convex position, then G admits a point-set embedding on S with the edges of G' drawn straight-line if and only if G has a planar embedding ψ such that the embedding of G' induced by ψ is outerplanar and each vertex of $V \setminus V'$ is on the external face of G' in ψ. Furthermore we show that, when the drawing exists, it can be computed in such a way that the edges of $E \setminus E'$ have at most 4 bends each.
- We extend the above investigation to sets of points in general position. If either G' is a face of G or if it is a simple path, a point-set embedding of G on any set of points in general position exists such that the edges of G' are straight-line edges and the edges of $E \setminus E'$ have at most 8 bends each.

– We finally consider the situation where G' is a set of k disjoint paths and S is any set of points in general position. In this case a point-set embedding exists where the edges of G' are straight-line edges and the edges of $E \setminus E'$ have $O(2^k)$ bends each.

In the remainder of the paper some proofs are omitted for reasons of space.

2   Preliminaries

Let $G = (V, E)$ be a graph. A *drawing* Γ of G maps each vertex v of G to a distinct point $p(v)$ of the plane and each edge $e = (u, v)$ of G to a simple Jordan curve connecting $p(u)$ and $p(v)$. Drawing Γ is *planar* if no two distinct edges intersect except at common end-vertices. Graph G is *planar* if it admits a planar drawing. A planar drawing Γ of G partitions the plane into topologically connected regions called the *faces* defined by Γ. The unbounded face is called the *external face*. The *boundary* of a face is its delimiting cycle described by the circular list of its edges and vertices. A face is *simple* if its boundary is a simple cycle.

An *embedding* of a planar graph G is an equivalence class of planar drawings that define the same set of faces, that is, the same set of face boundaries. A planar graph G together with the description of a set of faces F is called an *embedded planar graph*. A *maximal* embedded planar graph is such that all faces are triangles, that is, the boundary of each face has three vertices and three edges. Given any embedded planar graph G, it is easy to add edges that split the faces of G in order to obtain a maximal embedded planar graph that includes G. A graph G is *outerplanar* if it admits an embedding such that all vertices of G belong to a common face, which we can always choose as the external face.

A subdivision of a graph G is a graph obtained from G by replacing each edge by a path with at least one edge. Internal vertices on such a path are called *division vertices*, the edges of such path are called *sub-edges*.

Let $G = (V, E)$ be a planar graph with n vertices and let S be a set of n points in the plane. We say that the points of S are in *general position* if no three points of S lie on the same line. A *point-set embedding of G on S*, denoted as $\Gamma(G, S)$, is a planar drawing of G such that each vertex is mapped to a distinct point of S. $\Gamma(G, S)$ is called a *geometric point-set embedding* if all edges are drawn straight-line. Let $G' = (V', E')$ be a subgraph of G. A *point-set embedding of G on S with straight-line subgraph G'*, denoted as $\Gamma(G, G', S)$, is a point-set embedding of G on S such that the edges of G' are drawn as straight-line segments. The edges of G can be partitioned, with respect to G', in the following four sets:

Blue Edges. $E_{blue} = E'$;
Red Edges. $E_{red} = \{(u, v) \in E \setminus E' \mid u \notin V' \land v \notin V'\}$;
Black Edges. $E_{black} = \{(u, v) \in E \setminus E' \mid u \in V' \land v \notin V'\}$;
Green Edges. $E_{green} = \{(u, v) \in E \setminus E' \mid u \in V' \land v \in V'\}$.

Let $G = (V, E)$ be a planar graph. A *2-page book embedding* of G is a planar drawing of G such that all vertices of G are represented as points of a horizontal line ℓ called the *spine* and each edge is drawn in one of the two half-planes defined by ℓ. The half-plane above ℓ is called the *top page*, while the half-plane below the spine is called the *bottom page*. An edge of a 2-page book embedding is completely contained either in the top page or in the bottom page. Let $e_1 = (u_1, v_1)$ and $e_2 = (u_2, v_2)$ be two edges in the same page of a 2-page book embedding. Assume, without loss of generality, that u_1 is to the left of v_1 on ℓ, that u_2 is to the left of v_2 on ℓ, and that u_1 is to the left of u_2 on ℓ. A crossing between e_1 and e_2 can be avoided only if u_1, u_2, v_1, and v_2 do not appear in this order along the spine. On the other hand, if u_1, u_2, v_1, and v_2 do not appear in this order along the spine the two edges can easily be drawn without crossing (for example they can be drawn as circular arcs). Thus, the fact that two edges in a same page cross or not depends only on the relative position of their endvertices.

A *monotone topological book embedding* of a planar graph G is a 2-page book embedding of a subdivision of G such that each edge (u, v) of G has at most one division vertex d and: (i) u, d, and v appear in this order along the spine; (ii) sub-edge (u, d) is in the bottom page; (iii) sub-edge (d, v) is in the top page.

In [7] an algorithm to compute a monotone topological book embedding of a planar graph was presented. The next lemma, that will be used as a basic tool for the algorithms presented in the next sections, can be proved by exploiting the algorithm of [7] for computing a monotone topological book embedding.

Lemma 1. *Let $G = (V, E)$ be an embedded planar graph. If G contains a cycle $C = v_1, v_2, \ldots, v_h$ whose interior is triangulated with edges (v_1, v_i) for $3 \leq i \leq h - 1$, then there exists a monotone topological book embedding γ such that the linear order of the vertices of C in γ is $v_1, v_h, v_{h-1}, \ldots, v_2$; also, all edges of C and all edges (v_1, v_i) $(3 \leq i \leq h - 1)$ are in the bottom page of γ.*

3 Points in Convex Position

Let G be a planar graph and let G' be a subgraph of G. In this section we give a necessary and sufficient condition on the embedding of G and G' that guarantees that G has a point-set embedding on any set of points in convex position with straight-line subgraph G'. We start by giving some technical lemmas.

Lemma 2. *Let $G = (V, E)$ be an embedded simply connected outerplanar graph. There exists a graph $G_a = (V_a, E_a)$ such that: (i) $V_a = V$ and $E \subset E_a$; (ii) G_a is biconnected and outerplanar; (iii) every edge in $E_a \setminus E$ is on the external face.*

Lemma 3. *Let S be a set of points in convex position. It is possible to connect any two points p and q of S with a 1-bend polyline that does not intersect the interior of the convex hull of S.*

Lemma 4. *Let $G = (V, E)$ be a planar graph, let S be any set of points in convex position and let $G' = (V', E')$ be a biconnected subgraph of G. Let G admit a planar embedding ψ such that: (i) the embedding of G' induced by ψ is outerplanar; and (ii) all vertices of $V \setminus V'$ are on the external face of G' in the embedding ψ. Then G admits a point-set embedding on S with straight-line subgraph G'.*

Proof. Refer to Fig. 1 for an illustration. Since G' is biconnected, the boundary of its external face is a cycle C. Maintaining the embedding ψ we change the external face in such a way that an edge (u, v) of C is on the external face. Since the embedding of G' induced by ψ is outerplanar, then no vertex of G' is inside C; also no vertex of $V \setminus V'$ is inside C. Thus, C contains only blue edges. Let $v_1 = u, v_2 = v, \ldots, v_h$ be the vertices of C according to their circular ordering. Remove all blue edges inside C and replace them with blue edges of the type (v_1, v_i) for $3 \leq i \leq h - 1$. We denote the graph obtained from this transformation as $\overline{G}$ and its subgraph consisting of C and of the blue edges inside it as $\overline{G'}$. By Lemma 1, the graph $\overline{G}$ admits a monotone topological book embedding $\overline{\gamma}$ such that the linear order of the vertices of C in $\overline{\gamma}$ is $v_1, v_h, v_{h-1}, \ldots, v_2$ and all the edges of C along with the blue edges inside it are in the bottom page of $\overline{\gamma}$.

Let $u_1, u_2, \ldots, u_{n'}$ be the linear ordering of the (real and division) vertices of $\overline{G}$ in $\overline{\gamma}$. We enrich the set S with extra points so that the resulting set $\overline{S}$ has n' points in convex position. Let $p_1, p_2, \ldots, p_{n'}$ be the points of $\overline{S}$ according to their clockwise circular ordering. The points of $\overline{S} \setminus S$ are positioned in such a way that p_i is an extra point if and only if u_i is a division vertex ($1 \leq i \leq n'$). Vertex u_i is mapped to p_i ($1 \leq i \leq n$). All the subedges of $\overline{G}$ that are in the bottom page of $\overline{\gamma}$ will be drawn inside the convex hull $CH(\overline{S})$ of $\overline{S}$ as straight-line segments between their end-vertices. The subedges of $\overline{G}$ that are in the top page of $\overline{\gamma}$ are drawn outside $CH(\overline{S})$ with one bend. More precisely, let E_t be the set of sub-edges in the top page of $\overline{\gamma}$. We start drawing an edge $e_1 \in E_t$ such that, when drawn with the technique of Lemma 3, it leaves the points of $\overline{S}$ representing the end-vertices of the edges of E_t not yet drawn, on the convex hull of $\overline{S} \cup \{q\}$, where q is the bend point of e_1. We now can apply the same drawing technique to the set of subedges $E_t \setminus \{e_1\}$ and to the set of points $\overline{S} \cup q$.

Let $\overline{\Gamma}$ be the drawing obtained after replacing the division vertices with bends. We prove now that $\overline{\Gamma}$ is a point-set embedding of $\overline{G}$ on S with straight-line subgraph $\overline{G'}$. By construction, each vertex of $\overline{G}$ is mapped to a point of S. Let $e_1 = (u_1, v_1)$ and $e_2 = (u_2, v_2)$ be two sub-edges of $\overline{G}$; we prove that e_1 and e_2 do not cross. If they are in different pages in $\overline{\gamma}$ then one of them is inside $CH(\overline{S})$ and the other one is outside $CH(\overline{S})$ and therefore they do not cross. Suppose they are both in the bottom page; since they do not cross in $\overline{\gamma}$ then the order of their end-vertices in $\overline{\gamma}$ is either u_1, v_1, u_2, v_2 or u_1, u_2, v_2, v_1. Since the circular ordering of the points representing u_1, v_1, u_2, v_2 is coherent with the linear ordering that these vertices have in $\overline{\gamma}$, then e_1 and e_2 do not cross. If e_1 and e_2 are both in the top page, then they are drawn with one bend. Without loss of generality assume that e_1 is drawn before e_2. By construction, the polyline representing e_2 does not intersect the interior of the convex hull of $\overline{S} \cup \{q_i \mid q_i \text{ is the bend point of an edge drawn before } e_2\}$ and therefore it does not cross e_1. By Lemma 1 all edges of $\overline{G'}$ are in the bottom page of $\overline{\gamma}$ and therefore they are drawn straight-line.

We prove now that the drawing Γ obtained from $\overline{\Gamma}$ by restoring the original blue edges inside C is a point-set embedding of G on S with straight-line subgraph G'. Since C is drawn straight-line and the points are in convex position, then C is drawn as a convex polygon P. Also, by Lemma 1 the linear order of the vertices $v_1, v_2, \ldots, v_h$ of C in $\overline{\gamma}$ is $v_1, v_h, v_{h-1}, \ldots, v_2$; it follows that their circular order on the boundary of

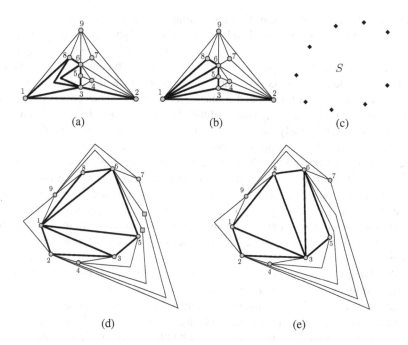

Fig. 1. (a) A maximal embedded planar graph G; the bold edges highlight a biconnected outerplanar subgraph G'. (b) The subgraph G' is transformed into a new subgraph $\overline{G'}$ with all chords incident on vertex 1. (c) A set S of 9 points in convex position. (d) A point-set embedding of $\overline{G}$ on $\overline{S}$ with straight-line subgraph $\overline{G'}$; $\overline{S}$ is a set of points obtained from S by adding two extra points. (d) A point-set embedding of G on S with straight-line subgraph G'.

P is $v_1, v_2, \ldots, v_h$, i.e. the same ordering they have in the embedding of ψ. Hence the original blue edges can be restored without creating any crossing.   □

Theorem 1. *Let $G = (V, E)$ be a planar graph, let S be any set of points in convex position and let $G' = (V', E')$ be a connected subgraph of G. G admits a point-set embedding $\Gamma(G, G', S)$ on S with straight-line subgraph G' if and only if G admits a planar embedding ψ such that: (i) the embedding of G' induced by ψ is outerplanar; (ii) all vertices of $V \setminus V'$ are on the external face of G' in the embedding ψ.*

Also, if $\Gamma(G, G', S)$ exists, it can be computed so that: If G' is biconnected, then every edge not in G' has at most 2 bends per edge; If G' is simply connected, then every edge not in G' has at most 4 bends per edge.

Proof. Let $\Gamma(G, G', S)$ be a point-set embedding of G on S with straight-line subgraph G'. Consider the drawing Γ' of G' in $\Gamma(G, G', S)$. Since the points of S are in convex position and Γ' is a straight-line drawing, all vertices of G' are on the external face of Γ' and therefore G' is outerplanar. Also, all points distinct from those representing the vertices of V' are on the external face of Γ'. It follows that the embedding defined by $\Gamma(G, G', S)$ is the desired embedding ψ.

Assume now that G does admit an embedding ψ that satisfies the statement; we prove that G admits a point-set embedding on S with straight-line subgraph G'. If G' is biconnected, then this is true by Lemma 4. If G' is not biconnected we apply the biconnection procedure described in Lemma 2 to make it biconnected. The dummy edges that are added to make G' biconnected can cross black and green edges. Every crossing between a dummy edge and a black/green edge is replaced with a division vertex that splits both the dummy edge and the black/green edge. Each black edge can be split at most once. Namely, let $e = (u, v)$ be a black edge and assume that $u \in V'$ and $v \in V \setminus V'$. If a dummy edge e' crosses e, then u is encountered between the two end-vertices of e' when walking clockwise along the external boundary of G'; we say that e' *covers* u. If e was crossed more than once then there would be at least two dummy edges that cover u. In this case, however, one of the two dummy edges cannot be on the external face of G', which is impossible by Lemma 2. Each green edge $e = (u, v)$ can be split at most twice. Namely, with an analogous argument as in the case of black edges, we have that there is at most one edge that covers u and at most one edge that covers v.

All the dummy (sub-)edges are considered to be blue edges. Let $e = (u, v)$ be a black edge split by a division vertex d and assume that $u \in V'$ and $v \in V \setminus V'$ sub-edge (u, d) is considered a blue edge, while sub-edge (d, v) is considered a black edge. Analogously, if e is a green edge split by two division vertices d_1 and d_2, the sub-edges (u, d_1) and (d_2, v) are considered to be blue edges, while sub-edge (d_1, d_2) is considered to be a green edge. The graph obtained after this transformation will be denoted as $\hat{G}$. This graph has as a subgraph the graph $\hat{G}'$ that consists of G' plus the blue edges added as dummy edges or obtained by splitting a black/green edge of G. $\hat{G}'$ is outerplanar and biconnected and by Lemma 2 $\hat{G}$ admits a point-set embedding on a suitably augmented set of points $\hat{S}$ with straight-line subgraph $\hat{G}'$. Since no edge of G' is split with a division vertex, removing dummy edges and replacing division vertices with bends we obtain a point-set embedding of G on S with straight-line subgraph G'.

We count now the number of bends along red, black, and green edges. If G' is biconnected, then each red, black, or green edge e is split by at most one division vertex in the monotone book embedding $\bar{\gamma}$. The two sub-edges obtained from e are in different pages and thus one of them is drawn straight-line and the other one is drawn with one bend. Since the division vertex that splits e is replaced at the end by an additional bend, then the number of bends along e is two.

Suppose now that G' is not biconnected and let e be a black edge. Edge e can be split into a blue sub-edge e_1 and a black sub-edge e_2 during the biconnection procedure. The drawing of the transformed graph $\hat{G}$ is computed like in the case when G' is biconnected, and therefore e_1 has no bends while e_2 has at most two bends. Since the division vertex that splits e into e_1 and e_2 is replaced by an additional bend, the total number of bends along e is at most three. Let e' be a green edge; edge e can be split into a blue sub-edge e_1, a green sub-edge e_2, and a blue sub-edge e_3 during the biconnection procedure. The drawing of the transformed graph $\hat{G}$ is computed like in the case when G' is biconnected, and therefore e_1 and e_3 have no bends while e_2 has at most two bends. Since the division vertices that split e into e_1, e_2, and e_3 are replaced by two additional bends, the total number of bends along e is at most four.   □

4 Points in General Position

In this section we consider points in general position and consider planar graphs whose subgraph is either a cycle or a path or a set of disjoint paths. Before giving the main results of this section we need to give some additional definitions and to prove a pair of lemmas that will be used subsequently.

Let Γ be a drawing of a graph G and let v be a vertex of G; we say that v is visible from below in Γ if Γ does not intersect the open vertical halfline below v. Analogously, v is visible from above in Γ if Γ does not intersect the open vertical halfline above v. Throughout this section we assume that the points of S have distinct x-coordinate; if this is not the case we can achieve this condition by a suitable rotation of the plane.

Lemma 5. *Let C be a cycle and let S be a set of points in general position. C admits a geometric point-set embedding Γ on S such that each vertex of C is visible in Γ either from above or from below.*

Proof. Let $CH(S)$ be the convex hull of S and let p_l and p_r be the leftmost and the rightmost point of $CH(S)$, respectively. In $CH(S)$ there are two paths from p_l to p_r: the first one, that we call *upper hull*, has no point of S above, the other one, that we call *lower hull*, has no point of S below. Let $p_l = p_1, p_2, \ldots, p_h = p_r$ be the points in the lower hull ordered from left to right. Let $q_1, q_2, \ldots, q_k$ be the remaining points ordered from right to left.

Let $v_1, v_2, \ldots, v_n$ be the vertices of C ordered clockwise. Notice that $n = h + k$. Vertex v_i, for $1 \le i \le h$, is mapped to point p_i; vertex v_{h+i}, for $1 \le i \le k$, is mapped to point q_i. Cycle C can be divided into two paths with end-vertices in common: $\pi_1 = v_1, v_2, \ldots, v_k$ and $\pi_2 = v_k, v_{k+1}, v_{k+2}, \ldots, v_{k+m}, v_1$. Both π_1 and π_2 are represented by a x-monotone polyline and therefore neither the drawing of π_1 nor the drawing of π_2 has a crossing. On the other hand, all points of π_2 are above those of π_1 except for the end-vertices that are in common. Hence there is no crossing between edges of π_1 and edges of π_2. Due to the monotonicity of π_1 and π_2, the vertices of π_1 are visible from below, and those of π_2 are visible from above. □

Lemma 6. *Let $G = (V, E)$ be an embedded planar graph and let S be a set of points in general position. If G contains a cycle $C = v_1, v_2, \ldots, v_h$ whose interior is triangulated with edges (v_1, v_i) for $3 \le i \le h - 1$, then there exists a point-set embedding of G on S such that: (i) each edge has at most two bends; and (ii) after removing the edges of C and the edges (v_1, v_i) ($3 \le i \le h - 1$) all vertices of C are visible from below.*

Proof. Let γ be a monotone topological book embedding of G computed according to Lemma 1. Let $w_1, w_2, \ldots, w_{n'}$ be the real and division vertices of G in the order they appear along the spine of γ. We enrich the set S with extra points so that the resulting set S' has n' points and, denoted as $p_1, p_2, \ldots, p_{n'}$ the points of S' according to their order along the x-direction, p_i is an extra point if and only if w_i is a division vertex ($1 \le i \le n'$).

We compute a point-set embedding of G on S' with the following technique introduced by Kaufmann and Wiese [14]. Vertex w_i is mapped to point p_i ($1 \le i \le n'$); the edges between vertices that are consecutive along the spine of γ (and therefore drawn on consecutive points of S') are drawn straight-line. Let σ be a value greater than the maximum slope of a segment $\overline{p_i p_{i+1}}$ ($1 \leq i \leq n' - 1$); the remaining edges are drawn as polylines with two segments (and hence one bend) with slope σ and $-\sigma$. Using segments with slope σ and $-\sigma$ it is possible to draw an edge either above or below the polyline $\pi = \overline{p_1 p_2}, \overline{p_2 p_3}, \ldots, \overline{p_{n'-1} p_{n'}}$. If a sub-edge e is in the top page of γ, then it is drawn above π; if e is in the bottom page of γ, then it is drawn below π. The resulting drawing is planar except that 1-bend edges that are incident on the same vertex may contain overlapping segments. To eliminate these overlaps, we perturb the overlapping edges by decreasing the absolute value of their segment slopes by slightly different amounts. The slope changes are chosen to be small enough to avoid creating edge crossings while preserving the planar embedding. See [14] for details.

Via this technique, each sub-edge of G has at most one bend. Since γ is a monotone topological book embedding, each edge of G is split into at most two sub-edges. Let $e = (u, v)$ be an edge of G subdivided into two sub-edges e_1 and e_2 by a division vertex $d(u, v)$. One of the two sub-edges is in the top page in γ and the other one is in the bottom page in γ. Thus, the point p representing $d(u, v)$ has two segments incident on it: one from above and the other one from below; also, these two segments have the same slope. Since these two segments are the only two incident on $d(u, v)$ no rotation is performed to remove overlaps and therefore, when $d(u, v)$ is removed, no extra bend is created at p. It follows that each edge of G has at most two bends.

By Lemma 1, all edges of C and all edges inside C are in the bottom page. Since point-set embedding of G on S preserves the embedding of G, then there is no vertex or edge inside the drawing of C. Thus, if all edges of C and all edges inside C are removed, the vertices of C are visible from below.   □

Theorem 2. *Let G be an embedded planar graph, let S be any set of points in general position and let C be the boundary of a simple face of G. G admits a point-set embedding on S with straight-line subgraph C such that every edge not in C has at most 8 bends per edge.*

Proof. We describe how to compute a point-set embedding of G on S with straight-line C. Maintaining the embedding of G we change the external face so that some edge $e = (u, v)$ of C is on the external face. Let $v_1 = u, v_2 = v, \ldots, v_h$ be the vertices of C according to their circular ordering. We add blue edges (v_1, v_i) ($3 \leq i < h$) inside C.

Let $S' \subseteq S$ be the set containing the first h points along the x-direction. A point-set embedding Γ' of C on S' is computed as described in the proof of Lemma 5. The drawing Γ' is computed so that the circular clockwise order of the vertices is preserved and vertex v_1 is placed at the leftmost point of S.

Let γ be a monotone topological book embedding of G computed according to Lemma 1. Let $w_1, w_2, \ldots, w_{n''}$ be the real and division vertices of G in the order they appear along the spine of γ. We enrich the set $S \setminus S'$ with extra points so that the resulting set S'' has n'' points and, denoted as $p_1, p_2, \ldots, p_{n''}$ the points of S'' according to their order along the x-direction, p_i is an extra point if and only if w_i is a division vertex or a vertex of C ($1 \leq i \leq n''$). We compute a point-set embedding of G on S'' according to Lemma 6. After this point-set embedding is computed we remove the edges of C and the blue edges inside C. Denote as Γ'' the resulting drawing.

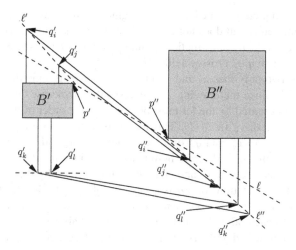

Fig. 2. An illustration for the proof of Theorem 2

Let v_i ($1 \leq i \leq h$) be a vertex of C. Let p_i' be the point representing v_i in Γ', and let p_i'' be the point representing v_i in Γ''. Notice that p_i' is a point of S, while p_i'' is not a point of S; therefore v_i will be represented by p_i' in the final drawing. All black/green edges incident to v_i are incident on point p_i''. We add a polyline from p_i' to p_i'', in order to make these edges incident to p_i'.

Let B' be the bounding box of Γ' and let B'' be the bounding box of Γ''. Let p' be the top-right corner of B' and let p'' be the bottom-left corner of B''. Refer to Fig. 2 for an illustration. Let ℓ be the straight line through p' and p''. Let ℓ' be a half-line that is completely above ℓ and intersects all vertical lines that intersect B'; let ℓ'' be a half-line that is completely below ℓ and intersects all vertical lines that intersect B''. Finally, let ℓ_H be a horizontal line below B'.

Let q_i' be the projection of point p_i' either on line ℓ', if p_i' is visible from above, or on line ℓ_H, if p_i' is visible from below. Notice that the leftmost and rightmost points of Γ' are visible both from above and from below. We project these two vertices on ℓ'. Let q_i'' be the projection of p_i'' on ℓ''. The polyline connecting p_i' to p_i'' is $\pi_i = \overline{p_i'q_i'}, \overline{q_i'q_i''}, \overline{q_i''p_i''}$.

Notice that, if there are $k > 1$ black/green edges incident to a vertex v_i, then all these edges are represented with a portion in common: the polyline π_i. It is possible to separate these edges by replacing each of the three points q_i', q_i'', p_i' with a set of k points coherently ordered.

Since both Γ' and Γ'' are planar, planarity can be proved by showing that the polylines π_i ($1 \leq i \leq h$) cross neither Γ' nor Γ'' and do not cross each other.

Consider a polyline π_i ($1 \leq i \leq h$) and assume first that q_i' is above B'. Segment $\overline{p_i'q_i'}$ does not cross Γ' because p_i' is visible from above and it does not cross Γ'' because it is completely to the left of B''. Segment $\overline{q_i''p_i''}$ does not cross Γ'' because p_i'' is visible from below and does not cross Γ' because it is completely to the right of B'. Since point q_i' is above ℓ and point q_i'' is below ℓ, segment $\overline{q_i'q_i''}$ crosses ℓ between p' and p'' and therefore it crosses neither Γ' nor Γ''. Suppose now that q_i' is below B'. In this case segment $\overline{p_i'q_i'}$ does not cross Γ' because p_i' is visible from below and it does not cross

Γ'' because it is completely to the left of B''. Segment $\overline{q_i'' p_i''}$ does not cross Γ'' because p_i'' is visible from below and does not cross Γ' because it is completely to the right of B'. Point q_i' is below B' by construction; point q_i'' is below ℓ and therefore below B''. It follows that segment $\overline{q_i' q_i''}$ crosses neither Γ' nor Γ''.

Consider now two polylines π_i and π_j ($1 \leq i, j \leq h$). Assume first that q_i' and q_j' are both above B'. Since the drawing Γ' preserves the clockwise circular order of C and since v_1 is represented by the leftmost point of Γ', the order of the projections on ℓ' is $q_1', q_h', q_{h-1}', \ldots, q_{h-k}'$ ($k \geq 1$). This means that the order of q_i' and q_j' on ℓ' is the same as the order of q_i'' and q_j'' on ℓ'' and therefore π_i and π_j do not cross each other. Suppose now that q_i' and q_j' are both below B'. Since the drawing Γ' preserves the clockwise circular order of C and since v_1 is represented by the leftmost point of Γ', the order of the projections on ℓ_H is $q_2', q_3', \ldots, q_{h-k-1}'$ ($k \geq 1$). This means that the order of q_i' and q_j' on ℓ_H is opposite to the order of q_i'' and q_j'' on ℓ'' and therefore π_i and π_j do not cross each other. Finally, suppose that q_i' is above B' and q_j' is below B'. In this case q_j'' is below and to the right of q_i'', and therefore a crossing is not possible.

We count now the number of bends per edge. Clearly the blue edges are drawn with zero bends in Γ'. The red edges are drawn with two bends in Γ''. Consider a black edge e. Edge e is first drawn with two bends in Γ'' with one end-vertex at a point p_i'' ($1 \leq i \leq h$); the drawing of this edge is then modified with the addition of the polyline π_i. Since polyline π_i has two bends and an extra bend can exist at point p_i'', then the total number of bends on e is at most 5. Let e' be a green edge. The two end-vertices of e' are vertices of C but e' is not an edge of C. Thus edge e' is first drawn with two bends in Γ'' with one end-vertex at a point p_i'' and the other one at a point p_j'' ($1 \leq i, j \leq h$); the drawing of this edge is then modified with the addition of the two polylines π_i and π_j. Each of these polylines has 2 bends and two extra bends can exist at p_i'' and p_j''. It follows that e' has at most 8 bends per edge. □

With techniques similar to that of Theorem 2 the following theorems can be proved.

Theorem 3. *Let G be an embedded planar graph, let S be any set of points in general position and let P be a subgraph of G that is a simple path. G admits a point-set embedding on S with straight-line subgraph P such that every edge not in C has at most 8 bends per edge.*

Theorem 4. *Let G be an embedded planar graph, let S be any set of points in general position and let $P_1, P_2, \ldots, P_k$ be a set of disjoint subgraphs of G each being a simple path. G admits a point-set embedding on S with straight-line subgraph $\bigcup_{i=1}^{k} P_i$ such that every edge not in $\bigcup_{i=1}^{k} P_i$ has $O(2^k)$ bends per edge.*

5 Open Problems

1. Theorem 1 shows that for every set of points in convex position and for every outerplanar subgraph a point-set embedding can be computed such that the subgraph has straight-line edges and the remaining edges of the graph have at most four bends each. How hard is it to compute a point-set embedding with the minimum number of bends in total?

2. Extend Theorems 2 and 3 by considering the case that G' is either a tree or a general outerplanar graph. Note that the outerplanar graphs are the largest family of graphs that admit a straight-line point-set embedding without mapping on any set of points [9].

3. We have studied constrained point-set embeddability on *any* given set of points. Given a planar graph G with n vertices, a non outerplanar subgraph G' of G, and a set S of n distinct points in convex position, how hard is it to test whether G admits a point-set embedding on S such that G' is drawn with straight-line edges and the remaining edges have constant curve complexity?

References

1. Badent, M., Di Giacomo, E., Liotta, G.: Drawing colored graphs on colored points. Theoret. Comput. Sci. 408(2-3), 129–142 (2008)
2. Bose, P.: On embedding an outer-planar graph on a point set. Comput. Geom. Theory Appl. 23, 303–312 (2002)
3. Bose, P., McAllister, M., Snoeyink, J.: Optimal algorithms to embed trees in a point set. J. Graph Algorithms Appl. 2(1), 1–15 (1997)
4. Di Giacomo, E., Didimo, W., Liotta, G., Meijer, H., Trotta, F., Wismath, S.K.: k-colored point-set embeddability of outerplanar graphs. J. Graph Algorithms Appl. 11(1), 29–49 (2008)
5. Di Giacomo, E., Liotta, G., Trotta, F.: On embedding a graph on two sets of points. Internat. J. Found. Comput. Sci. 17(5), 1071–1094 (2006)
6. Di Giacomo, E., Didimo, W., Liotta, G., Meijer, H., Wismath, S.K.: Point-set embeddings of trees with edge constraints. In: Hong, S.-H., Nishizeki, T., Quan, W. (eds.) GD 2007. LNCS, vol. 4875, pp. 113–124. Springer, Heidelberg (2008)
7. Di Giacomo, E., Didimo, W., Liotta, G., Wismath, S.K.: Curve-constrained drawings of planar graphs. Comput. Geom. Theory Appl. 30(1), 1–23 (2005)
8. Di Giacomo, E., Liotta, G., Trotta, F.: Drawing colored graphs with constrained vertex positions and few bends per edge. In: Hong, S.-H., Nishizeki, T., Quan, W. (eds.) GD 2007. LNCS, vol. 4875, pp. 315–326. Springer, Heidelberg (2008)
9. Gritzmann, P., Mohar, B., Pach, J., Pollack, R.: Embedding a planar triangulation with vertices at specified points. Amer. Math. Monthly 98(2), 165–166 (1991)
10. Halton, J.H.: On the thickness of graphs of given degree. Inform. Sci. 54, 219–238 (1991)
11. Ikebe, Y., Perles, M., Tamura, A., Tokunaga, S.: The rooted tree embedding problem into points in the plane. Discrete Comput. Geom. 11, 51–63 (1994)
12. Kaneko, A., Kano, M.: Straight line embeddings of rooted star forests in the plane. Discrete Appl. Math. 101, 167–175 (2000)
13. Kaneko, A., Kano, M.: Semi-balanced partitions of two sets of points and embeddings of rooted forests. Internat. J. Comput. Geom. Appl. 15(3), 229–238 (2005)
14. Kaufmann, M., Wiese, R.: Embedding vertices at points: Few bends suffice for planar graphs. J. Graph Algorithms Appl. 6(1), 115–129 (2002)
15. Pach, J., Törőcsik, J.: Layout of rooted trees. DIMACS Series in Discrete Math. and Theoretical Comput. Sci. 9, 131–137 (1993)
16. Pach, J., Wenger, R.: Embedding planar graphs at fixed vertex locations. Graphs and Combinatorics 17, 717–728 (2001)
17. Sugiyama, K.: Graph Drawing and Applications for Software and Knowledge Engineers. World Scientific, Singapore (2002)

Tree Drawings on the Hexagonal Grid

Christian Bachmaier, Franz J. Brandenburg, Wolfgang Brunner,
Andreas Hofmeier, Marco Matzeder, and Thomas Unfried

University of Passau, Germany
{bachmaier,brandenb,brunner,hofmeier,matzeder,
unfried}@fim.uni-passau.de

Abstract. We consider straight-line drawings of trees on a hexagonal grid. The hexagonal grid is an extension of the common grid with inner nodes of degree six. We restrict the number of directions used for the edges from each node to its children from one to five, and to five patterns: straight, Y, ψ, X, and full. The ψ–drawings generalize hv- or strictly upward drawings to ternary trees.

We show that complete ternary trees have a ψ–drawing on a square of size $\mathcal{O}(n^{1.262})$ and general ternary trees can be drawn within $\mathcal{O}(n^{1.631})$ area. Both bounds are optimal. Sub–quadratic bounds are also obtained for X–pattern drawings of complete tetra trees, and for full–pattern drawings of complete penta trees, which are 4–ary and 5–ary trees. These results parallel and complement the ones of Frati [8] for straight–line orthogonal drawings of ternary trees.

Moreover, we provide an algorithm for compacted straight–line drawings of penta trees on the hexagonal grid, such that the direction of the edges from a node to its children is given by our patterns and these edges have the same length. However, drawing trees on a hexagonal grid within a prescribed area or with unit length edges is $\mathcal{NP}$–hard.

1 Introduction

Drawing trees is one of the best studied areas in graph drawing. It has been initiated forty years ago by D. E. Knuth, who posed the question "How shall we draw a tree?" [12]. He proposed the hierarchical style, drawing binary trees level by level and left–to–right, and used a typewriter as a drawing tool. This idea has become the most common tree drawing technique, and has been turned into practice by the Reingold–Tilford algorithm [13] and its generalizations [16, 2].

Another approach are radial drawings, introduced by Eades [7]. Here trees are displayed in a centralized view with the root in the center and the nodes at depth d on the d–th ring. This approach is used for tree drawings in social sciences.

The third major approach was motivated by VLSI design and the theory of graph embeddings. Orthogonal drawings are obtained using the common grid as a host. The most important cost measure for such drawings is the area of the smallest surrounding rectangle.

I.G. Tollis and M. Patrignani (Eds.): GD 2008, LNCS 5417, pp. 372–383, 2009.
© Springer-Verlag Berlin Heidelberg 2009

Orthogonal drawings are restricted to graphs of degree at most four. This suffices for binary and ternary trees, which can be drawn on $\mathcal{O}(n)$ area, if bends of the edges are permitted [5,15]. However, the drawings obtained by these algorithms are not really pleasing. Even complete binary trees are deterred, as illustrations in these papers display. The typical tree structure is not visible, since the algorithms wind edges and recursively fold subtrees. This poor behavior also holds for other tree drawing algorithms achieving good area bounds [3,9,10,14].

The readability improves with the restriction to upward and hv–drawings, which use only three resp. two out of four directions. Now the hierarchical structure of the tree is preserved, and in case of hv–drawings the parent is in the upper left corner above its subtrees (with the Y–axis directed downwards).

For binary trees in general there is one degree of freedom on the orthogonal grid. Only three directions at a grid point are used in the drawing. This suffices for compact tree drawings on an almost linear area. In particular, complete binary trees can be drawn straight–line on a square of size less than $2n$ in the H–layout and on $\mathcal{O}(n)$ area in the hv–layout. Moreover, $\mathcal{O}(n \log n)$ is the upper and lower bound for the area of straight–line orthogonal upward drawings of arbitrary binary trees [3].

Ternary trees need all four directions on the orthogonal grid. For straight–line drawings this enforces more area and even fails when restricted to upward drawings [8]. Frati's upper bounds are $\mathcal{O}(n^{log_3 2})^2 = \mathcal{O}(n^{1.262})$ for complete ternary trees and $\mathcal{O}(n^{1.631})$ for arbitrary ternary trees. There are no non–trivial lower bounds for these types of tree drawings, yet.

The paper is organized as follows: In Sect. 2 we introduce hexagonal grid drawings and review previous work on straight–line orthogonal drawings of binary and ternary trees. In Sect. 3 we provide upper and lower bounds of ψ–drawings of ternary trees, and establish sub–quadratic upper bounds of X–pattern drawings of complete tetra trees and full–pattern drawings of complete penta trees. In Sect. 4 we introduce an algorithm for straight–line drawings of penta trees with patterns for the directions of the edges from the nodes to their children, and in Sect. 5 we establish $\mathcal{NP}$–hardness results on minimal area and minimal edge length drawings.

2 Preliminaries

We consider straight–line drawings on the *hexagonal grid* which consists of equilateral triangles as defined, e. g., in [11]. It defines three directions, called *grid lines*. The X–axis is directed to the east, the Y–axis has an angle of $2\pi/3$ and the diagonal one of $\pi/3$ clockwise against the X–axis. They are directed to the south–west, and to the south–east, respectively, see Fig. 1(a). The hexagonal grid can be sheared by a counter–clockwise rotation of the diagonal axis by $\pi/12$ and of the Y–axis by $\pi/6$ creating the *sheared grid*. Then, the Y–axis is as usual and directs downward, see Fig. 1(b). We switch between these representations whenever it is appropriate. The latter representation shows that hexagonal drawings are an extension of orthogonal drawings, where the orthogonal grid is underlying. A grid point v is defined by its $x-$ and y-coordinate (v_x, v_y).

We define the *distance* between two points a and b lying on the same grid line or on the same bisecting line between two grid lines as $d(a,b) = max(|a_x - b_x|,$ $|a_y - b_y|, |(a_x - a_y) - (b_x - b_y)|)$. The *distance* $d_D(a, L)$ of a point a to a segment of a bisecting line or a grid line L with respect to a *direction* D (along a grid line or a bisecting line), is defined by the distance of a to the intersection point c of L and the parallel of D through a. The distance is set to ∞, if this direction line does not intersect L. Note that in the sheared grid the bisecting lines are not really bisecting the angle between the grid lines.

See Fig. 1 for an example of distances: $d(u, w) = 4$, $d(u, v) = 2$, $d(v, w) = 4$, $d_{(v,c)}(v, (u, w)) = 2$ and $d_{(v,w)}(v, (u, w)) = 4$.

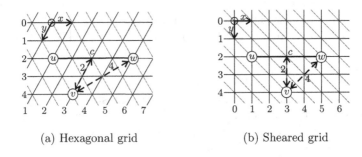

(a) Hexagonal grid   (b) Sheared grid

Fig. 1. The two grid versions

Let $T = (V, E)$ be a rooted tree. *Tetra* and *penta trees* are trees with outdegree at most 4 or 5, respectively. The *height* $h(T)$ of a rooted tree T is the maximum length (number of edges) of a path from the root to a leaf. Let T be a penta tree. A *straight–line drawing* $\Gamma(T)$ of T on the hexagonal grid is an embedding of the nodes of T to grid points. The edges are mapped to segments on the grid lines s. t. the embedding is planar. The *area* of a tree drawing $\Gamma(T)$ on the sheared grid is the size of the smallest surrounding rectangle. Let $width_\Gamma(T)$ and $height_\Gamma(T)$ denote the width and the height of this rectangle, whose quotient is the *aspect ratio*. The rectangle corresponds to an enclosing parallelogram in the hexagonal grid. We call a drawing $\Gamma(T)$ *globally uniform*, if all outgoing edges of nodes at the same depth have the same length. It is *locally uniform*, if outgoing edges for each node have the same length.

We first consider globally uniform ψ–drawings of ternary trees. In ψ–drawings three directions are used: to the east, south–east and south. ψ–drawings are extensions of hv–drawings of binary trees [4]. They correspond to upward drawings with three possible directions and are used only for binary and ternary trees.

We introduce another drawing style, *pattern drawings*. In pattern drawings the edge directions of the outgoing edges of each node $v \in V$ have a fixed angle to the direction of the incoming edge, see Fig. 2(a) – (e).

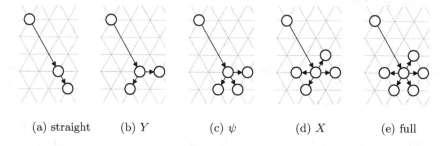

| (a) straight | (b) Y | (c) ψ | (d) X | (e) full |

Fig. 2. The five patterns of the drawings

Our investigations on hexagonal grid drawings are motivated by the fact, that they allow nice drawings of up to 5–ary trees. In particular, for ternary trees they provide a canonical generalization of upward and hv–drawings, which gives pleasing pictures. The pattern drawings can be seen as a step towards a discretization of radial drawings with a bounded number of slopes [6]. Let us first recall the state–of–the–art on straight–line orthogonal drawings of binary (see Tab. 1) and ternary trees (see Tab. 2).

Table 1. Area bounds for binary trees

Drawing Style	Complete Trees	Arbitrary Trees	Source
hv or strictly upward	$\Theta(n)$	$\Theta(n \log n)$	[4]
T or upward	$\Theta(n)$	$\Theta(n \log n)$	[3, 4]
H, all four directions	$\Theta(n)$	$\mathcal{O}(n \log \log n)$	[3, 14]

Straight–line orthogonal drawings of ternary trees were recently investigated by Frati [8]. Here the picture changes. We loose one degree of freedom, which increases the area and may even lead to non–drawability. Moreover, there are no non–trivial lower bounds, yet.

Table 2. Area bounds for ternary trees

Drawing Style	Complete Trees	Arbitrary Trees	Source
hv or strictly upward	non–drawable	non–drawable	trivial
T or upward	non–drawable	non–drawable	trivial
H, all four directions	$\mathcal{O}(n^{1.262})$	$\mathcal{O}(n^{1.631})$	[8]
ψ (on hexagonal grid)	$\Theta(n^{1.262})$	$\Theta(n^{1.631})$	Th. 1, Th. 2

3   Hexagonal Tree Drawings

In the following we consider upper and lower bounds with respect to the required area in ψ–drawings of complete and arbitrary ternary trees.

Theorem 1. *There is a linear time algorithm to draw a complete ternary tree with n nodes by a globally uniform ψ–drawing on the hexagonal grid in $\mathcal{O}(n^{1.262})$ area and with aspect ratio 1.*

Proof. We construct the drawing recursively, such that the root is in the upper left corner. Let the three subtrees of height h each be drawn inside a square with width and height $S(h)$. We move the three subtrees $1 + S(h)$ units away from the root following the X– and Y–axis and the diagonal creating a drawing of the tree with width and height $S(h + 1) = 2S(h) + 1$. This leads to $S(h) = 2^h - 1$ and to globally uniform edge lengths. Since $h = \log_3 n$ we obtain for a complete n–node ternary tree a drawing with width and height each in $\mathcal{O}(n^{0.631})$ and an area of $\mathcal{O}(n^{1.262})$.   □

Theorem 2. *There is a linear time algorithm to draw an unordered arbitrary ternary tree with n nodes by a ψ–drawing on the hexagonal grid in $\mathcal{O}(n^{1.631})$ area.*

Proof. Minimize the width and let the height be arbitrary, i. e., the height is $\mathcal{O}(n)$. Let $T_1, T_2,$ and T_3 be the subtrees of T with root r, with $width(T_1) \leq width(T_2) \leq width(T_3)$. Recursively construct the drawing with the root in the upper left corner. Relative to r place T_1 one unit diagonally under r, attach T_2 horizontally to the right at distance $2 + width(T_1)$, and attach T_3 by a vertical line underneath, see Fig. 3. Then $width(T) = \max\{2 + width(T_1) + width(T_2), width(T_3)\}$, which results in $width(T) = \mathcal{O}(n^{\log_3 2}) = \mathcal{O}(n^{0.631})$. This is shown by calculations as in the proof of Theorem 5 in [8].   □

See Fig. 4 for an example of a ψ–drawing of a complete ternary tree. We now turn to lower bounds between $n \log n$ and n^2. These are the lower bounds for the area of unordered and ordered hv–drawings of binary trees [4].

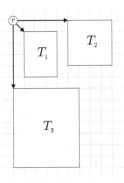

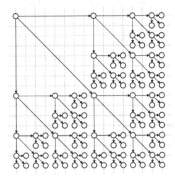

Fig. 3. Sketch for Theorem 2   **Fig. 4.** Complete ψ–drawing

Lemma 1. *Any ψ-drawing of a complete ternary tree with n nodes has a width (and a height) of $\Omega(n^{0.631})$.*

Proof. Consider ψ-drawings on the sheared grid. Let $\Gamma(T)$ be a ψ-drawing of a complete ternary tree of height h. We claim that the extreme grid points at $(0, 2^h - 1)$ and $(2^h - 1, 0)$ are occupied by the drawing. There is a node at these grid points or they are passed by some edge. The proof is by induction on the height h. The claim is clearly true for $h = 1$.

Assume for contradiction that there exists a minimal h, s. t. a complete subtree of height h does not occupy the grid points as described above. Let r be the root of a tree with height h placed at $(0, 0)$ and let T_1, T_2, and T_3 be the subtrees of r. By induction, every ψ-drawing of a tree with height $h - 1$ occupies points $(2^{h-1} - 1, 0)$ and $(0, 2^{h-1} - 1)$ relative to its root.

There is a vertical line from r to T_3, a diagonal line from r to T_1, and a horizontal line from r to T_2. If T_3 does not occupy $(0, 2^h - 1)$, then it must occupy the diagonal at a point (p, p) with $p \leq 2^{h-1} - 1$ and there is no space left for T_1. With a symmetric argument T_2 occupies $(2^h - 1, 0)$. □

From Theorem 1 and Lemma 1 we directly obtain.

Theorem 3. *The upper and lower bound for the area of ψ-drawings of complete ternary trees with n nodes in the sheared grid is $\Theta(n^{1.262})$.*

Theorem 4. *The upper and lower bound for the area of ψ-drawings of unordered arbitrary ternary trees with n nodes is $\Theta(n^{1.631})$.*

Proof. The upper bound follows directly from Theorem 2. For the lower bound consider a ternary tree consisting of a path of length $n/2$ followed by a complete ternary subtree of size $n/2$. Then, the path needs $\Omega(n)$ in at least one dimension, and the complete subtree needs $\Omega(n^{0.631})$ in any dimension. □

These results parallel the ones for hv-drawings of binary trees on the orthogonal grid, where the bounds are $\Theta(n)$ for complete and $\Theta(n \log n)$ for arbitrary binary trees.

ψ-drawings use less area than radial tree drawings, where the nodes are placed on concentric rings around the center. Suppose that two nodes must have at least unit distance. Consider the binary case; the general case is similar. Then, the outermost ring containing the leaves must have a circumference of at least $n/2$ for complete trees and, thus, the area is $\Omega(n^2)$.

We now turn to complete penta and tetra trees, and their pattern drawings on the hexagonal grid with five and four directions towards the children. We call the pattern drawings of complete penta trees *full-pattern drawings* and of complete tetra trees *X-pattern drawings*.

Theorem 5. *There is a linear time algorithm to draw a complete penta tree with n nodes by a globally uniform full-pattern drawing on the hexagonal grid in $\mathcal{O}(n^{1.37})$ area and with aspect ratio 1.*

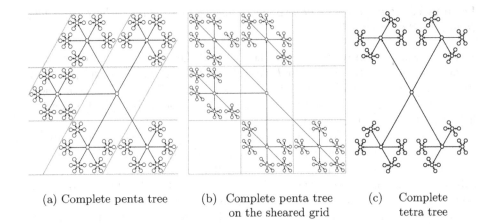

(a) Complete penta tree   (b) Complete penta tree   (c)   Complete
                                on the sheared grid   tetra tree

Fig. 5. Drawings of complete trees

Proof. Construct the drawings recursively. By an expansion by the factor three in each dimension one can draw a new tree of height $h + 1$ in a planar way. This can easily be seen in the sheared grid, see Fig. 5(b). Thus, the area is in $\mathcal{O}(9^h)$, where h is the height of the tree. All edges of the same depth have the same length. Since $h = \log_5 n$, the area is $\mathcal{O}(n^{\log_5 9})$, which is $\mathcal{O}(n^{1.37})$.   □

For an example see Fig. 5(a). In the same way we draw complete tetra trees, which need $\mathcal{O}(n^{\log_4 9})$ area, see Fig. 5(c).

Theorem 6. *There is a linear time algorithm to draw a complete tetra tree with n nodes by a globally uniform X–pattern drawing on the hexagonal grid in $\mathcal{O}(n^{1.58})$ area and with aspect ratio 1.*

4   Pattern Drawings of Penta Trees

In this section we introduce an algorithm for compacted pattern drawings of ordered penta trees on the hexagonal grid, e.g., see Fig. 7.

Once the directions of the outgoing edges of the root are fixed, the directions of all edges of the tree are predetermined. Thus, all edge directions can be computed in linear time by a top–down traversal of the tree. The only free and computable parameter is the length of the edges. We produce locally uniform drawings, i.e., the edge length is the same for all outgoing edges of a node. The goal is to keep it small which is achieved by a compaction method.

Our algorithm has some similarities with the Reingold–Tilford algorithm [13]. However, it uses simpler contours, which are convex hexagons, and it attempts to minimize the edge length and not the width of the drawing.

Definition 1. *The convex contour of a subtree T is defined by six coordinates: $min_x, min_y,$ and min_{x-y} are the smallest coordinates of the nodes of the subtree in x, y, and $(x-y)$ directions, respectively. The values $max_x, max_y,$ and max_{x-y} are defined analogously.*

The six corner points and the six segments of each contour can be computed in linear time obviously. The trivial convex contour of a leaf v consists of the values $min_x = max_x = v_x$, $min_y = max_y = v_y$, and $min_{x-y} = max_{x-y} = v_x - v_y$. In this case, the values of the contour match the absolute position of the leaf in the current drawing of the tree. We construct the contour C_r of an inner node r by merging the contours $C_1, \ldots, C_l$ of its children $s_1, \ldots, s_l$. We set the value $min_x(C_r) = min\{r_x, min_x(C_1), \ldots, min_x(C_l)\}$. The remaining five values are computed analogously. As an example see the contour C_1 of the drawing of the subtree rooted at r_1 in Fig. 6.

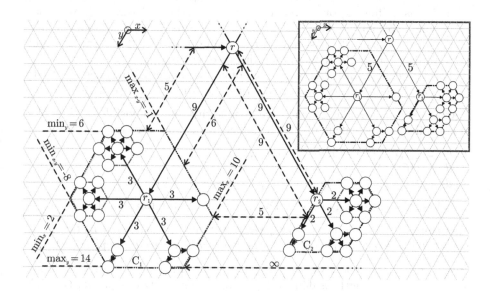

Fig. 6. Before and after trimming the outgoing edges of r

Definition 2. *Let C be a contour and let x be a point, a segment, or a contour. Let D be a direction. We define $d_D(x, C)$ as the length of the shortest segment parallel to D such that one end point lies on x and the other on C. We set $d_D(x, C) = \infty$, if such a segment does not exist.*

Note that all these distances can be computed in $\mathcal{O}(1)$ time using the distance between a point and a segment only, as each contour has at most six segments.

Algorithm 1 first produces a drawing of a penta tree T with sufficiently long edges s.t. the drawing is planar. Therefore, *drawTreeOnHexagonalGrid* is called, which uses the value *edgeLengthToChildren* for each node. These edge lengths suffice to get a planar pattern drawing of a penta tree, see Theorem 5. Then, it computes the compacted edge lengths which are used for the final drawing. As an example see Fig. 7.

To compact the subtree of a node r (see Algorithm 2), we create the trivial contour C of the node r (line 1), call the algorithm recursively for its children to compute their contours (lines 3 to 5), compute the trim of all outgoing edges of r (lines 6 to 19), move only the contours of the children (for efficiency reasons not the complete subtrees) towards r, and merge them with C (lines 20 to 22). The *edge trim* is the value the outgoing edges of r can be shortened. It is computed satisfying the following conditions:

1. For each pair of children r_i and r_j of r, the contour C_i of r_i does not cross the edge (r, r_j) (line 8).
2. For each pair of children r_i and r_j of r, the contours of r_i and r_j do not cross. Here we distinguish two cases: The angle α between the edges to r_i and r_j is $\frac{\pi}{3}$ (line 11) or the remaining cases $\alpha \in \left\{ \frac{2\pi}{3}, \pi \right\}$ (line 13). Note that moving two contours with $\alpha = \frac{\pi}{3}$ one unit towards their parent reduces their distance by one, whereas moving two contours with $\alpha \in \left\{ \frac{2\pi}{3}, \pi \right\}$ one unit towards their parent reduces their distance by two.
3. For each child r_i of r, the contour C_i of r_i does not cross the edge of r to its parent (line 16) or does not cross r (if there is no parent) (line 19).

As an example see Fig. 6, where we assume that the non visible children of r do not influence the calculations. The following distances are used:

1. $d_{(r,r_1)}((r, r_2), C_1) = 6$, $d_{(r,r_2)}((r, r_1), C_2) = 9$
2. $d_{(r_1,r_2)}(C_1, C_2) = 5$ ($\alpha = \frac{\pi}{3}$)
3. $d_{(r,r_1)}((r, r.parent), C_1) = 5$, $d_{(r,r_2)}((r, r.parent), C_2) = 9$

Therefore, the edge trim of r is 4. For the result see the top right box of Fig. 6.

Theorem 7. *Let T be an arbitrary penta tree with n nodes and root r. Algorithm 1 (drawTreeCompactedOnHexagonalGrid) has time complexity $\mathcal{O}(n)$.*

Proof. The time complexity of each line in Algorithm 1 is $\mathcal{O}(n)$, as the calculation of the initial edge lengths is done in $\mathcal{O}(n)$, and $drawTreeOnHexagonalGrid(T)$ and $computeCompactedEdgeLength(r)$ each have linear time complexity. For the latter one, the distance between a point and a segment is computed in $\mathcal{O}(1)$. As each node has at most five children and each convex contour consists of at most six segments, the edge trim is computed in $\mathcal{O}(1)$. Moving a contour is in $\mathcal{O}(1)$ as well. Thus, the time complexity for one node is $\mathcal{O}(1)$ and $\mathcal{O}(n)$ for the tree T. $\qquad\square$

5   NP–Completeness Results

Finally we establish some $\mathcal{NP}$–hardness results for the area and the edge length of drawings on the hexagonal grid. In contrast to the previous section the trees are unordered, i.e., the children of a node can be permuted. In the drawing this is a rotation or a flip.

Algorithm 1. drawTreeCompactedOnHexagonalGrid

Input: An ordered penta tree $T = (V, E)$ with root r and height h
Output: A compacted drawing $\Gamma(T)$ of T

1 **foreach** $v \in V$ **do** $v.edgeLengthToChildren \leftarrow 3^{h-depth(v)-1}$
2 $drawTreeOnHexagonalGrid(T)$
3 $computeCompactedEdgeLength(r)$
4 $drawTreeOnHexagonalGrid(T)$

Algorithm 2. computeCompactedEdgeLength

Input: A node r of an ordered penta tree
Output: $edgeLengthToChildren$ of each node in the subtree of r

1 Contour $C \leftarrow Contour(r)$
2 Set $\mathcal{C} \leftarrow \emptyset$
3 **foreach** *child* r_i *of* r **do**
4     $C_i \leftarrow computeCompactedEdgeLength(r_i)$
5     $\mathcal{C} = \mathcal{C} \cup \{C_i\}$
6 $edgeTrim \leftarrow \infty$
7 **foreach** $C_i, C_j (i \neq j) \in \mathcal{C}$ **do**
8     $edgeTrim \leftarrow \min \left(edgeTrim, d_{(r,r_i)}((r,r_j), C_i) - 1 \right)$
9     $\alpha \leftarrow$ angle between (r, r_i) and (r, r_j)
10   **if** $\alpha = \frac{\pi}{3}$ **then**
11        $edgeTrim \leftarrow \min \left(edgeTrim, d_{(r_i,r_j)}(C_i, C_j) - 1 \right)$
12   **else**
13        $edgeTrim \leftarrow \min \left(edgeTrim, \left\lfloor \frac{d_{(r_i,r_j)}(C_i,C_j)-1}{2} \right\rfloor \right)$
14 **if** $r.parent \neq nil$ **then**
15   **foreach** $C_i \in \mathcal{C}$ **do**
16        $edgeTrim \leftarrow \min \left(edgeTrim, d_{(r,r_i)}((r,r.parent), C_i) - 1 \right)$
17 **else**
18   **foreach** $C_i \in \mathcal{C}$ **do**
19        $edgeTrim \leftarrow \min \left(edgeTrim, d_{(r,r_i)}(r, C_i) - 1 \right)$
20 **foreach** $C_s \in \mathcal{C}$ **do**
21     $move(C_s, edgeTrim)$
22     $C.merge(C_s)$
23 $r.edgeLengthToChildren \leftarrow r.edgeLengthToChildren - edgeTrim$
24 **return** C

Theorem 8. *Let* $T = (V, E)$ *be an unordered penta tree. The following problems are* $\mathcal{NP}$*–hard:*

- *Given an integer* K*, does* T *have a straight–line drawing on the hexagonal grid with an area at most* K*?*
- *Does* T *have a straight–line drawing on the hexagonal grid with unit length edges?*

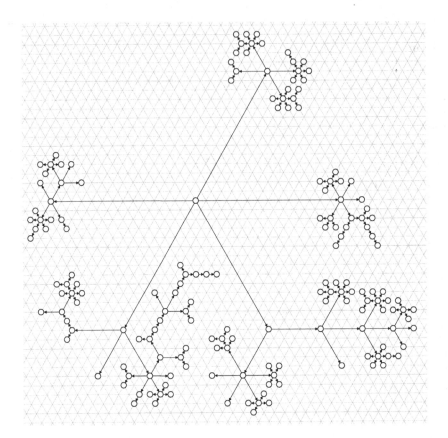

Fig. 7. Compacted drawing of a penta tree

Proof. (Sketch). We follow the Bhatt–Cosmadakis technique [1] and reduce from the NOT–ALL–EQUAL–3SAT problem. Suppose we draw on the sheared grid with the two axis and the diagonal. Then, the construction is made s. t. the diagonal cannot be used by the drawing, if the area bound or the unit length edges are preserved.   ☐

Corollary 1. *It is $\mathcal{NP}$–hard to draw unordered penta trees on a hexagonal grid within minimal area or with minimal edge length.*

6   Summary and Open Problems

In this paper we have shown upper and lower bounds for ψ–drawings of ternary trees and upper bounds for tetra and penta trees on the hexagonal grid. We have introduced a compaction algorithm for penta trees, which produces pleasing drawings in linear time. Finally, we have shown the $\mathcal{NP}$–hardness of drawing unordered penta trees with minimal area or minimal edge length.

As open problems remain finding a tree drawing algorithm, which adopts as much as possible from the Reingold–Tilford algorithm, establishing upper and lower bounds for the area of X-pattern and full–pattern drawings of ternary trees and considering the extension to the octagrid, which is the orthogonal grid with both diagonals.

References

1. Bhatt, S.N., Cosmadakis, S.S.: The complexity of minimizing wire lengths in VLSI layouts. Inf. Process. Lett. 25(4), 263–267 (1987)
2. Bloesch, A.: Aestetic layout of generalized trees. Softw. Pract. Exper. 23(8), 817–827 (1993)
3. Chan, T.M., Goodrich, M.T., Kosaraju, S.R., Tamassia, R.: Optimizing area and aspect ratio in straight-line orthogonal tree drawings. Comput. Geom. Theory Appl. 23(2), 153–162 (2002)
4. Crescenzi, P., Di Battista, G., Piperno, A.: A note on optimal area algorithms for upward drawings of binary trees. Comput. Geom. Theory Appl. 2, 187–200 (1992)
5. Dolev, D., Trickey, H.W.: On linear area embedding of planar graphs. Tech. Rep. STAN-CS-81-876, Stanford University, Stanford, CA, USA (1981)
6. Dujmović, V., Suderman, M., Wood, D.R.: Really straight graph drawings. In: Pach, J. (ed.) GD 2004. LNCS, vol. 3383, pp. 122–132. Springer, Heidelberg (2005)
7. Eades, P.: Drawing free trees. Bulletin of the Institute of Combinatorics and its Applications 5, 10–36 (1992)
8. Frati, F.: Straight-line orthogonal drawings of binary and ternary trees. In: Hong, S.-H., Nishizeki, T., Quan, W. (eds.) GD 2007. LNCS, vol. 4875, pp. 76–87. Springer, Heidelberg (2008)
9. Garg, A., Goodrich, M.T., Tamassia, R.: Planar upward tree drawings with optimal area. Int. J. Comput. Geometry Appl. 6(3), 333–356 (1996)
10. Garg, A., Rusu, A.: Straight-line drawings of binary trees with linear area and arbitrary aspect ratio. J. Graph Algo. App. 8(2), 135–160 (2004)
11. Kant, G.: Hexagonal grid drawings. In: Mulkers, A. (ed.) Live Data Structures in Logic Programs. LNCS, vol. 675, pp. 263–276. Springer, Heidelberg (1993)
12. Knuth, D.E.: The Art of Computer Programming, vol. 1. Addison-Wesley, Reading (1968)
13. Reingold, E.M., Tilford, J.S.: Tidier drawing of trees. IEEE Trans. Software Eng. 7(2), 223–228 (1981)
14. Shin, C.S., Kim, S.K., Chwa, K.Y.: Area-efficient algorithms for straight-line tree drawings. Comput. Geom. Theory Appl. 15(4), 175–202 (2000)
15. Valiant, L.G.: Universality considerations in VLSI circuits. IEEE Trans. Computers 30(2), 135–140 (1981)
16. Walker, J.Q.W.: A node-positioning algorithm for general trees. Softw. Pract. Exper. 20(7), 685–705 (1990)

Isometric Diamond Subgraphs

David Eppstein

Computer Science Department, University of California, Irvine
eppstein@uci.edu

Abstract. We test in polynomial time whether a graph embeds in a distance-preserving way into the hexagonal tiling, the three-dimensional diamond structure, or analogous higher-dimensional structures.

1 Introduction

Subgraphs of square or hexagonal tilings of the plane form nearly ideal graph drawings: their angular resolution is bounded, vertices have uniform spacing, all edges have unit length, and the area is at most quadratic in the number of vertices. For induced subgraphs of these tilings, one can additionally determine the graph from its vertex set: two vertices are adjacent whenever they are mutual nearest neighbors. Unfortunately, these drawings are hard to find: it is NP-complete to test whether a graph is a subgraph of a square tiling [2], a planar nearest-neighbor graph, or a planar unit distance graph [5], and Eades and Whitesides' *logic engine* technique can also be used to show the NP-completeness of determining whether a given graph is a subgraph of the hexagonal tiling or an induced subgraph of the square or hexagonal tilings.

With stronger constraints on subgraphs of tilings, however, they are easier to construct: one can test efficiently whether a graph embeds *isometrically* onto the square tiling, or onto an integer grid of fixed or variable dimension [7]. In an isometric embedding, the unweighted distance between any two vertices in the graph equals the L_1 distance of their placements in the grid. An isometric embedding must be an induced subgraph, but not all induced subgraphs are isometric. Isometric square grid embeddings may be directly used as graph drawings, while planar projections of higher dimensional embeddings can be used to draw any *partial cube* [6], a class of graphs with many applications [11].

Can we find similar embedding algorithms for other tilings or patterns of vertex placements in the plane and space? In this paper, we describe a class of d-dimensional patterns, the *generalized diamond structures*, which include the hexagonal tiling and the three-dimensional molecular structure of the diamond crystal. As we show, we can recognize in polynomial time the graphs that have isometric embeddings onto generalized diamonds of fixed or variable dimension.

2 Hexagons and Diamonds from Slices of Lattices

The three-dimensional points $\{(x,y,z) \mid x+y+z \in \{0,1\}\}$, with edges connecting points at unit distance, form a 3-regular infinite graph (Fig. 1, left) in which every vertex has

I.G. Tollis and M. Patrignani (Eds.): GD 2008, LNCS 5417, pp. 384–389, 2009.

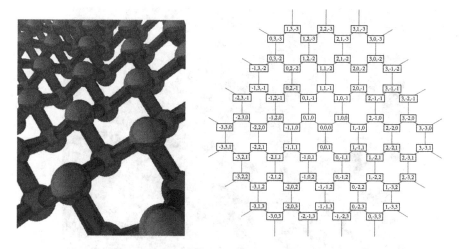

Fig. 1. Left: The unit distance graph formed by the integer points $\{(x,y,z) \mid x+y+z \in \{0,1\}\}$. Right: The same graph projected onto the plane $x+y+z = 0$ to form a hexagonal tiling.

three perpendicular edges [10]. Its projection onto the plane $x+y+z = 0$ is a hexagonal tiling (Fig. 1, right). In one higher dimension, the points $\{(w,x,y,z) \mid w+x+y+z \in \{0,1\}\}$, with edges connecting points at unit distance, projected into the three-dimensional subspace $w+x+y+z = 0$, form an infinite 4-regular graph embedded in space with all edges equally long and forming equal angles at every vertex (Fig. 2). This pattern of point placements and edges is realized physically by the crystal structure of diamonds, and is often called the *diamond lattice*, although it is not a lattice in the mathematical definition of the word; we call it the *diamond graph*.

Analogously, define a k-dimensional *generalized diamond graph* as follows. Form the set of $(k+1)$-dimensional integer points such that the sum of coordinates is either zero or one, connect pairs of points at unit distance, and project this graph onto the hyperplane in which the coordinate sum of any point is zero. The result is a highly symmetric infinite $(k+1)$-regular graph embedded in k-dimensional space. The generalized diamond graph is an isometric subset of the $(k+1)$-dimensional integer lattice, so any finite isometric subgraph of the generalized diamond graph is a partial cube. However, not every partial cube is an isometric subgraph of a generalized diamond: for instance, squares, cubes, or hypercubes are not, because these graphs contain four-cycles whereas the generalized diamonds do not. Thus we are led to the questions of which graphs are isometric diamond subgraphs, and how efficiently we may recognize them.

3   Coherent Cuts

A *cut* in a graph is a partition of the vertices into two subsets C and $V \setminus C$; an edge *spans* the cut if it has one endpoint in C and one endpoint in $V \setminus C$. If $G = (U,V,E)$ is a bipartite graph, we say that a cut $(C,(U \cup V) \setminus C)$ is *coherent* if, for every edge (u,v) that spans the cut (with $u \in U$ and $v \in V$), u belongs to C and v belongs to $(U \cup V) \setminus C$.

Fig. 2. The three-dimensional diamond graph

That is, if we color the vertices black and white, all black endpoints of edges spanning the cut are on one side of the cut, and all white endpoints are on the other side.

The *Djokovic–Winkler relation* of a partial cube G determines an important family of cuts. Define a relation $\sim$ on edges of G by $(p,q) \sim (r,s)$ if and only if $d(p,r) + d(p,s) = d(q,r) + d(q,s)$; then G is a partial cube if and only if it is bipartite and $\sim$ is an equivalence relation [4, 14]. Each equivalence class of G spans a cut $(C, V \setminus C)$; we call V and $V \setminus C$ *semicubes* [7]. One may embed G into a hypercube by choosing one coordinate per Djokovic–Winkler equivalence class, set to 0 within C and to 1 within $V \setminus C$. Since this embedding is determined from the distances in G, the isometric embedding of G into a hypercube is determined uniquely up to symmetries of the cube.

As an example, The *Desargues graph* (Fig. 3) is a symmetric graph on 20 vertices, the only known nonplanar 3-regular cubic partial cube [8]; it is used by chemists to model configuration spaces of molecules [1, 13]. The left view is a more standard symmetrical view of the graph while the right view has been rearranged to show more clearly the cut formed by one of the Djokovic–Winkler equivalence classes. As can be seen in the figure, this cut is coherent: each edge spanning the cut has a blue endpoint in the top semicube and a red endpoint in the bottom semicube. The Djokovic–Winkler relation partitions the edges of the Desargues graph into five equivalence classes, each forming a coherent cut.

Theorem 1. *A partial cube is an isometric subgraph of a generalized diamond graph if and only if all cuts formed by Djokovic–Winkler equivalence classes are coherent.*

Proof. In the generalized diamond graph itself, each semicube consists of the set of points in which some coordinate value is above or below some threshold, and each edge spanning a Djokovic–Winkler cut connects a vertex below the threshold to one above

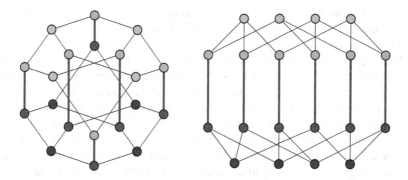

Fig. 3. Two views of the Desargues graph and of a coherent cut formed by a Djokovic–Winkler equivalence class

it. The bipartition of the generalized diamond graph consists of one subset of vertices for which the coordinate sum is zero and another for which the coordinate sum is one. In an edge spanning the cut, the endpoint on the semicube below the threshhold must have coordinate sum zero and the other endpoint must have coordinate sum one, so the cut is coherent. The Djokovic–Winkler relation in any isometric subset of a generalized diamond graph is the restriction of the same relation in the generalized diamond itself, and so any isometric diamond subgraph inherits the same coherence property.

Conversely let G be a partial cube in which all Djokovic–Winkler cuts are coherent; color G black and white. Choose arbitrarily some white base vertex v of G to place at the origin of a d-dimensional grid, where d is the number of Djokovic–Winkler equivalence classes, and assign a different coordinate to each equivalence class, where the ith coordinate value for a vertex w is zero if v and w belong to the same semicube of the ith equivalence class, $+1$ if v belongs to the white side and w belongs to the black side of the ith cut, and -1 if v belongs to the black side and w belongs to the white side of the cut. This is an instance of the standard embedding of a partial cube into a hypercube by its Djokovic–Winkler relationship, and (by induction on the distance from v) every vertex has coordinate sum either zero or one. Thus, we have embedded G isometrically into a d-dimensional generalized diamond graph.   □

For example, the Desargues graph is an isometric subgraph of a five-dimensional generalized diamond.

4   The Diamond Dimension

Theorem 1 leads to an algorithm for embedding any isometric diamond subgraph into a generalized diamond graph, but possibly of unnecessarily high dimension. Following our previous work on *lattice dimension*, the minimum dimension of an integer lattice into which a partial cube may be isometrically embedded [7], we define the *diamond dimension* of a graph G to be the minimum dimension of a generalized diamond graph into which G may be isometrically embedded. The diamond dimension may be as low as the lattice dimension, or (e.g., in the case of a path) as large as twice the lattice dimension. We now show how to compute the diamond dimension in polynomial time. The technique is similar to that for lattice dimension, but is somewhat simpler for the diamond dimension.

Color the graph black and white, and let $(C, V \setminus C)$ and $(C', V \setminus C')$ be two cuts determined by equivalence classes of the Djokovic–Winkler relation, where C and C' contain the white endpoints of the edges spanning the cut and the complementary sets contain the black endpoints. Partially order these cuts by the set inclusion relationship on the sets C and C': $(C, V \setminus C) \leq (C', V \setminus C')$ if and only if $C \subseteq C'$. The choice of which coloring of the graph to use affects this partial order only by reversing it. A *chain* in a partial order is a set of mutually related elements, an *antichain* is a set of mutually unrelated elements, and the *width* of the partial order is the maximum size of an antichain. By Dilworth's theorem [3] the width is also the minimum number of chains into which the elements may be partitioned. Computing the width of a given partial order may be performed by transforming the problem into graph matching, but even more efficient algorithms are possible, taking time quadratic in the number of ordered elements and linear in the width [12].

Theorem 2. *The diamond dimension of any isometric diamond subgraph G, plus one, equals the width of the partial order on Djokovic–Winkler cuts.*

Proof. First, the diamond dimension plus one is greater than or equal to the width of the partial order. For, suppose that G is embedded as an isometric subgraph of a d-dimensional generalized diamond graph; recall that this graph may itself be embedded isometrically into a $(d+1)$-dimensional grid. We may partition the partial order on cuts into $d+1$ chains, by forming one chain for the cuts corresponding to sets of edges parallel to each of the $d+1$ coordinate axes. The optimal chain decomposition of the partial order can only use at most as many chains.

In the other direction, suppose that we have partitioned the partial order on cuts into a family of $d+1$ chains. To use this partition to embed G isometrically into a d-dimensional generalized diamond graph, let each chain correspond to one dimension of a $(d+1)$-dimensional integer lattice, place an arbitrarily-chosen white vertex at the origin, and determine the coordinates of each vertex by letting traversal of an edge in the direction from white to black increase the corresponding lattice coordinate by one unit. Each other vertex is connected to the origin by a path that either has equal numbers of white-to-black and black-to-white edges (hence a coordinate sum of zero) or one more white-to-black than black-to-white edge (hence a coordinate sum of one). Thus, the diamond dimension of G is at most d. As we have upper bounded and lower bounded the diamond dimension plus one by the width, it must equal the width. □

Thus, we may test whether a graph may be embedded into a generalized diamond graph of a given dimension, find the minimum dimension into which it may be embedded, and construct a minimum dimension embedding, all in polynomial time. To do so, find a partial cube representation of the graph, giving the set of Djokovic–Winkler cuts [9], form the partial order on the cuts, compute an optimal chain decomposition of this partial order [12], and use the chain decomposition to form an embedding as described in the proof.

It would be of interest to find more general algorithms for testing whether a graph may be isometrically embedded into any periodic tiling of the plane, or at least any periodic tiling that forms an infinite partial cube. Currently, the only such tilings for which we have such a result are the square tiling [7] and (by the dimension two case of Theorem 2) the hexagonal tiling.

Acknowledgements

This work was supported in part by NSF grant 0830403. All figures are by the author, remain the copyright of the author, and are used by permission.

References

1. Balaban, A.T.: Graphs of multiple 1, 2-shifts in carbonium ions and related systems. Rev. Roum. Chim. 11, 1205 (1966)
2. Bhatt, S., Cosmodakis, S.: The complexity of minimizing wire lengths in VLSI layouts. Inform. Proc. Lett. 25, 263–267 (1987)
3. Dilworth, R.P.: A decomposition theorem for partially ordered sets. Annals of Mathematics 51, 161–166 (1950)
4. Djokovic, D.Z.: Distance preserving subgraphs of hypercubes. J. Combinatorial Theory, Ser. B 14, 263–267 (1973)
5. Eades, P., Whitesides, S.: The logic engine and the realization problem for nearest neighbor graphs. Theoretical Computer Science 169(1), 23–37 (1996)
6. Eppstein, D.: Algorithms for drawing media. In: Pach, J. (ed.) GD 2004. LNCS, vol. 3383, pp. 173–183. Springer, Heidelberg (2005)
7. Eppstein, D.: The lattice dimension of a graph. Eur. J. Combinatorics 26(5), 585–592 (July 2005), http://dx.doi.org/10.1016/j.ejc.2004.05.001
8. Eppstein, D.: Cubic partial cubes from simplicial arrangements. Electronic J. Combinatorics 13(1), R79 (2006)
9. Eppstein, D.: Recognizing partial cubes in quadratic time. In: Proc. 19th Symp. Discrete Algorithms, pp. 1258–1266. ACM and SIAM (2008)
10. Eppstein, D.: The topology of bendless three-dimensional orthogonal graph drawing. In: Proc. 16th Int. Symp. Graph Drawing (2008)
11. Eppstein, D., Falmagne, J.C., Ovchinnikov, S.: Media Theory: Applied Interdisciplinary Mathematics. Springer, Heidelberg (2008)
12. Felsner, S., Raghavan, V., Spinrad, J.: Recognition algorithms for orders of small width and graphs of small Dilworth number. Order 20(4), 351–364 (2003)
13. Mislow, K.: Role of pseudorotation in the stereochemistry of nucleophilic displacement reactions. Acc. Chem. Res. 3(10), 321–331 (1970)
14. Winkler, P.: Isometric embeddings in products of complete graphs. Discrete Applied Mathematics 7, 221–225 (1984)

Non-convex Representations of Graphs*

Giuseppe Di Battista, Fabrizio Frati, and Maurizio Patrignani

Dip. di Informatica e Automazione – Roma Tre University

Abstract. We show that every plane graph admits a planar straight-line drawing in which all faces with more than three vertices are non-convex polygons.

1 Introduction

In a *straight-line planar drawing* of a graph each edge is drawn as a segment and no two segments intersect. A *convex drawing* is a straight-line planar drawing in which each face is a convex polygon. While every plane graph admits a planar straight-line drawing [6], not every plane graph admits a convex drawing. Tutte showed that every triconnected plane graph admits such a drawing with its outer face drawn as an arbitrary convex polygon [12]. Thomassen [11] characterized the graphs admitting a convex drawing and Chiba *et al.* [3] presented a linear-time algorithm for producing such drawings. Convex drawings can be efficiently constructed in small area [4,2].

Hong and Nagamochi proved that for every triconnected plane graph whose boundary is a star-shaped polygon a drawing in which every internal face is a convex polygon can be obtained in linear time [8]. The same authors also investigated drawings where the outer face is a convex polygon and the internal faces are star-shaped [9,10].

As opposed to traditional convex graph drawing, some research works explored the properties of drawings with non-convexity requirements [7,1]. A *pointed drawing* is such that each vertex is incident to an angle larger than π. In [7] it is shown that a planar graph admits a straight-line pointed drawing iff it is minimally rigid or a subgraph of a minimally rigid graph. Since there exist planar graphs that do not admit straight-line pointed drawings, algorithms that construct pointed drawings with tangent-continuous biarcs, circular arcs, or parabolic arcs are studied in [1].

In this paper we address the problem of producing *non-convex drawings*, i.e., drawings where all faces with more than three vertices are non-convex. This can be considered as the opposite of the classic problem of constructing convex drawings. Also, it can be seen as the dual of the problem of constructing pointed drawings, since faces, and not vertices, are constrained to have an angle greater than π. We prove the following:

Theorem 1. *Every plane graph admits a non-convex drawing.*

In Sect. 3 we prove the previous theorem for biconnected graphs by means of a constructive algorithm whose inductive approach is reminiscent of Fary's construction [6], although it applies to non-triangulated graphs and relies on a more complex case study. Due to space restrictions, some proofs are omitted and can be found in [5]. In Sect. 4 we discuss how to extend the result to general plane graphs.

---

* Work partially supported by MUR under Project "MAINSTREAM: Algoritmi per strutture informative di grandi dimensioni e data streams."

I.G. Tollis and M. Patrignani (Eds.): GD 2008, LNCS 5417, pp. 390–395, 2009.

2   Preliminaries

A graph G is *simple* if it has no multiple edges and no self-loops. A planar drawing of G determines a circular ordering of the edges incident to each vertex. Two drawings of G are *equivalent* if they determine the same circular ordering around each vertex. A *planar embedding* is an equivalence class of planar drawings. A planar drawing partitions the plane into *faces*. The unbounded face is the *outer face* and is denoted by $f(G)$. A *chord* of $f(G)$ is an edge connecting two non-adjacent vertices of $f(G)$. A graph with a planar embedding and an outer face is called a *plane graph*. A *non-convex drawing* of a plane graph G is a planar straight-line drawing of G in which each internal (external) face with more than three vertices has an angle greater than π (smaller than π).

Consider a face f of G and two non-adjacent vertices u and v incident to f. The *contraction of u and v inside f* leads to a graph G' in which u and v are replaced by a single vertex w, connected to all vertices u and v are connected to in G. If u and v are connected to the same vertex y, then G' contains two edges (w, y) (hence G' is not simple), unless (u, y) and (y, v) are incident to f (in this case there is only one edge (w, y)). A contraction can also be performed on two adjacent vertices u and v, by removing edge (u, v) and contracting u and v inside the face created by the removal.

3   Proof of Theorem 1

In this section we assume that graph G is biconnected. The following lemmata hold.

Lemma 1. *Suppose that G has a face f with at least four incident vertices. Then there exist two vertices u and v incident to f, such that edge (u, v) can be added to G inside f, so that the resulting plane graph G' is simple.*

Lemma 2. *Let G be a plane graph with a face f having more than four incident vertices. Let G' be the plane graph obtained by inserting inside f an edge e between two non-adjacent vertices of f. Suppose that a non-convex drawing Γ' of G' exists. Then the drawing Γ obtained by removing e from Γ' is a non-convex drawing of G.*

In order to prove Theorem 1 for biconnected plane graphs, we prove the following:

Theorem 2. *Let G be a biconnected plane graph such that all the faces of G have three or four incident vertices. Let $f(G)$ be the outer face of G. If $f(G)$ has three vertices, then, for every triangle T in the plane, G admits a non-convex drawing in which $f(G)$ is represented by T. If $f(G)$ has four vertices and no chord, then, for every non-convex quadrilateral Q in the plane, G admits a non-convex drawing in which $f(G)$ is represented by Q. The mapping of the vertices of $f(G)$ to the vertices of T or Q is arbitrary, provided that the circular ordering of the vertices of $f(G)$ is respected.*

Theorem 2, together with Lemmata 1 and 2, implies Theorem 1. Namely, let G be any biconnected plane graph. While G has any face f with at least five vertices, add, by Lemma 1, a dummy edge inside f so that the augmented graph is still plane and simple. The obtained graph G' has (at least) one face with four incident vertices for each face of G with more than three incident vertices. In order to apply Theorem 2, consider

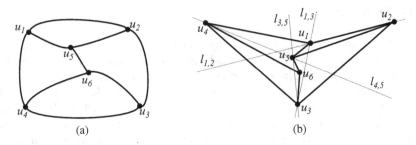

Fig. 1. (a) G_6^*. (b) Construction of a non-convex drawing of G_6^*.

$f(G')$ and, if $f(G')$ has four incident vertices and has a chord, insert a dummy edge inside $f(G')$ between the two vertices not incident to the chord, turning $f(G')$ into a triangular face. Construct a non-convex drawing of G', as in the proof of Theorem 2. By Lemma 2, removing the dummy edges inserted leaves the drawing non-convex.

Before proving Theorem 2, we give two more lemmata. Let G_5^* be the plane graph having a 4-cycle (u_1, u_2, u_3, u_4) as outer face and one internal vertex u_5 connected to u_1, u_2, u_3, and u_4. Let G_6^* be the plane graph having a 4-cycle (u_1, u_2, u_3, u_4) as outer face and two connected internal vertices u_5 and u_6 with u_5 connected to u_1 and u_2, and u_6 connected to u_3 and u_4 (see Fig. 1.a).

Lemma 3. *For any non-convex quadrilateral Q in the plane, there exists a non-convex drawing of G_5^* such that $f(G_5^*)$ is represented by Q.*

Lemma 4. *For any non-convex quadrilateral Q in the plane, there exists a non-convex drawing of G_6^* such that $f(G_6^*)$ is represented by Q (see Fig. 1.b).*

Proof of Theorem 2. Let G be a biconnected plane graph having all faces of size three or four and such that, if $f(G)$ has size four, then it has no chord.

The proof is by induction on the number of internal vertices of G. In the base case, either G has no internal vertex and the statement is trivially true, or $G = G_5^*$, or $G = G_6^*$. In the latter cases the statement follows by Lemmata 3 and 4, respectively. Inductively assume that the statement holds for any biconnected plane graph with less than n internal vertices. Suppose that G has n internal vertices. Three are the cases.

G has a separating 3-cycle C. Denote by G_1 the graph obtained by removing from G all vertices internal to C. Denote by G_2 the subgraph of G induced by the vertices internal to and on the border of C. If $f(G_1)$ has three (four) vertices, then consider any triangle T (resp. any non-convex quadrilateral Q) and construct a non-convex drawing Γ_1 of G_1 having T (resp. Q) as outer face. Consider the triangle T' representing C in Γ_1. Construct a non-convex drawing Γ_2 of G_2 having T' as outer face and insert Γ_2 inside Γ_1 by gluing the two drawings along the common face C represented by T' in both drawings. The resulting drawing Γ is a non-convex drawing of G.

G has no separating 3-cycle and G has a separating 4-cycle C. Denote by G_1 the graph obtained by removing from G all vertices internal to C. Denote by G_2 the subgraph of G induced by the vertices internal to and on the border of C. Notice that $f(G_2)$ has no chords, otherwise G_2 (and then G) would have a separating triangle. If

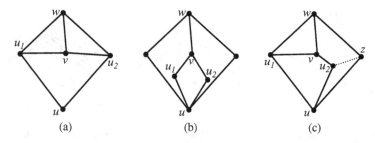

Fig. 2. (a) Both u_1 and u_2 are incident to $f(G)$. (b) Both u_1 and u_2 are internal vertices of G. (c) Vertex u_1 is incident to $f(G)$ and vertex u_2 is not.

$f(G_1)$ has three (four) vertices, then consider any triangle T (resp. any non-convex quadrilateral Q) and construct a non-convex drawing Γ_1 of G_1 having T (resp. Q) as outer face. Consider the non-convex quadrilateral Q' representing C in Γ_1. Construct a non-convex drawing Γ_2 of G_2 having Q' as outer face and insert Γ_2 inside Γ_1 by gluing the two drawings along the common face C represented by Q' in both drawings. The resulting drawing Γ is a non-convex drawing of G.

G has no separating 3-cycle and no separating 4-cycle. We are going to contract two vertices of G to obtain a graph with $n-1$ vertices in order to apply induction.

Consider any internal vertex v of G. First, we show that at least one of the following statements holds: (i) there exist two triangular faces incident to v and sharing an edge (v, u), such that contracting edge (v, u) does not create chords of $f(G)$; (ii) there exists a quadrilateral face $f = (v, u_1, u, u_2)$, such that either contracting v and u or contracting u_1 and u_2 does not create chords of $f(G)$; (iii) G is G_5^*; (iv) G is G_6^*.

First, suppose that two triangular faces (u, v, u_1) and (u, v, u_2) exist. If also u is internal to G, then contracting (u, v) does not create chords of $f(G)$ and statement (i) holds. Otherwise, assume that u is incident to $f(G)$. Then, statement (i) does not hold only if both the following are true: (a) $f(G)$ is a 4-cycle and (b) there exists an edge connecting v and the vertex w of $f(G)$ not adjacent to u. In fact, if (a) does not hold, then $f(G)$ is a 3-cycle and no chord can be generated by contracting any two vertices. Also, if (b) does not hold, then contracting v to u does not create chords of $f(G)$.

Then, suppose that $f(G)$ is a 4-cycle (u, w_1, w, w_2) and that edge (v, w) exists. If at least one of u_1 and u_2 is internal to G, then either (u, v, w, w_1) or (u, v, w, w_2) is a separating 4-cycle, contradicting the hypotheses. Hence, assume that both u_1 and u_2 are incident to $f(G)$. The four 3-cycles (u, v, u_1), (u, v, u_2), (w, v, u_1), and (w, v, u_2) do not have internal vertices by hypothesis. Hence, G contains no internal vertex other than v, $G = G_5^*$, and statement (iii) holds.

If two triangular faces incident to v and sharing an edge do no exist, then there exists a face $f = (v, u_1, u, u_2)$. Edge (u, v) does not belong to G, otherwise either (v, u, u_1) or (v, u, u_2) would be a separating 3-cycle. Analogously, edge (u_1, u_2) does not belong to G. If u is internal to G, then contracting v and u in f does not create chords of $f(G)$ and statement (ii) holds. Otherwise, with similar arguments as above, statement (ii) does not hold for v and u only if both the following are true: (a) $f(G)$ is a 4-cycle and (b) there exists an edge connecting v and the vertex w of $f(G)$ not adjacent to u.

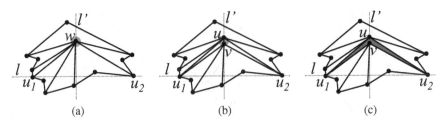

Fig. 3. (a) and (b): Inductive construction of Γ, when statement (i) holds. The light shaded region represents disk D. (a) and (c): Inductive construction of Γ, when statement (ii) holds. The dark shaded region represents face (u, u_1, v, u_2) in Γ.

Then, suppose that $f(G)$ is a 4-cycle and that edge (v, w) exists. We distinguish three cases: (1) If both u_1 and u_2 are incident to $f(G)$ (see Fig. 2.a), then G contains 3-cycles (u_1, v, w) and (u_2, v, w). Since G has no separating triangle, such cycles are faces of G, contradicting the hypothesis that two triangular faces incident to v and sharing an edge do no exist. (2) If both u_1 and u_2 are internal vertices of G (see Fig. 2.b), then contracting u_1 and u_2 does not create chords of $f(G)$ and statement (ii) holds. (3) If one out of u_1 and u_2, say u_1, is incident to $f(G)$ and the other, say u_2, is not (see Fig. 2.c), consider the fourth vertex z of $f(G)$. If edge (u_2, z) does not exist, vertices u_1 and u_2 can be contracted without creating chords of $f(G)$ and statement (ii) holds. If edge (u_2, z) exists, there exist 3-cycles (u_1, v, w) and (u, u_2, z), and 4-cycles (v, u_1, u, u_2) and (v, w, z, u_2). Such cycles contain no vertices in their interior, since G has no separating 3-cycle or 4-cycle. Hence, $G = G_6^*$ and statement (iv) holds.

We prove that whichever of the statements holds, a non-convex drawing of G can be constructed. Suppose that statement (i) holds, with G having faces (v, u, w_1) and (v, u, w_2). Contract edge (u, v) to a vertex w. The resulting graph G' is simple. If not, then there would exist a vertex w_3 adjacent to both u and v in G, with $w_3 \neq w_1, w_2$. This would imply that (u, v, w_3) is a separating triangle, contradicting the hypotheses.

Inductively construct a non-convex drawing Γ' of G' (see Fig. 3.a). Now consider the point p where vertex w is drawn in Γ'. There exists a small disk D (see Fig. 3.b) centered at p such that moving p to any point inside D leaves Γ' a non-convex drawing. Consider the line l' through p and orthogonal to the line l connecting u_1 and u_2. Remove w and its incident edges. Insert vertices u and v on l', so that both are inside D. Connect u and v to their neighbors. It is easy to see that all faces that had an angle greater than π in Γ' still have an angle greater than π in Γ and that the drawing is still planar.

Now, suppose that statement (ii) holds. More precisely, suppose that contracting u and v inside a face $f = (u, u_1, v, u_2)$ does not create chords of $f(G)$. Contract u and v inside f to a vertex w. The resulting graph G' is simple. If not, then there would exist a vertex w_3 adjacent to both u and v in G, with $w_3 \neq u_1, u_2$. This would imply that either (u, v, u_1, w_3) or (u, v, u_2, w_3) is a separating 4-cycle, contradicting the hypotheses.

Inductively construct a non-convex drawing Γ' of G' (see Fig. 3.a). Perturb the vertices so that they are in general position and that Γ' is still a non-convex drawing. Consider the point p where w is drawn in Γ' and consider the line l' through p and orthogonal to the line l connecting u_1 and u_2. There exists a small disk D (see Fig. 3.c) centered at p such that D does not intersect l and such that moving p to any point inside

D leaves Γ' non-convex. Remove w and its incident edges. Insert vertices u and v on l', so that both are inside D. Connect u and v to their neighbors. It is easy to see that all faces that had an angle greater than π in Γ' still have an angle greater than π in Γ and that the drawing is still planar. Further, face (u, v, u_1, u_2) is non-convex, as well, since the angle incident to the one of u and v farther from l is greater than π.

Finally, suppose that statement (iii) or (iv) holds. We are in one of the base cases and the claim directly follows from Lemmata 3 or 4, respectively. $\square$

4 Conclusions

We have proved that every biconnected plane graph admits a drawing in which each face with more than three vertices has an angle greater than π. This result can be extended to general plane graphs as follows. Any simply-connected graph G can be suitably augmented to a biconnected graph G' by adding dummy edges in such a way that for each face of G with more than three vertices there exists a corresponding face of G' with more than three vertices. Removing dummy edges from a non-convex drawing of G' yields a non-convex drawing of G. We believe it is of interest to determine whether non-convex drawings can be realized on a polynomial-size grid.

References

1. Aichholzer, O., Rote, G., Schulz, A., Vogtenhuber, B.: Pointed drawings of planar graphs. In: Bose, P. (ed.) CCCG 2007, pp. 237–240 (2007)
2. Bárány, I., Rote, G.: Strictly convex drawings of planar graphs. Doc. Math. 11, 369–391 (2006)
3. Chiba, N., Yamanouchi, T., Nishizeki, T.: Linear algorithms for convex drawings of planar graphs. In: Bondy, J.A., Murty, U.S.R. (eds.) Progress in Graph Theory, pp. 153–173. Academic Press, New York (1984)
4. Chrobak, M., Goodrich, M.T., Tamassia, R.: Convex drawings of graphs in two and three dimensions. In: Symposium on Computational Geometry, pp. 319–328 (1996)
5. Di Battista, G., Frati, F., Patrignani, M.: Non-convex representations of graphs. Tech. Report RT-DIA-134-2008, Dip. Informatica e Automazione, Univ. Roma Tre (2008)
6. Fary, I.: On straight line representations of planar graphs. Acta. Sci. Math. 11, 229–233 (1948)
7. Haas, R., Orden, D., Rote, G., Santos, F., Servatius, B., Servatius, H., Souvaine, D., Streinu, I., Whiteley, W.: Planar minimally rigid graphs and pseudo-triangulations. Comput. Geometry Theory Appl. 31, 31–61 (2005)
8. Hong, S., Nagamochi, H.: Convex drawings of graphs with non-convex boundary. In: Fomin, F.V. (ed.) WG 2006. LNCS, vol. 4271, pp. 113–124. Springer, Heidelberg (2006)
9. Hong, S., Nagamochi, H.: Star-shaped drawings of planar graphs. In: Brankovich, L., Lin, Y., Smyth, W.F. (eds.) IWOCA 2007. College Publications (2007)
10. Hong, S., Nagamochi, H.: Star-shaped drawing of planar graphs with fixed embedding and concave corner constraints. In: Hu, X., Wang, J. (eds.) COCOON 2008. LNCS, vol. 5092, pp. 405–414. Springer, Heidelberg (2008)
11. Thomassen, C.: Plane representations of graphs. In: Progress in Graph Theory, pp. 43–69. Academic Press, London (1984)
12. Tutte, W.T.: Convex representations of graphs. Proc. London Math. Soc. 10, 304–320 (1960)

Subdivision Drawings of Hypergraphs*

Michael Kaufmann[1], Marc van Kreveld[2], and Bettina Speckmann[3]

[1] Institut für Informatik, Universität Tübingen
mk@informatik.uni-tuebingen.de
[2] Department of Computer Science, Utrecht University
marc@cs.uu.nl
[3] Department of Mathematics and Computer Science, TU Eindhoven
speckman@win.tue.nl

Abstract. We introduce the concept of subdivision drawings of hypergraphs. In a subdivision drawing each vertex corresponds uniquely to a face of a planar subdivision and, for each hyperedge, the union of the faces corresponding to the vertices incident to that hyperedge is connected. Vertex-based Venn diagrams and concrete Euler diagrams are both subdivision drawings. In this paper we study two new types of subdivision drawings which are more general than concrete Euler diagrams and more restricted than vertex-based Venn diagrams. They allow us to draw more hypergraphs than the former while having better aesthetic properties than the latter.

1 Introduction

A graph G is a pair $G = (V, E)$, where V is a set of elements or vertices and $E \subseteq V \times V$ is a set of pairs of vertices, called edges. A *hypergraph* $H = (V, E)$ is a generalization of a graph, where again V is a set of elements or vertices, but E is a set of non-empty subsets of V, called *hyperedges* [1]. The set $E \subseteq \mathcal{P}(V)$ of hyperedges is a subset of the powerset of V.

Hypergraphs are not as common as graphs, but they do arise in many application areas. In relational databases there is a natural correspondence between database schemata and hypergraphs, with attributes corresponding to vertices and relations to hyperedges [7]. Hypergraphs are used in VLSI design for circuit visualization [6,14] and also appear in computational biology [10,12] and social networks [3].

Drawings of hypergraphs are less well-understood than drawings of graphs. There is no single "standard" method of drawing hypergraphs, comparable to the point-and-arc drawings for graphs. When drawing hypergraphs, vertices are usually depicted as points or regions in the plane, but hyperedges can have very varied forms, including Steiner trees, closed curves in the plane, faces of subdivisions, and points. As a result, there is no unique definition of planarity for hypergraphs—different drawing methods imply different, non-equivalent planarity definitions. In the following, we describe some of the drawing methods for hypergraphs in more detail.

---

* This research was initiated during the Bertinoro Workshop on Graph Drawing, 2008. Bettina Speckmann is supported by the Netherlands Organisation for Scientific Research (NWO) under project no. 639.022.707.

I.G. Tollis and M. Patrignani (Eds.): GD 2008, LNCS 5417, pp. 396–407, 2009.

Hypergraph Drawings. The two most common methods to draw hypergraphs are the subset-based and the edge-based method. A subset-based drawing highlights the fact that a hypergraph can be interpreted as a set system: Vertices are drawn as points in the plane and hyperedges are drawn as simple closed curves that contain exactly those vertices that they are incident to (see Fig. 1(a)). It is easy to see that any hypergraph can be drawn this way. Subset-based drawings neither know any concept of or experience any problems with planarity. Bertault and Eades [2] show how to create subset-based hypergraph drawings.

An edge-based hypergraph drawing resembles standard drawings of graphs more closely [11]. Vertices are again drawn as points, but hyperedges are drawn as Steiner trees (see Fig. 1(b)). Edge-based drawings imply the most common definition for hypergraph planarity: A hypergraph is *planar* if and only if it has an edge-based drawing without hyperedge crossings.

Another way to draw hypergraphs is the so-called *Zykov representation*. Vertices are again drawn as points, but hyperedges are visualized by faces of a subdivision. The vertices around a face of the subdivision are the ones connected by the corresponding hyperedge (see Fig. 1(c)). To distinguish faces that represent hyperedges from faces that do not, a background color is needed to fill all faces of the subdivision that do not correspond to hyperedges. A hypergraph is *Zykov-planar* if it has a Zykov representation. Zykov-planarity is equivalent to hypergraph planarity as induced by edge-based drawings.

A hypergraph H can also be drawn as a bipartite graph where one set of vertices corresponds to the vertices of H and the other set corresponds to the hyperedges (see Fig. 1(d)). The edges of the bipartite graph represent vertex-hyperedge incidences. Also this representation immediately implies a definition of planarity, which is equivalent to Zykov-planarity and hence to hypergraph planarity as induced by edge-based drawings [17].

Many hypergraphs are not (Zykov-)planar and hence have neither an edge-based drawing, nor a Zykov representation, nor a drawing as a planar bipartite incidence graph. Motivated by this fact, Pollak and Johnson [9] study alternative definitions of hypergraph planarity which are implied by yet two more methods to draw hypergraphs— *hyperedge-based Venn diagrams* and *vertex-based Venn diagrams*. (The choice of these

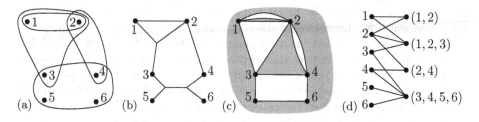

Fig. 1. Four drawings of the hypergraph $H = (V, E)$ with $V = \{1, 2, 3, 4, 5, 6\}$ and $E = \{(1, 2), (1, 2, 3), (2, 4), (3, 4, 5, 6)\}$: (a) subset-based drawing; (b) edge-based drawing; (c) Zykov representation; (c) incidence representation (bipartite graph)

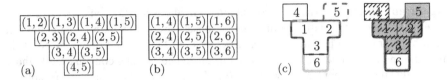

Fig. 2. A hyperedge-based Venn diagram for (a) K_5 and (b) $K_{3,3}$; (c) two different drawings of a vertex-based Venn diagram for H

names is somewhat unfortunate, since the drawings have little in common with standard Venn diagrams.)

In a hyperedge-based Venn diagram, hyperedges correspond uniquely to the faces of a planar subdivision in such a way that for any vertex v, the union of the faces corresponding to hyperedges that contain v, is connected. In a vertex-based Venn diagram, vertices correspond uniquely to faces of a planar subdivision in such a way that for any hyperedge, the union of the faces corresponding to its incident vertices is connected. Note that the subdivision itself does not show the hyperedges directly, curves that enclose unions of faces must still be drawn to visualize them. A hypergraph is *hyperedge-planar* or *vertex-planar* if a planar subdivision with the required properties exists. It is easy to see that both K_5 and $K_{3,3}$ are hyperedge-planar (see Fig. 2(a, b)). Concerning vertex-planarity, consider the following hypergraph H: H has six vertices and three hyperedges $(1, 2, 3, 4)$, $(1, 2, 3, 5)$, and $(1, 2, 3, 6)$. H is vertex-planar but not (Zykov-)planar. Figure 2(c) shows two drawings of H which showcase different methods to draw the hyperedges in a vertex-based Venn diagram.

Subdivision Drawings. Vertex-based Venn diagrams are a particular type of hypergraph drawings which we call *subdivision drawings*. In a subdivision drawing each vertex corresponds uniquely to a face of a planar subdivision. Furthermore, for any hyperedge, the union of the faces corresponding to the vertices incident to that hyperedge, the *hyperedge region*, is connected. Hypergraph drawings which are based on Euler diagrams are subdivision drawings as well. Below we discuss them in some more detail.

To draw a hypergraph H as an Euler diagram we again need to interpret H as a set system. Euler diagrams represent a collection of sets by simple, closed curves in the plane, such that the interior of each curve represents the elements of the corresponding set. Any face induced by the collection of curves is called a *zone*, which lies in the interior of some curves and in the exterior of the rest. In an Euler diagram, a zone z is only present if an element exists that is in exactly those sets whose curves have z as their common interior, and z is in the common exterior of all other curves [13]. No two zones may represent the same intersections of sets. Often, certain well-formedness conditions are considered part of the definition of Euler diagrams. These specify that no point may be the intersection of three of more curves, and all intersections of two curves are proper intersections (no two curves partially overlap). Flower and Howse call diagrams satisfying these conditions *concrete Euler diagrams* [8].

Extended Euler diagrams [16] allow curves to intersect in more general ways: They may partially coincide, and multiple intersections are allowed. Also, zones exist for all possible subsets that can be obtained by intersections. For example, the set system

$\mathcal{S}: \{ \{1, 2, 3\}, \{1, 2, 4\}, \{1, 3, 4\}, \{2, 3, 4\} \}$ generates all four units, all six pairs, and all four triples of elements as intersections of one or more sets, so an extended Euler diagram will have 14 zones for these intersections. Strictly speaking, an extended Euler diagram is not a subdivision drawing, since there is no unique mapping between the vertices and the faces of the subdivision anymore. (There is, however, an injective mapping.)

There is no concrete Euler diagram for $\mathcal{S}$, but if we interpret $\mathcal{S}$ as a hypergraph, then a vertex-based Venn diagram of $\mathcal{S}$ with four faces exists. On the other hand, hyperedge regions of a concrete Euler diagram are simply connected, whereas hyperedge regions of a vertex-based Venn diagram need only be connected, that is, they can have holes. In this paper we study two new types of subdivision drawings for hypergraphs—*simple* and *compact* subdivision drawings—which are more general than concrete Euler diagrams and more restricted than vertex-based Venn diagrams. They allow us to draw more hypergraphs than the former while having better aesthetic properties than the latter. An interesting connection to traditional graph drawing is established by the observation that drawing simple subdivisions corresponds to the problem of overlapping cluster planarity [4] when single edges are missing.

Results. In Sect. 2 we define simple and compact subdivision drawings. We also show that there are hypergraphs which have a subdivision drawing, but not a simple subdivision drawing, and hypergraphs which have a simple, but not a compact subdivision drawing. Pollack and Johnson [9] proved that it is NP-complete to decide if a given hypergraph has a subdivision drawing. Nevertheless, there are classes of graphs that always have a subdivision drawing. In Sect. 3 we prove that hypergraphs which correspond to a particular hierarchy (when viewed as a set system) have a compact subdivision drawing where each face of the subdivision is convex. In the full paper we also show that hypergraphs which are reduced line graphs of complete graphs have a compact subdivision drawing.

2 Subdivision Drawings

Before we can define simple and compact subdivision drawings we first need to introduce some notation and state some assumptions. Let $H = (V, E)$ be a hypergraph. Two vertices u and v of H are *equivalent* with respect to E, if every hyperedge contains either both or none of u and v. To simplify the following discussion we assume that no two vertices of H are equivalent. (Equivalent vertices can easily be removed in a preprocessing step and can be added to the final drawing in an equally easy post-processing step.)

Recall that in a subdivision drawing each vertex corresponds uniquely to a face of a planar subdivision D. Furthermore, the hyperedge region of each hyperedge (the union of the faces corresponding to the vertices incident to that hyperedge) is connected. We assume that the subdivision D has only vertices of degree three. Not every bounded face of D has to correspond to a vertex of H. We call a face that does correspond to a vertex of H a *vertex face*.

A subdivision drawing is *simple* if every hyperedge region is simple, that is, bounded by one simple closed curve. A subdivision drawing is *compact* if it is simple and each

$H = (V, E)$ $V = \{1, 2, 3, 4, 5, 6\}$ $E = \{(1, 2, 3), (2, 4), (3, 4, 5, 6)\}$

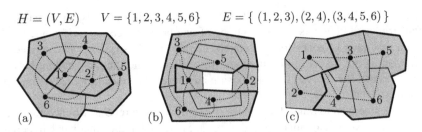

Fig. 3. Subdivision drawing of a hypergraph H, only the curves bounding the hyperedge $(3, 4, 5, 6)$ are indicated (a); simple subdivision drawing of H (b); compact subdivision drawing of H (c). Dotted edges indicate a planar support.

bounded face of D is a vertex face. That is, the complement of all vertex faces is connected. Note that collapsing a face that is not a vertex face will never destroy the connectivity of hyperedge regions, but it may create non-simple hyperedge regions. Consider the white face in Fig. 3(b): Removing it will make the region of either hyperedge $(1, 2, 3)$ or hyperedge $(3, 4, 5, 6)$ non-simple.

A concrete Euler diagram is a compact subdivision drawing which has only proper intersections between the simple closed curves which bound hyperedge regions: No three curves have a common point and no two curves intersect over a stretch of positive length. In Flower and Howse's terminology, not even the set system $\{\{1, 2\}, \{2, 3\}, \{1, 3\}\}$ has a concrete Euler diagram.

A graph G is a *support* for a hypergraph H if the vertices of G correspond to the vertices of H such that for each hyperedge e the subgraph of G induced by e is connected. G is a *planar support* if it is planar. A planar support G is *simple* if G has a planar embedding where each cycle in a subgraph induced by a hyperedge e does not have any other vertex of G on the inside. Hence the planar support in Fig. 4(a) is a simple planar support if every hyperedge that induces the cycle c is also incident to vertex v. Intuitively, the planar support is a subgraph of the dual graph of any subdivision drawing of H (see Fig. 3).

Observation 1. *A hypergraph H*

(i) has a simple subdivision drawing if and only if it has a simple planar support.
(ii) has a compact subdivision drawing if and only if it has a simple planar support with an embedding where all bounded faces are triangulated.

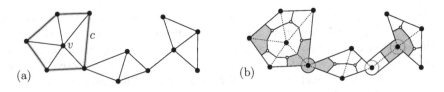

Fig. 4. A (simple) planar support (a); turning a support into a subdivision (b)

Subdivisions and their dual graphs have been extensively studied in, for example, graph theory and VLSI design, and there are several methods that can turn a planar support into a dual subdivision. For completeness, we sketch an easy approach that creates such a subdivision. Use any straight-line embedding of the support. (i) Trace a "race track" around every edge and cut it in the middle. Each vertex face is formed by the union of all half race tracks adjacent to the corresponding vertex. (ii) Place a vertex inside each triangle and connect it to the middle of each edge, forming three quadrilaterals. Place a small circle around every cut vertex, trace a race track around every bridge edge, and cut it in the middle (see Fig. 4(b)). Each vertex face is now the union of all quadrilaterals, circles, and half race tracks adjacent to the corresponding vertex in the support.

We now show that there are hypergraphs which have a subdivision drawing, but not a simple subdivision drawing, and hypergraphs which have a simple but not a compact subdivision drawing. First, consider the hypergraph H_1 on four vertices. The hyper-edges of H_1 are the six pairs and the four triples of vertices. Since the hyperedges include all pairs of vertices all vertex faces of the subdivision must be adjacent. Hence, modulo re-labeling of vertices and the removal of white non-vertex faces, we must have a subdivision drawing as the one depicted in Fig. 5(a) with a non-simple hyperedge region for hyperedge $(2, 3, 4)$.

Second, consider the hypergraph H_2 that is schematically depicted in Fig. 5(b). The hyperedges of H_2 are all black edges plus the two hyperedges with nine vertices each, which are indicated by the gray contours. The black edges form a planar support which is uniquely defined, up to the choice of the outer face. Each hyperedge region is simply connected, so the black edges even form a simple planar support. However, we can not triangulate either of the quadrilaterals without making one hyperedge non-simple. Since at most one of the quadrilaterals can be the outer face, H_2 does not have a compact subdivision drawing.

3   Hierarchies and Subdivision Drawings

In this section we characterize certain hypergraphs which have a compact subdivision drawing. In fact, they even have a compact subdivision drawing where each face of the subdivision is convex. We again view hypergraphs as set systems and study a partic-ular *hierarchy* defined on these set systems. Any hypergraph H with n vertices and k

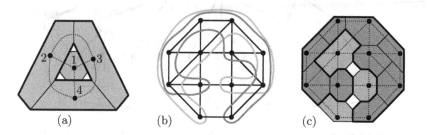

Fig. 5. A hypergraph which has no simple subdivision drawing (a); the hypergraph H_2 (b); a simple but not compact subdivision drawing of H_2 (c)

hyperedges can be interpreted as a set system $\mathcal{S} = \{S_1, \dots, S_k\}$ on a base set M of n elements. (Here we again assume that no two vertices of H are equivalent.) Subdivision drawings naturally visualize set containment well: The region of a set (hyperedge) S_i is contained in the region of a set (hyperedge) S_j if and only if $S_i \subset S_j$.

A *hierarchy* $\mathcal{H}$ is a directed acyclic graph induced by a set system or hypergraph. A hierarchy has two types of vertices: *base vertices*, which represent a singleton set for each element in the base set M, and *set vertices*, which represent each set in the set system $\mathcal{S}$. Hence $\mathcal{H}$ has $n + k$ vertices. With slight abuse of notation we refer to the vertices of $\mathcal{H}$ by the names of the sets they represent. $\mathcal{H}$ has a directed edge (S_1, S_2) with $S_1, S_2 \in \mathcal{S} \cup \{\{v\} : v \in M\}$ if and only if $S_2 \subset S_1$ and for no set $S_3 \in \mathcal{S}$, we have $S_2 \subset S_3 \subset S_1$. That is, edges are directed from larger to smaller sets and represent direct containment—our hierarchies do not contain transitive edges. The base nodes are the *leaves* of the hierarchy (with only incoming edges) and the set nodes are the *internal nodes* of the hierarchy (each internal node has at least two outgoing edges). A hierarchy $\mathcal{H}$ corresponds uniquely to a hypergraph H and vice versa. $\mathcal{H}$ is a *planar hierarchy* if it is planar. We say that a hierarchy is *based* if it has a set vertex S_M that represents the complete base set M. S_M is necessarily the root of the hierarchy and has only outgoing edges. See Fig. 6 for an example.

In the following we prove that a hypergraph H has a compact drawing where each face of the subdivision is convex if the corresponding hierarchy $\mathcal{H}$ is based and planar. In particular we describe an algorithm that transforms $\mathcal{H}$ into an outerplanar support for H by "sliding" certain edges of $\mathcal{H}$ down to the leaf level. The complete process comprises the following four steps:

1. Fix an embedding and construct a depth-first search spanning tree of $\mathcal{H}$.
2. Slide all non-tree edges of $\mathcal{H}$ down to the leaf level.
3. Remove all internal nodes of $\mathcal{H}$ to create an outerplanar support for H.
4. Construct a compact subdivision drawing from the outerplanar support.

We now describe each step in more detail.

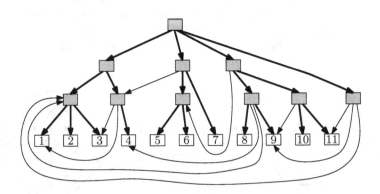

Fig. 6. A based planar hierarchy $\mathcal{H}$ defined by $M = \{1, 2, 3, 4, 5, 6, 7, 8, 9, 10, 11\}$ and the sets $\{1, 2, 3\}$, $\{1, 3, 4\}$, $\{5, 6\}$, $\{1, 2, 3, 4, 8, 9\}$, $\{9, 10, 11\}$, $\{1, 2, 3, 9, 11\}$, $\{1, 2, 3, 4\}$, $\{1, 3, 4, 5, 6, 7\}$, $\{1, 2, 3, 4, 5, 6, 8, 9, 10, 11\}$, $S_M = \{1, 2, 3, 4, 5, 6, 7, 8, 9, 10, 11\}$

Step 1: *Embedding $\mathcal{H}$ and constructing a depth-first search spanning tree.*

We embed $\mathcal{H}$ such that the root S_M lies on the outer face. Recall that all edges are directed from the root to the leaves. The embedding defines a left to right order (counterclockwise) of the children of each node. We traverse the hierarchy in a depth-first search (DFS) manner, creating an ordered DFS spanning tree, where every edge that reaches some node for the first time defines the *true parent* of that node. The true parent structure is a tree which categorizes the edges of $\mathcal{H}$ into *tree edges* and *non-tree edges*. W.l.o.g. we relabel the vertices in the base set with $1, 2, \ldots, n$ in the order in which they are encountered in the DFS traversal.

Step 2: *Slide all non-tree edges down to the leaf level.*

We now transform the based planar hierarchy $\mathcal{H}$ into another based planar hierarchy $\mathcal{H}'$ where all non-tree edges point to leaves. We do this in such a way that the outerplanar support for $\mathcal{H}'$ which we create in Step 3 is also an outerplanar support for $\mathcal{H}$.

Let $\vec{a}$ be any non-tree edge that points to an internal node S_i. We distinguish five different cases, depending on the next edge in the clockwise order around S_i, see Fig. 7. If the next edge in clockwise order is (i) an outgoing tree edge, then we can slide $\vec{a}$ to point to the child of that tree edge. If it is (ii) an incoming non-tree edge $\vec{c}$, then we can proceed with $\vec{c}$ instead. If it is (iii) an outgoing non-tree edge $\vec{c}$, then we can slide $\vec{a}$ to point to the destination of $\vec{c}$. Finally, the next clockwise edge can be an incoming tree edge from the parent. In this case we consider the counterclockwise neighbor of $\vec{a}$ at S_i. Due to the construction of the DFS tree, this can be only an outgoing tree edge (iv) or an incoming non-tree edge $\vec{c}$ (v). We can treat (iv) symmetric to (i) and (v) symmetric to (ii).

The following lemma shows that this transformation preserves the adjacencies captured by the original hierarchy.

Lemma 1. *For any based planar hierarchy $\mathcal{H}$ there exists another based planar hierarchy $\mathcal{H}'$ where every non-tree edge points to a leaf, and a planar support for $\mathcal{H}'$ is also a planar support for $\mathcal{H}$.*

Proof. We first argue that the transformation described above terminates and results in a hierarchy $\mathcal{H}'$ where all non-tree edges point to leaves. The five cases of the transformation can be grouped as follows: In cases (i), (iii), and (iv) the non-tree edge $\vec{a}$ slides

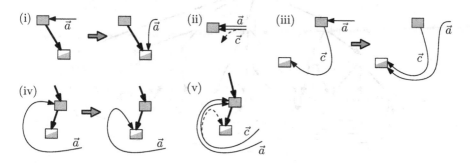

Fig. 7. Cases of the transformation. Grey nodes are internal, half-grey nodes can be internal or leaf nodes.

to a node S_j with $S_j \subset S_i$. In cases (ii) and (v) no sliding is done, but it is easy to see that these cases cannot occur more often than there are non-tree edges pointing to S_i. Hence the process terminates when all non-tree edges point to leaves.

Now we prove that a planar support for $\mathcal{H}'$ is also a planar support for $\mathcal{H}$ by considering the corresponding hypergraphs (set systems). Let H be the hypergraph corresponding to $\mathcal{H}$. We argue that any sliding operation results in a hypergraph H' whose planar support is also a planar support for H. In particular, let (S_i, S_j) be a non-tree edge that is slid and which becomes edge (S_i', S_k). The sets of H' are precisely the sets of H with the exception of S_i which is replaced by S_i'. We know that $S_k \subset S_j$ and $S_i' \subseteq S_i$. Consider a planar support G' for H'. The base elements in S_j, S_k, and S_i' are connected in G'. We have to show that the base elements of S_i are also connected in G', although S_i is not a set of H' and hence not a node of $\mathcal{H}'$. We have $S_i = S_i' \cup S_j$ and also $S_k = S_i' \cap S_j$ which implies $S_i' \cap S_j \neq \emptyset$. Hence we can conclude that the base elements of S_i are connected in G'. Since this argument holds for every sliding operation the lemma follows. $\qquad\square$

Step 3: *Remove all internal nodes and create an outerplanar support.*

After Step 2 we have a hierarchy where every non-tree edge points directly to a leaf (see Fig. 8). We now embed this hierarchy in the plane such that all leaves (base elements) lie on a horizontal line ℓ. Non-tree edges are drawn as curves that connect an internal node to a leaf, either without crossing ℓ or by crossing it exactly once. Such a planar embedding always exists—its can be obtained directly from the planar embedding used in Step 1 of our algorithm (see Fig. 8).

We say that a base element *encounters* ℓ if either its leaf is intersected by ℓ or if ℓ intersects a non-tree edge that is directed to its leaf. Let $W = w_1, \dots, w_s$ be the sequence of the base elements as they encounter ℓ from left to right. In Fig. 8, the sequence is $W = 1, 2, 3, 1, 4, 5, 6, 7, 6, 8, 4, 1, 9, 10, 11, 9, 1$. The base elements in any set (internal node) form a subinterval of W. Hence, any planar graph on the base elements that realizes all adjacencies of W is a planar support. Therefore we call W the *support sequence*.

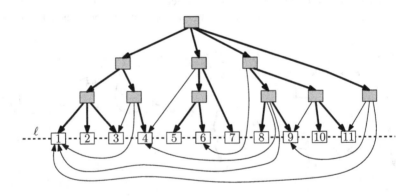

Fig. 8. Hierarchy after the transformation

Lemma 2. *A support sequence W does not have a subsequence ..a..b..a..b.., where a and b represent any two distinct base elements.*

(In other words, W is a Davenport-Schinzel sequence of order 2 [15].)

Proof. By construction, the first occurrences of a and b in W correspond to leaves, the later ones to crossings of non-tree edges with ℓ. A subsequence ..a..b..a..b.. implies that the non-tree edges to a and b cross below ℓ, but the initial hierarchy was planar and the transformations preserved planarity.   □

Lemma 3. *There is an outerplanar graph G that realizes all adjacencies of W.*

Proof. Assume that the base elements are numbered $1, \ldots, n$. We compute G from W as follows. Create a node for every base element and connect them into a path using $n - 1$ edges $(i, i + 1), 1 \le i \le n - 1$. Scan W from left to right, and for any repeated occurrence of i in $\ldots j, i, k \ldots$, create edges (i, j) and (i, k) such that the path edges $(i, i + 1)$ always keep the unbounded face to the left. The planarity of this construction follows directly from the planarity of the transformed hierarchy $\mathcal{H}'$. G is outerplanar since all path edges $(i, i + 1)$ bound the outer face and hence all nodes are incident to the outer face, see Fig. 9.   □

Step 4: *Construct a compact subdivision drawing from the outerplanar support.*

The outerplanar graph G that results from Step 3 forms the basis of the planar support. We add an edge $(1, n)$ if it is not present yet and triangulate all bounded faces (in Fig. 9, only the faces $1, 4, 8, 9$ and $4, 5, 6, 8$).

An easy way to construct the regions is the following: Use any straight-line embedding of the planar support with triangulated bounded faces. For each triangle, choose any point in its interior, for instance the center of mass. For any edge, place an extra point in the middle. Partition each triangle into three quadrilaterals by drawing edges between the center and three edge midpoints. The region of each node is the union of the incident quadrilaterals (see Fig. 10(a)).

We can ensure that each face of the subdivision is convex, by using a slightly more involved method based on Voronoi diagrams. Dillencourt [5] showed how to realize an outerplanar graph as the Delaunay triangulation of points. The dual of this Delaunay triangulation, clipped to lie within a bounding box, is a compact subdivision drawing where each face is convex, see Fig. 10(b).

Theorem 1. *A hypergraph that corresponds to a based planar hierarchy has a compact subdivision drawing where each face of the subdivision is convex.*

To conclude this section we note that hypergraphs corresponding to a (non-based) planar hierarchy need not have a compact subdivision drawing, or even a simple one. Consider again the hypergraph H_1 on four vertices. The hyperedges of H_1 are the six pairs

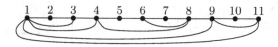

Fig. 9. Planar support for the hierarchy shown in Fig. 8

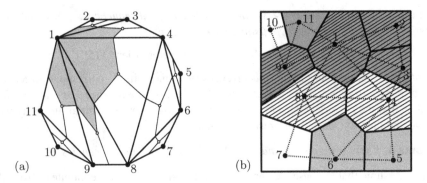

Fig. 10. Triangle-partition based method of constructing a subdivision for the planar support of Fig. 9, the face of node 1 is indicated (a); Voronoi diagram based method for the same support, showing 3 sets: $\{5, 6\}$, $\{1, 2, 3, 9, 11\}$, and $\{1, 2, 3, 4, 8, 9\}$ (b)

and the four triples of vertices. Its hierarchy is planar, as shown in Fig. 11(a), but—as argued in Sect. 2—H_1 has no simple subdivision drawing. It is easy to see, however, that each hypergraph corresponding to a planar hierarchy has a subdivision drawing, the planar support can be extracted from any planar embedding of the hierarchy by iteratively contracting edges incident to base vertices.

4   Conclusion and Open Problems

We introduced the concept of subdivision drawings of hypergraphs which comprises both vertex-based Venn diagrams and Euler diagrams. We studied two new types of subdivision drawings, simple and compact subdivision drawings, and established some of their basic properties. We also characterized certain hypergraphs that have a compact subdivision drawing.

It is NP-complete to decide if a hypergraph has a subdivision drawing, but it is not clear if the same result holds for simple and compact subdivision drawings as well. Of

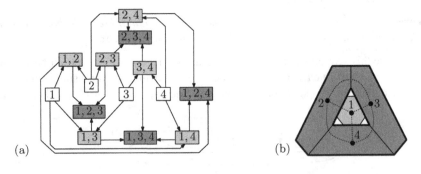

Fig. 11. Planar drawing of the hierarchy of H_1 (a); subdivision drawing of H_1 (b)

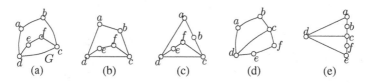

Fig. 1. (a) The graph G, (b) a drawing of G on seven segments, (c) a drawing of G on five segments, (d) another embedding of G, and (e) a minimum segment drawing of G

segments in a drawing [1], and the insightful results presented in their work have established a new line of research henceforth. However, as their results suggest, this problem is quite difficult for most of the non-trivial graph classes. For most of these cases, bounds have been given on the number of segments in a drawing, but no algorithm is known so far for computing a minimum segment drawing. For example, although Dujmović *et al.* have provided an algorithm for computing minimum segment drawings of trees, no algorithm is known for biconnected and triconnected planar graphs. The problem has also been studied for plane graphs. Although dealing with plane graphs is typically easier than dealing with planar graphs, no algorithm is known for computing minimum segment drawings of biconnected and triconnected plane graphs as well. Even for degree restricted cases of plane graphs, e.g., for plane triconnected cubic graphs, no algorithm has yet been devised for computing minimum segment drawings.

In this paper, we study the minimum segment drawing problem for series-parallel graphs with the maximum degree three. For such a graph G, we give linear-time algorithms for choosing such an embedding of G that admits a straight-line drawing on the minimum number of segments, and for computing a minimum segment drawing of G. The rest of this paper is organized as follows. In Section 2 we give some definitions and present our primary results. In Section 3 we give a linear-time algorithm for computing a minimum segment drawing of a biconnected series-parallel graph with the maximum degree three. In Section 4 we briefly illustrate how our idea from Section 3 can be extended to compute minimum segment drawing of a series-parallel graph which is not necessarily biconnected. Finally Section 5 is a conclusion.

2 Preliminaries

In this section we give some relevant definitions and present our preliminary results. For basic graph theoretic and graph drawing related definitions we refer to [4].

A graph $G = (V, E)$ is called a *series-parallel graph* (with *source* s and *sink* t) if either G consists of a pair of vertices s and t connected by a single edge, or there exist two series-parallel graphs $G_i = (V_i, E_i)$, $i = 1, 2$, with source s_i and sink t_i such that $V = V_1 \cup V_2, E = E_1 \cup E_2$, and either (*i*) $s = s_1, t_1 = s_2$ and $t = t_2$, or (*ii*) $s = s_1 = s_2$ and $t = t_1 = t_2$ [6]. A pair $\{u, v\}$ of vertices of a connected graph G is a *split pair* if there exist two subgraphs $G_1 = (V_1, E_1)$ and $G_2 = (V_2, E_2)$ such that: (*i*) $V = V_1 \cup V_2, V_1 \cap V_2 = \{u, v\}$; and (*ii*) $E = E_1 \cup E_2, E_1 \cap E_2 = \emptyset$, $|E_1| \geq 1, |E_2| \geq 1$. Thus every pair of adjacent vertices of G is a split pair of G. A *split component* of a split pair $\{u, v\}$ is either an edge (u, v) or a maximal connected subgraph H of G such that $\{u, v\}$ is not a split pair of H.

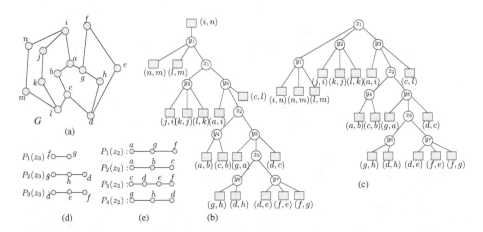

Fig. 2. (a) A biconnected series-parallel graph G with $\Delta(G) = 3$, (b) SPQ-tree T of G with respect to reference edge (i, n), (c) SPQ-tree T of G with P-node z_1 as the root, (d) the three core paths of node z_3, and (e) the four core paths of node z_2

Let G be a biconnected series-parallel graph. The SPQ-tree T of G with respect to a *reference edge* $e = (s, t)$ is a recursive decomposition of G induced by its split pairs [6]. T is a rooted ordered tree whose nodes are of three types: S, P and Q. Each node x of T corresponds to a subgraph of G, called its *pertinent graph* $G(x)$. Tree T is recursively defined as follows.

(i) *Trivial Case*: In this case, G consists of two parallel edges e and e' joining s and t. The tree T consists of a single Q-node x. The pertinent graph $G(x)$ consists of only the edge e'. (ii) *Parallel Case*: In this case, the split pair $\{s, t\}$ has three or more split components $G_0, G_1, \ldots, G_k$ ($k \geq 2$), and G_0 consists of only a reference edge $e = (s, t)$. The root of T is a P-node x. The pertinent graph $G(x) = G_1 \cup G_2 \cup \cdots \cup G_k$. ($iii$) *Series Case*: In this case, the split pair $\{s, t\}$ has exactly two split components, and one of them consists of the reference edge e. One may assume that the other split component has cut-vertices $c_1, c_2, \ldots c_{k-1}$ ($k \geq 2$), that partition the component into its blocks $G_1, G_2, \ldots, G_k$ in this order from s to t. Then the root of T is an S-node x. The pertinent graph $G(x)$ of node x is a union of $G_1, G_2, \ldots, G_k$. In Fig. 2 we have illustrated the concept of representing the recursive decomposition of a given biconnected series-parallel graph through an SPQ-tree. In each of the cases mentioned above, we call the edge e the *reference edge* of node x. Except for the trivial case, node x of T has children $x_1, x_2, \ldots, x_k$ in this order; x_i is the root of the SPQ-tree of graph $G(x_i) \cup e_i$ with respect to the reference edge $e_i, 1 \leq i \leq k$. We call edge e_i *the reference edge of node x_i*, and call the endpoints of edge e_i the *poles* of node x_i. The tree obtained so far has a Q-node associated with each edge of G, except the reference edge e. We complete the SPQ-tree T by adding a Q-node, representing the reference edge e, and making it the parent of x so that it becomes the root of T. One can easily modify T to an SPQ-tree T' with an arbitrary P-node as the root as illustrated in Fig. 2(e). In the remainder of this paper, we consider SPQ-trees having P-nodes as their roots. Based on the assumption that $\Delta(G) = 3$, the following facts were mentioned in [6].

Fact 1. *Let (s, t) be the reference edge of an S-node x of T, and let $x_1, x_2, \ldots, x_k$ be the children of x in this order from s to t. Then the following (i)–(iii) hold. (i) Each child x_i of x is either a P-node or a Q-node; (ii) both x_1 and x_k are Q-nodes; and (iii) x_{i-1} and x_{i+1} must be Q-nodes if x_i is a P-node where $2 \leq i \leq k - 1$.*

Fact 2. *The root P-node of T has exactly three children and each non-root P-node of T has exactly two children. For a non-root P-node x in T, either both the children of x are S-nodes, or one child of x is an S-node and the other child of x is a Q-node.*

A node x in T is *primitive* if x does not have any descendant P-node in T. We define the *height* of a primitive P-node to be zero. The *height* of any other P-node is $(i + 1)$ if the maximum of the heights of its descendant P-nodes is i. For two given P-nodes x and z in T, we say that z is a *child P-node* of x if there is an S-node y in T such that y is a child of x and z is a child of y in T.

Let G be a biconnected series-parallel graph with $\Delta(G) = 3$. Let G' be the plane graph corresponding to a straight line drawing Γ of G. Let T' be an SPQ-tree of G', and r be such a P-node in T' that the poles of r appear on the outerface of Γ. An SPQ-tree of G corresponding to Γ is the SPQ-tree obtained by considering T' rooted at r. We use T_Γ to denote an SPQ-tree of G corresponding to a drawing Γ of G. For a node x in T_Γ, let P_x and N_x denote the number of P-nodes and primitive P-nodes in the subtree of T_Γ rooted at x. If x is a non-root P-node, then let y and y' denote the two children of x in T_Γ. Let p and q denote the number of child P- and Q-nodes respectively of the node y in T_Γ. Similarly, let p' and q' denote the number of child P- and Q-nodes respectively of y' in T_Γ. Let z_i denote the i-th child P-node of y in T_Γ and e_i denote the edge corresponding to the i-th child Q-node of y in T_Γ. Similarly, let z'_i denote the i-th child P-node of y' in T_Γ and e'_i denote the edge corresponding to the i-th child Q-node of y' in T_Γ. For each non-root P-node x of T_Γ, we now define the *core paths* of $G(x)$ as follows. If x is a primitive P-node, then let $q \geq q'$. We then define three core paths $P_i(x)$ $(1 \leq i \leq 3)$ of $G(x)$ as $P_1(x) = e_1$, $P_2(x) = G(y')$ and $P_3(x) = \bigcup_{i=2}^{q} e_i$, as shown in Fig. 2(d). Otherwise, x is not primitive, and we consider the following two subcases. If either of the two nodes y and y' has at least two child P-nodes then we assume that y is such a node, otherwise we assume that $p \geq p'$ and proceed as follows. Let e_j and e_k denote the edges corresponding to the Q-nodes immediately preceding z_1 and z_p, respectively in T_Γ. We then define four core paths $P_i(x)$ $(1 \leq i \leq 4)$ of $G(x)$ as $P_1(x) = P_1(z_1) \cup \bigcup_{i=1}^{j} e_i$, $P_2(x) = \bigcup_{i=1}^{p'} P_2(z'_i) \cup \bigcup_{i=1}^{q'} e'_i$, $P_3(x) = P_3(z_p) \cup \bigcup_{i=k+1}^{q} e_i$, and $P_4(x) = \bigcup_{i=1}^{p} P_2(z_i) \cup \bigcup_{i=j+1}^{k} e_i$, as shown in Fig. 2(e). We define a straight line drawing Γ of G to be a *canonical drawing of G* if the following (a) and (b) hold for Γ. (a) For each non-root P-node x in T_Γ, each core path $P_i(x)$ of $G(x)$ is drawn on a different line segment $L_i(x)$; and (b) there is a primitive P-node w in T_Γ such that the poles of w appear on the outerface of Γ.

Let $L(\Gamma)$ denote the number of segments in the drawing Γ of G. We call a line segment l_1 in Γ to be *collinear* with another line segment l_2 in Γ if l_1 and l_2 have the same slope, and the perpendicular distance between l_1 and l_2 is zero. For a node x in T_Γ, we use $\Gamma(x)$ to denote the drawing of $G(x)$ in Γ, and $\Gamma \setminus \Gamma(x)$ to denote the drawing obtained by deleting $\Gamma(x)$ from Γ. If Γ is a canonical drawing of G, then we say that $\Gamma(x)$ is a *canonical drawing of $G(x)$*. We say that $\Gamma(x)$ *shares* a line segment

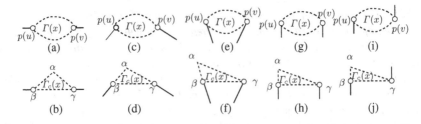

Fig. 3. All possible cases for computing $\Gamma_c(x)$ when x is a primitive P-node

with $\Gamma \setminus \Gamma(x)$ if there is a line segment l_1 in $\Gamma(x)$ and a line segment l_2 in $\Gamma \setminus \Gamma(x)$ such that l_1 and l_2 are collinear and have a common end point.

3   Biconnected Series-Parallel Graphs

In this section we give our algorithm for computing a minimum segment drawing of a biconnected series-parallel graph G with $\Delta(G) = 3$. We first show that any drawing Γ of G can be transformed into a canonical drawing Γ_c such that $L(\Gamma_c) \leq L(\Gamma)$. We then give a lower bound of $L(\Gamma)$, and describe our drawing algorithm.

We have the following lemma on transformation of a drawing into canonical drawing.

Lemma 1. *Let G be a biconnected series-parallel graph with $\Delta(G) = 3$. Then for any straight-line drawing Γ of G, a canonical drawing Γ_c of G can be computed such that $L(\Gamma_c) \leq L(\Gamma)$.*

Proof. Let x be a non-root P-node having poles u and v in $\mathcal{T}_\Gamma$. By Fact 1, there is a sibling Q-node of x preceding it and a sibling Q-node of x following it in $\mathcal{T}_\Gamma$. Let $e(x) = (u', u)$ and $e'(x) = (v', v)$ denote the two edges corresponding to these two Q-nodes respectively. Let $h(x)$ denote the height of x in $\mathcal{T}_\Gamma$. Using induction on $h(x)$ we now prove that for each non-root P-node x of $\mathcal{T}_\Gamma$, we can compute a canonical drawing $\Gamma_c(x)$ of $G(x)$ such that replacing $\Gamma(x)$ with $\Gamma_c(x)$ in Γ does not increase $L(\Gamma)$.

For $h(x) = 0$, we compute $\Gamma_c(x)$ by first drawing a triangle with three segments $L_i(x)$ $(1 \leq i \leq 3)$, and then drawing the core path $P_i(x)$ on $L_i(x)$ $(1 \leq i \leq 3)$. Considering all possible orientations of $l(e(x))$ and $l(e'(x))$, computation of $\Gamma_c(x)$ is shown in Fig. 3. In each case, we choose the line segment closed between α and β as $L_1(x)$, the one closed between β and γ as $L_2(x)$, and the one closed between α and γ as $L_3(x)$. We now show that, $L(\Gamma)$ will not increase if we replace $\Gamma(x)$ with $\Gamma_c(x)$. Since $G(x)$ is a simple cycle, any straight line drawing of $G(x)$ would require at least three line segments. Again, in any straight line drawing of G, $\Gamma(x)$ may share at most two line segments with $\Gamma \setminus \Gamma(x)$. Except for the case where $l(e(x))$ and $l(e'(x))$ are parallel (as in Fig. 3(g) and (i)) or diverging (as in Fig. 3(e)), we have not reduced the number of line segments that might have been shared between $\Gamma(x)$ and $\Gamma \setminus \Gamma(x)$ as shown in Fig. 3(a)–(d). If $l(e(x))$ and $l(e'(x))$ are parallel or diverging as illustrated in Fig. 3(e)–(j), our drawing might have reduced this number by at most one if $\Gamma(x)$ had shared both the line-segments $l(e(x))$ and $l(e'(x))$. However, if $\Gamma(x)$ had shared both

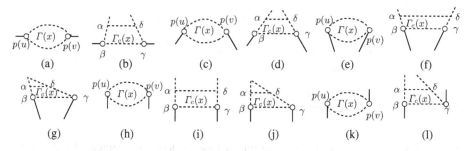

Fig. 4. All possible cases for computing $\Gamma_c(x)$ when $h(x) > 0$

the line segments $l(e(x))$ and $l(e'(x))$, then in every case, any straight line drawing of $G(x)$ would require at least four segments, and we have reduced this number by at least one. Hence, replacing $\Gamma(x)$ with $\Gamma_c(x)$ would not increase $L(\Gamma)$ in any of the cases.

We now assume that $h(x) > 0$ and $\Gamma_c(w)$ has been computed for all the descendant P-nodes w of x. To compute $\Gamma_c(x)$, we first draw a quadrangle with four line segments $L_i(x)$ $(1 \le i \le 4)$, in such a way that $L_2(x)$ is the line segment closed between $p(u)$ and $p(v)$. Based on different orientation of $l(e(x))$ and $l(e'(x))$, the quadrangle is illustrated in Fig. 4. In every case, we choose the line segment closed between α and β as $L_1(x)$, the one closed between β and γ as $L_2(x)$, the one between γ and δ as L_3 and the one between α and δ as $L_4(x)$. We then draw the core path $P_i(x)$ along $L_i(x)$ $(1 \le i \le 4)$. Finally, for each child P-node w of y, we add $\Gamma_c(w)$ by making $L_2(w)$ and $L_4(x)$ collinear. Similarly, for each child P-node w of y', we draw $\Gamma_c(w)$ by making $L_2(w)$ and $L_2(x)$ collinear. The fact that replacing $\Gamma(x)$ with $\Gamma_c(x)$ does not increase $L(\Gamma)$ can be understood as follows. Let $\Gamma'(x)$ denote the drawing obtained by considering $\Gamma_c(x)$ and the two line segments $l(e(x))$ and $l(e'(x))$. Let $G'(x)$ denote the underlying graph of $\Gamma'(x)$. If $l(e(x))$ and $l(e'(x))$ are collinear or converging as illustrated in Fig. 4(a)–(d), then for each degree two vertex v' of $G'(x)$, the two incident edges of v' are collinear in $\Gamma'(x)$ with the exception that for each primitive P-node, the incident edges of exactly one degree two vertex are non-collinear. Again for each degree three vertex v' of $G'(x)$, exactly two of the three incident edges are collinear in $\Gamma'(x)$. Thus, $\Gamma'(x)$ has the maximum possible sharing between the drawings of the edges of $G'(x)$, and replacing $\Gamma(x)$ with $\Gamma_c(x)$ will not increase $L(\Gamma)$. Similarly, if $l(e(x))$ and $l(e'(x))$ are parallel with the angle between them being $0°$ as shown in Fig. 4(h)–(j), and if x has a child S-node with at least two child P-nodes, then $\Gamma'(x)$ will have the maximum possible sharing between the drawings of the edges of $G(x)$, and replacing $\Gamma(x)$ with $\Gamma_c(x)$ will not increase $L(\Gamma)$. On the other hand, if each child S-node of x has at most one child P-node, then the three incident edges of v are pairwise non-collinear in $\Gamma'(x)$ as shown in Fig. 4(j) and Fig. 5(c). However, $L(\Gamma)$ will not increase even in this case. If both the line segments $l(e(x))$ and $l(e'(x))$ were shared by some line-segment in $\Gamma(x)$, then either of the following (a) and (b) will hold. (a) There is at least one descendant non-primitive P-node x' of x such that at either of the two poles of x', all the three incident edges are pairwise non-collinear as shown in Fig. 5(a); and (b) there is at least one descendant primitive P-node x'' of x such that $\Gamma(x'')$ uses four line-segments as shown in Fig. 5(b). In both the cases, replacing $\Gamma(x)$ with $\Gamma_c(x)$

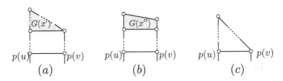

Fig. 5. (a) and (b) Two drawings of $G(x)$ that shares both the line segments $l(e(x))$ and $l(e'(x))$ with the rest of the drawing, (c) a canonical drawing $\Gamma_c(x)$ of $G(x)$

does not increase $L(\Gamma)$. The reasoning for the cases where line segments $l(e(x))$ and $l(e'(x))$ are parallel with the angle between them being $180°$ (as shown in Fig. 4(k) and (l)), or where $l(e(x))$ and $l(e'(x))$ are diverging (as shown in Fig. 4(e)–(g)) follows from similar arguments.

It now remains for us to show that we can obtain a primitive P-node w in $\mathcal{T}_\Gamma$ such that the poles of w appear on the outerface of Γ_c. One can observe that for each non-root P-node x of $\mathcal{T}_\Gamma$, there is a primitive P-node w in the subtree of $\mathcal{T}_\Gamma$ rooted at x such that the poles of w appear on the outerface of $\Gamma_c(x)$. Let r denote the root of $\mathcal{T}$. Let y_1, y_2 and y_3 denote the three children of r in $\mathcal{T}$. We first consider the nodes y_1 and y_2 as the two children of a temporary P-node x' and compute $\Gamma_c(x')$ in the same way as described in the inductive step above. We then replace $\Gamma(x')$ with $\Gamma_c(x')$, and this does not increase $L(\Gamma)$. We then take a single line segment and draw on it all the edges corresponding to the child Q-nodes of y_3 along with all the paths $P_2(z)$ for each child P-node z of y_3. Let $\Gamma'(y_3)$ denote this drawing of $G(y_3)$. We then compute Γ_c by merging $\Gamma_c(x')$ with $\Gamma'(y_3)$. The details of the proof that the merging of $\Gamma_c(x')$ and $\Gamma'(y_3)$ does not increase $L(\Gamma)$ is omitted in this extended abstract since the arguments are similar to those given in the induction step above. One can also observe that after performing the merging of $\Gamma_c(x')$ and $\Gamma'(y_3)$, we will obtain the poles of a primitive P-node in the outerface of Γ_c. $\qquad\Box$

Lemma 1 implies that any straight line drawing Γ of G requires at least $L(\Gamma_c)$ line segments where Γ_c is a canonical drawing obtained by transforming Γ. We therefore give here a lower bound of $L(\Gamma_c)$. For clarity of notations, we use $\mathcal{T}$ instead of $\mathcal{T}_{\Gamma_c}$ to denote an SPQ-tree corresponding to Γ_c. Since there is always a primitive P-node w in $\mathcal{T}$ such that the poles of w appear on the outerface of Γ_c, we assume that the root of $\mathcal{T}$ has two child S-nodes that are primitive in $\mathcal{T}$. We first have the following lemma.

Lemma 2. *Let G be a biconnected series-parallel graph with $\Delta(G) = 3$. Let Γ_c be a canonical drawing of G. Let $\mathcal{T}$ be an SPQ-tree of G corresponding to Γ_c. Then for a non-root P-node x in $\mathcal{T}$, $L(\Gamma_c(x)) \geq P_x + N_x + 1$.*

Proof. We use induction on P_x. In the basis case, $P_x = 1$, i.e., x is a primitive P-node. Hence $N_x = 1$ and $P_x + N_x + 1 = 3$. Since $G(x)$ is a simple cycle when x is primitive and any straight line drawing of a cycle requires at least three segments, the claim holds.

We now assume that $P_x > 0$ and the claim holds for every P-node w in $\mathcal{T}$ having $P_w < P_x$. Hence $L(\Gamma_c(w)) \geq P_w + N_w + 1$. We now take a child P-node w of x and delete the drawing $\Gamma_c(w)$ from Γ_c. Let G' denote the underlying graph of this drawing $\Gamma_c \setminus \Gamma_c(w)$. The graph G' is not necessarily a biconnected series-parallel graph. Let u and v be the two poles of w in $\mathcal{T}$. In order to ensure that we are working with a

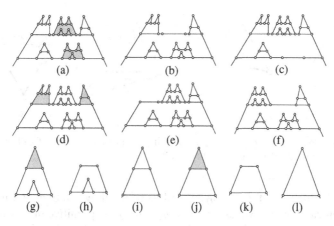

Fig. 6. Cases in the induction step of the proof of Lemma 2. $\Gamma_c(w)$ shown highlighted in each case.

biconnected series-parallel graph, we now add an edge (u, v) to G', and a new line segment between the points $p(u)$ and $p(v)$ in Γ_c. We then replace the node w in $\mathcal{T}$ with a Q-node representing the edge (u, v), and rename the node x as x'. Let Γ' denote this newly computed drawing of $G(x')$. Since $P_{x'} < P_x$, we have $L(\Gamma_c(x')) \geq P_{x'} + N_{x'} + 1$. We now have the following two cases to consider.

Case 1. Γ' *is canonical.*

This case may occur in either of the following two subcases. *(i)* $p > 2$ and $w = z_i$ $(2 \leq i \leq p - 1)$, as illustrated in Fig. 6(a) and (b); and *(ii)* $p' \geq 1$ and $w = z'_i$ $(1 \leq i \leq p')$, as illustrated in Fig. 6(a) and (c). In both the subcases, $\Gamma_c(w)$ had exactly one line segment shared with $\Gamma_c(x)$. Thus, $L(\Gamma') = L(\Gamma_c(x)) - L(\Gamma_c(w)) + 1$. Again, since Γ' is canonical, $L(\Gamma')) = L(\Gamma_c(x'))$. By induction hypothesis we have $L(\Gamma_c(x)) \geq P_{x'} + N_{x'} + 1 + P_w + N_w + 1 - 1 = P_x + N_x + 1$.

Case 2. Γ' *is not canonical.*

Here we have the following three subcases.
(i) $p > 2$ and either $w = z_1$ or $w = z_p$, as illustrated in Fig. 6(d) and (e), *(ii)* $p = p' = 1$ and $w = z_1$, as illustrated in Fig. 6(g) and (h); and *(iii)* $p = 1, p' = 0$ and $w = z_1$, as illustrated in Fig. 6(j) and (k). We omit the proofs for the second and third subcase in this extended abstract. For the subcase 2(i), $\Gamma_c(w)$ had exactly two line segments shared with $\Gamma_c(x)$. Thus, $L(\Gamma') = L(\Gamma_c(x)) - L(\Gamma_c(w)) + 2$. Since Γ' is not canonical, we now make it canonical by making $L_1(z_2)$ collinear with $L_1(x')$ if $w = z_1$ or, by making $L_3(z_{p-1})$ collinear with $L_3(x')$ if $w = z_p$ as illustrated in Fig. 6(f). One can observe that, in both the cases, the number of line segments decreases by exactly one in $\Gamma_c(x')$. Thus, $L(\Gamma_c(x')) = L(\Gamma') - 1$. By induction hypothesis we then have $L(\Gamma_c(x)) \geq P_{x'} + N_{x'} + 1 + P_w + N_w + 1 - 1 = P_x + N_x + 1$. □

We now have the following theorem.

Theorem 1. *Let* G *be a biconnected series-parallel graph with* $\Delta(G) = 3$. *Let* Γ_c *be a canonical drawing of* G. *Let* $\mathcal{T}$ *be an* SPQ-tree of G *corresponding to* Γ_c. *Let* $P_{\mathcal{T}}$ *and* $N_{\mathcal{T}}$ *denote the number of P-nodes and the number of primitive P-nodes respectively in* $\mathcal{T}$. *Then the following* (a) *and* (b) *hold.* (a) $L(\Gamma_c) \geq P_{\mathcal{T}} + N_{\mathcal{T}} + 2$, *if every S-node in* $\mathcal{T}$ *has at most one child P-node; and* (b) $L(\Gamma_c) \geq P_{\mathcal{T}} + N_{\mathcal{T}} + 1$, *otherwise.*

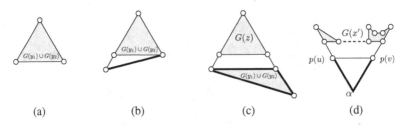

Fig. 7. (a) The drawing of $G(y_1) \cup G(y_2)$, (b) Γ_c if y_3 is primitive, (c) Γ_c if y_3 has exactly one child P-node, and (d) Γ_c if y_3 has at least two child P-nodes

Proof. Let r be the root of $\mathcal{T}$. Let y_1, y_2 and y_3 be the three children of r in $\mathcal{T}$. Since Γ_c is a canonical drawing, we assume that y_1 and y_2 are primitive in $\mathcal{T}$. Since G is a simple graph, exactly one of y_1, y_2 and y_3 can be a Q-node in $\mathcal{T}$. Thus, if there is a Q-node among y_1, y_2 and y_3, then we assume that y_2 is the Q-node. The proofs of the claims (a) and (b) are given below.

(a) We have the following two cases to consider here.

Case 1. y_3 is primitive in $\mathcal{T}$.

Since $G(y_1) \cup G(y_2)$ is a cycle, at least three line segments are required to draw $G(y_1) \cup G(y_2)$ in Γ_c as illustrated in Fig. 7(a). Since $G(y_3)$ is a path, at least one new line segment is required to draw $G(y_3)$ along with $G(y_1) \cup G(y_2)$ in Γ_c as illustrated through the thick line segment in Fig. 7(b). Since $P_\Gamma = 1, N_\Gamma = 1$, and $P_\Gamma + N_\Gamma + 2 = 4$, we thus have $L(\Gamma_c) = 4 \geq P_\Gamma + N_\Gamma + 2$.

Case 2. y_3 is not primitive in $\mathcal{T}$.

Let z denote the child P-node of y_3 in $\mathcal{T}$. By Lemma 2, $L(\Gamma_c(z)) \geq P_z + N_z + 1$. One can observe that $G(y_1) \cup G(y_2)$ is connected with $G(y_3)$ through the two edges incident to the two poles of $G(y_3)$. Hence, any drawing of $G(y_1) \cup G(y_2)$ can share at most two segments with the drawing $\Gamma_c(z)$. However, as shown in the proof of Lemma 1, since each S-node in $\mathcal{T}$ has at most one child P-node, we cannot draw the line segments $L_1(z)$ and $L_3(z)$ as converging in the exterior of $\Gamma_c(z)$ without increasing $L(\Gamma_c(z))$. Since $L_1(z)$ and $L_3(z)$ are converging in the interior of $\Gamma_c(z)$, at least two new segments are required to draw $G(y_1) \cup G(y_2)$ along with $G(y_3)$ as shown through the thick line segments in Fig. 7(c). Thus, $L(\Gamma_c) \geq P_z + N_z + 1 + 2$. Since $P_\Gamma = P_z + 1, N_\Gamma = N_z$, we thus have $L(\Gamma_c) \geq P_\Gamma + N_\Gamma + 1$.

(b) In this case y_3 has at least two child P-nodes in $\mathcal{T}$. We consider y_2 and y_3 as the two S-nodes of a temporary P-node x', and compute $\Gamma_c(x')$ as described in the proof of Lemma 1. By Lemma 2, $L(\Gamma_c(x')) \geq P_{x'} + N_{x'} + 1$. As shown in the proof of Lemma 1, since at least one S-node in $\mathcal{T}$ has two child P-nodes, we can draw the line segments $L_1(x')$ and $L_3(x')$ as converging in the exterior of $\Gamma_c(x')$ as illustrated in Fig. 7(d). Let α denote the point where $L_1(x')$ and $L_3(x')$ converges. Let u and v denote the poles of r. Since y_3 is not a Q-node, we can now complete the drawing of $G(y_3)$ on the two line segments closed between $p(u), \alpha$ and $\alpha, p(v)$ without requiring any new line segment. Since $P_{x'} = P_\Gamma, N_{x'} = N_\Gamma$, we have $L(\Gamma_c) \geq P_{x'} + N_{x'} + 1 = P_\Gamma + N_\Gamma + 1$. □

We now present our main result on minimum segment drawing in the following theorem.

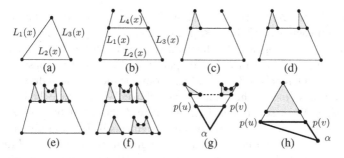

Fig. 8. (a) Drawing of $G(x)$ for a primitive P-node x, (b) the quadrangle for a non-root and non-primitive P-node x, (c)–(f) completing the drawing $\Gamma(x)$ for a non-root and non-primitive P-node x, (g) Γ when there is a suitable root in $\mathcal{T}$, and (h) Γ when there is no suitable root in $\mathcal{T}$

Theorem 2. *Let G be a biconnected series-parallel graph with $\Delta(G) = 3$. Then a minimum segment drawing of G can be computed in linear time.*

Proof. We first compute a planar straight line drawing Γ of G in linear time and then show that Γ is a minimum segment drawing of G.

Let $\mathcal{T}$ be an SPQ-tree of G rooted at an arbitrary P-node r. Let y_1, y_2 and y_3 denote the three children of r in $\mathcal{T}$. Since G is a simple graph, at most one of y_1, y_2 and y_3 can be a Q-node. Thus, if there is a Q-node among y_1, y_2 and y_3, then we assume that y_2 is the Q-node. In order to compute a minimum segment drawing of G, we want the following two conditions to hold for y_1, y_2 and y_3 in $\mathcal{T}$. *(a)* y_1 and y_2 are primitive in $\mathcal{T}$; and *(b)* y_3 has at least two child P-nodes in $\mathcal{T}$. If these conditions hold for the three children of our arbitrarily chosen root r, then we are done. Otherwise, we search for such a P-node r in $\mathcal{T}$. If there is such a P-node r in $\mathcal{T}$, then there is an S-node x in $\mathcal{T}$ such that x has at least two child P-nodes, one of which is primitive. We can search for such an S-node x in $\mathcal{T}$ in linear time. If we find such an S-node x in $\mathcal{T}$, then the child primitive P-node r of x will be our desired root of $\mathcal{T}$. If we fail to find any such S-node x in $\mathcal{T}$, then each S-node in $\mathcal{T}$ has at most one child P-node in $\mathcal{T}$. We then choose any primitive P-node r in $\mathcal{T}$ as the root of $\mathcal{T}$. We now compute Γ in a bottom up traversal of $\mathcal{T}$. At first, in each non-root P-node x of $\mathcal{T}$, we compute a canonical drawing $\Gamma(x)$ of $G(x)$ from the previously computed canonical drawings $\Gamma(w)$ of $G(w)$ for each child P-node w of x. Then we compute $\Gamma = \Gamma(r)$ for the root r of $\mathcal{T}$. We describe this construction inductively. For a primitive P-node x in $\mathcal{T}$, we draw the three core paths $P_i(x)$ $(1 \leq i \leq 3)$ of $G(x)$ on three line segments $L_i(x)$ $(1 \leq i \leq 3)$ such that each line segment $L_i(x)$ is closed between the end vertices of $P_i(x)$ $(1 \leq i \leq 3)$, as illustrated in Fig. 8(a). For a non-root and non-primitive P-node x in $\mathcal{T}$, we first draw the four core paths $P_i(x)$ $(1 \leq i \leq 4)$ on four line segments $L_i(x)$ $(1 \leq i \leq 4)$ such that each line segment $L_i(x)$ is closed between the end vertices of $P_i(x)$ $(1 \leq i \leq 4)$, as illustrated in Fig. 8(b). For each child P-node w of x, we now add $\Gamma(w)$ to this quadrangle and complete the drawing $\Gamma(x)$ as follows. At first, we draw $\Gamma(z_1)$ by making $L_1(z_1)$ collinear with $L_1(x)$, and $L_2(z_1)$ collinear with $L_4(x)$ as shown in Fig. 8(c). Next we draw $\Gamma(z_p)$ by making $L_3(z_p)$ collinear with $L_3(x)$ and $L_2(z_p)$ collinear with $L_4(x)$ as shown in Fig. 8(d). Finally, for each $w = z_i$

$(2 \leq i \leq p - 1)$, we draw $\Gamma(w)$ by making $L_2(w)$ collinear with $L_4(x)$ as shown in Fig. 8(e), and for each $w = z_i'$ $(1 \leq i \leq p')$, we draw $\Gamma(w)$ by making $L_2(w)$ collinear with $L_2(x)$ as shown in Fig. 8(f). We finally assume that x is the root P-node of $\mathcal{T}$. Let u and v denote the poles of x in $\mathcal{T}$. To compute $\Gamma = \Gamma(x)$, we first consider y_2 and y_3 as the children of a temporary P-node x' and compute the canonical drawing $\Gamma(x')$ in the same way as described in the inductive case above. We now have the following two cases to consider. We first consider the case where $\mathcal{T}$ has a suitable root as described earlier. By construction, we will have two line segments in $\Gamma(x')$ in this case, namely $L_1(x')$ and $L_3(x')$ that can be drawn as converging in the exterior of $\Gamma(x')$. Let α denote the point where $L_1(x')$ and $L_3(x')$ converge. We then draw the graph $G(y_3)$ on the two line segments closed between $p(u), \alpha$ and $\alpha, p(v)$, as shown in Fig. 8(g). We next consider the case where $\mathcal{T}$ does not have a suitable root r as described earlier. In this case, we take a point α in the exterior of $\Gamma(x')$ such that the points $p(u), p(v)$ and α form a triangle as shown in Fig. 8(h). We then draw the graph $G(y_3)$ on the two line segments closed between $p(u), \alpha$ and $\alpha, p(v)$. Clearly, the drawing Γ described above can be computed in linear time. We omit the details of this proof of time complexity in this extended abstract.

We now prove that Γ has the minimum number of segments. We first prove that for each non-root P-node x of $\mathcal{T}$, we draw $G(x)$ on $P_x + N_x + 1$ segments. We give here an inductive proof by taking induction on the height $h(x)$ of x. For $h(x) = 0$, $P_x + N_x + 1 = 3$. We have drawn $G(x)$ on three line segments, and our claim holds for $h(x) = 0$. We now consider $h(x) > 0$ and x is a non-root and non-primitive P-node. While computing $\Gamma(x)$, we have drawn $G(y')$ in such a way that all the edges corresponding to the child Q-nodes of y' were drawn on a single segment, and $L_2(z_i')$ for each $G(z_i')$ was drawn on the same segment. Thus the number of segments in this drawing of $G(y')$ is $P_y' + N_y' + p' - (p' - 1) = P_y' + N_y' + 1$. Similarly, $G(y)$ was first drawn on $P_y + N_y + 1$ segments and then the path $\bigcup_{i=1}^{j} e_i$ was drawn on the same segment as $L_1(z_1)$ and the path $\bigcup_{i=k+1}^{q} e_i$ was drawn on the same segment as $L_3(z_p)$. Here e_j and e_k are the two edges corresponding to the two Q-nodes immediately preceding z_1 and z_p respectively in $\mathcal{T}$. Since we had reused an already drawn segment, this last operation did not increase the number of segments. We finally had merged these drawings of $G(y)$ and $G(y')$ together to get a drawing of $G(x)$ on $P_y + N_y + 1 + P_y' + N_y' + 1 = (P_y + P_y' + 1) + N_x + 1 = P_x + N_x + 1$ segments. Finally, in the root node, we did not draw any new line segment if a suitable root r was found for $\mathcal{T}$, otherwise we had drawn exactly one new line segment. Thus we had drawn Γ on $P_x' + N_x' + 1$ segments in the first case, and on $P_x' + N_x' + 2$ segments in the second case. Since $P_\Gamma = P_{x'}, N_\Gamma = N_{x'}$, we have ultimately drawn Γ on $P_\Gamma + N_\Gamma + 2$ segments if each S-node in $\mathcal{T}$ had at most one child P-node, and on $P_\Gamma + N_\Gamma + 1$ segments otherwise. Both these quantities matches the bound given on $L(\Gamma)$ in Theorem 1, and this completes the proof.   □

4   Series-Parallel Graphs with Cut Vertices

So far we have dealt with biconnected series-parallel graphs with the maximum degree three. However, the same idea can be adopted to compute a minimum segment drawing of a series-parallel graph G that contains cut vertices. In this case, we first compute the blocks of G. Each block of G is either a single edge or a series-parallel graph G' which can be decomposed similarly to the pertinent subgraph $G(x)$ of a P-node x in the SPQ-tree of a biconnected series-parallel graph with the maximum degree three. For each such graph $G(x)$, we then compute a canonical drawing of $G(x)$. Next we add a single line segment aligned with the path $P_2(x)$ of each block $G(x)$ and complete the drawing of G.

5 Conclusion

In this paper we have given a linear-time algorithm for computing minimum segment drawings of series-parallel graphs with the maximum degree three. To the best of our knowledge, this is the first result in this problem focusing on an important subclass of planar graphs. It remains as our future work to achieve similar results for wider subclasses of planar graphs.

Acknowledgements

This work is based on the M.Sc. Engineering Thesis [7] completed in the Department of Computer Science and Engineering, Bangladesh University of Engineering and Technology (BUET). We acknowledge BUET for supporting this work throughout.

References

1. Dujmović, V., Eppstein, D., Suderman, M., Wood, D.R.: Drawings of planar graphs with few slopes and segments. Comput. Geom. 38(3), 194–212 (2007)
2. Fáry, I.: On straight line representation of planar graphs. Acta Sci. Math. Szeged 11, 229–233 (1948)
3. de Fraysseix, H., Pach, J., Pollack, R.: How to draw a planar graph on a grid. Combinatorica 10(1), 41–51 (1990)
4. Nishizeki, T., Rahman, M.S.: Planar Graph Drawing. Lecture Notes Series on Computing, vol. 12. World Scientific Publishing Co., Singapore (2004)
5. Purchase, H.C.: Which aesthetic has the greatest effect on human understanding? In: DiBattista, G. (ed.) GD 1997. LNCS, vol. 1353, pp. 248–261. Springer, Heidelberg (1997)
6. Rahman, M.S., Egi, N., Nishizeki, T.: No-bend orthogonal drawings of series-parallel graphs. In: Healy, P., Nikolov, N.S. (eds.) GD 2005. LNCS, vol. 3843, pp. 409–420. Springer, Heidelberg (2006)
7. Samee, M.A.H.: Minimum Segment Drawings of Series-Parallel Graphs with the Maximum Degree Three. M.Sc. Engineering Thesis, Department of Computer Science and Engineering, Bangladesh University of Engineering and Technology (July 2008)
8. Schnyder, W.: Embedding planar graphs on the grid. In: First ACM-SIAM Symp. on Discrete Algorithms, pp. 138–148 (1990)
9. Stein, K.S.: Convex maps. Amer. Math. Soc. 2, 464–466 (1951)
10. Wagner, K.: Bemerkungen zum vierfarbenproblem. Jahresber. Deutsch. Math-Verein. 46, 26–32 (1936)

Dunnart: A Constraint-Based Network Diagram Authoring Tool

Tim Dwyer, Kim Marriott, and Michael Wybrow

Clayton School of Information Technology, Monash University, 3800, Australia
{Tim.Dwyer,Kim.Marriott,Michael.Wybrow}@infotech.monash.edu.au

Abstract. We present a new network diagram authoring tool, Dunnart, that provides *continuous network layout*. It continuously adjusts the layout in response to user interaction, while still maintaining the layout style and, where reasonable, the current layout topology. The diagram author uses placement constraints, such as alignment and distribution, to tailor the layout style and can guide the layout by repositioning diagram components or rerouting connectors. The key to the flexibility of our approach is the use of topology-preserving constrained graph layout.

1 Introduction

Producing well laid out network diagrams is not easy and extremely tedious for any but the simplest networks. While automatic graph layout algorithms can provide high-quality layout [4], in many situations users would like the ability to interactively control and fine-tune the layout with similar flexibility to that provided in standard diagram authoring tools. Although some general purpose diagramming tools, such as Microsoft Visio[1] and Omnigraffle,[2] provide automatic graph layout, the integration of graph layout into these tools is quite unsatisfactory. Similar concerns apply to the network layout tool yEd.[3] The issue is that these tools use static graph layout algorithms which are not well-matched to the inherently interactive nature of diagramming tools. They provide only "once off" graph layout and allow little flexibility for the author to tailor the resulting layout by, say, requiring that certain nodes are aligned.

We believe that a better model for integrating automatic graph layout into diagramming tools is *continuous network layout*. In this model the graph-layout engine runs continuously to improve the layout in response to user interaction. The author uses placement constraints, such as alignment and distribution, to tailor the layout style and can guide the layout by repositioning diagram components or rerouting connectors. Importantly, layout should be fast enough to allow the diagram author to immediately see the effect of their changes. Thus, continuous network layout requires efficient dynamic graph layout techniques that support placement constraints.

Continuous network layout was introduced in GLIDE [13]. However, the spring-based layout algorithm used by GLIDE was not robust or powerful enough to truly

---

[1] "Layout Assistant for Visio", Tom Sawyer Soft., http://www.tomsawyer.com/lav/
[2] "Omnigraffle", The Omni Group, http://www.omnigroup.com/omnigraffle/
[3] "yEd", yWorks, http://www.yworks.com/products/yed/

I.G. Tollis and M. Patrignani (Eds.): GD 2008, LNCS 5417, pp. 420–431, 2009.

support the model. Here we present a new network diagram authoring tool, Dunnart,[4] that provides continuous network layout and which uses a recently developed topology preserving constrained graph layout algorithm [7]. This provides considerably more robust and powerful automatic layout than is possible with unconstrained optimisation techniques such as those underlying GLIDE.

Dunnart supports a variety of different layout styles, arbitrary clusters of nodes, and placement tools such as alignment, distribution and separation. Dunnart's layout engine continuously adjusts the layout in response to user interaction, ensuring that the diagram remains "tidy" by, for instance, removing object overlap, while still maintaining the layout style and user imposed placement constraints. Figure 1 illustrates the use of Dunnart.

One of the most interesting innovations in Dunnart is a simple, readily understood physical metaphor for *layout adjustment*: Poly-line connectors and cluster boundaries act like rubber-bands, trying to shrink in length and hence straighten. Like physical rubber-bands, the connectors and cluster boundaries are impervious in that nodes and other connectors cannot pass through them. This means that layout adjustment preserves the general structure of the network drawing, i.e. its *topology*, and so changes are smooth and predictable. Changes to the topology only occur as the result of explicit direction by the author and for common user editing actions, such as moving objects during direct manipulation or resizing a node, the diagram topology is preserved.

Usability concerns have guided the design of Dunnart from its beginning and we have carefully considered the design of the constraint-based placement tools including how to provide adequate feedback about constraint interaction (especially in the case of inconsistency) and how to ensure that the diagram is not too cluttered by visual representation of constraints (See Fig. 2). One important factor improving usability is that layout adjustment occurs in real-time, providing immediate feedback about the effect of user changes.

2   Related Work

Our work brings together research into constraint-based diagramming editors and research into graph drawing. Since the very infancy of diagram authoring tools there has been interest in allowing the author to specify persistent layout relationships on the diagram components, e.g. [8,12,14]. Previous systems have explored constraint solving techniques and user interaction for constraint-based placement tools. However, apart from GLIDE [13], these were not designed for network diagrams and none provided automatic network layout in the sense that we are discussing.

GLIDE was the first constraint-based diagramming tool explicitly designed for network diagrams. It introduced continuous network layout and provided high-level placement constraints (called VOFs) which the author could add to control the layout and which the layout engine endeavoured to satisfy during subsequent changes to the layout. However GLIDE had two serious limitations. The first was a lack of robustness. GLIDE used springs to approximately enforce layout constraints which effectively meant the constraints were solved by minimising a goal function that contains an error term for

---

[4] Dunnart, http://www.dunnart.org/

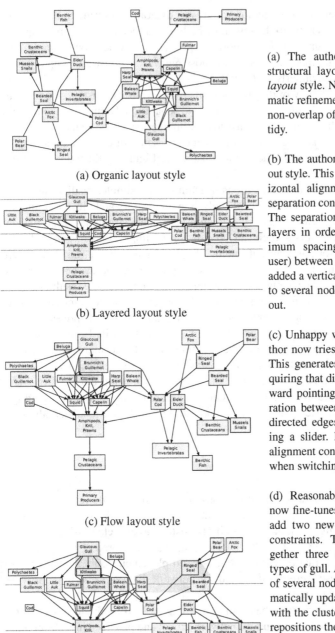

(a) Organic layout style

(b) Layered layout style

(c) Flow layout style

(d) Final layout

(a) The author initially calls the structural layout tool with *organic layout* style. Note how various automatic refinement constraints such as non-overlap of nodes keep the layout tidy.

(b) The author tries the *layered* layout style. This generates a set of horizontal alignment constraints with separation constraints between them. The separation constraints keep the layers in order and enforce a minimum spacing (adjustable by the user) between layers. The author has added a vertical alignment constraint to several nodes to improve the layout.

(c) Unhappy with the result, the author now tries the *flow* layout style. This generates style constraints requiring that directed edges be downward pointing. The minimum separation between nodes connected by directed edges can be adjusted using a slider. Note that the vertical alignment constraint was maintained when switching layout styles.

(d) Reasonably happy, the author now fine-tunes the layout. They first add two new horizontal alignment constraints. They then cluster together three seal species and two types of gull. As a result the position of several nodes and edges are automatically updated to remove overlap with the clusters. Finally, the author repositions the "Beluga" and "Arctic Fox" nodes to improve the clarity of the diagram.

Fig. 1. Example of interactive network layout with Dunnart. Network data from the Many Eyes "Arctic food chain" visualisation, http://www.many-eyes.com/.

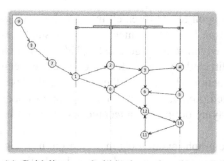

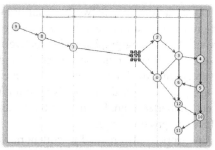

(a) Guidelines are bold in the regions between attached objects and are significantly faded outside those regions. This helps to reduce clutter while still allowing the user to attach objects to guidelines. The four vertical alignment constraints (represented by the guidelines) are involved in a horizontal distribution constraint which requires them to be equally spaced. The user is currently adjusting the distribution spacing (about the highlighted guideline) by dragging a handle on right-side of the distribution indicator.

(b) The six alignment constraints have horizontal separation constraints between them. The minimum separation distance may be adjusted by dragging a handle on the separation indicator. Note that three constraints are currently active (highlighted in red) while the other two have some slack. The grey band on the right edge of the page shows that a page-containment constraint is not satisfied. This results from the user indirectly pushing a shape outside the page while dragging another shape.

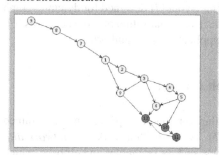

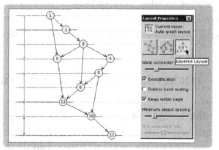

(c) Flow layout has been enabled, constraining all directed edges to point downward. A cycle of directed edges causes a constraint conflict. Dunnart drops one of the conflicting constraints, and highlights this as well as the set of affected nodes.

(d) The widgets in the "Layout Properties" window control the structural layout setting. They also allow the user to adjust the ideal edge length, and toggle generation of non-overlap and page-boundary constraints.

Fig. 2. Screenshots showing the visual representations of constraints in Dunnart

each constraint. In the case of conflicting forces, such as, for example, alignment of nodes in a network with springy connectors, the so called "constraints" would simply not be satisfied, or worse, the whole system could become unstable and not converge to a local minimum. The second limitation was that GLIDE provided little automatic network layout. While it did allow the user to manually impose VOFs to control edge length, when used in combination with other user-specified VOFs this led to conflicting forces and unsatisfied constraints.

Our techniques for network layout draw upon recent research into graph drawing. One relevant area of research is *dynamic graph layout* [2] which focuses on stable re-layout of changing graphs, or interactive navigation of large graphs [10]. Most such systems are based on unconstrained force-directed layout in which the forces between nodes are modified in response to user interaction. However, in these systems the level of user control over the layout is very limited (i.e., alignment or distribution of nodes is not supported) and, because of the underlying optimisation techniques, it would be difficult to provide more.

Our work is also related to collaborative graph-layout tools in which the user can interact with the optimisation engine to improve the layout and escape local minima by providing user hints [11], such as repositioning a node. This is also true in the continuous network layout model, since user interaction can guide the layout engine away from undesired local minima. The fundamental difference is that Dunnart is a generic network diagramming tool, while collaborative graph layout is intended to allow the user to improve the layout obtained with a single specialised layout engine. Thus, user hints are quite restrictive and depend on the underlying layout algorithm. For example, the systems of [11,1] are built on top of a layered graph-drawing algorithm for directed graphs, while the Giotto system [3] is built on top of an orthogonal graph layout engine.

Dunnart is based upon so called *constrained graph layout* algorithms which perform graph layout subject to various kinds of layout constraints [9,5]. It uses a recent algorithm for topology preserving constrained graph layout [7] designed for dynamic graph layout. This has previously been used for interactive visualisation of large networks [6]. Here we demonstrate its usefulness in a new application area: authoring.

3 Background: Constrained Graph Layout

In this section we briefly review the algorithm for topology preserving constrained graph layout. It is described more fully in [7]. The algorithm works on *network diagrams*. These can contain: basic graphic shapes, such as rectangles and ellipses, which are treated as rectangular nodes in the diagram; connectors, which form the edges in the diagram and may be directed; and container shapes, which contain a set of nodes and so specify node clusters in the diagram.

A *layout* for a network diagram gives a position for each node in the diagram and a route for the *paths*, i.e. edge routes and cluster boundaries, in the network.

Constrained graph layout allows constraints on the placement of nodes. These are required to be *separation constraints* in a single dimension.[5] The layout must also satisfy various *refinement constraints* to ensure that it is "tidy." The refinement constraints are:

- no two nodes overlap;
- the nodes inside the region defined by the boundary of each cluster are exactly the nodes in the cluster;

---

[5] Separation constraints have the form $u + d \leq v$ or $u + d = v$ where u and v are variables representing horizontal or vertical position of a node and d is a constant giving the minimum separation required between u and v.

– every path is *valid* and *tight* where a valid path is one in which no segment passes *through* a node and a tight path is one in which the path "wraps" tightly around each node corner in the path.

Topology-preserving constrained graph layout uses the *P-stress* (for path-stress) goal function to measure the quality of a layout. *P-stress* modifies the standard *stress* function to penalise nodes that are too close together, but not nodes that are more than their ideal distance apart, thus eliminating long range attraction which can cause issues in highly constrained problems. *P-stress* also tries to make the length of each path in the network no more than its ideal length. This has the effect of straightening edges and making clusters more compact and circular in shape.

The basic algorithm to find a layout that minimises *P-stress* and which satisfies the layout constraints is:

(1) Find a position for the nodes satisfying the layout constraints by projecting the current position of the nodes on to the placement constraints and then using a greedy heuristic to satisfy the non-overlap constraints and cluster containment constraints (modeled using a rectangular box).
(2) Perform edge routing using an incremental poly-line connector routing algorithm [15] to compute poly-line routes for each edge, which minimise edge length and amount of bend. The cluster boundary is obtained using the convex hull of the cluster.
(3) Optimise the layout by iteratively improving the current layout using gradient projection to reduce *P-stress*. This preserves the topology of the initial layout.

As noted previously, unlike force-directed layout, constrained graph layout techniques ensure that the generated layouts really do satisfy all of the layout constraints (unless the constraints are infeasible).

4   Dunnart

Dunnart is intended to be a generic diagramming tool that supports most diagram types, including network diagrams. The original motivation for Dunnart was to explore usability issues in constraint-based diagramming tools. Thus, usability has been a focus of its design from the beginning. Feedback from its use—for constructing a wide variety of diagrams including UML diagrams and biological networks—has greatly improved the interface design. We now look at its more novel aspects.

A primary usability consideration was when and how much the layout engine should change the layout in response to user interaction. Typically, when first constructing a network diagram, the user will try different layout styles and, for each style, wants the tool to automatically find a good layout. Then, once the basic layout and style is chosen, the user will fine-tune the layout. During fine-tuning, it is important that changes made by the layout engine are predictable and controllable by the author. To support these two use cases, Dunnart provides two kinds of network layout: *structural layout* and *layout adjustment*. We now look at these.

4.1   Structural Layout

Dunnart provides a structural layout tool which is free to completely rearrange the layout so long as the user-specified placement constraints remain satisfied. It is explicitly invoked by the author to re-layout the network. The other function of the structural layout tool is to impose a layout style on the diagram. Dunnart currently provides three layout styles: organic, flow and layered (shown in Fig. 1). It could be extended with other layout styles. The only requirement is that the aesthetic constraints imposed by the style must be able to be modelled using separation constraints so that the layout aesthetic can be maintained in subsequent interaction.

Organic layout is the most basic style since it does not impose any style constraints. It simply calls the constrained graph layout algorithm sketched in Sect. 3. Flow-style layout is the same except that the tool adds style constraints ensuring that the start node of each directed edge is above its end node.

Structural layout can also use external graph layout algorithms to find a layout and determine the style constraints. As an example of this, structural layout with the layered style uses the Graphviz[6] library implementation of the Sugiyama algorithm. This determines a layer for each object in the network, the ordering of objects on each layer and a routing for connectors through the layers which minimises crossings. An alignment placement constraint is generated for each layer and a separation constraint between each pair of layers keeps them a minimum distance apart and preserves the layer order. Currently, existing placement constraints are initially ignored in this style and only imposed in the subsequent layout adjustment step.

Style constraints behave like author specified placement constraints. Thus, the author is free to modify the layout by removing style constraints. Using constraints to model layout style is one of the reasons Dunnart is very flexible. It means that, unlike most previous diagramming tools, layout styles are not brittle and the author is free to tailor the layout style by adding placement constraints to the diagram before calling the structural layout tool, or by subsequently modifying the placement and style constraints.

4.2   Layout Adjustment

The second kind of automatic layout provided in Dunnart is called *layout adjustment*. This supports fine-tuning of the layout and runs continuously during interaction. Changes made by the layout engine during layout adjustment need to be predictable and (reasonably) continuous. Consequently, we believe layout adjustment should preserve the topology of the starting layout as far as possible.

We now describe how the layout is updated after the main kinds of user interaction provided in Dunnart. For most interactions this has two steps. First, find a new feasible layout satisfying the placement, style and refinement constraints that changes the topology of the current layout as little as possible. Second, perform step (3) of the layout algorithm (Sect. 3) to optimise the layout while preserving its topology. Table 1 gives details of how the new feasible layout is found for different kinds of user interaction. We make use of two techniques.

---

[6] Graphviz, AT&T Research, http://www.graphviz.org/

Table 1. Computation of new feasible layout after common kinds of user interaction. Note that this step is always followed by topology-preserving layout optimisation.

Add graphic object: Node repair followed by edge routing repair.
Delete graphic object: Edge routing repair.
Add connector: Automatically or manually route connector.
Delete connector: Nothing—layout remains feasible.
Add/modify cluster: Node repair followed by edge routing repair.
Delete cluster: Nothing—layout remains feasible.
Cut/Copy (to clipboard): Copy nodes to clipboard and perform edge routing repair. If cutting, delete graphic objects and connectors.
Paste (from clipboard): Add nodes to canvas and perform node repair. Then perform edge routing repair (based on connector routing in clipboard for pasted connectors).
Add a placement constraint: Node repair followed by edge routing repair. However, nodes that have moved too far because of the placement behave as if cut and pasted.
Delete a placement constraint: Nothing—layout remains feasible.

The first is *node position repair*. This is done using step (1) of the layout algorithm (Sect. 3) to compute new position for the nodes which satisfies the placement and style constraints as well as the non-overlap and cluster containment constraints.

The second technique, which we call *rubber-banding*, is for repairing edge routes. The issue is that the route may have become invalid because it now passes through a graphic object or is no longer tight and so should be shortened by straightening and merging some adjacent segments. As much as possible we want to preserve the current route. Rubber-banding finds a new edge route by tracing the original connector path—object corner by object corner—until the destination object is reached. At all stages the connector acts like a rubber-band, fitting snugly around objects encountered so far on the route. The rubber-banding implementation uses the connector routing algorithm to dynamically route from the start object to the current object corner while preserving as much of the previous route as possible. More exactly, the last vertex in the route is removed from the route whenever the bend angle around the vertex becomes 180° or more, and routing proceeds from the preceding vertex.

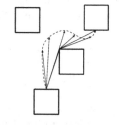

Rubber-banding is also used for manual specification of connector routes. Connectors are typically created by specifying their start and end object, in which case automatic connector routing is used to determine a shortest-path route. However, the author is also free to specify the *topological* route of a connector. The author starts from an object and then threads the connector through the objects to the destination object with rubber-banding computing the route to the current cursor location. This is shown in Fig. 3.

Fig. 3. Manual connector routing. The author "threads" the endpoint of the connector between the objects to specify the topological route for the connector.

The remaining user interactions are kinds of *direct manipulation* of the diagram. A strength of Dunnart is that the layout engine is fast enough to provide "live" feedback during direct manipulation. With live feedback, all objects and connectors in the

Table 2. Implementation of user actions providing live feedback during manipulation

---

Dragging objects: Simply add terms to the goal function for each node v being manipulated of form $(y_v - y_d)^2 + (x_v - x_d)^2$ where (x_d, y_d) is the new desired position of node v.

Horizontal resizing of a node: The node to be resized is internally replaced by two artificial nodes which correspond to the left and right boundary edges of the original node's bounding box. Separation constraints couched in terms of these nodes are generated to maintain non-overlap between the bounding box and the other nodes. The width is changed by dragging the two artificial nodes to the required width, and updating the appearance of the node.

Vertical resizing of a node: Analogous to horizontal resizing.

Simultaneous vertical and horizontal resizing of a node: Achieved by resizing in small horizontal and vertical increments.

Tuning of goal function: The user can use sliders to change parameters of the goal function, such as the desired edge length.

---

diagram have their position and routing updated immediately in response to user manipulation. Direct manipulation is guaranteed not to change the topology of the layout. Details of the process—essentially achieved by performing step (3) of the layout algorithm with a modified P-stress goal function—are given in Table 2.

Clearly topology-preservation means that when dragging objects the author cannot move objects through connectors or other objects, since this changes the topology. This may make it difficult or impossible for the author to achieve their objective of, say, snapping an object to an alignment guideline because the alignment guideline keeps moving away from the object being dragged. For this reason, Dunnart allows the author to temporarily escape from continuous layout adjustment during object dragging by depressing a modifier key. This suspends any current layout activity and causes those objects not being directly manipulated to maintain their current position. The user is now free to move objects through connectors and other objects or to add or remove an object from a container shape. This allows the user to quickly and easily modify the topology of the diagram.

Depressing the modifier key also breaks the selected objects free from placement and style constraints involving non-selected objects. Dunnart treats this as if the objects have been cut and pasted into their new location. The only difference is that connectors between the manipulated objects and non-manipulated objects are treated as new, automatically routed connectors.

4.3   Understanding Constraints

Placement constraints are the primary method for the author to tailor the layout without having to explicitly position objects. The placement tool sets up a persistent relationship that is maintained in subsequent interaction until the author explicitly removes it rather than a once-off position adjustment. Dunnart provides standard placement tools: horizontal and vertical alignment and distribution, horizontal and vertical separation (sequencing) that keeps objects a minimum distance apart horizontally or vertically while preserving their relative ordering, and an "anchor" tool that allows the user to fix the current position of a selected object or set of objects.

Table 3. Indicative running times on an average (Dual Core 2GHz) PC for various sized randomly generated directed networks with flow style. For each graph we give the number of nodes and edges. Note that the number of separation constraints imposing downward edges is $|E|$. We give the time to find (a) a feasible layout after adding a new alignment constraint; and (b) the average rate of layout updates during dragging of a random node and the time for the layout to converge following the movement.

(a) Feasibility			(b) Direct manipulation			
$|V|$	$|E|$	Feasibility repair (seconds)	$|V|$	$|E|$	Layout frame rate (frames/sec)	Time to converge (seconds)
59	62	0.19	59	62	15.83	0.94
105	117	0.84	105	117	11.72	1.75
156	167	1.96	156	167	8.59	4.50
230	276	5.16	230	276	2.21	7.26

Like most constraint-based diagramming tools, there is a graphical representation for each placement relation in the diagram. A potential usability issue for constraint-based layout tools that utilise such visual representations is that they clutter the diagram. To reduce clutter we have chosen to use an explicit visual representation only for user-created placement constraints and some style constraints but not for refinement constraints since the objects themselves and their behaviour during manipulation provide sufficient feedback. To further reduce clutter, the visual representation for constraints is by default very faded, leaving the actual diagram components clearly visible (see Fig. 2).

Another well-known usability issue of constraint-based layout tools is that users can find it difficult to understand interaction between the constraints. Immediate feedback during direct manipulation helps this considerably since it allows the author to quickly notice unexpected interaction between the objects being manipulated and other parts of the diagram. As a more sophisticated way to understand constraint interaction, Dunnart also provides a query tool dubbed "Information Mode." This tool finds the path of constraints between two objects and illustrates this to the user by highlighting the relevant constraint indicators.

The extreme kind of unexpected interaction between constraints is when the author tries to perform an action which will give rise to inconsistent constraints. For instance: the author may try to add a downward pointing connector which creates a cycle of downward edges; try to apply a placement tool which gives rise to an inconsistent constraint; or use the modifier key to move an object to an infeasible position. To allow the author to understand the problem, Dunnart highlights the placement and style constraints and objects associated with the subset of separation constraints causing the inconsistency.

5   Performance

One of the most important requirements of Dunnart is that the layout algorithms are fast enough for interactive layout. Table 3(a) lists for network diagrams of various sizes the time taken to complete node position and edge routing repair after the addition of a new alignment constraint. Up to a few seconds are required to layout networks of around

250 nodes. We have found that the dominating cost of this process is finding the initial connector routing.

Perhaps more interesting, is the speed of topology-preserving layout adjustment, especially during direct manipulation. Table 3(b) shows the average number of layout updates per second while the user drags a random node slowly to the four corners of the screen and back to the centre. It also shows the time taken for the layout to converge, once the user has stopped dragging the object. As expected, because the layout optimisation algorithm generally starts from a solution close to the optimal solution it converges quite quickly, allowing real-time feedback during manipulation of graphs with up to 100–150 nodes. It is worth noting that layout occurs in separate thread so that Dunnart is still responsive while layout adjustment is taking place. Furthermore, layout adjustment typically finds a near optimal solution very rapidly, and the majority of time is spent moving nodes only very slightly.

6   Conclusion

We have described Dunnart, a new network diagram authoring tool that provides powerful automatic graph layout, yet still allows the user total layout flexibility. Topology preserving constrained graph layout provides predictable behaviour during editing and allows the author to use placement constraints to control and improve the layout.

The underlying graph layout engine is fast enough to provide live update of the layout during direct manipulation for networks with up to 100 nodes. This is more than sufficient for the kind of diagrams that are typically created with interactive authoring tools. For larger networks we believe that a combination of fast layout techniques (for an overview layout) and topology preserving constrained graph layout (for the detailed view) is the right approach [6].

There are a number of extensions to Dunnart that we intend to investigate. One is orthogonal connector routing. We want to explore further use of Dunnart in particular application areas, such as biological networks and concept maps.

References

1. Böhringer, K.-F., Paulisch, F.N.: Using constraints to achieve stability in automatic graph layout algorithms. In: CHI 1990: Proceedings of the SIGCHI conference on Human Factors in Computing Systems, pp. 43–51. ACM Press, New York (1990)
2. Brandes, U., Wagner, D.: A bayesian paradigm for dynamic graph layout. In: DiBattista, G. (ed.) GD 1997. LNCS, vol. 1353, pp. 236–247. Springer, Heidelberg (1997)
3. Bridgeman, S.S., Fanto, J., Garg, A., Tamassia, R., Vismara, L.: InteractiveGiotto: An algorithm for interactive orthogonal graph drawing. In: DiBattista, G. (ed.) GD 1997. LNCS, vol. 1353, pp. 303–308. Springer, Heidelberg (1997)
4. Di Battista, G., Eades, P., Tamassia, R., Tollis, I.G.: Graph Drawing: Algorithms for the Visualization of Graphs. Prentice-Hall, Inc., Englewood Cliffs (1999)
5. Dwyer, T., Koren, Y., Marriott, K.: IPSep-CoLa: An incremental procedure for separation constraint layout of graphs. IEEE Transactions on Visualization and Computer Graphics 12(5), 821–828 (2006)

6. Dwyer, T., Marriott, K., Schreiber, F., Stuckey, P.J., Woodward, M., Wybrow, M.: Exploration of networks using overview+detail with constraint-based cooperative layout. In: IEEE Transactions on Visualization and Computer Graphics (InfoVis 2008) (to appear, 2008)
7. Dwyer, T., Marriott, K., Wybrow, M.: Topology preserving constrained graph layout. In: GD 2008. LNCS. Springer, Heidelberg (to appear, 2009)
8. Gleicher, M.: Briar: A constraint-based drawing program. In: CHI 1992: Proceedings of the SIGCHI conference on Human Factors in Computing Systems, pp. 661–662. ACM Press, New York (1992)
9. He, W., Marriott, K.: Constrained graph layout. Constraints 3, 289–314 (1998)
10. Huang, M.L., Eades, P., Lai, W.: Online visualization and navigation of global web structures. The International Journal of Software Engineering and Knowledge Engineering 13(1), 27–52 (2003)
11. do Nascimento, H.A.D., Eades, P.: User hints for directed graph drawing. In: Mutzel, P., Jünger, M., Leipert, S. (eds.) GD 2001. LNCS, vol. 2265, pp. 205–219. Springer, Heidelberg (2002)
12. Nelson, G.: Juno, a constraint-based graphics system. In: SIG-GRAPH 1985 Conference Proceedings. ACM Press, New York (1985)
13. Ryall, K., Marks, J., Shieber, S.M.: An interactive constraint-based system for drawing graphs. In: ACM Symposium on User Interface Software and Technology, pp. 97–104 (1997)
14. Sutherland, I.E.: Sketchpad: A Man-Machine Graphical Communication System. Ph.D. thesis, Massachusetts Institute of Technology (1963)
15. Wybrow, M., Marriott, K., Stuckey, P.J.: Incremental connector routing. In: Healy, P., Nikolov, N.S. (eds.) GD 2005. LNCS, vol. 3843, pp. 446–457. Springer, Heidelberg (2006)

Approximating the Crossing Number of Apex Graphs

Markus Chimani[1], Petr Hliněný[2,*], and Petra Mutzel[1]

[1] Faculty of CS, Dortmund University of Technology, Germany
{markus.chimani,petra.mutzel}@tu-dortmund.de
[2] Faculty of Informatics, Masaryk University, Brno, Czech Republic
hlineny@fi.muni.cz

Abstract. We show that the crossing number of an apex graph, i.e. a graph G from which only one vertex v has to be removed to make it planar, can be approximated up to a factor of $\Delta(G - v) \cdot d(v)/2$ by solving the *vertex inserting* problem, i.e. inserting a vertex plus incident edges into an optimally chosen planar embedding of a planar graph. Due to a recently developed polynomial algorithm for the latter problem, this establishes the first polynomial fixed-constant approximation algorithm for the crossing number problem of apex graphs with bounded degree.

Keywords: Crossing number, apex graph, vertex insertion.

1 Edge and Vertex Insertion Problems

We assume that the reader is familiar with the standard notation of terminology of graph theory, and especially with topological graphs, see [5]. A graph G is called an *apex graph* if there is a vertex v such that $G - v$ is planar. The *crossing number* $\mathrm{cr}(G)$ of a graph G is the minimum number of pairwise edge crossings in a drawing of G in the plane. Determining the crossing number of a given graph is an NP-complete problem, and exact crossing numbers are in general extremely difficult to compute.

A common heuristic way of finding a drawing of a graph G with few crossings starts with a planar subgraph of G, and then re-inserts the remaining edges one by one in a locally optimal way. The edge insertion problem can be solved to optimality by a linear-time algorithm [3]. A subsequent result [4] uses that algorithm to give an approximation of the crossing number of almost planar graphs (i.e. those made planar by removing one edge) up to a factor of $\Delta(G)$ (recently improved to the best possible $\Delta(G)/2$ in [1]).

A natural generalization of the previous results is to consider the problem of *inserting a vertex* with a specified neighbourhood into a planar embedding of a graph G, with the least number of crossings. Although this shows to be a much harder question than that of edge insertion, a very recent result of [2] reads:

Theorem 1 (Chimani, Gutwenger, Mutzel, and Wolf). *The vertex insertion problem for a planar graph can be optimally solved in polynomial time.*

---

* Supported by the Institute for Theoretical Computer Science ITI, project 1M0545.

I.G. Tollis and M. Patrignani (Eds.): GD 2008, LNCS 5417, pp. 432–434, 2009.

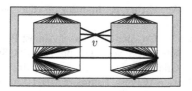

Fig. 1. An example of a vertex v insertion instance requiring many crossings, eventhough the crossing number of the graph is small. The gray regions denote dense subgraphs.

2   Crossing Number Approximation

We can apply Theorem 1 to approximate the crossing number of apex graphs.

Theorem 2. *Let G be a graph and v its vertex such that $G - v$ is planar, the maximum degree in $G - v$ is Δ, and v has degree d in G. Then the vertex insertion problem of v back into a planar embedding of $G - v$ has a solution with at most $d \cdot \lfloor \Delta/2 \rfloor \cdot \mathrm{cr}(G)$ crossings.*

This new result immediately gives us a polynomial approximation algorithm for the crossing number of an apex graph G up to factor $d \cdot \lfloor \Delta/2 \rfloor$. On the other hand, it is possible to construct examples for which optimal solutions to the vertex insertion problem require up to $d \cdot \Delta \cdot \mathrm{cr}(G)/4$ crossings, cf. Fig. 1.

The idea of the proof is as follows (compare to [4]): Assume Γ is a plane embedding of the graph $G - v$ achieving optimality in the vertex v insertion problem, Γ' is a crossing-optimal drawing of the graph G, and let F be a minimal edge set such that $\Gamma' - v - F$ is a plane embedding. Then $|F| \le \mathrm{cr}(G)$ and the embedding $\Gamma' - v - F$ can be turned into $\Gamma - F$ by a sequence of 1- and 2-flips (Whitney flips), which allows to re-embed the edges F without crossings in $G - v$. The central argument is that the number of new crossings introduced on the edges of v is limited by an iteration of the following claim over all $f \in F$:

Lemma 3. *Let H be an apex graph with a vertex v, having a drawing with ℓ crossings in which $H - v$ is connected and plane embedded. Let an edge f connect vertices of $H - v$. If $(H - v) + f$ is planar, then there is a drawing of $H + f$ with plane embedded $(H - v) + f$ having at most $\ell + d(v) \cdot \lfloor \Delta(H - v)/2 \rfloor$ crossings.*

In contrast to [4], establishing Theorem 2 using this lemma requires a careful consideration of non-biconnected graphs and the fact that the position of the newly introduced vertex v is unknown and probably different between Γ and Γ'.

References

1. Cabello, S., Mohar, B.: Crossing and weighted crossing number of near planar graphs. In: GD 2008. LNCS, vol. 5417. Springer, Heidelberg (to appear, 2009)
2. Chimani, M., Gutwenger, C., Mutzel, P., Wolf, C.: Inserting a vertex into a planar graph. In: ACM-SIAM Symposium on Discrete Algorithms (to appear, 2009)

3. Gutwenger, C., Mutzel, P., Weiskircher, R.: Inserting an edge into a planar graph. Algorithmica 41, 289–308 (2005)
4. Hliněný, P., Salazar, G.: On the Crossing Number of Almost Planar Graphs. In: Kaufmann, M., Wagner, D. (eds.) GD 2006. LNCS, vol. 4372, pp. 162–173. Springer, Heidelberg (2007)
5. Mohar, B., Thomassen, C.: Graphs on surfaces. Johns Hopkins Studies in the Mathematical Sciences. Johns Hopkins University Press, Baltimore MD, USA (2001)

Policy-Aware Visualization of Internet Dynamics

Luca Cittadini, Tiziana Refice, Alessio Campisano,
Giuseppe Di Battista, and Claudio Sasso

Dipartimento di Informatica e Automazione – Università di Roma Tre
{ratm,refice,gdb}@dia.uniroma3.it,
{alessio.campisano,claudio.sasso}@gmail.com

1 Why Visualizing Inter-domain Routing Dynamics

The *Internet* can be represented as a graph of *Autonomous Systems (ASes)*. Each AS dynamically selects the *(AS-)paths* to reach *destinations* on the Internet, according to inter-AS *customer-provider relationships*. Such relationships define a *hierarchy* of all the ASes.

The Internet is renowned to be highly dynamic as AS-paths to any destination may frequently change (from an *old path* to a *new path*). Such *AS-path changes* may impact the Internet operation and are usually debugged manually. Figure 1(a) shows the AS-paths (valid at a specific time) from a set of ASes to a specific destination, as displayed by *BGPlay* [2]. After a few seconds the state of the network can be significantly different. Due to the enormous amount of AS-path changes occurring in a short time period, it is very difficult to spot them and, thus, to locate their root causes. Hence, effectively visualizing an AS-path change can significantly help understand the Internet dynamics.

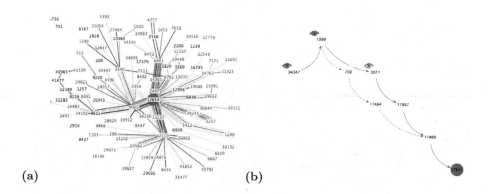

(a)　　　　　　　　　　　　　　　　　　　　　　　　(b)

Fig. 1. (a) AS-paths towards a specific destination. Each number represents an AS. The red node is the destination. (b) An AS-path change displayed by BGPATH. Solid edges belong to the new path, dashed edges belong to the old path.

I.G. Tollis and M. Patrignani (Eds.): GD 2008, LNCS 5417, pp. 435–436, 2009.

2   How-To Visualize Internet Dynamics to Support Analysis of Network Events

We propose to focus the visualization on a single AS-path change and to display it on the customer-provider hierarchy, as [6] shows that network events located at different levels of the hierarchy have usually significantly different impact on the Internet.

We now describe our algorithm to draw an AS-path change. First, we assign customer-provider relationships to the links of both old and new AS-paths, and we direct the links from providers to customers. As shown by [5], we can then classify ASes of both paths according to the *valley-free property* as follows:

type 1 : nodes in the "uphill" portion of the old path
type 2 : nodes in the "uphill" portion of the new path
type 3 : nodes in the "downhill" portion of the old path
type 4 : nodes in the "downhill" portion of the new path

We assign vertical coordinates using a topological sort of the graph. We then compute the horizontal coordinates. Namely, we first split edges spanning over multiple vertical layers by adding extra nodes and edges. Further, we add extra edges between nodes on the same layer, from nodes with lower type values to nodes with higher values. Finally, the topological sort of this augmented graph provides us with the horizontal coordinates, such that nodes in the new path are placed right of nodes in the old path, according to the common intuition of time flowing left-to-right.

In [3,4] we detail how our visualization paradigm supports the analysis of network events. We also developed a publicly available tool, BGPATH [1], which visualizes a user-specified AS-path change according to the approach described above and provides useful information to help identify its root cause. Figure 1(b) shows how BGPATH displays a sample AS-path change.

References

1. BGPath, http://nero.dia.uniroma3.it/rca/
2. BGPlay, http://www.ris.ripe.net/bgplay/
3. Campisano, A., Cittadini, L., Di Battista, G., Refice, T., Sasso, C.: Tracking Back the Root Cause of a Path Change in Interdomain Routing. In: IEEE/IFIP NOMS (2008)
4. Cittadini, L., Refice, T., Campisano, A., Di Battista, G., Sasso, C.: Measuring and Visualizing Interdomain Routing Dynamics with BGPath. In: IEEE ISCC (2008)
5. Gao, L.: On Inferring Autonomous System Relationships in the Internet. IEEE/ACM Transactions on Networking (2001)
6. Zhao, X., Zhang, B., Terzis, A., Massey, D., Zhang, L.: The Impacts of Link Failure Location on Routing Dynamics: A Formal Analysis. In: ACM SIGCOMM Asia Workshop (2005)

Enhancing Visualizations of Business Processes

Philip Effinger, Michael Kaufmann, and Martin Siebenhaller

Universität Tübingen, WSI für Informatik, Sand 13, 72076 Tübingen, Germany
{effinger,mk,siebenha}@informatik.uni-tuebingen.de

In today's business world, there exist multitudinous tools to model processes. For an automatic layout algorithm for those processes, it is important to take into account criteria like process flow and the semantic of the modeling notation. Most of the existing tools use standard layout approaches that often produce unsatisfying results since the calculated drawings are too dense and/or too large. This makes it difficult for users to understand the underlying process model.

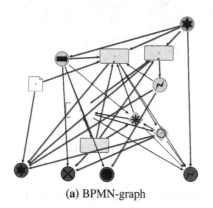
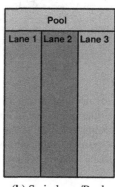

(a) BPMN-graph (b) Swimlanes/Pools

Fig. 1. (a) shows a graph representing a business process. It was drawn by starting with placing nodes and finally inserting straight-line edges. The different categories of BPMN elements [3] are represented by graph objects differing in color and shape.

The popular business process modeling notation (BPMN) [3] consists of the following categories of elements (see Fig. 1): *Flow objects* control the flow in a process and *connecting objects* are used to connect flow objects. *Swimlanes/Pools* partition flow objects into logical units, e.g. departments of a company. *Annotations* offer the possibility to add comments to flow and connecting objects.

For the core layout we use the orthogonal layout approach described in [2] that incorporates different constraints needed for the automatic layout of activity diagrams which are related to business process diagrams. The supported constraints include partitions, clusters as well as a common flow direction of edges which is especially important for such diagrams. To improve the layout quality we introduce two concepts - connectors and cuts.

Connectors replace edges by a pair of connector nodes, which are connected to the corresponding endpoints of the replaced edge (see Fig. 2). Connector nodes belonging

I.G. Tollis and M. Patrignani (Eds.): GD 2008, LNCS 5417, pp. 437–438, 2009.
© Springer-Verlag Berlin Heidelberg 2009

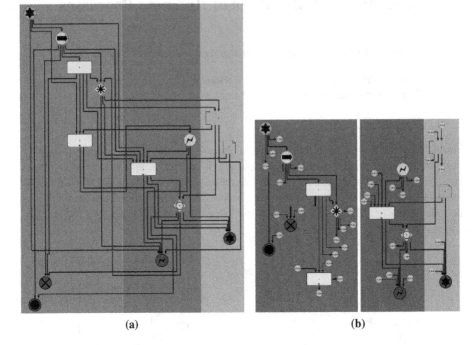

(a)   (b)

Fig. 2. Applying our approach to the BPMN-graph of Fig. 1. (a) shows the result after applying the core layout (the assignment of nodes to swimlanes is given as input). After inserting connectors and performing a vertical cut we obtain the sub-layouts shown in (b). Note, that the sum of the area of the resulting subgraphs is considerably smaller and the significant layout properties (embedding, shape and orientation) are maintained.

to the same edge get the same label to denote their correlation. Connectors offer a way to reduce the number of unaesthetic edges, i.e. edges with many bends and crossings. We determine candidates for a replacement by connectors by means of a badness function. In cases where process models become very complex and cannot easily be overlooked, it is desirable to split the resulting diagram into smaller pieces. The main objective for a *cut* is to split as few edges as possible and to obtain subgraphs of nearly equal size. We use the dual graph to find an appropriate route for a cut. The resulting subgraphs are relayouted using the sketch-driven approach described in [1]. It reduces the area of the drawing without changing the user's mental map (see Fig. 2).

References

1. Brandes, U., Eiglsperger, M., Kaufmann, M., Wagner, D.: Sketch-driven orthogonal graph drawing. In: Goodrich, M.T., Kobourov, S.G. (eds.) GD 2002. LNCS, vol. 2528, pp. 1–11. Springer, Heidelberg (2002)
2. Siebenhaller, M., Kaufmann, M.: Drawing activity diagrams. Technical Report WSI-2006-02, Wilhelm-Schickard-Institute, University of Tübingen (2006)
3. White, S.A.: Introduction to bpmn (2004), http://www.bpmn.org

A Robust Biclustering Method Based on Crossing Minimization in Bipartite Graphs

Cesim Erten and Melih Sözdinler

Computer Science and Engineering, Işık University

Clustering refers to the process of organizing a set of input vectors into clusters based on similarity defined according to some preset distance measure. In many cases it is more desirable to simultaneously cluster the dimensions as well as the vectors themselves. This special instance of clustering, referred to as *biclustering*, was introduced by Hartigan [3]. It has many applications in areas including data mining, pattern recognition, and computational biology. Considerable attention has been devoted to it from the gene expression data analysis; see [5] for a nice survey. Input is represented in a data matrix, where the rows and columns of the matrix correspond to genes and conditions respectively. Each entry in the matrix reflects the expression level of a gene under a certain condition. From a graph-teoretical perspective the data matrix can be viewed as a weighted bipartite graph, where the vertex set of one partition is the set of genes and the vertex set of the other partition is the set of conditions. An existing weighted edge incident on a *gene-condition* pair reflects the expression level of the gene under that specific experimental condition. The biclustering problem may then be described in terms of the various versions of the biclique extraction problem in bipartite graphs. Many interesting versions that directly apply to the biclustering problem are NP-hard [4]. Various graph-theoretical approaches employing heuristics have been suggested [1,4,6,7].

One drawback of these approaches is the assumption that the corresponding bipartite graph is unweighted. Of these approaches the one following a direct graph-drawing approach is [1]. A crossing minimization procedure is applied on the unweighted bipartite graph resulting from preprocessing the original input data matrix. Our approach is similar in essence. However we do not have a discretization/normalization step to convert the weighted bipartite graph into an unweighted one as this would cause some data loss and produce erroneous output. Instead we apply crossing minimization directly on the original weighted graph. Various efficient crossing minimization heuristics have been shown to work well on weighted bipartite graphs and a 3-approximation algorithm has been suggested [2]. Our algorithm consists mainly of three steps. *Initial placement* phase applies a two-sided crossing minimization on the weighted graph until there is no change on the node orders. To do this we employ algorithm 3-WOLF of [2] (one-sided crossing minimization procedure) repeatedly, each time alternating the fixed layer. We have verified that if the input data is noise-free then this initial placement is usually enough to identify bicliques and extract the biclusters. *Adaptive Noise Hiding* phase removes the weighted edges in the graph that correspond to noise in the original input data. Sliding a window around the primeter of each node pair, where $(i \pm 1, j), (i, j \pm 1), (i \pm 1, j \pm 1)$ constitutes the perimeter of a pair (i, j), we check whether the window satisfies a threshold density in terms of the number of nonzero

I.G. Tollis and M. Patrignani (Eds.): GD 2008, LNCS 5417, pp. 439–440, 2009.

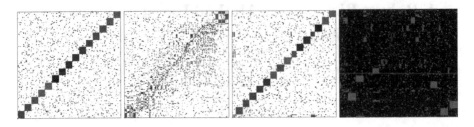

Fig. 1. Assumed noise is 0.05. (a) Initial artificial design with 15 biclusters of $K_{12,12}$; (b) Without noise removal; (c) Our complete algorithm; (d) Matlab Bioinformatics Box

weight edges. If it does not, the pairs on the perimeter are considered *suspicious*. Once sliding is finished we find the *most suspicious* weight and remove all the suspicious pairs with that weight. We adaptively apply our two-sided crossing minimization procedure on the new graph and continue noise hiding after incrementing the threshold density. The removal of the *suspicious* edges and the crossings couple each other in terms of noise removal. Each time the partitions of the graph are reordered to reduce crossings, new suspicious pairs are created. Once the noise removal phase is over, we finally gather the biclusters by applying a procedure similar to the one described in [1]. Details of the algorithm are left for the final paper. We note that different from previous approaches we directly apply weighted crossing minimization on the original input data, not to lose possibly important data that can not be considered noise. Secondly our application of the crossing minimization is two-folds. Besides providing a good initial placement, crossing minimization is also used to handle noise removal. Our preliminary experiments provided better results than the *clustergram* function of the Matlab Bioinformatics Box. Figure 1 provides a sample visualization from our initial tests.

References

1. Abdullah, A., Hussain, A.: A new biclustering technique based on crossing minimization. Neurocomputing 69(16-18), 1882–1896 (2006)
2. Çakiroglu, O., Erten, C., Karatas, Ö., Sözdinler, M.: Crossing minimization in weighted bipartite graphs. In: Demetrescu, C. (ed.) WEA 2007. LNCS, vol. 4525, pp. 122–135. Springer, Heidelberg (2007)
3. Hartigan, J.A.: Direct clustering of a data matrix. Journal of the American Statistical Association 67(337), 123–129 (1972)
4. Lonardi, S., Szpankowski, W., Yang, Q.: Finding biclusters by random projections. Theor. Comput. Sci. 368(3), 217–230 (2006)
5. Madeira, S.C., Oliveira, A.L.: Biclustering algorithms for biological data analysis: A survey. IEEE/ACM Trans. on Comp. Biol. and Bioinf.(TCBB) 1(1), 24–45 (2004)
6. Mishra, N., Ron, D., Swaminathan, R.: On finding large conjunctive clusters. In: COLT, pp. 448–462 (2003)
7. Tanay, A., Sharan, R., Shamir, R.: Discovering statistically significant biclusters in gene expression data. Bioinformatics 18(supagesl. 1), 136–144 (2002)

Visualizing the Results of Metabolic Pathway Queries

Allison P. Heath[1], George N. Bennett[2], and Lydia E. Kavraki[1,3,4]

[1]Department of Computer Science, [2]Department of Biochemistry and Cell Biology,
[3]Department of Bioengineering, Rice University, Houston, TX 77005, USA
[4]Structural and Computational Biology and Molecular Biophysics, Baylor College of Medicine,
Houston, TX, 77005, USA

1 Introduction and Problem Definition

Biology contains a wealth of network data, such as metabolic, transcription, signaling and protein-protein interaction networks. Our research currently focuses on metabolic networks, although similar ideas may be applied to other biological networks. Metabolic networks consist of the chemical compounds and reactions necessary to support life. Traditionally, series of successive metabolic reactions have been organized into simple metabolic pathways and manually drawn. However, as we move into the era of systems biology, it is becoming apparent that automated ways of processing and visualizing metabolic networks must be developed.

Our main goal is to create helpful visualizations of a large number of small metabolic networks or paths for biological researchers. This is a distinct problem from previous work on visualizing large metabolic networks and single pathways [1,2,3,4,5]. As a concrete example, a common query is to find all of the paths between two chemicals in the network. Using a methodology we developed, we find 27,912 paths of length 13 from L-2-aminoadipate to L-lysine in KEGG. The number of paths quickly scales with length; there are 693,943 paths of length 15 between the same two compounds. Displaying a list of all of these pathways or merging all of these pathways together produces an unsatisfactory visualization.

2 Approach and Results

While merging all of the pathways together produces a poor visualization for biological researchers, it does reveal that the main variation between the pathways is the reactions, not the compounds. Therefore, we investigated clustering the results. We define a distance measure based on the similarity of the chemical compounds in the pathways: $\frac{|c(X) \oplus c(Y)|}{|c(X)| + |c(Y)|}$, where $c(X)$ and $c(Y)$ are the set of chemical compounds in path X and Y. Using this distance measure, we cluster using a simple leader algorithm. This algorithm builds a list of the paths in random order, then selects the first path and designates it a cluster center. It then iterates over the remaining paths. If the next path is less than a predetermined maximum distance from a cluster center, it is added to the nearest cluster. Otherwise, it is designated as a new cluster center. This algorithm is fast and does not require knowledge of the number of clusters. While it is a localized algorithm, it can be run multiple times and the clusters can be compared to see how consistent they are.

I.G. Tollis and M. Patrignani (Eds.): GD 2008, LNCS 5417, pp. 441–442, 2009.

For our example data consisting of 27,912 paths of length 13 from L-2-aminoadipate to L-lysine this method appears to work relatively well. At a distance cutoff of 0 we get 81 clusters ranging in size from 8 to 140 nodes. At a distance cutoff of 0.2 we get 35 clusters ranging in size from 8 to 146. Visualizing these clusters using Cytoscape produce qualitatively decent results. However, further investigation of distance measures and clustering technique will likely be needed for larger or more dissimilar result sets.

3   Discussion

We demonstrate that simple clustering methods can help reveal the structure of the data and create simpler, more useful visualizations. In addition to information obtained from clustering the results, there is a wealth of external biological information that can assist and enrich the visualization. In order to be fully useful, the display should enable the user to interact with the results. We have begun work on a Cytoscape plugin which should enable interactive features. We hope to combine these ideas together to create useful visualizations of many small metabolic networks or pathways. However, many open questions remain to be investigated on visualizing biological pathway data.

Acknowledgements

This work is supported in part by a Hamill Innovation Grant and NSF CCF 0523908. APH is supported by a NSF Graduate Research Fellowship.

References

1. Bourqui, R., Cottret, L., Lacroix, V., Auber, D., Mary, P., Sagot, M.F., Jourdan, F.: Metabolic network visualization eliminating node redundance and preserving metabolic pathways. BMC Syst. Biol. 1, 29 (2007)
2. Brandes, U., Dwyer, T., Schreiber, F.: Visual understanding of metabolic pathways across organisms using layout in two and a half dimensions. Journal of Integrative Bioinformatics 1, 1 (2004)
3. Goesmann, A., Haubrock, M., Meyer, F., Kalinowski, J., Giegerich, R.: PathFinder: reconstruction and dynamic visualization of metabolic pathways. Bioinformatics 18, 124–129 (2002)
4. Kojima, K., Nagasaki, M., Miyano, S.: Fast grid layout algorithm for biological networks with sweep calculation. Bioinformatics 24(12), 1433–1441 (2008)
5. Schreiber, F.: High quality visualization of biochemical pathways in biopath. Silico. Biol. 2(2), 59–73 (2002)

Visual Specification of Layout

Sonja Maier, Steffen Mazanek, and Mark Minas

Universität der Bundeswehr München, Germany

Abstract. We give an overview of a drawing approach that combines the concepts constraint satisfaction, attribute evaluation and transformation. The approach is tailored to an editor for visual languages, which supports structured editing as well as free-hand editing. In this paper, we focus on the visual specification of such a layout algorithm. As a running example, deterministic finite automata are used.

1 Visual Specification

When implementing an editor for a visual language, a challenging task is layout. The drawing approach should produce a good-looking result and should support the user. Additionally, the layout specification should be very easy. In the following, we introduce an approach that aims to achieve these concurrent goals.

Implementation. The approach was implemented and tested in DIAMETA [1], an editor generation framework. Figure 1(a) shows a DIAMETA editor for DFA's.

Aspects. The approach supports structured editing as well as free-hand editing. Structured editors offer the user some operations that transform correct diagrams into (other) correct diagrams. Free-hand editors allow to arrange diagram components from a language-specific set on the screen without any restrictions, thus giving the user more freedom.

The approach is best suited for layout refinement, which starts with an initial layout and performs minor changes to improve it while still preserving the "mental map" [2] of the original layout.

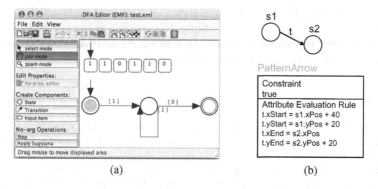

(a)   (b)

Fig. 1. (a) Editor for DFA's. (b) Visual Rule.

I.G. Tollis and M. Patrignani (Eds.): GD 2008, LNCS 5417, pp. 443–444, 2009.

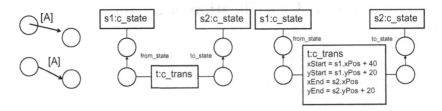

Fig. 2. Transitions (Editor, Rule (HGM))

Algorithm. We have introduced a dynamic layout algorithm, which combines the concepts constraint satisfaction and attribute evaluation [1]. This approach provides us with all we need for layout refinement. To simplify layout specification, we combined graph transformation with this dynamic drawing approach. With the approach, layout specification may be done on the abstract or on the concrete syntax level of a diagram language. In DIAMETA, this means changing an EMF model (abstract syntax level) or a hypergraph (concrete syntax level). Up to now, the layout specification is textual. To allow a more intuitive description of layout, we introduce a visual language for layout specification.

Example. We demonstrate our approach by specifying a rather simple layout for DFA's. Generally, a layout algorithm consists of a set of rules that modify the hypergraph. Each rule either changes attributes or the structure of the hypergraph. From this set, the drawing facility is generated and automatically included in the editor. The layout algorithm may be triggered manually (click "Apply Sugiyama", Fig. 1(a)) or automatically.

Figure 2 shows a sample layouting rule that modifies attributes. Similarly, also rules that change the structure of the graph may be specified. On the left side, a DFA before and after applying rules that update the attributes *xStart*, *yStart*, *xEnd* and *yEnd* of the arrow *t* is shown. On the right side, the rule that is responsible for updating these attributes is presented. The rule checks for two states connected by a transition, if the arrow exactly starts and ends at the borderline of the circle. If this is not the case, the attributes are updated. Figure 1(b) shows this rule specified visually. An editor for specifying rules visually is automatically generated.

Conclusion. In this paper, we gave an overview of a drawing approach that supports structured editing as well as free-hand editing. It is possible to specify the layout algorithm visually. With this approach, an environment was created that allows us to conduct experiments easily and to identify the "best" layouting strategy.

References

1. Maier, S., Minas, M.: A generic layout algorithm for meta-model based editors. In: Schürr, A., Nagl, M., Zündorf, A. (eds.) AGTIVE 2007. LNCS, vol. 5088, pp. 66–81. Springer, Heidelberg (2008)
2. Purchase, H.C., Hoggan, E.E., Görg, C.: How important is the "mental map"? - an empirical investigation of a dynamic graph layout algorithm. In: Kaufmann, M., Wagner, D. (eds.) GD 2006. LNCS, vol. 4372, pp. 184–195. Springer, Heidelberg (2007)

Spine Crossing Minimization in Upward Topological Book Embeddings

Tamara Mchedlidze and Antonios Symvonis

Dept. of Mathematics, National Technical University of Athens, Athens, Greece
{mchet,symvonis}@math.ntua.gr

1 Introduction

An *upward topological book embedding* of a planar st-digraph G is an upward planar drawing of G such that its vertices are aligned along the vertical line, called the *spine*, and each edge is represented as a simple Jordan curve which is divided by the intersections with the spine (*spine crossings*) into segments such that any two consecutive segments are located at opposite sides of the spine. When we treat the problem of obtaining an upward topological book embedding as an optimization problem, we are naturally interested in embeddings with the minimum possible number of spine crossing.

We define the problem of *HP-completion with crossing minimization problem* (for short, *HPCCM*) as follows: Given an embedded planar graph $G = (V, E)$, directed or undirected, one non-negative integer c, and two vertices s, $t \in V$, the HPCCM problem asks whether there exists a superset E' containing E and a drawing $\Gamma(G')$ of graph $G' = (V, E')$ such that (i) G' has a hamiltonian path from vertex s to vertex t, (ii) $\Gamma(G')$ has at most c edge crossings, and (iii) $\Gamma(G')$ preserves the embedded planar graph G. When the input digraph G is acyclic, we can insist on HP-completion sets which leave the HP-completed digraph G' also acyclic. We refer to this version of the problem as the *Acyclic-HPCCM problem*.

2 Results

A detailed presentation of our results (including technical proofs) is available as a Technical Report through arXiv [1]. The following theorem establishes the equivalence between the Acyclic-HPCCM problem and the problem of spine crossing minimization in upward topological book embedding for st-digraphs.

Theorem 1. *Let $G = (V, E)$ be an n node st-digraph. G has a crossing-optimal HP-completion set E_c with Hamiltonian path $P = (s = v_1, v_2, \ldots, v_n = t)$ such that the corresponding optimal drawing $\Gamma(G')$ of $G' = (V, E \cup E_c)$ has c crossings if and only if G has an optimal (wrt the number of spine crossings) upward topological book embedding with c spine crossings where the vertices appear on the spine in the order $\Pi = (s = v_1, v_2, \ldots, v_n = t)$.*

I.G. Tollis and M. Patrignani (Eds.): GD 2008, LNCS 5417, pp. 445–446, 2009.

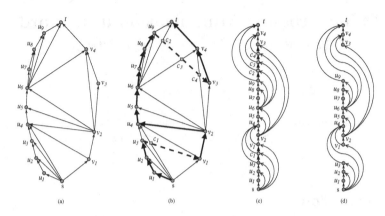

Fig. 1. Example of the construction of an optimal topological book embedding

For an outerplanar triangulated st-digraph G, we define the *st-polygon decomposition of G* and, based on the decomposition's properties, we develop a linear-time dynamic programming algorithm that solves the Acyclic-HPCCM problem with at most one crossing per edge. This is summarized in the following theorem.

Theorem 2. *Given an n node outerplanar triangulated st-digraph G, a crossing-optimal HP-completion set for G with at most one crossing per edge can be computed in $O(n)$ time.*   □

Our main result follows from the Theorems 1&2:

Theorem 3. *Given an n node outerplanar triangulated st-digraph G, an upward topological book embedding for G with minimum number of spine crossings and at most one spine crossing per edge can be computed in $O(n)$ time.*   □

Figure 1.a shows an upward planar st-digraph G that is not hamiltonian. In Figure 1.b graph G is augmented by the edges of an optimal HP-completion set (bold dashed edges) produced by our algorithm. The created Hamiltonian path is drawn with bold edges. By splitting the crossing edges we obtain graph G_c. Figure 1.c shows an upward topological book embedding of G_c with its vertices placed on the spine in the order they appear on a hamiltonian path of G_c. The edges appearing on the left (resp. right) side of the Hamiltonian path (as traveling from s to t) are placed on the left (resp. right) half-plane. Figure 1.d shows the optimal upward topological book embedding of G created from the drawing in Figure 1.c by deleting c_1, c_2, c_3, c_4 and merging the split edges of G.

References

1. Mchedlidze, T., Symvonis, A.: Optimal acyclic hamiltonian path completion for outerplanar triangulated st-digraphs (with application to upward topological book embeddings). arXiv:0807.2330, http://arxiv.org/abs/0807.2330

ILOG Elixir

Georg Sander and The ILOG Elixir team

ILOG SA, 9 rue de Verdun - BP 85, 94253 Gentilly Cedex, France
sander@ilog.fr
http://elixir.ilog.com

1 Introduction

The Adobe technology platform including Adobe© Flex© [2] and Adobe AIR™ [3] deliver portability, high performance and rich graphical UI to internet and desktop applications. ILOG Elixir [1] enhances this platform by adding advanced data visualization displays. It includes ready-to-use schedule displays, map displays, dials, gauges, 3D and radar charts, treemap charts and organization charts. ILOG Elixir is completely integrated with Adobe Flex Builder and fully supports Adobe Flexs data-binding and event models.

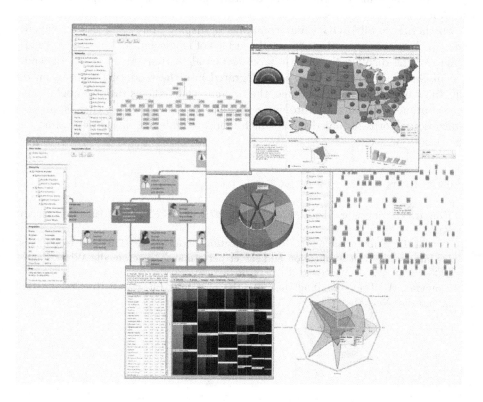

Fig. 1. Sample Applications of ILOG Elixir

I.G. Tollis and M. Patrignani (Eds.): GD 2008, LNCS 5417, pp. 447–448, 2009.

2   Highlights

Some of the display components are traditional business displays: The charts components display data series in radial or linear manner, optionally with a 3D look. The maps component is suitable for the cartography domain. It allows to import the ESRI Shapefile format, to style the display and to display arbitrary symbols on top of the map. These symbols can be charts, gauges, dials or custom components. The treemap can display large hierarchical data sets for the purpose of detecting data trends and outliers. Treemaps combine data clustering algorithms with advanced rendering techniques to help users identify clusters of particularly significant data. For example, a treemap can be used to depict the health of the global economy with visual attributes tied to a countrys size and gross domestic product (GDP).

Directly related to graph layout technology is the organization chart component. It depicts the interrelationships between people, equipment, or functions. It contains an intelligent tree layout algorithm specialized for this business domain that places the nodes in top-down or tip-over style while optimizing the available space. The component allows zooming, partial views, and dynamic level of details. When changing the partial view or the level of detail, the layout algorithm automatically reorganizes the diagram incrementally to adapt to the new situation.

Furthermore, all ILOG Elixir components display labels and use intelligent label decluttering algorithms to avoid overlaps of labels and to increase the readability of the display. No generic label layout algorithm can be used. Instead, specilialized labeling technology is integrated into the rendering mechanism of the different displays. For instance, the treemap chart allows various label placement options including an automatic visibility control of labels, that is, an optimization algorithm that decides which labels must be displayed depending on the situation.

References

1. Kim, E.: Getting Started with Flex 3, section 7.1: ILOG Elixir. O'Reilly, Sebastopol (2008)
2. Purcell, B., Subramanian, D.: Flex application performance: Tips and techniques for improving client application and server performance, macromedia White Paper (2004)
3. Simmons, A.: Understanding the potential of Adobe Integrated Runtime (AIR). integration New Media, Inc. - White Paper (2008)

DAGmap View

Vassilis Tsiaras and Ioannis G. Tollis

Institute of Computer Science, FORTH, and
Department of Computer Science, University of Crete
Heraklion, Greece
{tsiaras,tollis}@ics.forth.gr

Abstract. DAGmap view is a program written in Java that draws directed acyclic graphs using space filling techniques. In DAGmap view the layout function and the hierarchy presentation function have been decomposed to improve the stability of the layout during navigation and zooming.

1 Description

Among the many alternative ways to visualize a tree, space filling visualizations, such as treemaps, have become very popular due to their efficiency, their scalability, and their easiness of navigation and user interaction [1]. Recently, we investigated space filling visualizations for hierarchies that are modeled by Directed Acyclic Graphs (DAG) and we defined the constraints for such a visualization [2].

DAGmap view is a program written in Java that draws specific classes of DAGs, such as Two Terminal Series Parallel digraphs (TTSP), layered planar st-graphs and trees, using space filling techniques (Fig. 1). Additionally, it implements the vertex duplication heuristic and allows the user to specify a set of vertices to be duplicated. The tool has implemented many novel ideas such as drawing of vertices and of edges of a DAG, separate layout and hierarchy presentation functions, zooming without changing the size of the nesting borders and keeping the layout of the rectangles constant during zooming and navigation.

The two main functions of DAGmap are the layout and the hierarchy presentation functions. The layout function assigns rectangles to vertices and edges of a DAG G while the hierarchy presentation function illustrates the structure of a hierarchy using a number of techniques including the cushion, the nested, and the cascaded presentations. In DAGmap view we implemented nested presentation and we plan to implement cascaded and cushion presentations in the next release of the program.

In treemaps, nesting is trivial and is done along with the layout. The drawing rectangle R_u of a node u is shrunk and the resulting rectangle R'_u is located inside R_u. Then the border $R_u \setminus R'_u$ is used for displaying information concerning u and R'_u is used for drawing the children of u. And this procedure is repeated recursively. In DAGmaps, layout and nesting should better be implemented as two separate functions. First, the layout function assigns rectangles to vertices

I.G. Tollis and M. Patrignani (Eds.): GD 2008, LNCS 5417, pp. 449–450, 2009.
© Springer-Verlag Berlin Heidelberg 2009

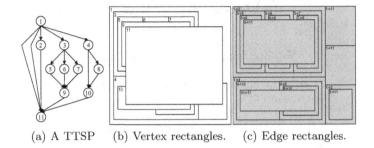

(a) A TTSP   (b) Vertex rectangles.   (c) Edge rectangles.

Fig. 1. Example of a TTSP digraph DAGmap drawing using the squarified layout

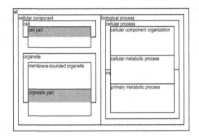

Fig. 2. In this example a subgraph of the Gene Ontology (GO) DAG is drawn. The color refers to the relationship between two GO terms. We use white color for *is_a* relationship and light brown color for *part_of* relationship. The term "cell part" *is_a* "cellular component" and *part_of* "cell".

and/or to edges of a DAG G, in case that such an assignment is possible. Then the nesting function shrinks the rectangles proportionally to their longest path distance from the sources, in order to reveal the hierarchy structure. Decomposing the nesting from the layout greatly facilitates the implementation of the layout algorithm although slightly complicates the implementation of the nesting algorithm.

References

1. Bederson, B.B., Shneiderman, B., Wattenberg, M.: Ordered and quantum treemaps: Making effective use of 2d space to display hierarchies. ACM Transactions on Graphics 21(4), 833–854 (2002)
2. Tsiaras, V., Triantafilou, S., Tollis, I.G.: Treemaps for directed acyclic graphs. In: Hong, S.-H., Nishizeki, T., Quan, W. (eds.) GD 2007. LNCS, vol. 4875, pp. 377–388. Springer, Heidelberg (2008)

Brain Network Analyzer

Vassilis Tsiaras[1,2], Ioannis G. Tollis[1,2], and Vangelis Sakkalis[1]

[1] Institute of Computer Science, FORTH
[2] Department of Computer Science, University of Crete
Heraklion, Greece
{tsiaras,tollis,sakkalis}@ics.forth.gr

Abstract. Brain Network Analyzer is an application, written in Java, that displays and analyzes synchronization networks from brain signals. The program implements a number of network indices and visualization techniques. The program has been used to analyze networks produced by electroencephalogram data of alcoholic and control patients.

1 Introduction

One of the major issues in neuroscience is to describe how different brain areas communicate with each other during perception, cognition, action as well as during spontaneous activity in the default or resting state. Data acquired using noninvasive techniques [like functional magnetic resonance imaging (fMRI); electroencephalography (EEG); magnetoencephalography (MEG)] may be used to estimate functional connectivity, which is defined as the statistical dependence between the activations of distinct and often well separated neuronal populations. Network models provide a common framework for describing functional connectivity. However defining what nodes and edges should be is a challenging problem. Network nodes can easily be identified with fMRI data but the dependence between nodes can be measured only for low frequencies (< 0.2 Hz) due to the limited time resolution of fMRI. On the other hand the time resolution of EEG/MEG is excellent but the mapping from generators in the brain to the sensors on the scalp is complex and the topology of a network in sensors space is different from the topology in generators space [1]. To identify generators from EEG/MEG data one has to solve the inverse problem which is ill posed and sensitive to noise. In the following we assume that connectivity networks associated with EEG/MEG data have been defined as adjacency matrices.

2 Program Presentation

Brain Network Analyzer is a program that visualizes and analyzes functional connectivity networks as well as connectivity networks at sensors space. The input consists of a series of adjacency matrices W_n, $n = 1, 2, \ldots, T$. To improve the signal to noise ratio the user may use hard or soft thresholds. Using hard thresholds each matrix W_n is transformed into a binary matrix with entries in $\{0, 1\}$.

I.G. Tollis and M. Patrignani (Eds.): GD 2008, LNCS 5417, pp. 451–452, 2009.

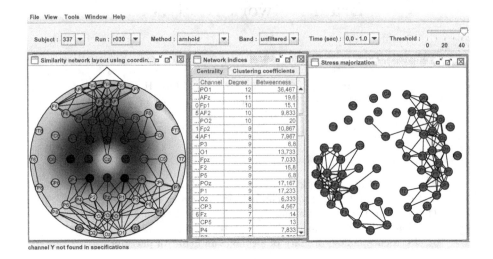

Fig. 1. Synchronization network and topographic map are calculated at sensors space from EEG data. Network indices and multidimensional scaling windows are also shown.

Soft thresholds, such as the power function $f(w) = w^{\beta}$, are functions from $[0,1]$ to $[0,1]$. Entry $W_n(i,j)$ is a measure of how similar (or synchronous) is the dynamics of node j to the dynamics of node i. To turn this similarity measure into a dissimilarity (or distance) measure we use the transformation $d(w) = -log(w)$. Distance measures are needed for calculating multi-dimensional scaling and displaying, in the plane, nodes with similar functionality close together (see Fig. 1). In the program's menu there are options to visualize raw signals, potential maps, scalp topographies, and networks where node coordinates are defined either by the position of the generators in the brain or the sensors on the scalp or by an algorithm that groups functionally similar nodes. The user may also produce video with the evolution of networks during a cognitive task.

Finally, most of network indices, that appeared in complex networks literature, have been implemented. Using this program we found differences in clustering coefficient indices among the alcoholic and control subjects beta band (13-30 Hz) networks at sensors space [2].

References

1. Ioannides, A.A.: Dynamic functional connectivity. Current Opinion in Neurobiology 17, 161–170 (2007)
2. Sakkalis, V., Tsiaras, V., Zervakis, M., Tollis, I.G.: Optimal brain network synchrony visualization: Application in an alcoholism paradigm. In: IEEE EMBS, pp. 4285–4288 (2007)

Graph Drawing Contest Report

Ugur Dogrusoz[1], Christian A. Duncan[2], Carsten Gutwenger[3], and Georg Sander[4]

[1] Bilkent University and Tom Sawyer Software
ugur@cs.bilkent.edu.tr
[2] Louisiana Tech University, Ruston, LA 71272, USA
duncan@latech.edu
[3] University of Dortmund, Germany
carsten.gutwenger@cd.uni-dortmund.de
[4] ILOG, 94253 Gentilly Cedex, France
sander@ilog.fr

Abstract. This report describes the 15th Annual Graph Drawing Contest, held in conjunction with the 2008 Graph Drawing Symposium in Heraklion, Crete, Greece. The purpose of the contest is to monitor and challenge the current state of graph-drawing technology.

1 Introduction

This year's Graph Drawing Contest had five distinct categories: four special graph categories, and the Graph Drawing Challenge. The special graph categories provided four real world graphs from different application domains: a mystery graph, a graph from electric engineering, a graph from social sciences, and a biological network. The mystery graph had 71 nodes and 145 directed and labeled edges and represents a series of social or cultural events. The task was to determine which events are represented by this graph and to create a drawing of its logical structure. For the remaining categories, the task was to provide a visualization typical for the corresponding domain. The Graph Drawing Challenge, which took place during the conference, focused on minimizing the number of crossings of upward grid drawings of graphs with edge bends. We received 18 submissions: 7 for the four special graph categories, and 11 for the Graph Drawing Challenge. Unfortunately, we did not receive any submissions for the biological network, which represented the mTOR signalling pathways with 90 entities, 54 interactions and 85 inclusions.

2 Mystery Graph

Honoring this year's conference location (Greece), the mystery graph represents the torch relay routes of all Olympic Summer and Winter Games. The nodes are countries, and the edges are labeled with the year of the games when the torch traveled from one country to the next. The data was collected from Wikipedia [5], but the order of the nodes and edges was randomized. All three teams that submitted a drawing found the correct solution.

I.G. Tollis and M. Patrignani (Eds.): GD 2008, LNCS 5417, pp. 453–458, 2009.

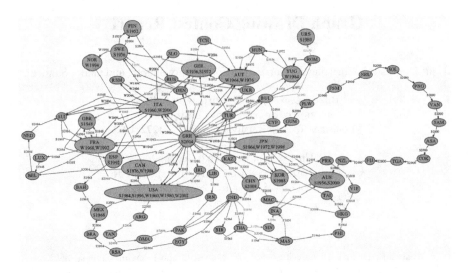

Fig. 1. First place, Mystery Graph: Olympic Torch Relay Routes (original in color)

Since the nodes represent countries, we received two submissions that depicted the graph on top of a geographic map. However, the judges felt that this did not illustrate the logical structure of the graph clear enough. The winning submission by Yifan Hu and Emden Gansner (Fig. 1) does not use geographic locations. Instead it places the "Greece" node (the starting point of all torch relay routes) in the middle and groups the other country nodes around it. The different years of the routes are displayed by colors on the edges. The initial layout of this submission used the sfdp code in a developmental version of GraphViz[1]. It was then hand-tuned and fed through a spline routine.

3   Graph from Electrical Engineering

This electrical network represents the architecture of the FR500 VLIW processor. The graph data is inspired by a diagram published in [4]. The original diagram (Fig. 2) contains a mix of directed and undirected edges, but the contest graph is simplified (directions are removed, multi-edges are collapsed, an auxiliary node is introduced to detangle multi-edges). The resulting data consists of 35 deeply nested nodes and 48 undirected edges. The task was to produce a fully automatic drawing without manual tuning.

There were three submissions for the electric diagram. One submission was an energy-based compound straight-line drawing, and another submission used a 3D layout of the graph. The winning submission by Melanie Badent and Pietro Palladino used orthogonal edges with bends, which fits well for diagrams in the electrical application domain. The drawing (Fig. 3) was obtained by implementing a module `Orthogonal-GroupLayouter` for the freely available graph editor YEd [6]. This new module is based on the topology-shape-metrics approach with three phases (planar embedding with the introduction of dummy nodes for crossings, calculating the edge shapes using bends, and finding the metrics for the shapes to obtain the final coordinates).

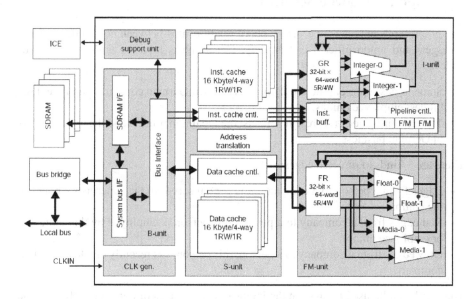

Fig. 2. FR500 VLIW processor, Original Diagram

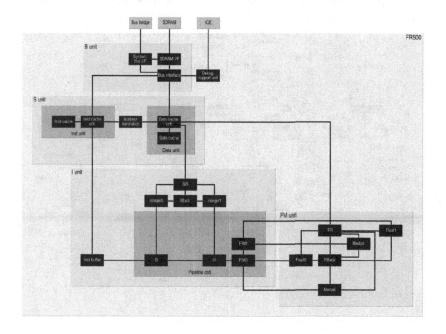

Fig. 3. First place, FR500 VLIW processor

4   Graph from Social Sciences

The graph represents the supervisory board relationships between companies and top managers and union officers in Germany. The graph is bipartite with two kinds of nodes:

- Some nodes represent the top 25 publicly traded German companies and the three biggest employee unions.
- The remaining nodes represent persons: the top union officers and the top managers in German business.

There are 3 kinds of directed edges:

- Edges from a person to a company: this person serves in the supervisory board of the company.
- Red edges from a company to a person: this person is employed by that company or employee union.
- Gray edges from a company to a person: this person was recently employed by that company.

The graph can be used to analyze the interest dependencies of companies. If a person is employed by a company and sits in the supervisory board of another company, the supervised company is partially controlled by the employing company. If a person was formerly employed, the situation is however less clear: in some cases, a former CEO

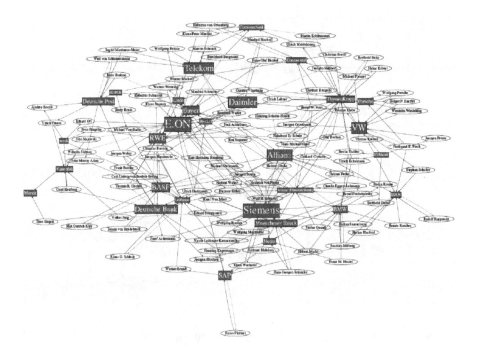

Fig. 4. First place, Supervisory Board Dependencies (original in color)

moved into the supervisory board simply to free the space for a new CEO but still kept strong ties to that company. In other cases, a former CEO parted completely from a company (e.g., got fired) and now acts on own interests.

The first prize was awarded to the only submission in this category by Yifan Hu and Emden Gansner (Fig. 4). They use a color coding on the nodes and edges. The company label size is proportional to the market capitalization of that company. The layout is obtained using the sfdp code of GraphViz without hand-tuning.

5   Graph Drawing Challenge

This year's challenge dealt with minimizing the number of crossings of upward grid drawings of graphs with edge bends. This is a subproblem of the popular layered lay-out technique by Sugiyama e.a. [3] which is known to be NP-hard. It requires that all nodes be placed on grid positions, that nodes and edge bends don't overlaps, and that all edge segments point strictly upwards. At the start of the one-hour on-site compe-tition, the contestants were given six nonplanar, acyclic, directed graphs with a legal upward layout that however had a huge number of crossings. The goal was to re-arrange the layout to reduce the number of crossings. Only the number of crossings was judged; other aesthetic criteria such as the number of edge bends or the area were ignored.

We partitioned the challenge into two subcategories: automated and manual. The seven manual teams solved the problems by hand using ILOG's Simple Graph Editing Tool provided by the committee. They received graphs ranging in size from 19 nodes / 32 edges to 148 nodes / 200 edges. The four automated teams were allowed to use their own sophisticated software tools with specialized algorithms for the problem. They re-ceived graphs ranging in size from 24 nodes / 46 edges to 993 nodes / 1383 edges. Both subcategories were judged independently by summing up the scores of each graph. The score of a graph was determined by dividing the crossing number of the best submission by the crossing number of the current submission (hence, the best submission receives 1 point and the other submissions receive a fraction of 1).

The winner in the manual subcategory was the team of University Konstanz (Melanie Badent, Martin Mader, Christian Pich). They had the best manual result for 3 graphs and obtained an overall score of 4.6. The other manual teams obtained scores between 2 and 4.2. The winner in the automated subcategory was the team of TU Dortmund (Hoi-Ming Wong, Markus Chimani, Karsten Klein) using software based on a recently published algorithm [2]. They had the best automated result for all 6 graphs, hence obtaining the maximum possible score of 6 points.

Some graphs used in both subcategories were constructed in a way so that the optimal crossing number was known. While some automated and manual teams reached the optimal crossing number for the smaller graphs, the optimal crossing number for larger graphs was neither reached by any manual nor by any automated team. However, the automated teams usually obtained better results than the teams that solved the challenge manually. Figure 5 shows the optimal result of a graph with 99 nodes/157 edges (4 crossings) and the corresponding best results of the manual teams (100 crossings) and of the automated teams (15 crossings).

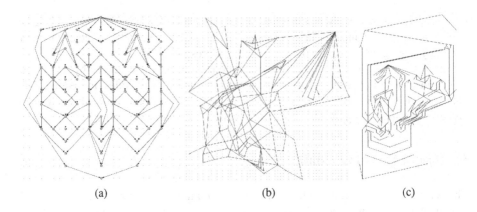

| (a) | (b) | (c) |

Fig. 5. Challenge graph with 99 nodes / 157 edges: (a) the optimal solution: 4 crossings, (b) the best manually obtained result by team Konstanz: 100 crossings, (c) the best automated result by team Dortmund: 15 crossings

Acknowledgments

We thank Emek Demir of Memorial Sloan-Kettering Cancer Center, New York, for providing the data of the contest graph in the biological domain, and Lothar Krempel of Max Planck Institute for the Study of Societies, Germany, for the idea of the graph in the social sciences domain. Finally, we thank the generous sponsors of the symposium and all the contestants.

References

1. AT&T: Graphviz, http://www.graphviz.org
2. Chimani, M., Gutwenger, C., Mutzel, P., Wong, H.: Layer-free upward crossing minimization. In: McGeoch, C.C. (ed.) WEA 2008. LNCS, vol. 5038, pp. 55–68. Springer, Heidelberg (2008)
3. Sugiyama, K., Tagawa, S., Toda, M.: Methods for visual understanding of hierarchical systems. IEEE Trans. Sys. Man, and Cyb. SMC 11(2), 109–125 (1981)
4. Sukemura, T.: FR500 VLIW-architecture high-performance embedded microprocessor. FUJITSU Sci. Tech. J. 36(1), 31–38 (2000)
5. Wikipedia: Olympic flame (accessed, March 2008), http://en.wikipedia.org/wiki/Olympic_Flame
6. yWorks: yFiles, http://www.yworks.com

Author Index